RENEWALS 458-4574

DATE DUE

GAYLORD			PRINTED IN U.S.A.

Biotechnology
Second Edition

Cumulative Index

 WILEY-VCH

Biotechnology

Second Edition

Fundamentals

Volume 1
Biological Fundamentals

Volume 2
Genetic Fundamentals and
Genetic Engineering

Volume 3
Bioprocessing

Volume 4
Measuring, Modelling and Control

Products

Volume 5a
Recombinant Proteins, Monoclonal
Antibodies and Therapeutic Genes

Volume 5b
Genomics and Bioinformatics

Volume 6
Products of Primary Metabolism

Volume 7
Products of Secondary Metabolism

Volumes 8a and b
Biotransformations I and II

Special Topics

Volume 9
Enzymes, Biomass, Food and Feed

Volume 10
Special Processes

Volumes 11a–c
Environmental Processes I–III

Volume 12
Legal, Economic and
Ethical Dimensions

All volumes are also displayed on our Biotech Website:
http://www.wiley-vch.de/books/biotech

A Multi-Volume Comprehensive Treatise

Biotechnology

Second, Completely Revised Edition

Edited by
H.-J. Rehm and G. Reed
in cooperation with
A. Pühler and P. Stadler

Cumulative Index

 **WILEY-VCH**

Weinheim · New York · Chichester · Brisbane · Singapore · Toronto

Series Editors:

Prof. Dr. H.-J. Rehm
Institut für Mikrobiologie
Universität Münster
Corrensstraße 3
D-48149 Münster
FRG

Dr. G. Reed
1029 N. Jackson St. #501-A
Milwaukee, WI 53202-3226
USA

Prof. Dr. A. Pühler
Biologie VI (Genetik)
Universität Bielefeld
Postfach 100131
D-33501 Bielefeld
FRG

Prof. Dr. P. I W. Stadler
Artemis Pharmaceuticals
Geschäftsführung
Pharmazentrum Köln
Neurather Ring 1
D-51063 Köln
FRG

Library of Congress Card No.: applied for

British Library Cataloguing-in-Publication Data:
A catalogue record for this book is available from the British Library

Die Deutsche Bibliothek – CIP-Einheitsaufnahme

A catalogue record for this book
Is available from Der Deutschen Bibliothek
  ISBN 3-527-28337-4

Scientific Advisory Board

Contents

Contents of Volumes 1 to 12

Volume 1: Biological Fundamentals

Volume 2: Genetic Fundamentals and Genetic Engineering

Introduction 1
A. Pühler

I. Classical Genetics

1 Mutagenesis 5
A. Crueger
2 Genetic Exchange Processes for Prokaryotes 47
B. W. Holloway
3 Genetic Exchange Processes in Lower Eukaryotes 73
U. Stahl, K. Esser
4 Cell Fusion 93
J. Anné
5 Gene Mapping in Animals and Plants 141
G. Wricke, H. Geldermann, W. E. Weber

II. Molecular Genetics

6 Structure and Function of DNA 191
F. Götz
7 Principles of Gene Expression 233
R. Mattes
8 DNA Sequencing 257
G. Volckaert, P. Verhasselt, M. Voet, J. Robben
9 DNA Synthesis 317
J. W. Engels, B. Sprunkel, E. Uhlmann

III. Genetic Engineering of Microorganisms

10 Principles of Genetic Engineering for *Escherichia coli* 373
H. Schwab

11 Genetic Engineering of Gram Negative Bacteria 427
U. B. Priefer
12 Genetic Engineering of Gram Positive Bacteria 455
W. Wohlleben, G. Muth, J. Kalinowski
13 Genetic Engineering of Yeast 507
P. E. Sudbery
14 Genetic Engineering of Filamentous Fungi 529
G. Turner

IV. Genetic Engineering of Plants

15 Genetic Engineering of Plant Cells 547
G. Kahl, K. Weising
16 Transgenic Plants 627
L. Willmitzer

V. Genetic Engineering of Animals

17 Genetic Engineering of Animal Cells 663
M. Wirth, H. Hauser
18 Transgenic Animals 745
G. Brem

VI. Biosafety Concepts

19 Safety Aspects in Biotechnology 835
R. Simon, W. Frommer

Index 855

Volume 3: Bioprocessing

Introduction   1
G. Stephanopoulos

I. Nature and Issues of Bioprocessing

1 Fermentation – An Overview   7
B. C. Buckland, M. D. Lilly

2 Industrial Animal Cell Culture   23
B. D. Kelley, T.-W. Chiou,
M. Rosenberg, D. I. C. Wang

3 Overview of Downstream
Processing   39
R. Spears

4 Proteins and Peptides as Drugs: Sources
and Methods of Purification   57
S. E. Builder, R. L. Garnick,
J. C. Hodgdon, J. R. Ogez

II. Product Formation (Upstream Processing)

5 Bioreactors: Description and
Modelling   77
J. Nielsen, J. Villadsen

6 Cell Culture Bioreactors   105
A. Sambanis, W.-S. Hu

7 Media for Microbial Fermentations   127
R. L. Greasham

8 Media for Cell Culture   141
R. A. Wolfe

9 Media and Air Sterilization   157
G. K. Raju, C. L. Cooney

10 Oxygen Transfer and Mixing:
Scale-Up Implications   185
M. Reuss

11 Oxygen Transfer in Cell Culture
Bioreactors   219
J. G. Aunins, H.-J. Henzler

12 Strategies for Fermentation with
Recombinant Organisms   283
T. Imanaka

13 Anaerobic Fermentations   295
L. E. Erickson, D. Y. C. Fung,
P. Tuitemwong

14 Fermentation Monitoring and
Control   319
T. Chattaway, G. A. Montague,
A. J. Morris

15 Fermentation Data Analysis for
Diagnosis and Control   355
G. Stephanopoulos, K. Konstantinov,
U. Saner, T. Yoshida

16 Design of Aseptic, Aerated
Fermentors   401
M. Charles, J. Wilson

17 Biotransformations and Enzyme
Reactors   427
A. S. Bommarius

III. Product Recovery and Purification

18 Cell and Cell Debris Removal:
Centrifugation and Crossflow
Filtration   469
R. V. Datar, C.-G. Rosén

19 Cell Disruption and Isolation of
Non-Secreted Products   505
H. Schütte, M.-R. Kula

20 *In vitro* Protein Refolding   527
J. L. Cleland, D. I. C. Wang

21 Liquid-Liquid Extraction (Small
Molecules)   557
K. Schügerl

22 Protein Purification by Liquid-Liquid
Extraction   593
B. D. Kelley, T. A. Hatton

23 Protein Separation and Purification   617
J.-C. Janson, L. Rydén

24 Affinity Separations   643
S. Sundaram, M. L. Yarmush

25 Electrokinetic Separations   679
A. J. Grodzinsky, M. L. Yarmush

26 Final Recovery Steps:
Lyophilization, Spray-Drying   695
C. F. Gölker

IV. Process Validation, Regulatory Issues

27 Analytical Protein Chemistry   717
*S. Sundaram, D. M. Yarmush,
M. L. Yarmush*

28 Biotechnology Facility Design and
Process Validation   739
M. G. Beatrice

29 Treatment of Biological Waste   769
D. F. Liberman

Index   789

Volume 4: Measuring, Modelling, and Control

Introduction   1

I. Instruments

1 Common Instruments for Process
   Analysis and Control   5
   K. Schügerl
2 Methods and Instruments in Fermentation
   Gas Analysis   27
   E. Heinzle, I. J. Dunn
3 Biosensors   75
   B. Mattiasson

II. Measurements in Bioreactor Systems

4 Characterization of Bioreactors   107
   A. Lübbert
5 On-Line Analysis of Broth   149
   K. Schügerl
6 Determination of Cell Concentration
   and Characterization of Cells   179
   K. F. Reardon, T. H. Scheper
7 Bioreactor State Estimation   225
   G. Stephanopoulos, S. Park
8 Optimization of Sampling   251
   A. Munack

III. Modelling of Bioreactor Systems

9 Cell Models   267
   K.-H. Bellgardt

10 Stirred Tank Models   299
    M. Reuss, R. Bajpai
11 Tower Reactor Models   349
    J. C. Merchuk
12 Process Models: Optimization of Yeast
    Production – A Case Study   383
    K.-H. Bellgardt, J. Yuan
13 Aerobic Waste Water Process
    Models   407
    G. F. Andrews
14 Anaerobic Waste Water Process
    Models   441
    D. Schürbüscher, C. Wandrey
15 Bioprocess Kinetics and Modelling
    of Recombinant Fermentation   485
    D. D. Y. Ryu, J.-Y. Kim, S. B. Lee

IV. Control and Automation

16 Control of Bioreactor Systems   509
    H. C. Lim, K.-S. Lee
17 Automation in Biotechnology   561
    A. Lübbert
18 Modelling, Design, and Control of
    Downstream Processing   603
    S. Shioya, K.-I. Suga
19 Expert Systems for Biotechnology   625
    A. Halme, N. Karim

Index   637

Volume 5a: Recombinant Proteins, Monoclonal Antibodies and Therapeutic Genes

Introduction   XV
U. Ney, D. Schomburg

Structure and Function of Proteins

1 Types and Function of Proteins   1
   S. Wolf, H. G. Gassen

2 Sequence and Structure of Proteins   43
   F. Eisenhaber, P. Bork

3 Protein Interactions   87
   M. Meyer, D. Schomburg

Recombinant Proteins

4 *In vitro* Folding
   of Inclusion Body Proteins
   on an Industrial Scale   111
   R. Rudolph, H. Lilie, E. Schwarz

5 Medical Applications
   of Recombinant Proteins
   in Humans and Animals   125
   G. D. Wetzel

6 Enzymes for Industrial Applications   189
   W. Aehle, O. Misset

Monoclonal Antibodies

7 Antibody Engineering
   and Expression   219
   J. R. Adair

8 Manufacture of Therapeutic
   Antibodies   245
   A. J. Racher, J. M. Tong, J. Bonnerjea

9 Use of Antibodies for Immunopurification   275
   D. J. King

10 Preclinical Testing of Antibodies:
   Pharmacology, Kinetics,
   and Immunogenicity   289
   R. Foulkes, S. Stephens

11 Preclinical Testing of Antibodies:
   Safety Aspects   303
   R. W. James

12 Therapeutic Applications
   of Monoclonal Antibodies:
   A Clinical Overview   311
   M. Sopwith

Case Studies

13 Antibodies for Sepsis:
   Some Lessons Learnt   329
   S. M. Opal

14 An Engineered Human Antibody
   for Chronic Therapy: CDP571   343
   M. Sopwith, S. Stephens

15 Antibody Targeted Chemotherapy   355
   M. S. Berger, P. R. Hamann, M. Sopwith

Volume 5b: Genomics and Bioinformatics

Introduction   1
C. W. Sensen

Application Domains

1 Genome Projects of Model Organisms   5
A. Pühler, D. Jording, J. Kalinowski,
D. Buttgereit, R. Renkawitz-Pohl,
L. Altschmied, A. E. Danchin,
H. Feldmann,H.-P. Klenk, M. Kröger

2 The Human Genome Project   41
L. C. Tsui, S. W. Scherer

3 Genetic Disease   61
G. Dellaire

4 Genomics and Human Disease   105
R. C. Green

5 Pharmaceutical Bioinformatics and Drug
Discovery   123
C. S. Ramanathan, D. B. Davison

6 Agri-Food and Genomics   145
I. A. P. Parkin, S. J. Robinson, A. G. Sharpe,
K. Rozwadowski, D. Hegedus, D. J. Lydiate

DNA Technologies

7 Genomic Mapping and Positional Cloning,
with Emphasis on Plant Science   165
A. Vanavichit, S. Tragoonrung,
T. Toojinda

8 Sequencing Technology   193
L. R. Middendorf, P. G. Humphrey,
N. Narayanan, S. C. Roemer

9 A DNA Microarrays Fabrication Strategy
for Research Laboratories   227
D. C. Tessier, D. Y. Thomas, R. Brousseau

Protein Technologies

10 Two-Dimensional Gel Electrophoresis
and Mass Spectrometry for Proteomic
Studies: State-of-the-Art   241
D. Figeys

11 Proteome Analysis by Capillary Electro-
phoresis   269
N. J. Dovichi, S. Hu, D. Michels,
Z. Zhang, S. N. Krylov

Bioinformatics

12 Using the Molecular Biology Data   281
E. M. Zdobnov, R. Lopez, R. Apweiler,
T. Etzold

13 Tools for DNA Technologies   301
P. Rice

14 Tools for Protein Technologies   325
D. S. Wishart

15 Structure Information   345
M. Cygler, A. Matte, J. D. Schrag

16 Automated Genome Annotation and
    Comparative Genomics   361
    S. Gopal, T. Gaasterland
17 Genomic Data Representation through
    Images – MAGPIE as an Example   379
    P. Gordon, T. Gaasterland, C. W. Sensen
18 Interaction Databases   399
    G. D. Bader, C. W. V. Hogue

Ethical, Legal and Social Issues

19 Ethical Aspects of Genome Research and
    Banking   417
    B. M. Knoppers
20 Genomics – Five Years from Now   431
    C. W. Sensen

Index   587

Volume 6: Products of Primary Metabolism

Introduction 1
M. Roehr

I. Raw Materials and Raw Material Strategies

1a Sugar-Based Raw Materials for
Fermentation Applications 5
E. Stoppok, K. Buchholz

1b Starch-Based Raw Materials for
Fermentation Applications 31
J.-C. de Troostembergh

2 Raw Material Strategies – Economical
Problems 47
F. Schneider, H. Steinmüller

II. Products of Primary Metabolism

3 Ethanol – Classical Methods 59
T. Senn, H. J. Pieper

4 Ethanol – Potential Source of Energy
and Chemical Products 121
N. Kosaric

5 Microbial Production of Glycerol and
Other Polyols 205
H.-J. Rehm

6 Microbial Production of
Acetone/Isopropanol 229
P. Dürre, H. Bahl

7 Microbial Production of
2,3-Butanediol 269
I. S. Maddox

8 Lactic Acid 293
J. S. Kaščák, J. Komínek, M. Roehr

9 Citric Acid 307
M. Roehr, C. P. Kubicek, J. Komínek

10 Gluconic Acid 347
M. Roehr, C. P. Kubicek, J. Komínek

11 Further Organic Acids 363
M. Roehr, C. P. Kubicek

12 Acetic Acid 381
H. Ebner, S. Sellmer, H. Follmann

13 PHB and Other Polyhydroxyalkanoic
Acids 403
A. Steinbüchel

14a Amino Acids – Technical Production
and Use 465
W. Leuchtenberger

14b Enzymology of Amino Acid
Production 503
*N. Esaki, S. Nakamori, T. Kurihara,
S. Furuyoshi, K. Soda*

15 Nucleotides and Related
Compounds 561
A. Kuninaka

16 Extracellular Polysaccharides 613
I. W. Sutherland

17 Biosurfactants 659
N. Kosaric

Index 719

Volume 7: Products of Secondary Metabolism

Introduction
H. von Döhren, H. Kleinkauf

1 General Aspects of Secondary
  Metabolism   1
  H. von Döhren, U. Gräfe

2 Regulation of Bacterial Antibiotic
  Production   57
  K. Chater, M. Bibb

3 Screening of Novel Receptor-Active
  Compounds of Microbial Origin   107
  H. Tanaka, S. Ōmura

4 Microbial Lipids   133
  C. Ratledge

5 Microbial Siderophores   199
  G. Winkelmann, H. Drechsel

6 Advances in the Molecular Genetics of
  β-Lactam Antibiotic Biosynthesis   247
  P. L. Skatrud, T. Schwecke,
  H. v. Liempt, M. B. Tobin

7 Peptide Antibiotics   277
  H. Kleinkauf, H. von Döhren

8 Lantibiotics   323
  R. Jack, F. Götz, G. Jung

9 Glycopeptide Antibiotics
  (Dalbaheptides)   369
  G. Lancini, B. Cavalleri

10 Aminoglycosides and Sugar
   Components in Other Secondary
   Metabolites   397
   W. Piepersberg, J. Distler

11 Products from Basidiomycetes   489
   G. Erkel, T. Anke

12 Cyclosporins: Recent Developments in
   Biosynthesis, Pharmacology and
   Biology, and Clinical Applications   535
   J. Kallen, V. Mikol, V. F. J. Quesniaux,
   M. D. Walkinshaw, E. Schneider-Scherzer,
   K. Schörgendorfer, G. Weber, H. Fliri

13 Secondary Products from Plant Cell
   Cultures   593
   J. Berlin

14 Biotechnical Drugs as Antitumor
   Agents   641
   U. Gräfe, K. Dornberger, H.-P. Saluz

Index   707

Volume 8a: Biotransformations I

Introduction   1
D. R. Kelly

1 Perspectives in Biotransformation   5
   M. Turner
2 Biotransformations – Practical Aspects   25
   D. R. Kelly

Hydrolases

3 Biotransformations with Lipases   37
   R. Kazlauskas, U. Bornscheuer
4 Esterases   193
   S. Phythian
5 Cleavage and Formation
   of Amide Bonds   243
   D. Hoople
6 Nitriles   277
   A. Bunch

Redox Enzymes

7 Alkaloids   327
   N. Bruce
8 Yeast   363
   S. Servi
9 Alcohol Dehydrogenases – Characteristics,
   Design of Reaction Conditions,
   and Applications   391
   J. Peters
10 Hydroxylation and Dihydroxylation   475
   H. Holland
11 Flavin Monooxygenases – Uses as Catalysts
   for Baeyer–Villiger Ring Expansion
   and Heteroatom Oxidation   535
   D. R. Kelly, P. Wan, J. Tsang

Index   589

Volume 8b: Biotransformations II

Introduction   1
D. R. Kelly

Carbon–Carbon Bond Formation, Addition, Elimination and Substitution Reactions

1 Carbon–Carbon Bond Formation Using Enzymes   5
   A. D. M. Curtis
2 Lyases   41
   D. R. Kelly, J. Mahdi
3 Halocompounds   173
   G. Robinson, S. Jackman, J. Stratford
4 Phosphorylation   221
   S. Jones
5 Enzymes in Carbohydrate Chemistry: Formation of Glycosidic Linkages   243
   S. Flitsch, G. M. Watt

Applications

6 Industrial Biotransformations   277
   U. T. Bornscheuer

7 Regioselective Oxidation of Aminosorbitol with *Gluconobacter oxydans*, A Key Reaction in the Industrial Synthesis of 1-Deoxynojirimycin   295
   M. Schedel
8 Engineering Microbial Pathways for Amino Acid Production   313
   I. Fotheringham
9 Biotechnological Production of Natural Aroma Chemicals by Fermentation Processes   333
   J. Rabenhorst
10 Synthetic Applications of Enzyme-Catalyzed Reactions   351
   J. McGregor-Jones

The Future of Biotransformations

11 Catalytic Antibodies   403
   G. M. Blackburn, A. Garçon
12 Synthetic Enzymes: Artificial Peptides in Stereoselective Synthesis   491
   P. Bentley

Index   519

Volume 9: Enzymes, Biomass, Food and Feed

Introduction 1
G. Reed, T. W. Nagodawithana

I. Enzymes

1 The Nature of Enzymes   5
G. M. Smith
2 Production of Enzymes as Fine
Chemicals   73
K. A. Foster, S. Frackman, J. F. Jolly
3 Kinetics of Multi-Enzyme Systems   121
A. Cornish-Bowden
4 Analytical Uses of Enzymes   137
G.-B. Kresse

II. Biomass

5 Production of Microbial Biomass   167
H. Boze, G. Moulin, P. Galzy
6 Nutritional Value and Safety of "Single
Cell Protein"   221
N. S. Scrimshaw, E. B. Murray

III. Food Fermentations

7 Baked Goods   241
G. Spicher, J.-M. Brümmer
8 Commercial Production of Baker's Yeast
and Wine Yeast   321
C. Caron
9 Cheese   353
N. F. Olson

10 Other Fermented Dairy Products   385
R. C. Chandan, K. M. Shahani
11 Brewing   419
J. Russell, G. G. Stewart
12 Wine and Brandy   463
H. H. Dittrich
13 Indigenous Fermented Foods   505
L. R. Beuchat
14 Cocoa Fermentation   561
A. S. Lopez, P. S. Dimick
15 Vinegar   579
H. Ebner, H. Follmann, S. Sellmer
16 Olive Fermentations   593
*A. Garrido Fernández, P. García García,
M. Brenes Balbuena*
17 Vegetable Fermentations   629
H. P. Fleming, K. H. Kyung, F. Breidt
18 Use of Enzymes in Food Processing   663
H. S. Olsen
19 Carbohydrate-Based Sweeteners   737
R. E. Hebeda

IV. Fermented Feeds

20 Fermented Feeds and Feed
Products   769
R. D. Shaver, K. K. Batajoo

Index 785

Volume 10: Special Processes

Introduction   1
H.-J. Rehm

Energy by Microbial Processes

1 Bioelectrochemical Fuel Cells   5
D. Sell

Special Microorganisms of Biotechnological Interest

2 Maintenance of Cell Cultures – with Special Emphasis on Eukaryotic Cells   27
C. Bardouille

3 New Methods of Screening in Biotechnology   41
K. Frobel, S. Metzger

4 Biocatalysis under Extreme Conditions   61
C. Bertoldo, R. Grote, G. Antranikian

5 Biotechnology with Cyanobacteria and Microalgae   105
O. Pulz, K. Scheibenbogen, W. Groß

6 Biotechnology with Protozoa   137
A. Tiedtke

"Inorganic" Biotechnology

7 Biotechnology of Coal   153
J. Klein, R. Fakoussa, U. Hölker, M. Hofrichter, H. Schmiers, C. Sinder, A. Steinbüchel

8 Microbial Leaching of Metals   191
H. Brandl

9 Accumulation and Transformation of Metals by Microorganisms   225
G. M. Gadd

10 Microbial Corrosion and its Inhibition   265
W. Sand

Special Substances

11 Vitamins and Related Compounds: Microbial Production   319
S. Shimizu

12 Biochemistry of Polyketide Synthases   341
R. S. Gokhale, D. Tuteja

13 Biotechnological Production of Terpenoid Flavor and Fragrance Compounds   373
J. Schrader, R. G. Berger

14 New Degradable Resins   423
A. Steinbüchel

Biotechnology in Various Areas and Special Processes

15 Marine Biotechnology   441
U. Lindequist, T. Schweder

16 Biotechnology in Plant Protection   485
F. Niepold, K. Rudolph

Volume 11a: Environmental Processes I

Introduction   1
J. Winter

I General Aspects

1 Historical Development of Wastewater
Treatment Processes   5
L. Hartmann

2 Bacterial Metabolism in Wastewater
Treatment Systems   17
C. Gallert, J. Winter

3 Nitrification and Denitrification –
Microbial Fundamentals and
Consequences for Application   55
D. Zart, R. Stüven, E. Bock

4 Autoaggregation of Microorganisms:
Flocs and Biofilms   65
J. Wingender, H.-C. Flemming

5 Nucleic Acid-Based Techniques
for Analyzing the Diversity, Structure,
and Dynamics of Microbial Communities
in Wastewater Treatment   85
A. Schramm, R. Amann

6 Analytical Parameters for Monitoring
of Wastewater Treatment Processes   109
H. Kroiss, K. Svardal

7 Monitoring of Environmental Processes
with Biosensors   125
M. Reiss, W. Hartmeier

8 Laws, Statutory Orders and Directives
on Waste and Wastewater Treatment   141
P. Nisipeanu

II Processes of Wastewater Treatment
Waste Water Sources and Composition

9 Municipal Wastewater
and Sewage Sludge   161
P. Koppe, A. Stozek, V. Neitzel

10 Industrial Wastewater Sources
and Treatment Strategies   191
K.-H. Rosenwinkel, U. Austermann-Haun, H. Meyer

11 Agricultural Waste and Wastewater
Sources and Management   217
P. Weiland

Aerobic Carbon, Nitrogen, and Phosphate Removal

12 Biological Processes in Wetland Systems
for Wastewater Treatment   241
R. Kuschk, A. Wießner, U. Stottmeister

13 Activated Sludge Process   253
R. Kayser

14 Biological and Chemical Phosphorus
Removal   285
G. Schön, N. Jardin

15 Continuous Flow and Sequential
Processes in Municipal Wastewater
Treatment   321
E. Morgenroth, P. A. Wilderer

16 Trickling Filter Systems   335
    P. Baumann, B. Dorias

17 Submerged Fixed-Bed Reactors   349
    J. Schulz genannt Menningmann

18 Experience with Biofilters
    in Wastewater Treatment   365
    C. Sieker, M. Barjenbruch

19 Special Aerobic Wastewater
    and Sludge Treatment Processes   373
    U. Wiesmann, J. Libra

20 Modeling of Aerobic Wastewater
    Treatment Process   417
    M. Henze

Metal Ion Removal

21 Metal Removal by Biomass:
    Physico-Chemical Elimination
    Methods   431
    G. Bunke, P. Götz, R. Buchholz

Anaerobic Processes

22 Anaerobic Metabolism
    and its Regulation   455
    M. McInerney

23 CSTR-Reactors and Contact Processes
    in Industrial Wastewater Treatment   479
    H. Kroiss, K. Svardal

24 Fixed Film Stationary Bed
    and Fluidized Bed Reactors   493
    H.-J. Jördening, K. Buchholz

25 Possibilities and Potential of Anaerobic
    Wastewater Treatment Using Anaerobic
    Sludge Bed (ASB) Reactors   517
    *G. Lettinga, L. W. Hulshoff Pol,
    J. B. van Lier, G. Zeemann*

26 Modeling of Biogas Reactors   527
    H. Märkl

27 Future Aspects –
    Cleaner Production   561
    N. Räbiger

Index   579

Volume 11b: Environmental Processes II

Introduction   1
J. Klein

I General Aspects

1 Contaminated Soil Areas, Different
Countries and Contaminants, Monitoring
of Contaminants   5
W. Ulrici

2 Characterization of the Geological and
Hydrogeological Situation, Effects on
Natural Geochemical Barriers and
Remediation   43
W. G. Coldewey, C. Klinger

3 Bioavailability of Contaminants   61
B. Mahro

4 "Humification" Process or Formation of
Refractory Soil Organic Matter   89
M. Kästner

5 Ecotoxicological Assessment   127
A. Eisenträger, K. Hund

II Microbiological Aspects

6 Aerobic Degradation by Microorganisms
145
W. Fritsche, M. Hofrichter

7 Principles of Anaerobic Degradation of
Organic Compounds   169
B. Schink

8 Bacterial Degradation of Aliphatic
Hydrocarbons   193
J. E. T. van Hylckama Vlieg, D. B. Janssen

9 Degradation of Aromatic and
Polyaromatic Compounds   211
M. Kästner

10 Degradation of Chlorinated Compounds
241
C. Wischnak, R. Müller

11 Microbial Degradation of Compounds
with Nitro Functions   273
K.-H. Blotevogel, T. Gorontzy

III Processes for Soil Clean-Up

12 Thermal Processes, Scrubbing/Extraction,
Bioremediation and Disposal   305
M. Koning, K. Hupe, R. Stegmann

13 Bioremediation with Heap Technique
319
V. Schulz-Berendt

14 Bioreactors   329
R. H. Kleijntjens, K. Ch. A. M. Luyben

15 *In situ* Remediation   349
T. Held, H. Dörr

16 Degradation by Plants – Phytoremediation of Organics   371
J. L. Schnoor

17 Phytoremediation of Metals   385
D. E. Salt, A. J. M. Baker

18 Advanced *in situ* Bioremediation –
A Hierarchy of Technology Choices   399
R. Unterman, M. F. DeFlaun, R. J. Steffan

Volume 11c: Environmental Processes III

Introduction   1
J. Winter

I Solid Waste Treatment

1 Bio- and Pyrotechnology of Solid Waste
   Treatment   5
   C. Gallert, J. Winter

2 Microbiology of Composting   35
   H. J. Kutzner

3 Composting of Plant Residues and Waste
   Plant Materials   101
   F. Schuchardt

4 Technology and Strategies of
   Composting   127
   U. Krogmann, I. Körner

5 Anaerobic Fermentation of Wet or
   Semi-Dry Garbage Waste Fractions   151
   N. Rilling

6 Landfill Systems, Sanitary Landfilling of
   Solid Wastes – Long-Term Problems with
   Leachates   167
   K.-U. Heyer, R. Stegmann

7 Sanitary Landfills – Long-Term Stability
   and Environmental Implications   191
   M. Switzenbaum

8 Combined Mechanical and Biological
   Treatment of Municipal Solid Waste   203
   P. Schalk

9 Hygienic Considerations of Aerobic/
   Anaerobic Treatment of Wastewater,
   Sludge, Biowaste, and Compost   213
   D. Strauch, W. Philipp, R. Böhm

10 Future Settlement Structures with
   Minimized Waste and Wastewater
   Generation   249
   R. Otterpohl

II Waste Gas Treatment

General Aspects

11 Process Engineering of Biological Waste
   Gas Purification   259
   *M. Waweru, V. Herrygers,
   H. Van Langenhove, W. Verstraete*

12 Microbiological Aspects of Biological
   Waste Gas Purification   275
   K.-H. Engesser, T. Plaggemeier

Processes

13 Bioscrubbers   305
   E. Schippert, H. Chmiel

14 Biofilters   321
   K. Fischer

15 Treatment of Waste Gas Pollutants in
   Trickling Filters   333
   T. Plaggemeier, O. Lämmerzahl

16 Membrane Processes and Alternative
   Techniques   345
   M. Reiser

Applications

17 Commercial Applications of Biological
   Waste Gas Purification   357
   D. E. Chitwood, J. S. Devinny

III Drinking Water Preparation

Index   489

Volume 12: Legal, Economic and Ethical Dimensions

Introduction   1
D. Brauer

I. Modern Biotechnology – What Is New About It?

1 The Evaluation of Technology as an Interactive Commitment-Building Process – The Failure of Technology Assessment   5
G. Van Steendam

2 Concepts of Risk Assessment: The "Process versus Product" Controversy Put to the Rest   39
H. I. Miller

3 Biosafety in rDNA Research and Production   63
D. Brauer, M. Bröker, C. Kellermann, E.-L. Winnacker

4 Biotechnology and Bioethics: What is Ethical Biotechnology?   115
D. R. J. Macer

II. Product Development and Legal Requirements

5 Structured Risk Assessment of rDNA Products and Consumer Acceptance of These Products   157
C. T. Verrips

6 Strategic Regulations for Safe Development of Transgenic Plants   197
T. L. Medley, S. L. McCammon

7 Biomedicinal Product Development   213
J.-P. Gregersen

8 Regulations for Recombinant DNA Research, Product Development and Production in the US, Japan and Europe   239
D. Brauer, H. D. Schlumberger

III. Intellectual Property and Bioinformatics

9 Biotechnology and Intellectual Property   281
J. Straus

10 Patent Applications for Biomedicinal Products   299
J.-P. Gregersen

11 Databases in Biotechnology   323
E. Poetzsch

IV. Biotechnology in a Developing World

12 Commercial Biotechnology: Developing World Prospects   339
G. T. Tzotzos, M. Leopold

13 Biotechnology in the Asian-Pacific Region   369
R. D. Schmid, B.-H. Chung, A. J. Jones, S. Saono, J. Scriven, J. H. J. Tsai

14 Biotechnology and Biological Diversity   433
M. Kawai

Cumulative Author Index
of Volumes 1 to 12

The figures in bold give the volume number (before the point) and the chapter number (after the point), respectively. The figure thereafter is the page number.

Adair, J. R.
– Antibody Engineering and Expression **5a**.7, 219

Adams, Gregor B.
– Gene Therapy for HIV Infection **5a**.21, 471

Aehle, W.
– Enzymes for Industrial Applications **5a**.6, 189

Alfermann, A. W.
– Plant Cell Cultures **1**.17, 577

Altschmied, L.
– Genome Projects of Model Organisms **5b**.1, 5

Amann, R.
– Nucleic Acid-Based Techniques for Analyzing the Diversity, Structure, and Dynamics of Microbial Communities in Wastewater Treatment **11a**.5, 85

Andrews, G. F.
– Aerobic Waste Water Process Models **4**.13, 407

Anke, T.
– Products from Basidiomycetes **7**.11, 489

Anné, J.
– Cell Fusion **2**.4, 93

Antranikian, G.
– Biocatalysis under Extreme Conditions **10**.4, 61

Apweiler, R.
– Using the Molecular Biology Data **5b**.12, 281

Auling, G.
– Pseudomonads **1**.12, 401

Aunins, J. G.
– Oxygen Transfer in Cell Culture Bioreactors **3**.11, 219

Austermann-Haun, U.
– Industrial Wastewater Sources and Treatment Strategies **11a**.10, 191

Bader, G. D.
– Interaction Databases **5b**.18, 399

Bahl, H.
– Clostridia **1**.9, 285

Bahl, H.
– Microbial Production of Acetone/Isopropanol **6**.6, 229

Bajpai, R.
– Stirred Tank Models **4**.10, 299

Baker, A. J. M.
– Phytoremediation of Metals **11b**.17, 385

Bardouille, C.
– Maintenance of Cell Cultures – with Special Emphasis on Eukaryotic Cells **10**.2, 27

Barjenbruch, M.
– Experience with Biofilters in Wastewater Treatment **11a**.18, 365

Batajoo, K. K.
– Fermented Feeds and Feed Products **9**.20, 769

Baumann, P.
– Trickling Filter Systems **11a**.16, 335

Beatrice, M. G.
– Biotechnology Facility Design and Process Validation **3**.28, 739

Bellgardt, K.-H.
– Cell Models **4**.9, 267

Bellgardt, K.-H.
– Process Models: Optimization of Yeast Production – A Case Study **4**.12, 383

Bentley, P.
– Synthetic Enzymes: Artificial Peptides in Stereoselective Synthesis **8b**.12, 491

Berger, M. S.
– Antibody Targeted Chemotherapy **5a**.15, 355

Berger, R. G.
– Biotechnological Production of Terpenoid Flavor and Fragrance Compounds **10**.13, 373

Berlin, J.
– Secondary Products from Plant Cell Cultures **7**.13, 593

Bertoldo, C.
– Biocatalysis under Extreme Conditions
   10.4, 61
Beuchat, L. R.
– Indigenous Fermented Foods   **9**.13, 505
Bibb, M.
– Regulation of Bacterial Antibiotic Production
   7.2, 57
Blackburn, G. M.
– Catalytic Antibodies   **8b**.11, 403
Blotevogel, K.-H.
– Microbial Degradation of Compounds with Nitro
   Functions   **11b**.11, 273
Bock, E.
– Nitrification and Denitrification – Microbial
   Fundamentals and Consequences for Application
   11a.3, 55
Bommarius, A. S.
– Biotransformations and Enzyme
   Reactors   **3**.17, 427
Bonnerjea, J.
– Manufacture of Therapeutic Antibodies   **5a**.8,
   245
Bork, P.
– Sequence and Structure of Proteins   **5a**.2, 43
Bornscheuer, U. T.
– Biotransformations with Lipases   **8a**.3, 37
Bornscheuer, U. T.
– Industrial Biotransformations   **8b**.6, 277
Boze, H.
– Production of Microbial Biomass   **9**.5, 167
Böhm, R.
– Hygienic Considerations of Aerobic/Anaerobic
   Treatment of Wastewater, Sludge, Biowaste, and
   Compost   **11c**.9, 213
Brandl, H.
– Microbial Leaching of Metals   **10**.8, 191
Brauer, D.
– Biosafety in rDNA Research and Production
   12.3, 63
Brauer, D.
– Introduction   **12**, 1
Brauer, D.
– Regulations for Recombinant DNA Research,
   Product Development and Production in the US,
   Japan and Europe   **12**.8, 239
Breidt, F.
– Vegetable Fermentations   **9**.17, 629
Brem, G.
– Transgenic Animals   **2**.18, 745
Brenes Balbuena, M.
– Olive Fermentations   **9**.16, 593
Brousseau, R.
– A DNA Microarrays Fabrication Strategy for
   Research Laboratories   **5b**.9, 227
Bröker, M.
– Biosafety in rDNA Research and Production
   12.3, 63

Bruce, N.
– Alkaloids   **8a**.7, 327
Brümmer, J.-M.
– Baked Goods   **9**.7, 241
Buchholz, K.
– Fixed Film Stationary Bed and Fluidized Bed
   Reactors   **11a**.24, 493
Buchholz, K.
– Sugar-Based Raw Materials for Fermentation
   Applications   **6**.1a, 5
Buchholz, R.
– Metal Removal by Biomass: Physico-Chemical
   Elimination Methods   **11a**.21, 431
Buckland, B. C.
– Fermentation – An Overview   **3**.1, 7
Builder, S. E.
– Proteins and Peptides as Drugs: Sources and
   Methods of Purification   **3**.4, 57
Bunch, A.
– Nitriles   **8a**.6, 277
Bunke, G.
– Metal Removal by Biomass: Physico-Chemical
   Elimination Methods   **11a**.21, 431
Buttgereit, D.
– Genome Projects of Model Organisms   **5b**.1, 5
Cantley, M. F.
– The Regulation of Modern Biotechnology: A
   Historical and European Perspective   **12**.18,
   505
Caron, C.
– Commercial Production of Baker's Yeast and
   Wine Yeast   **9**.8, 321
Carter, B. J.
– Viral Vectors for Gene Therapy   **5a**.18, 395
Cavalleri, B.
– Glycopeptide Antibiotics (Dalbaheptides)   **7**.9,
   369
Chandan, R. C.
– Other Fermented Dairy Products   **9**.10, 385
Charles, M.
– Design of Aseptic, Aerated Fermentors   **3**.16,
   401
Chater, K.
– Regulation of Bacterial Antibiotic Production
   7.2, 57
Chattaway, T.
– Fermentation Monitoring and Control   **3**.14,
   319
Chiou, T.-W.
– Industrial Animal Cell Culture   **3**.2, 23
Chitwood, D. E.
– Commercial Applications of Biological Waste
   Gas Purification   **11c**.17, 357
Chmiel, H.
– Bioscrubbers   **11c**.13, 305
Chung, B.-H.
– Biotechnology in the Asian-Pacific Region
   12.13, 369

Cleland, J. L.
– *In vitro* Protein Refolding   **3**.20, 527
Coldewey, W. G.
– Characterization of the Geological and
  Hydrogeological Situation, Effects on Natural
  Geochemical Barriers and Remediation   **11b**.2,
  43
Cooney, C. L.
– Growth of Microorganisms   **1**.3, 111
Cooney, C. L.
– Media and Air Sterilization   **3**.9, 157
Cornish-Bowden, A.
– Kinetics of Multi-Enzyme Systems   **9**.3, 121
Cramer, H.
– Microencapsulation-Based Cell
  Therapy   **10**.19, 547
Crueger, A.
– Mutagenesis   **2**.1, 5
Curtis, A. D. M.
– Carbon–Carbon Bond Formation Using
  Enzymes   **8b**.1, 5
Cygler, M.
– Structure Information   **5b**.15, 345
Danchin, A. E.
– Genome Projects of Model Organisms   **5b**.1, 5
Datar, R. V.
– Cell and Cell Debris Removal: Centrifugation
  and Crossflow Filtration   **3**.18, 469
Davison, D. B.
– Pharmaceutical Bioinformatics and Drug
  Discovery   **5b**.5, 123
DeFlaun, M. F.
– Advanced *in situ* Bioremediation – A Hierarchy
  of Technology Choices   **11b**.18, 399
Dellaire, G.
– Genetic Disease   **5b**.3, 61
Devinny, J. S.
– Commercial Applications of Biological Waste
  Gas Purification   **11c**.17, 357
Dijkhuizen, L.
– Methylotrophs   **1**.8, 265
Dimick, P. S.
– Cocoa Fermentation   **9**.14, 561
Distler, J.
– Aminoglycosides and Sugar Components in
  Other Secondary Metabolites   **7**.10, 397
Dittrich, H. H.
– Wine and Brandy   **9**.12, 463
Dorias, B.
– Trickling Filter Systems   **11a**.16, 335
Dornberger, K.
– Biotechnical Drugs as Antitumor Agents   **7**.14,
  641
Dovichi, N. J.
– Proteome Analysis by Capillary Electrophoresis
  5b.11, 269
Dörr, H.
– *In situ* Remediation   **11b**.15, 349

Drechsel, H.
– Microbial Siderophores   **7**.5, 199
Dunn, I. J.
– Methods and Instruments in Fermentation
  Gas Analysis   **4**.2, 27
Dürre, P.
– Clostridia   **1**.9, 285
Dürre, P.
– Microbial Production of Acetone/Isopropanol
  6.6, 229
Easter, J. P.
– Bacterial Activity Enhancement and Soil
  Decontamination   **11b**.20, 425
Easter, J. P.
– Genetically Engineered Microorganisms and
  Bioremediation   **11b**.21, 441
Ebner, H.
– Acetic Acid   **6**.12, 381
Ebner, H.
– Vinegar   **9**.15, 579
Ehmig, S. C.
– Press Coverage of Genetic Engineering in
  Germany: Facts, Faults and Causes   **12**.17, 495
Eisenhaber, F.
– Sequence and Structure of Proteins   **5a**.2, 43
Eisenträger, A.
– Ecotoxicological Assessment   **11b**.5, 127
Engels, J. W.
– DNA Synthesis   **2**.9, 317
Engesser, K.-H.
– Microbiological Aspects of Biological Waste Gas
  Purification   **11c**.12, 275
Erickson, L. E.
– Anaerobic Fermentations   **3**.13, 295
Erkel, G.
– Products from Basidiomycetes   **7**.11, 489
Esaki, N.
– Enzymology of Amino Acid Production   **6**.14b,
  503
Esser, K.
– Filamentous Fungi   **1**.15, 515
Esser, K.
– Genetic Exchange Processes in Lower Eukary-
  otes   **2**.3, 73
Etzold, T.
– Using the Molecular Biology Data   **5b**.12, 281
Fakoussa, R.
– Biotechnology of Coal   **10**.7, 153
Feldmann, H.
– Genome Projects of Model Organisms   **5b**.1, 5
Figeys, D.
– Two-Dimensional Gel Electrophoresis and Mass
  Spectrometry for Proteomic Studies: State-of-
  the-Art   **5b**.10, 241
Fischer, K.
– Biofilters   **11c**.14, 321
Fleming, H. P.
– Vegetable Fermentations   **9**.17, 629

Flemming, H.-C.
– Autoaggregation of Microorganisms: Flocs and Biofilms   **11a**.4, 65

Flemming, H.-C.
– Biofilm in Biofiltration   **11c**.21, 445

Fliri, H.
– Cyclosporins: Recent Developments in Biosynthesis, Pharmacology and Biology, and Clinical Applications   **7**.12, 535

Flitsch, S.
– Enzymes in Carbohydrate Chemistry: Formation of Glycosidic Linkages   **8b**.5, 243

Follmann, H.
– Acetic Acid   **6**.12, 381

Follmann, H.
– Vinegar   **9**.15, 579

Foster, K. A.
– Production of Enzymes as Fine Chemicals   **9**.2, 73

Fotheringham, I.
– Engineering Microbial Pathways for Amino Acid Production   **8b**.8, 313

Foulkes, J.
– Regulation of Antibodies and Recombinant Proteins   **5a**.22, 495

Foulkes, R.
– Preclinical Testing of Antibodies: Pharmacology, Kinetics, and Immunogenicity   **5a**.10, 289

Frackman, S.
– Production of Enzymes as Fine Chemicals   **9**.2, 73

Fritsche, W.
– Aerobic Degradation by Microorganisms   **11b**.6, 145

Frobel, K.
– New Methods of Screening in Biotechnology   **10**.3, 41

Frommer, W.
– Safety Aspects in Biotechnology   **2**.19, 835

Fuhr, G.
– Microencapsulation-Based Cell Therapy   **10**.19, 547

Fung, D. Y. C.
– Anaerobic Fermentations   **3**.13, 295

Furuyoshi, S.
– Enzymology of Amino Acid Production   **6**.14b, 503

Gaasterland, T.
– Automated Genome Annotation and Comparative Genomics   **5b**.16, 361

Gaasterland, T.
– Genomic Data Representation through Images – MAGPIE as an Example   **5b**.17, 379

Gadd, G. M.
– Accumulation and Transformation of Metals by Microorganisms   **10**.9, 225

Gallert, C.
– Bacterial Metabolism in Wastewater Treatment Systems   **11a**.2, 17

Gallert, C.
– Bio- and Pyrotechnology of Solid Waste Treatment   **11c**.1, 5

Gallert, C.
– Perspectives of Waste, Wastewater, Off-Gas, and Drinking Water   **11c**.23, 479

Galzy, P.
– Production of Microbial Biomass   **9**.5, 167

García García, P.
– Olive Fermentations   **9**.16, 593

Garçon, A.
– Catalytic Antibodies   **8b**.11, 403

Garnick, R. L.
– Proteins and Peptides as Drugs: Sources and Methods of Purification   **3**.4, 57

Garrido Fernández, A.
– Olive Fermentations   **9**.16, 593

Gassen, H. G.
– Types and Function of Proteins   **5a**.1, 1

Geldermann, H.
– Gene Mapping in Animals and Plants   **2**.5, 141

Gimbel, R.
– Potable Water Treatment   **11c**.18, 371

Gokhale, R. S.
– Biochemistry of Polyketide Synthases   **10**.12, 341

Gopal, S.
– Automated Genome Annotation and Comparative Genomics   **5b**.16, 361

Gordon, P.
– Genomic Data Representation through Images – MAGPIE as an Example   **5b**.17, 379

Gorontzy, T.
– Microbial Degradation of Compounds with Nitro Functions   **11b**.11, 273

Gölker, C. F.
– Final Recovery Steps: Lyophilization, Spray-Drying   **3**.26, 695

Götz, F.
– Lantibiotics   **7**.8, 323

Götz, F.
– Structure and Function of DNA   **2**.6, 191

Götz, P.
– Metal Removal by Biomass: Physico-Chemical Elimination Methods   **11a**.21, 431

Gräfe, U.
– Biotechnical Drugs as Antitumor Agents   **7**.14, 641

Gräfe, U.
– General Aspects of Secondary Metabolism   **7**.1, 1

Greasham, R. L.
– Media for Microbial Fermentations   **3**.7, 127

Green, R. C.
– Genomics and Human Disease   **5b**.4, 105

Gregersen, J.-P.
– Biomedicinal Products Development   **12**.7, 213
Gregersen, J.-P.
– Patent Applications for Biomedicinal Products **12**.10, 299
Grodzinsky, A. J.
– Electrokinetic Separations   **3**.25, 679
Groß, W.
– Biotechnology with Cyanobacteria and Microalgae   **10**.5, 105
Grote, R.
– Biocatalysis under Extreme Conditions **10**.4, 61
Halme, A.
– Expert Systems for Biotechnology   **4**.19, 625
Hamann, P. R.
– Antibody Targeted Chemotherapy   **5a**.15, 355
Hartmann, L.
– Historical Development of Wastewater Treatment Processes   **11a**.1, 5
Hartmeier, W.
– Monitoring of Environmental Processes with Biosensors   **11a**.7, 125
Hasse, C.
– Microencapsulation-Based Cell Therapy   **10**.19, 547
Hatton, T. A.
– Protein Purification by Liquid-Liquid Extraction   **3**.22, 593
Hauser, H.
– Genetic Engineering of Animal Cells   **2**.17, 663
Hebeda, R. F.
– Carbohydrate-Based Sweeteners   **9**.19, 737
Hegedus, D.
– Agri-Food and Genomics   **5b**.6, 145
Heinisch, J. J.
– Yeasts   **1**.14, 469
Heinzle, E.
– Methods and Instruments in Fermentation Gas Analysis   **4**.2, 27
Held, T.
– *In situ* Remediation   **11b**.15, 349
Henze, M.
– Modeling of Aerobic Wastewater Treatment Processes   **11a**.20, 417
Henzler, H.-J.
– Oxygen Transfer in Cell Culture Bioreactors **3**.11, 219
Herrygers, V.
– Process Engineering of Biological Waste Gas Purification   **11c**.11, 259
Heyer, K.-U.
– Landfill Systems, Sanitary Landfilling of Solid Wastes – Long-Term Problems with Leachates **11c**.6, 167
Hodgdon, J. C.
– Proteins and Peptides as Drugs: Sources and Methods of Purification   **3**.4, 57

Hofrichter, M.
– Aerobic Degradation by Microorganisms   **11b**.6, 145
Hofrichter, M.
– Biotechnology of Coal   **10**.7, 153
Hogue, C. W. V.
– Interaction Databases   **5b**.18, 399
Holland, H.
– Hydroxylation and Dihydroxylation   **8a**.10, 475
Hollenberg, C. P.
– Yeasts   **1**.14, 469
Holloway, B. W.
– Genetic Exchange Processes for Prokaryotes **2**.2, 47
Hoople, D.
– Cleavage and Formation of Amide Bonds   **8a**.5, 243
Hölker, U.
– Biotechnology of Coal   **10**.7, 153
Hu, S.
– Proteome Analysis by Capillary Electrophoresis **5b**.11, 269
Hu, W.-S.
– Cell Culture Bioreactors   **3**.6, 105
Hulshoff Pol, L. W.
– Possibilities and Potential of Anaerobic Wastewater Treatment Using Anaerobic Sludge Bed (ASB) Reactors   **11a**.25, 517
Humphrey, P. G.
– Sequencing Technology   **5b**.8, 183
Hund, K.
– Ecotoxicological Assessment   **11b**.5, 127
Hupe, K.
– Thermal Processes, Scrubbing/Extraction, Bioremediation and Disposal   **11b**.12, 305
Huttner, S. L.
– Government, Researchers and Activists: The Critical Public Policy Interface   **12**.16, 459
Imanaka, T.
– Strategies for Fermentation with Recombinant Organisms   **3**.12, 283
Jack, R.
– Lantibiotics   **7**.8, 323
Jackman, S.
– Halocompounds   **8b**.3, 173
James, R. W.
– Preclinical Testing of Antibodies: Safety Aspects **5a**.11, 303
Janson, J.-C.
– Protein Separation and Purification   **3**.23, 617
Janssen, D. B.
– Bacterial Degradation of Aliphatic Hydrocarbons   **11b**.8, 193
Jardin, N.
– Biological and Chemical Phosphorus Removal **11a**.14, 285

Jolly, J. F.
– Production of Enzymes as Fine Chemicals   **9**.2, 73

Jones, A. J.
– Biotechnology in the Asian-Pacific Region   **12**.13, 369

Jones, S.
– Phosphorylation   **8b**.4, 221

Jording, D.
– Genome Projects of Model Organisms   **5b**.1, 5

Jork, A.
– Microencapsulation-Based Cell Therapy   **10**.19, 547

Jördening, H.-J.
– Fixed Film Stationary Bed and Fluidized Bed Reactors   **11a**.24, 493

Jung, G.
– Lantibiotics   **7**.8, 323

Kahl, G.
– Genetic Engineering of Plant Cells   **2**.15, 547

Kalinowski, J.
– Genetic Engineering of Gram Positive Bacteria   **2**.12, 455

Kalinowski, J.
– Genome Projects of Model Organisms   **5b**.1, 5

Kallen, J.
– Cyclosporins: Recent Developments in Biosynthesis, Pharmacology and Biology, and Clinical Applications   **7**.12, 535

Karim, N.
– Expert Systems for Biotechnology   **4**.19, 625

Kästner, M.
– "Humification" Process or Formation of Refractory Soil Organic Matter   **11b**.4, 89

Kästner, M.
– Degradation of Aromatic and Polyaromatic Compounds   **11b**.9, 211

Kaščák, J. S.
– Lactic Acid   **6**.8, 293

Kawai, M.
– Biotechnology and Biological Diversity   **12**.14, 433

Kayser, R.
– Activated Sludge Process   **11a**.13, 253

Kazlauskas, R.
– Biotransformations with Lipases   **8a**.3, 37

Kellermann, C.
– Biosafety in rDNA Research and Production   **12**.3, 63

Kelley, B. D.
– Industrial Animal Cell Culture   **3**.2, 23

Kelley, B. D.
– Protein Purification by Liquid-Liquid Extraction   **3**.22, 593

Kelly, D. R.
– Biotransformations – Practical Aspects   **8a**.2, 25

Kelly, D. R.
– Flavin Monooxygenases – Uses as Catalysts for Baeyer–Villiger Ring Expansion and Heteroatom Oxidation   **8a**.11, 535

Kelly, D. R.
– Introduction   **8a**, 1

Kelly, D. R.
– Introduction   **8b**, 1

Kelly, D. R.
– Lyases   **8b**.2, 41

Kepplinger, H. M.
– Press Coverage of Genetic Engineering in Germany: Facts, Faults and Causes   **12**.17, 495

Kim, J.-Y.
– Bioprocess Kinetics and Modelling of Recombinant Fermentation   **4**.15, 485

King, D. J.
– Use of Antibodies for Immunopurification   **5a**.9, 275

Kleijntjens, R. H.
– Bioreactors   **11b**.14, 329

Klein, J.
– Biotechnology of Coal   **10**.7, 153

Klein, J.
– Introduction   **11b**, 1

Klein, J.
– Possibilities, Limits, and Future Developments of Soil Bioremediation   **11b**.22, 465

Kleinkauf, H.
– Introduction   **7**, 1

Kleinkauf, H.
– Peptide Antibiotics   **7**.7, 277

Klenk, H.-P.
– Genome Projects of Model Organisms   **5b**.1, 5

Klinger, C.
– Characterization of the Geological and Hydrogeological Situation, Effects on Natural Geochemical Barriers and Remediation   **11b**.2, 43

Knoppers, B. M.
– Ethical Aspects of Genome Research and Banking   **5b**.19, 417

Komínek, J.
– Citric Acid   **6**.9, 307

Komínek, J.
– Gluconic Acid   **6**.10, 347

Komínek, J.
– Lactic Acid   **6**.8, 293

Koning, M.
– Thermal Processes, Scrubbing/Extraction, Bioremediation and Disposal   **11b**.12, 305

Konstantinov, K.
– Fermentation Data Analysis for Diagnosis and Control   **3**.15, 355

Koppe, P.
– Municipal Wastewater and Sewage Sludge   **11a**.9, 161

Kosaric, N.
– Biosurfactants   **6**.17, 659

Kosaric, N.
– Ethanol – Potential Source of Energy and
  Chemical Products   **6**.4, 121
König, H.
– Methanogens   **1**.7, 251
Körner, J.
– Technology ancl Strategies of Composting
  11c.4, 127
Krämer, R.
– Metabolism   **1**.2, 47
Kresse, G.-B.
– Analytical Uses of Enzymes   **9**.4, 137
Krogmann, U.
– Technology ancl Strategies of Composting
  11c.4, 127
Kroiss, H.
– Analytical Parameters for Monitoring of Waste-
  water Treatment Processes   **11a**.6, 109
Kroiss, H.
– CSTR-Reactors and Contact Processes in
  Industrial Wastewater Treatment   **11a**.23, 479
Kröger, M.
– Genome Projects of Model Organisms   **5b**.1, 5
Krylov, S. N.
– Proteome Analysis by Capillary Electrophoresis
  5b.11, 269
Kubicek, C. P.
– Citric Acid   **6**.9, 307
Kubicek, C. P.
– Further Organic Acids   **6**.11, 363
Kubicek, C. P.
– Gluconic Acid   **6**.10, 347
Kula, M.-R.
– Cell Disruption and Isolation of Non-Secreted
  Products   **3**.19, 505
Kuninaka, A.
– Nucleotides and Related Compounds   **6**.15, 561
Kurihara, T.
– Enzymology of Amino Acid Production   **6**.14b,
  503
Kuschk, P.
– Biological Processes in Wetland Systems for
  Wastewater Treatment   **11a**.12, 241
Kutzner, H. J.
– Microbiology of Composting   **11c**.2, 35
Kyung, K. H.
– Vegetable Fermentations   **9**.17, 629
Lämmerzahl, O.
– Treatment of Waste Gas Pollutants in Trickling
  Filters   **11c**.15, 333
Lancini, G.
– Glycopeptide Antibiotics (Dalbaheptides)   **7**.9,
  369
Lee, K.-S.
– Control of Bioreactor Systems   **4**.16, 509
Lee, S. B.
– Bioprocess Kinetics and Modelling of
  Recombinant Fermentation   **4**.15, 485

Leopold, M.
– Commercial Biotechnology: Developing World
  Prospects   **12**.12, 339
Lettinga, G.
– Possibilities and Potential of Anaerobic
  Wastewater Treatment Using Anaerobic Sludge
  Bed (ASB) Reactors   **11a**.25, 517
Leuchtenberger, W.
– Amino Acids – Technical Production and Use
  6.14a, 465
Liberman, D. F.
– Treatment of Biological Waste   **3**.29, 769
Libra, J.
– Special Aerobic Wastewater and Sludge
  Treatment Processes   **11a**.19, 373
Liempt, H. v.
– Advances in the Molecular Genetics of β-Lactam
  Antibiotic Biosynthesis   **7**.6, 247
Lilie, H.
– In vitro Folding of Inclusion Body Proteins
  on an Industrial Scale   **5a**.4, 111
Lilly, M. D.
– Fermentation – An Overview   **3**.1, 7
Lim, H. C.
– Control of Bioreactor Systems   **4**.16, 509
Lindequist, U.
– Marine Biotechnology   **10**.15, 441
Lopez, A. S.
– Cocoa Fermentation   **9**.14, 561
Lopez, R.
– Using the Molecular Biology Data   **5b**.12, 281
Lübbert, A.
– Automation in Biotechnology   **4**.17, 561
Lübbert, A.
– Characterization of Bioreactors   **4**.4, 107
Luyben, K. Ch. A. M.
– Bioreactors   **11b**.14, 329
Lydiate, D. J.
– Agri-Food and Genomics   **5b**.6, 145
Macer, D. R. J.
– Biotechnology and Bioethics: What is Ethical
  Biotechnology?   **12**.4, 115
Maddox, I. S.
– Microbial Production of 2,3-Butanediol   **6**.7,
  269
Mahdi, J.
– Lyases   **8b**.2, 41
Mahro, B.
– Bioavailability of Contaminants   **11b**.3, 61
Mälzer, H.-J.
– Potable Water Treatment   **11c**.18, 371
Märkl, H.
– Modeling of Biogas Reactors   **11a**.26, 527
Matte, A.
– Structure Information   **5b**.15, 345
Mattes, R.
– Principles of Gene Expression   **2**.7, 233
Mattiasson, B.

– Biosensors   **4**.3, 75

Mayer, F.
– Cell Structure   **1**.1, 5

McCammon, S. L.
– Strategic Regulations for Safe Development of Transgenic Plants   **12**.6, 197

McClure, Myra
– Gene Therapy for HIV Infection   **5a**.21, 471

McGregor-Jones, J.
– Synthetic Applications of Enzyme-Catalyzed Reactions   **8b**.10, 351

McInerney, M.
– Anaerobic Metabolism and its Regulation   **11a**.22, 455

Medley, T. L.
– Strategic Regulations for Safe Development of Transgenic Plants   **12**.6, 197

Meinhardt, F.
– Filamentous Fungi   **1**.15, 515

Menn, Fu-Min
– Bacterial Activity Enhancement and Soil Decontamination   **11b**.20, 425

Menn, Fu-Min
– Genetically Engineered Microorganisms and Bioremediation   **11b**.21, 441

Merchuk, J. C.
– Tower Reactor Models   **4**.11, 349

Metzger, S.
– New Methods of Screening in Biotechnology   **10**.3, 41

Meyer, H.
– Industrial Wastewater Sources and Treatment Strategies   **11a**.10, 191

Meyer, J.
– Bacteriophages   **1**.16, 543

Meyer, M.
– Protein Interactions   **5a**.3, 87

Michels, D.
– Proteome Analysis by Capillary Electrophoresis   **5b**.11, 269

Middendorf, L. R.
– Sequencing Technology   **5b**.8, 183

Mikol, V.
– Cyclosporins: Recent Developments in Biosynthesis, Pharmacology and Biology, and Clinical Applications   **7**.12, 535

Miller, H. I.
– Concepts of Risk Assessment: The "Process versus Product" Controversy Put to the Rest   **12**.2, 39

Millet, A.
– "Oui" or "Non" to Biotechnology: The Other French Referendum   **12**.15, 449

Misset, O.
– Enzymes for Industrial Applications   **5a**.6, 189

Montague, G. A.
– Fermentation Monitoring and Control   **3**.14, 319

Morgenroth, E.
– Continuous Flow and Sequential Processes in Municipal Wastewater Treatment   **11a**.15, 321

Morris, A. J.
– Fermentation Monitoring and Control   **3**.14, 319

Moulin, G.
– Production of Microbial Biomass   **9**.5, 167

Mountain, A.
– Overview of Gene Therapy   **5a**.17, 383

Müller, R.
– Degradation of Chlorinated Compounds   **11b**.10, 241

Munack, A.
– Optimization of Sampling   **4**.8, 251

Murray, E. B.
– Nutritional Value and Safety of "Single Cell Protein"   **9**.6, 221

Muth, G.
– Genetic Engineering of Gram Positive Bacteria   **2**.12, 455

Nagodawithana, T. W.
– Introduction   **9**, 1

Nakamori, S.
– Enzymology of Amino Acid Production   **6**.14b, 503

Narayanan, N.
– Sequencing Technology   **5b**.8, 183

Neijssel, O. M.
– Overproduction of Metabolites   **1**.4, 163

Neitzel, V.
– Municipal Wastewater and Sewage Sludge   **11a**.9, 161

Ney, U.
– Introduction   **5a**, XV

Nicholson, I. J.
– Economic Considerations   **5a**.24, 531

Nielsen, J.
– Bioreactors: Description and Modelling   **3**.5, 77

Niepold, F.
– Biotechnology in Plant Protection   **10**.16, 485

Nisipeanu, P.
– Laws, Statutory Orders and Directives on Waste and Wastewater Treatment   **11a**.8, 141

Ōmura, S.
– Screening of Novel Receptor-Active Compounds of Microbial Origin   **7**.3, 107

Ogez, J. R.
– Proteins and Peptides as Drugs: Sources and Methods of Purification   **3**.4, 57

Olsen, H. S.
– Use of Enzymes in Food Processing   **9**.18, 663

Olson, N. F.
– Cheese   **9**.9, 353

Omar, S. H.
– Special Morphological and Metabolic Behavior of Immobilized Micro-organisms   **1**.6, 223

Opal, S. M.
– Antibodies for Sepsis: Some Lessons Learnt
  5a.13, 329
Otterpohl, R.
– Future Settlement Structures with Minimized
  Waste and Wastewater Generation   **11c**.10, 249
Park, S.
– Bioreactor State Estimation   **4**.7, 225
Parker, J.
– Regulation of Human Gene Therapy   **5a**.23, 517
Parkin, I. A. P.
– Agri-Food and Genomics   **5b**.6, 145
Peters, J.
– Alcohol Dehydrogenases – Characteristics,
  Design of Reaction Conditions, and Applications
  8a.9, 391
Petersen, M.
– Plant Cell Cultures   **1**.17, 577
Philipp, W.
– Hygienic Considerations of Aerobic/Anaerobic
  Treatment of Wastewater, Sludge, Biowaste, and
  Compost   **11c**.9, 213
Phythian, S.
– Esterases   **8a**.4, 193
Pieper, H. J.
– Ethanol – Classical Methods   **6**.3, 59
Piepersberg, W.
– Aminoglycosides and Sugar Components in
  Other Secondary Metabolites   **7**.10, 397
Piepersberg, W.
– Streptomycetes and Corynebacteria   **1**.13, 433
Plaggemeier, T.
– Microbiological Aspects of Biological Waste Gas
  Purification   **11c**.12, 275
Plaggemeier, T.
– Treatment of Waste Gas Pollutants in Trickling
  Filters   **11c**.15, 333
Poetzsch, E.
– Databases in Biotechnology   **12**.11, 323
Posten, C. H.
– Growth of Microorganisms   **1**.3, 111
Poznansky, M.
– Gene Therapy for HIV Infection   **5a**.21, 471
Preuß, G.
– Artificial Groundwater Recharge and Bank
  Filtration   **11c**.20, 425
Priefer, U. B.
– Genetic Engineering of Gram Negative Bacteria
  2.11, 427
Priest, F. C.
– Bacillus   **1**.11, 367
Pühler, A.
– Genome Projects of Model Organisms   **5b**.1, 5
Pühler, A.
– Introduction   **2**, 1
Pulz, O.
– Biotechnology with Cyanobacteria and
  Microalgae   **10**.5, 105

Quesniaux, V. F. J.
– Cyclosporins: Recent Developments in
  Biosynthesis, Pharmacology and Biology, and
  Clinical Applications   **7**.12, 535
Rabenhorst, J.
– Biotechnological Production of Natural Aroma
  Chemicals by Fermentation Processes   **8b**.9, 333
Räbiger, N.
– Future Aspects – Cleaner Production   **11a**.27,
  561
Racher, A. J.
– Manufacture of Therapeutic Antibodies   **5a**.8,
  245
Raju, G. K.
– Media and Air Sterilization   **3**.9, 157
Ramanathan, C. S.
– Pharmaceutical Bioinformatics and Drug
  Discovery   **5b**.5, 123
Ratledge, C.
– Microbial Lipids   **7**.4, 133
Reardon, K. F.
– Determination of Cell Concentration and
  Characterization of Cells   **4**.6, 179
Reed, G.
– Introduction   **9**, 1
Rehm H.-J.
– Introduction   **10**, 1
Rehm, H.-J.
– Application of Immobilized Microorganisms in
  Soil Decontamination   **11b**.19, 415
Rehm, H.-J.
– Comments on Other Fields of
  Biotechnology   **10**.20, 573
Rehm, H.-J.
– Microbial Production of Glycerol and Other
  Polyols   **6**.5, 205
Rehm, H.-J.
– Special Morphological and
  Metabolic Behavior of Immobilized Micro-
  organisms   **1**.6, 223
Reiser, M.
– Membrane Processes and Alternative Techniques
  11c.16, 345
Reiss, M.
– Monitoring of Environmental Processes with
  Biosensors   **11a**.7, 125
Renkawitz-Pohl, R.
– Genome Projects of Model Organisms   **5b**.1, 5
Reuss, M.
– Oxygen Transfer and Mixing: Scale-Up Implica-
  tions   **3**.10, 185
Reuss, M.
– Stirred Tank Models   **4**.10, 299
Rice, P.
– Tools for DNA Technologies   **5b**.13, 301
Rilling, N.
– Anaerobic Fermentation of Wet or Semi-Dry
  Garbage Waste Fractions   **11c**.5, 151

Robben, J.
– DNA Sequencing   **2**.8, 257
Robinson, G.
– Halocompounds   **8b**.3, 173
Robinson, S. J.
– Agri-Food and Genomics   **5b**.6, 145
Roehr, M.
– Citric Acid   **6**.9, 307
Roehr, M.
– Further Organic Acids   **6**.11, 363
Roehr, M.
– Gluconic Acid   **6**.10, 347
Roehr, M.
– Introduction   **6**, 1
Roehr, M.
– Lactic Acid   **6**.8, 293
Roemer, M.
– Sampling and Investigation of Solid Matter
  11b.23, 477
Roemer, S. C.
– Sequencing Technology   **5b**.8, 183
Rosén, C.-G.
– Cell and Cell Debris Removal: Centrifugation
  and Crossflow Filtration   **3**.18, 469
Rosenberg, M.
– Industrial Animal Cell Culture   **3**.2, 23
Rosenwinkel, K.-H.
– Industrial Wastewater Sources and Treatment
  Strategies   **11a**.10, 191
Rothmund, M.
– Microencapsulation-Based Cell
  Therapy   **10**.19, 547
Rozwadowski, K.
– Agri-Food and Genomics   **5b**.6, 145
Rudolph, K.
– Biotechnology in Plant Protection   **10**.16, 485
Rudolph, R.
– *In vitro* Folding of Inclusion Body Proteins
  on an Industrial Scale   **5a**.4, 111
Russell, J.
– Brewing   **9**.11, 419
Rydén, L.
– Protein Separation and Purification   **3**.23, 617
Ryu, D. D. Y.
– Bioprocess Kinetics and Modelling of
  Recombinant Fermentation   **4**.15, 485
Sahm, H.
– Metabolic Design   **1**.5, 189
Sahm, H.
– Introduction   **1**, 1
Salt, D. E.
– Phytoremediation of Metals   **11b**.17, 385
Saluz, H.-P.
– Biotechnical Drugs as Antitumor Agents   **7**.14,
  641
Sambanis, A.
– Cell Culture Bioreactors   **3**.6, 105

Sand, W.
– Microbial Corrosion and its
  Inhibition   **10**.10, 265
Sandmeier, H.
– Bacteriophages   **1**.16, 543
Saner, U.
– Fermentation Data Analysis for Diagnosis and
  Control   **3**.15, 355
Saono, S.
– Biotechnology in the Asian-Pacific Region
  12.13, 369
Sayler, G. S.
– Bacterial Activity Enhancement and Soil
  Decontamination   **11b**.20, 425
Sayler, G. S.
– Genetically Engineered Microorganisms and
  Bioremediation   **11b**.21, 441
Schalk, P.
– Combined Mechanical and Biological Treatment
  of Municipal Solid Waste   **11c**.8, 203
Schedel, M.
– Regioselective Oxidation of Aminosorbitol with
  Gluconobacter oxydans, A Key Reaction in the
  Industrial Synthesis of 1-Deoxynojirimycin
  8b.7, 295
Scheibenbogen, K.
– Biotechnology with Cyanobacteria and
  Microalgae   **10**.5, 105
Scheper, T. H.
– Determination of Cell Concentration and
  Characterization of Cells   **4**.6, 179
Scherer, S. W.
– The Human Genome Project   **5b**.2, 41
Schink, B.
– Principles of Anaerobic Degradation of Organic
  Compounds   **11b**.7, 169
Schippert, E.
– Bioscrubbers   **11c**.13, 305
Schleef, M.
– Issues for Large-Scale Plasmid DNA
  Manufacturing   **5a**.20, 443
Schlumberger, H. D.
– Regulations for Recombinant DNA Research,
  Product Development and Production in the US,
  Japan and Europe   **12**.8, 239
Schmid, R. D.
– Biotechnology in the Asian-Pacific Region
  12.13, 369
Schmiers, H.
– Biotechnology of Coal   **10**.7, 153
Schnabl, H.
– Biotechnology in Space   **10**.17, 507
Schneider, F.
– Raw Material Strategies – Economical Problems
  6.2, 47

Schneider-Scherzer, E.
– Cyclosporins: Recent Developments in Biosynthesis, Pharmacology and Biology, and Clinical Applications **7**.12, 535

Schnoor, J. L.
– Degradation by Plants – Phytoremediation of Organics **11b**.16, 371

Schoenen, D.
– Hygienic Aspects of Drinking Water **11c**.19, 413

Schomburg, D.
– Introduction **5a**, XV

Schomburg, D.
– Protein Interactions **5a**.3, 87

Schön, G.
– Biological and Chemical Phosphorus Removal **11a**.14, 285

Schörgendorfer, K.
– Cyclosporins: Recent Developments in Biosynthesis, Pharmacology and Biology, and Clinical Applications **7**.12, 535

Schrader, J.
– Biotechnological Production of Terpenoid Flavor and Fragrance Compounds **10**.13, 373

Schrag, J. D.
– Structure Information **5b**.15, 345

Schramm, A.
– Nucleic Acid-Based Techniques for Analyzing the Diversity, Structure, and Dynamics of Microbial Communities in Wastewater Treatment **11a**.5, 85

Schuchardt, F.
– Composting of Plant Residues and Waste Plant Materials **11c**.3, 101

Schügerl, K.
– Common Instruments for Process Analysis and Control **4**.1, 5

Schügerl, K.
– Liquid-Liquid Extraction (Small Molecules) **3**.21, 557

Schügerl, K.
– On-Line Analysis of Broth **4**.5, 149

Schulte-Ebbert, U.
– Artificial Groundwater Recharge and Bank Filtration **11c**.20, 425

Schulz, J., genannt Menningmann
– Submerged Fixed-Bed Reactors **11a**.17, 349

Schulz-Berendt, V.
– Bioremediation with Heap Technique **11b**.13, 319

Schürbüscher, D.
– Anaerobic Waste Water Process Models **4**.14, 441

Schütte, H.
– Cell Disruption and Isolation of Non-Secreted Products **3**.19, 505

Schwab, H.
– Principles of Genetic Engineering for *Escherichia coli* **2**.10, 373

Schwarz, E.
– *In vitro* Folding of Inclusion Body Proteins on an Industrial Scale **5a**.4, 111

Schwecke, T.
– Advances in the Molecular Genetics of β-Lactam Antibiotic Biosynthesis **7**.6, 247

Schweder, T.
– Marine Biotechnology **10**.15, 441

Scrimshaw, N. S.
– Nutritional Value and Safety of "Single Cell Protein" **9**.6, 221

Scriven, J.
– Biotechnology in the Asian-Pacific Region **12**.13, 369

Sell, D.
– Bioelectrochemical Fuel Cells **10**.1, 5

Sellmer, S.
– Acetic Acid **6**.12, 381

Sellmer, S.
– Vinegar **9**.15, 579

Senn, T.
– Ethanol – Classical Methods **6**.3, 59

Sensen, C. W.
– Introduction **5b**, 1

Sensen, C. W.
– Genomic Data Representation through Images – MAGPIE as an Example **5b**.17, 379

Sensen, C. W.
– Genomics – Five Years from Now **5b**.20, 431

Servi, S.
– Yeast **8a**.8, 363

Shahani, K. M.
– Other Fermented Dairy Products **9**.10, 385

Sharpe, A. G.
– Agri-Food and Genomics **5b**.6, 145

Shaver, R. D.
– Fermented Feeds and Feed Products **9**.20, 769

Shimizu, S.
– Vitamins and Related Compounds: Microbial Production **10**.11, 319

Shioya, S.
– Modelling, Design, and Control of Downstream Processing **4**.18, 603

Sieker, C.
– Experience with Biofilters in Wastewater Treatment **11a**.18, 365

Simon, R.
– Safety Aspects in Biotechnology **2**.19, 835

Sinder, C.
– Biotechnology of Coal **10**.7, 153

Skatrud, P. L.
– Advances in the Molecular Genetics of β-Lactam Antibiotic Biosynthesis **7**.6, 247

Smith, G. M.
– The Nature of Enzymes **9**.1, 5

Teuber, M.
- Lactic Acid Bacteria **1**.10, 325

Thomas, D. Y.
- A DNA Microarrays Fabrication Strategy for Research Laboratories **5b**.9, 227

Thürmer, F.
- Microencapsulation-Based Cell Therapy **10**.19, 547

Tiedtke, A.
- Biotechnology with Protozoa **10**.6, 137

Tobin, M. B.
- Advances in the Molecular Genetics of β-Lactam Antibiotic Biosynthesis **7**.6, 247

Tong, J. M.
- Manufacture of Therapeutic Antibodies **5a**.8, 245

Toojinda, T.
- Genomic Mapping and Positional Cloning, with Emphasis on Plant Science **5b**.7, 165

Tragoonrung, S.
- Genomic Mapping and Positional Cloning, with Emphasis on Plant Science **5b**.7, 165

Traynor, G.
- Regulation of Antibodies and Recombinant Proteins **5a**.22, 495

Troostembergh, J.-C. de
- Starch-Based Raw Materials for Fermentation Applications **6**.1b, 31

Tsai, J. H. J.
- Biotechnology in the Asian-Pacific Region **12**.13, 369

Tsang, J.
- Flavin Monooxygenascs – Uses as Catalysts for Baeyer–Villiger Ring Expansion and Heteroatom Oxidation **8a**.11, 535

Tsui, L. C.
- The Human Genome Project **5b**.2, 41

Tuitemwong, P.
- Anaerobic Fermentations **3**.13, 295

Turner, G.
- Genetic Engineering of Filamentous Fungi **2**.14, 529

Turner, M.
- Perspectives in Biotransformation **8a**.1, 5

Tuteja, D.
- Biochemistry of Polyketide Synthases **10**.12, 341

Tzotzos, G. T.
- Commercial Biotechnology: Developing World Prospects **12**.12, 339

Uhl, W.
- Biofiltration Processes for Organic Matter Removal **11c**.22, 457

Uhlmann, E.
- DNA Synthesis **2**.9, 317

Ulrici, W.
- Contaminated Soil Areas, Different Countries and Contaminants, Monitoring of Contaminants **11b**.1, 5

Unterman, R.
- Advanced *in situ* Bioremediation – A Hierarchy of Technology Choices **11b**.18, 399

van Hylckama Vlieg, J. E. T.
- Bacterial Degradation of Aliphatic Hydrocarbons **11b**.8, 193

Van Langenhove, H.
- Process Engineering of Biological Waste Gas Purification **11c**.11, 259

van Lier, J. B.
- Possibilities and Potential of Anaerobic Wastewater Treatment Using Anaerobic Sludge Bed (ASB) Reactors **11a**.25, 517

Van Steendam, G.
- The Evaluation of Technology as an Interactive Commitment-Building Process – The Failure of Technology Assessment **12**.1, 5

Vanavichit, A.
- Genomic Mapping and Positional Cloning, with Emphasis on Plant Science **5b**.7, 165

Verhasselt, P.
- DNA Sequencing **2**.8, 257

Verrips, C. T.
- Structured Risk Assessment of rDNA Products and Consumer Acceptance of These Products **12**.5, 157

Verstraete, W.
- Process Engineering of Biological Waste Gas Purification **11c**.11, 259

Viikari, L.
- Biotechnology in the Pulp and Paper Industry **10**.18, 523

Villadsen, J.
- Bioreactors: Description and Modelling **3**.5, 77

Voet, M.
- DNA Sequencing **2**.8, 257

Volckaert, G.
- DNA Sequencing **2**.8, 257

von Döhren, H.
- General Aspects of Secondary Metabolism **7**.1, 1

von Döhren, H.
- Introduction **7**, 1

von Döhren, H.
- Peptide Antibiotics **7**.7, 277

Walkinshaw, M. D.
- Cyclosporins: Recent Developments in Biosynthesis, Pharmacology and Biology, and Clinical Applications **7**.12, 535

Wan, P.
- Flavin Monooxygenascs – Uses as Catalysts for Baeyer–Villiger Ring Expansion and Heteroatom Oxidation **8a**.11, 535

Wandrey, C.
- Anaerobic Waste Water Process Models **4**.14, 441

Wang, D. I. C.
- Industrial Animal Cell Culture **3**.2, 23

Wang, D. I. C.
– *In vitro* Protein Refolding   **3**.20, 527
Watt, G. M.
– Enzymes in Carbohydrate Chemistry: Formation of Glycosidic Linkages   **8b**.5, 243
Waweru, M.
– Process Engineering of Biological Waste Gas Purification   **11c**.11, 259
Weber, G.
– Cyclosporins: Recent Developments in Biosynthesis, Pharmacology and Biology, and Clinical Applications   **7**.12, 535
Weber, W. E.
– Gene Mapping in Animals and Plants   **2**.5, 141
Weiland, P.
– Agricultural Waste and Wastewater Sources and Management   **11a**.11, 217
Weir, N.
– Non-Viral Vectors for Gene Therapy   **5a**.19, 427
Weising, K.
– Genetic Engineering of Plant Cells   **2**.15, 547
Weisman, H. F.
– ReoPro Clinical Development: A Case Study   **5a**.16, 365
Wetzel, G. D.
– Medical Applications of Recombinant Proteins in Humans and Animals   **5a**.5, 125
Wiesmann, U.
– Special Aerobic Wastewater and Sludge Treatment Processes   **11a**.19, 373
Wießner, A.
– Biological Processes in Wetland Systems for Wastewater Treatment   **11a**.12, 241
Wilderer, A.
– Continuous Flow and Sequential Processes in Municipal Wastewater Treatment   **11a**.15, 321
Willmitzer, L.
– Transgenic Plants   **2**.16, 627
Wilson, J.
– Design of Aseptic, Aerated Fermentors   **3**.16, 401
Wingender, J.
– Autoaggregation of Microorganisms: Flocs and Biofilms   **11a**.4, 65
Winkelmann, G.
– Microbial Siderophores   **7**.5, 199
Winnacker, E.-L.
– Biosafety in rDNA Research and Production   **12**.3, 63
Winter, J.
– Bacterial Metabolism in Wastewater Treatment Systems   **11a**.2, 17
Winter, J.
– Bio- and Pyrotechnology of Solid Waste Treatment   **11c**.1, 5
Winter, J.
– Introduction   **11a**, 1

Winter, J.
– Introduction   **11c**, 1
Winter, J.
– Perspectives of Waste, Wastewater, Off-Gas, and Drinking Water   **11c**.23, 479
Wirth, M.
– Genetic Engineering of Animal Cells   **2**.17, 663
Wischnak, C.
– Degradation of Chlorinated Compounds   **11b**.10, 241
Wishart, D. S.
– Tools for Protein Technologies   **5b**.14, 325
Wohlleben, W.
– Genetic Engineering of Gram Positive Bacteria   **2**.12, 455
Wolf, S.
– Types and Function of Proteins   **5a**.1, 1
Wolfe, R. A.
– Media for Cell Culture   **3**.8, 141
Wricke, G.
– Gene Mapping in Animals and Plants   **2**.5, 141
Yarmush, D. M.
– Analytical Protein Chemistry   **3**.27, 717
Yarmush, M. L.
– Affinity Separations   **3**.24, 643
Yarmush, M. L.
– Analytical Protein Chemistry   **3**.27, 717
Yarmush, M. L.
– Electrokinetic Separations   **3**.25, 679
Yoshida, T.
– Fermentation Data Analysis for Diagnosis and Control   **3**.15, 355
Yuan, J.
– Process Models: Optimization of Yeast Production – A Case Study   **4**.12, 383
Zart, D.
– Nitrification and Denitrification – Microbial Fundamentals and Consequences for Application   **11a**.3, 55
Zdobnov, E. M.
– Using the Molecular Biology Data   **5b**.12, 281
Zeemann, G.
– Possibilities and Potential of Anaerobic Wastewater Treatment Using Anaerobic Sludge Bed (ASB) Reactors   **11a**.25, 517
Zhang, Z.
– Proteome Analysis by Capillary Electrophoresis   **5b**.11, 269
Zimmermann, H.
– Microencapsulation-Based Cell Therapy   **10**.19, 547
Zimmermann, U.
– Microencapsulation-Based Cell Therapy   **10**.19, 547

Cumulative Subject Index
of Volumes 1 to 12

Each entry is followed either by the volume number on boldface and page numers(s) where the subject is discussed, or by reference to another entry

A

AA *see Aspergillus* acylase I
AAAD *see* aromatic-L-amino acid decarboxylase
AAD *see* acetoacetate decarboxylase
AαD *see* aspartate α-decarboxylase
abattori wastewater **11A**, 206ff
– concentrations **11A**, 207
– specific amounts **11A**, 207
– treatment **11A**, 207
ABC (Association of Biotechnology Companies) **12**, 535
ABCDE system **10**, 161f
Abciximab *see* ReoPro
ABRAC (Agricultural Biotechnology Research Advisory Committee) **12**, 207, 470
absorption, terminology **10**, 228
absorption column, air-water distribution coefficients **11C**, 310ff
– design of **11C**, 310f
– – basic equation **11C**, 311
– mass balance in **11C**, 310
– two-film model for mass transfer **11C**, 310
abzymes **5B**, 357, **8A**, 269f
– – *see also* catalytic antibodies
– industrial future of **8B**, 446
– screening via cell growth **8B**, 445
– semi-synthetic – **8B**, 420
abzymes (catalytic antibodies) **5A**, 221, 233
acarbose **7**, 438f
acarbose family, structure of – **7**, 472
Accident Prevention Regulations for Biotechnology, in Germany **12**, 102
accumulation, of metals by microorganisms **10**, 225f
accuracy, definition of **4**, 45
ACeDB **5B**, 30, 285
acenaphthene, biodegradation of **11A**, 389
acenaphthylene, structure of **11B**, 216
acetal, hydrolysis of, use of catalytic antibodies **8B**, 432ff, 465
acetaldehyde, formation in wine making **9**, 475f

acetamidase, gene (*amdS*) **1**, 536
acetate, as a thermodynamic regulator **11A**, 465
– as substrate for methane production **11A**, 460
– use by methanogenic bacteria **11A**, 459f
acetate kinase, in industrial biotransformations **3**, 459
– phosphorylation by **8B**, 224
acetate oxidation, by *Acetobacter* **6**, 387
acetate production, by clostridia **1**, 309
acetc acid, formation in wine making **9**, 476
acetic acid **6**, 318ff
– – *see also* vinegar
– converson of **11A**, 538ff
– degradation in wastewater purification, by *Methanosarcina barkeri* **4**, 457
– – energy balances **4**, 446
– – mathematical model **4**, 463ff
– distribution coefficients, butoxy group effects **3**, 566
– – concentration effects **3**, 566
– – in liquid-liquid extraction **3**, 561ff
– extraction **3**, 561f, 566ff
– – by secondary amines **3**, 568
– – by tertiary amines **3**, 568
– – costs **3**, 569
– – performance **3**, 570
– – with Cyanex **3**, 575, , 923
– – with Hostarex A **3**, 327, 574
– formation in sour doughs **9**, 295, 297
– gas analysis in bioreactors **4**, 60
– in vinegar, detection of **9**, 585f
acetic acid bacteria **6**, 383ff
– classification **6**, 384
– effect of environmental factors **6**, 389
– gluconic acid production **6**, 351
– immobilized – **6**, 395
– lack of ethanol **6**, 389
– nitrogen requirements **6**, 387
– overoxidation **6**, 389
– oxygen demand **6**, 388
– specific growth rate **6**, 389
– specific product formation **6**, 389

– vitamin requirements **6**, 388
– wine spoilage by – **9**, 493
acetic acid fermentation **6**, 381ff
– bacteriophage problem **6**, 386
– biochemistry **6**, 386ff
– cell recycle **6**, 393
– continuous culture **6**, 392f
– continuous fermentation **6**, 390
– liquid surface process **6**, 394
– mathematical modeling **6**, 392
– physiology **6**, 388f
– raw materials **6**, 383f
– required nutrients **6**, 384
– semicontinuous fermentation **6**, 390
– the Frings Acetator® **6**, 391f
– trickling process **6**, 394f
– two-stage processes **6**, 390
acetic acid production, by anaerobic fermentation **6**, 396f
– – *Clostridium thermoaceticum* **6**, 396
– environmental protection **6**, 395f
– inoculation vinegar **6**, 385
– production volumes **6**, 396
– use of immobilized acetic acid bacteria **6**, 395
– with *Acetobacter* **6**, 384ff
– with *Frateuria* **6**, 384f
– with *Gluconobacter* **6**, 384f
acetoacetate decarboxylase (AAD) **8B**, 58ff
– *Clostridium acetobutylicum* **8B**, 59ff
– decarboxylation of acetoacetate, mechanism **8B**, 60
acetoacetic acid, decarboxylation of, abiotic mechanism **8B**, 59
– – by acetoacetate decarboxylase **8B**, 60
Acetobacter, formation of acetic acid **6**, 384
– hexose monophosphate pathway **6**, 387
– oxidation of acetate **6**, 387
Acetobacter europaeus **6**, 385f
– plasmid profile **6**, 386
Acetobacterium, anaerobic wastewater process **4**, 447
Acetobacter methanolicus, gluconic acid production **6**, 352, 356
Acetobacter strains, genetic improvement **6**, 386
– plasmid profile analysis **6**, 385f
– specific bacteriophages **6**, 386
Acetobacter suboxydans see Gluconobacter oxydans
Acetobacter xylinum, cellulose formation **6**, 634
acetogenic bacteria **11C**, 153
acetogenic bateria **11A**, 457
Acetogenium, anaerobic wastewater process **4**, 447
Acetogenium kivui, growth kinetics **1**, 133
acetoin **8B**, 63f
– gas analysis in bioreactors **4**, 60
α-acetolactate, biosynthesis of **8B**, 64

– decarboxylation of, by acetolactate decarboxylase **8B**, 63
α-acetolactate decarboxylase, use in brewing **9**, 685
acetolactate decarboxylase **8B**, 62ff
– of *Bacillus*, sp. **8B**, 63f
acetolactate synthase **8B**, 91
acetone, anaerobic degradation of **11B**, 175
– bacterial formation of **6**, 231ff
– formation by clostridia **6**, 236
– – pathway **6**, 238
– gas analysis in bioreactors **4**, 60
– physical properties **6**, 230
– production by *Clostridium acetobutylicum* **8B**, 61f
– production volumes 1910–1929 **8A**, 9
– removal by trickling filters **11C**, 336
acetone-butanol fermentation **3**, 308, **6**, 229ff
– batch culture **6**, 245
– by mixed cultures **3**, 308
– cell immobilization **6**, 252ff
– cell recycling **6**, 253f
– continuous culture **6**, 248ff
– – two-stage process **6**, 249, 252
– enzymes involved in solvent formation **6**, 243ff
– flow diagram **6**, 256
– history **6**, 230ff
– in clostridia **1**, 309
– industrial plants **6**, 232
– microorganisms **6**, 233ff
– – patented bacteria **6**, 234
– product extraction, *in situ* – **3**, 585
– product recovery, by distillation **6**, 253
– – by solvent extraction **6**, 254f
– regulation of product formation **6**, 245f
– strain improvement **6**, 246f
– – isolation of mutants **6**, 247
– – recombinant DNA techniques **6**, 247
– traditional commerical batch process **6**, 247f
– whey substrate **6**, 255
– yield **3**, 308
acetone production **6**, 229ff
– by clostridia **6**, 236
acetylcholine, hydrolysis by acetylcholinesterase **8A**, 212
acetylcholine receptor agonist **7**, 123
acetylcholinesterase **8A**, 212ff
– asymmetrization of cyclic *meso*-diesters **8A**, 212f
– – products **8A**, 213
acetyl-CoA **1**, 72
acetyl coenzyme A **9**, 31
– determination **9**, 143
N-acetyl cysteine, resolution of **8A**, 18
acetylenes, antifungal activity **7**, 503
O-acetylhomoserine sulfhydrylase, reactions catalyzed by – **6**, 530

acetyl phosphate, phosphorylation potential **8B**, 231

ACGM (Advisory Committee for Genetic Modification) **12**, 575f

Achromobacter obae, α-amino- -caprolactam racemase **6**, 514f

acid anhybrides, in lipase-catalyzed acylations **8A**, 58

acid hydrolases, formation by *Tetrahymena* **10**, 142ff

acidic tomato broth (ATB), for *Leuconostoc oenos* isolation **1**, 348

Acidonomas **1**, 270

acidophiles, characteristics of **10**, 68

acidophilic bacilli *see* bacilii

acid phosphatase, production **1**, 239

Acinetobacter, as reagent for biological Baeyer-Villiger reactions **8A**, 546

– Baeyer-Villiger biotransformations **8A**, 555ff

– cyclohexanone monooxygenase, Criegee intermediates **8A**, 553

– – mechanism of the Baeyer-Villiger reactions **8A**, 548

– immobilization of **8A**, 298

– isolation from elective culture **8A**, 547

– nitrilase activity **8A**, 298

Acinetobacter baumannii, acetinobactin **7**, 219

Acinetobacter calcoaceticus, bioemulsifier production **6**, 665

– emulsan formation **6**, 650

Acinetobacter haemolyticus, acetinoferrin **7**, 221f

Acinetobacter junii, Baeyer-Villiger biotransformations **8A**, 560

Acinetobacter sp., polyphosphate accumulation **11A**, 41

– role in enhanced biological phosphorus removal **11A**, 296

Acinetobacter sp. NCIMB 9871, sulfoxidations by **8A**, 574

acinetobactin, siderophore of *Acinetobacter baumannii* **7**, 219

– structure **7**, 219

acinetoferrin, dihydroxamate siderophore of *Acinetobacter haemolyticus* **7**, 221f

Acnuc **5B**, 292

aconitase **8B**, 109ff

cis-aconitate, decarboxylation of, by aconitate decarboxylase **8B**, 65

aconitate decarboxylase **8B**, 64f

– decarboxylation of *cis*-aconitate **8B**, 65f

– of *Aspergillus terreus* **8B**, 64f

acoustic methods, for biomass determination **4**, 195

acoustic resonance densitometry (ARD) **4**, 195, 431

ACRE (Advisory Committee on Releases to the Environment) **12**, 577

Acremonium chrysogenum see Cephalosporium acremonium

acridine **2**, 26f

acridine orange **2**, 26

acrolein, formation in distillers' fermentations **6**, 99

– in wine spoilage **9**, 495

acrylamide, production of **10**, 426

acrylamide production, enzyme process **8A**, 302

acrylic acid production, enzyme process **8A**, 302

– from acrylonitrile **8A**, 311f

acrylonitrile, hydrolysis to acrylamide **8A**, 284f

actagardine, action on cell wall of susceptible bacteria **7**, 357

– primary structure **7**, 337f

actin filaments **5A**, 31

Actinobacillus succinogenes, succinate formation by **10**, 581f

actinoidin A **7**, 375

– structure **7**, 373

Actinomadura, chemotaxonomic marker **1**, 435

– G + C content **1**, 435

– morphology **1**, 435

Actinomadura madurae, maduraferrin **7**, 219f

Actinomucor elegans, in meitauza fermentation **9**, 525

– in sufu fermentation **9**, 524

actinomycetes, as respiratory allergens and pathogens **11C**, 76

– as sources of nonribosomal pepetides **7**, 304ff

– degradation of, biopolymers **11C**, 69

– – cellulose **11C**, 69

– – lignin **11C**, 69

– – lignocellulose **11C**, 69

– in composting **11C**, 60ff

– lateral gene tansfer to Bacillaceae **7**, 457

– mesophilic – **11C**, 60

– oligosacharide structures **7**, 35

– product application **1**, 449

– production of antibiotics **1**, 445

– thermophilic – **11C**, 60

– – in compost of stable manure **11C**, 62

– – in self-heated material **11C**, 63

– – isolation from compost **11C**, 61

– – isolation from self-heated hay **11C**, 61

– – municipal waste compost **11C**, 62

– – temperature optimum **11C**, 40

actinomycin D **7**, 650ff

– antitumor agent **7**, 651

– structure **7**, 651

actinomycins, antitumor activity **7**, 650ff

actinophage, of *Streptomyces* **1**, 568

Actinoplanes, chemotaxonomic marker **1**, 435

– G + C content **1**, 435

– morphology **1**, 435

actinorhodin **1**, 215

– biosynthetic pathway **10**, 349

activated carbon absorbers, bacteria in **11C**, 466f

– removal of organic matter **11C**, 458ff
– – in drinking water treatment **11C**, 458ff
activated carbon biofilters, pathogens in **11C**, 468
activated flocs, role of extracellular polymeric substances **11A**, 74f
activated sludge *see also* sludge, waste activated sludge (WAS)
– aeration tanks, biofilm supports **11A**, 257f
– aerobic degradation in, carbon flow **11A**, 21
– biomass content **11A**, 261
– chemical oxygen demand (COD) **11A**, 261
– concentration of microorganisms **11A**, 72f
– *Cytophaga* in **11A**, 97f
– definition of **11A**, 254, 323
– enhanced biological phosphorus removal **11A**, 290ff
– – aerobic phosphorus uptake **11A**, 291ff
– – anaerobic phosphorus release **11A**, 291f
– – mainstream process **11A**, 297ff
– – pilot plant **11A**, 304f
– – polyphosphate accumulating bacteria **11A**, 296f
– *Flavobacterium* in **11A**, 97f
– metal desorption from **10**, 234
– NH$_4$ oxidation by *Nitrosomonas* sp. **11A**, 397
– NO$_2$ oxidation by *Nitrobacter* sp. **11A**, 397
– phosphorus uptake **11A**, 290ff
– plants **11A**, 255, 322f
– removal of organic matter, model matrix **11A**, 419
activated sludge flocs, composition of **11A**, 69f
– floc forming and filamentous organisms **11A**, 73
– microcolonies in **11A**, 69
– model of aggregation mechanisms and structure **11A**, 69f
– size of **11A**, 69
– structure of **11A**, 69
activated sludge models (ASM), activated sludge model No. 1 **11A**, 420f
– – calibration steps **11A**, 424
– – model calculations **11A**, 425f
– – processes for heterotrophic and nitrifying bacteria **11A**, 422
– – process matrix **11A**, 421
– activated sludge model No. 2 **11A**, 297f, 420ff
– – COD fractionation **11A**, 423
– – wastewater components **11A**, 424
– activated sludge model No. 3, processes for heterotrophic organisms **11A**, 422
– aeration **11A**, 419
– component participation **11A**, 420
– components **11A**, 419
– computer programs **11A**, 425
– hydraulic patterns **11A**, 419
– mass balances **11A**, 420
– model calibration **11A**, 424f
– processes **11A**, 418f

– rates **11A**, 420
– use of model calculations **11A**, 425f
activated sludge process **11A**, 252ff, 400ff
– alternating nitrification and denitrification process **11A**, 276f
– Bio-Denitro process **11A**, 276
– biofilm module **11A**, 258
– bulking **11A**, 73, 367f
– carbon dioxide removal **11A**, 266
– carbon removal **11A**, 259ff, 269
– definition of **11A**, 254
– denitrification **11A**, 264f
– – pre-anoxic zone – **11A**, 270f
– design loads **11A**, 279
– development of **11A**, 10ff, 254ff
– effect of, alkalinity **11A**, 266
– – dissolved oxygen **11A**, 265f
– – pH **11A**, 266
– effluent quality **11A**, 279
– excess sludge production **11A**, 260
– final clarifier **11A**, 278f
– floc formation **11A**, 73
– flow diagram **11A**, 254, 256, 276
– food-microorganism ratio **11A**, 260
– high-rate reactors **11A**, 258
– intermittent nitrification and denitrification **11A**, 264ff
– – plant configurations **11A**, 274ff
– – process control **11A**, 275
– Jülich wastewater treatment process (JARV) **11A**, 276f
– membrane separation technique **11A**, 258
– microbiological aspects **11A**, 259ff
– mixed liquor **11A**, 267f
– model **11A**, 401ff
– nitrification **11A**, 262ff
– – effect of pH **11A**, 263f
– – growth rate of nitrifiers **11A**, 262f
– – inhibition by ammonia **11A**, 263f
– nitrogen removal **11A**, 256f, 269f
– – pre-anoxic zone denitrification **11A**, 270f
– – processes for low COD/TKN ratio **11A**, 277
– – step-feed denitrification **11A**, 271ff
– oxygenation capacity (OC value) **11A**, 11
– phosphate removal **11A**, 257
– plant configurations **11A**, 268ff
– – aeration tanks **11A**, 268
– – closed loop tanks **11A**, 268f
– – mixing tanks **11A**, 268
– post-denitrification **11A**, 277f
– process configuration **11A**, 280
– pure oxygen – **11A**, 257
– role of floc forming and filamentous bacteria **11A**, 73
– simultaneous nitrification and denitrification **11A**, 269, 273f
– – (ORP) controller **11A**, 274
– – plant configurations **11A**, 273f

– – process control **11A**, 273f
– single-stage – **11A**, 254f
– sludge volume index (SVI) **11A**, 267
– solid retention time (SRT) **11A**, 260
– surplus sludge **11A**, 11
– technological aspects **11A**, 259ff
– toxic substances **11A**, 267
– Tri-Cycle process **11A**, 276
– two-stage – **11A**, 255f
– use for post-denitrification **11A**, 345
activated sludge system, mass balance equations, for a stirred tank **4**, 431ff
– mass transfer resistance, in flocs **4**, 431
– modelling sludge settling **4**, 435
– wastewater treatment **4**, 430
activated sludge systems, scheme **11A**, 400
– use for pre-denitrification **11A**, 343f
activated sludge tanks, design of **11C**, 307ff
– dimensions of **11C**, 307f
– in bioscrubbing plants and in aerobic biological wastewater treatment, comparison of **11C**, 309f
– kinetics of biodegradation **11C**, 307
activated zones, process scheme **11B**, 361
activation energy **1**, 66
active site model, for pig liver esterase **8A**, 207ff
– for sulfoxidation, by *Acinetobacter* **8A**, 570ff
– of cyclohexanone monooxygenase **8A**, 551
active sites, prediction of **5B**, 333f
active transport, substrate uptake, ligand "taxi" **9**, 191
– – primary – **9**, 190f
– – secondary – **9**, 191
– – vectorial partitioning **9**, 191
activin **5A**, 165
acute myelogenous leukemia (AML), phase I study, preliminary results – **5A**, 376
– treatment of **5A**, 358ff
ACVS *see* δ-(L-α-aminoadipyl)-L-cysteinyl-D-valine synthetase
ACV-synthetase, gene **1**, 529
acylamide, production of **8B**, 287f
acylase I *see* aminoacylase
acylases **8A**, 244ff
– L-stereospecificity **8B**, 356
– microbial –, lactam hydrolysis **8A**, 257
– production of non-natural amino acids **8B**, 285f
acylation, enantioselective –, tripeptide benzamide catalyzed – **8B**, 510ff
– of alcohols, use tetrapeptide catalysts **8B**, 512f
– of racemic hydroxyacid esters **8A**, 418f
acylation reactions, use of acyl donors **8A**, 56ff
acyl-carrier protein domain, of polyketide synthases **10**, 361
acyl cleavage reaction, transition state analogs **8B**, 411
acyl-coenzyme A:isopenicillin N acyltransferase (AT) **7**, 257ff

acyl donors, in lipase-catalyzed acylations, enol esters **8A**, 56ff
acyl lactylates, use in cosmetics **6**, 674
acyl migration, in mono- and diacylglycerides **8A**, 139
acyloin condensation **8B**, 393
– pyruvate decarboxylase mediated – **8B**, 49
– yeast catalyzed – **8B**, 366
acyl transferase domains, of polyketide synthases **10**, 360f
acyl transfer reactions, tetrahedral intermediates **8B**, 410
– use of catalytic antobodies **8B**, 474ff
ADA *see* adenosine deaminase deficiency
adamantanes, hydroxylation of **8A**, 487
ADAPT (antibody-dependend abzyme prodrug therapy) **5A**, 221
adaptive control **4**, 551ff
– schematic diagram **4**, 521
– self-tuning control **4**, 521
adaptive optimization **4**, 551ff
– definition **4**, 551
– multivariable on-line optimization **4**, 556
– of batch and fedbatch bioreactors **4**, 557f
– of continuous bioreactors, baker's yeast culture **4**, 552ff
– – implementation **4**, 554f
– – parameter estimation **4**, 553f
– – phases **4**, 555ff
– – process model **4**, 553
– procedures **4**, 554
– schematic diagram **4**, 552
additives, use in heap technology **11B**, 322
adechlorin, inhibition of adenosine deaminase **7**, 648
adenine **2**, 194, 320
adenine arabinoside, formation by transarabinosylation, by *Enterobacter aerogenes* **6**, 596
adenine dinucleotides, fluorescence **4**, 185ff
adeno-associated virus (AAV), biology **5A**, 413f
– genome **5A**, 414f
– integration site **5A**, 414
– life cycle **5A**, 413f
– serotypes **5A**, 414
– structure **5A**, 414f
– type 2 **5A**, 414
adeno-associated virus vectors **5A**, 397f, 413ff
– cell entry **5A**, 415
– clinical studies **5A**, 421
– design **5A**, 416
– gene cassettes, design rules **5A**, 417
– – limitations **5A**, 417
– genome **5A**, 416
– host responses **5A**, 420
– host tropism **5A**, 415
– in HIV gene therapy **5A**, 480f
– manufacturing **5A**, 417ff
– – complementation systems **5A**, 417

– – recombination events **5A**, 417
– of cystic fibrosis **5A**, 421
– packaging limit **5A**, 417
– packaging systems **5A**, 417f
– particle structure effects **5A**, 417
– persistence and integration **5A**, 415f
– preclinical studies **5A**, 419ff
– – *in vitro* **5A**, 419ff
– – *in vivo* **5A**, 420
– properties **5A**, 417
– purification **5A**, 419
– regulatory elements, design rules **5A**, 417
– – limitations **5A**, 417
– replication **5A**, 415f
– replication competent – **5A**, 418
– target cells **5A**, 420
– toxicity **5A**, 420
– wild-type generation **5A**, 418
adenomatous polyposis coli, familial – **5B**, 81, 86
adenomatous polyposis coli gene (APC) **5B**, 81
adenomycin, cyclitol moieties **7**, 454
adenosine, production by *Bacillus* sp., mutant strains **6**, 591
– structure **6**, 518
– therapeutic potential **6**, 564f
adenosine deaminase deficiency (ADA) **5A**, 472
– gene therapy of **5A**, 444
adenosine deaminase inhibitors, treatment of leukemia **7**, 648
adenosine diphosphate (ADP), formation, by cell-free phosphorylase **6**, 601
– – by immobilized phosphorylase **6**, 601
adenosine monophosphate (AMP), fermentative production **6**, 578
– formation, by cell-free phosphorylase **6**, 601
– – by immobilized phosphorylase **6**, 601
adenosine triphosphate (ATP), and phosphodiester formation **8B**, 232
– ATP generating systems, microbial – **10**, 330
– formation, by *Candida boidinii* **6**, 599
– – by cell-free phosphorylase **6**, 601
– – by enzymatic conversion form adenosine **6**, 600
– – by *Erwinia carotovora* **6**, 600f
– – by immobilized phosphorylase **6**, 601
– – by *Saccharomyces cerevisiae* **6**, 599
– formation by methane bacteria **4**, 449f
– microbial production of **10**, 330f
– monitoring, bioluminescence assays **4**, 211
– – ³¹P-MR **4**, 211
– production in aerobic metabolism **4**, 414
– structure of **8B**, 222
adenosine 3',5'-monophosphate *see* cyclic adenosine 3',5' monophosphate (cAMP)
S-adenosyl-L-homocysteine hydrolase **6**, 518f
adenoviral transfer **5B**, 85
adenovirus, biology **5A**, 406
– genome **5A**, 406f
– latency **5A**, 408
– life cycle **5A**, 407f
– persistence **5A**, 408
– replication competent – **5A**, 410f
– structure **5A**, 406f
– type 5 genome **5A**, 407
adenovirus vectors **5A**, 397ff, 406ff
– animal models **5A**, 411
– cell targeting **5A**, 411f
– clinical studies **5A**, 413
– design **5A**, 408f
– expression *in vivo* **5A**, 412
– Food and Drug Administration requirements **5A**, 411
– genome **5A**, 408
– immune responses **5A**, 410, 412
– inflammation **5A**, 412
– manufacturing **5A**, 409ff
– – complementation systems **5A**, 409
– packaging cells **5A**, 409f
– persistence *in vivo* **5A**, 412
– preclinical studies **5A**, 411ff
– – *in vitro* **5A**, 411
– – *in vivo* **5A**, 411f
– production systems **5A**, 409
– properties **5A**, 408f
– purification **5A**, 411
– RCA contamination **5A**, 410f
– recombination **5A**, 410
– scale-up of production **5A**, 411
– therapy of cystic fibrosis **5A**, 413
– therapy of tumors **5A**, 413
– toxicity **5A**, 412
– transgene expression cassettes **5A**, 409
adenylate kinase, phosphorylation by **8B**, 226
adenylate system **1**, 54
ADEPT/ADAPT **5A**, 233
ADEPT (antibody-directed enzyme prodrug therapy) **5A**, 221, 321
ADH *see* alcohol dehydrogenase
– – *see* alcohol dehydrogenases
adherent cells **3**, 107
adhesin **1**, 18
adhesives, biodeterioration of **10**, 311
adhumulone **9**, 423
adipic acid, production of polymers **10**, 426
adjuvants, use of emulsions **6**, 680f
adsorbable organic chlorinated compounds *see* AOX
adsorption, of lipases **8A**, 51f
– of microorganisms **11B**, 417f
– terminology **10**, 228
adsorption processes, in drinking water treatment **11C**, 395ff
– – granular activated carbon **11C**, 397f
– – powdered activated carbon **11C**, 396f
adsorption technology, in wastewater treatment **11A**, 566

adult polycystic kidney disease (APKD) **5B**, 68
adult respiratory distress syndrome (ARDS), use
  of liposomes in gene therapy **5A**, 433
adventitious agents, in cell cultures **10**, 33f
aerated pile composting, with automatic turning
  machines **11C**, 140f
aeration *see* bioreactor aeration
– in stirred tanks, effect on circulation time **4**,
  312
– – effect on mixing time **4**, 311ff
aeration in biofilm reactors **11A**, 354f
aeration membranes **3**, 245ff
– – *see also* diffusion membranes
– Accurel – **3**, 249, 255
– concentration gradients **3**, 245
– mass transfer **3**, 245, 249
– – Accurel membranes **3**, 245
– – external modules **3**, 254f
– microporous – **3**, 245, 249
– – bubble point **3**, 246
– – mass transfer coefficients **3**, 245
– – materials **3**, 246
– – properties **3**, 246
– – tortuosity **3**, 246
– open pore – **3**, 245
– polypropylene – **3**, 245
– polytetrafluoroethylene (PTFE) – **3**, 245
– silicone diffusion – **3**, 245
– structures **3**, 245
– transport properties **3**, 245ff
– tubing design **3**, 252
aereolysin **10**, 85
Aerobacter, in ogi fermentation **9**, 545
Aerobacter aerogenes see Klebsiella pneumoniae
– aerobactin **7**, 22f
– 2,3-butanediol productivity **6**, 276
– 2,3-butanediol yield **6**, 276
aerobactin, siderophore of *Aerobacter aerogenes*
  7, 222f
aerobic degradation *see* bacterial degradation,
  biodegradation
– of biopolymers **11A**, 20ff
– of glucose, mass and energy balance **11A**, 21ff
– of xenobiotics **11A**, 44ff
aerobic fermentations, bioreactor design **3**, 404
aerobic metabolism, adenosine triphosphate formation **11A**, 20
– coenzyme A formation **11A**, 20
aerobic thermophilic stabilization of sludge, with
  subsequent anaerobic digestion **11C**, 221f
aerobic wastewater treatment processes, modeling
  of **11A**, 417ff
Aeromonas hydrophila, amonabactins **7**, 206
aeropyrolysin **10**, 87
aerosols, containing microorganisms **10**, 287
– formation aerosol droplets **10**, 287
– from composting plants, actinomycetes **11C**, 74
– – bacterial population of **11C**, 73f

– – health risks **11C**, 72ff
– in composting plants, occupational health risks
  11C, 237ff
A-factor **7**, 27
– of *Streptomyces griseus*, cascade **7**, 87f
– structure **7**, 80
affinity chromatography, in enzyme purification
  9, 56, 106ff
affinity chromatography (AFF) *see* chromatography
affinity ligands, for protein purification **3**, 598ff
affinity partitioning, metal ion binding **3**, 601
– of proteins **3**, 598ff
affinity reagents **9**, 41f
aflatoxin, in indigenous fermented foods **9**, 552
AFLP *see* amplified fragment length polymorphism
AFNOR (Association Française pour la Normalisation) **12**, 591f, 641
African fermented foods **9**, 542ff
– banku **9**, 544f
– dawadawa **9**, 542f
– gari **9**, 543f
– injera **9**, 546
– kaffir beer **9**, 546
– merissa **9**, 547
– ogi **9**, 545f
agar **6**, 628
– as products of marine biotechnology **10**, 446
– production by algae **10**, 130
Agaricus, strain improvement **2**, 120
Agaricus bisporus **1**, 534f
– cultivation **1**, 536
– endogenous siderophore **7**, 228
agarose gel simulation **5B**, 394
Agency for Technology Assessment and Application *see* BPPT
Agenda 21 **12**, 628f
agglutinins, in budding yeast **1**, 33
aggregation models of humic substances **11B**,
  104, 106
agitation *see* bioreactor agitation
agitation system, measurement of power input **4**,
  126f
agitator *see* bead mill, impeller
AGRICOLA **5B**, 283
agricultural biotechnology, development of transgenic plants **12**, 197ff, 358f
– in developing countries **12**, 358ff
– – crops **12**, 360ff
– – entry barriers **12**, 344f, 361
– product labeling **12**, 465f
– public opinion **12**, 484
– technology assessment **12**, 358ff
– technology transfer **12**, 363ff
agricultural wastes **11A**, 217ff
– aerobic treatment **11A**, 232ff
– – inactivation of Enterobacteria **11A**, 232

– – thermophilic processes **11A**, 232f
– anaerobic treatment **11A**, 225ff
– – classification of processes **11A**, 225
– – digester systems **11A**, 226, 230
– – dry fermentation **11A**, 230f
– – process conditions **11A**, 231
– biogasification **11A**, 230ff
– – biogas yield **11A**, 231f
– – mass balance **11A**, 231
– characteristics **11A**, 220ff
– co-fermentation plant, scheme **11A**, 228f
– composting **11A**, 233f
– – windrows **11A**, 233
– definition of **11A**, 218
– ensiling juice **11A**, 223
– from crop processing **11A**, 224
– from livestock farming **11A**, 219ff
– from slaughterhouses **11A**, 221f
– – amount of solid and liquid wastes per livestock unit **11A**, 222
– harvesting residues **11A**, 218, 222f
– land application of digested residues, limits for
11A, 229
– liquid –, value of **11C**, 225
– management **11A**, 217ff
– manure treatment plants **11A**, 227f
– plant materials **11A**, 219
– quantities **11A**, 220ff
– solids treatment **11A**, 229ff
– – anaerobic – **11A**, 229ff
– – wet fermentation **11A**, 229ff
– sources **11A**, 217ff
– types of **11A**, 218f
– upgrading to higher value products **11A**, 234ff
– – examples **11A**, 235
– utilization of **11A**, 225ff
agricultural wastewater, aerobic treatment **11A**,
232ff
– – inactivation of Enterobacteria **11A**, 232
– – thermophilic processes **11A**, 232f
– anaerobic treatment **11A**, 225ff
– – classification of processes **11A**, 225
– – digester systems **11A**, 226
– from dairy plants, characteristics **11A**, 221
– – quantities **11A**, 221
– management **11A**, 217ff
– sources **11A**, 217ff, 433f
Agricus bisporus genes **10**, 577
agrifood, and genomics **5B**, 145ff
– ESTs and microarrays **5B**, 149f
– genome sequencing **5B**, 148
agrobacteria, amidase activity **8A**, 294
– nitrile hydratase activity **8A**, 294f
– synthesis of indoleacetic acid **8A**, 294
agrobacterial signal molecule **2**, 551ff
– aromatic phenolics **2**, 551ff
– flavonoids **2**, 554
– monosaccharides **2**, 551f

– opines **2**, 553
agrobacterial T-DNA **2**, 549f
– circle **2**, 555
– double-stranded **2**, 555
– expression **2**, 556
– gene products **2**, 556
– gene suppression in hosts **2**, 596
– genomic integration **2**, 556f
– integration pattern **2**, 556
– mobilization **2**, 551
– origin of transfer (*oriT*) **2**, 554
– single-stranded (T-strand) **2**, 554
– transfer **2**, 555
– vector **2**, 557ff
– – binary **2**, 559
– – design **2**, 557, 560ff
– – disarmed **2**, 557
agrobacterial virulence gene (*vir*) **2**, 551
– inducers **2**, 551f
– organization **2**, 551
– regulation **2**, 551f
Agrobacterium **2**, 549
– root loci gene (*rol*) **2**, 557
– signal molecule recognition **2**, 554
Agrobacterium azotophilum, in pozol fermentation **9**, 549
Agrobacterium-mediated gene transfer **2**, 573ff
– efficiency **2**, 575
– host range extension **2**, 574
– limitations **2**, 573, 576
– strategies **2**, 573
– to plant tissue **2**, 573
Agrobacterium-mediated transformation **5B**, 153
Agrobacterium-mediated virus infection **2**, 574f
Agrobacterium radiobacter, D-hydantoinase/carba-
moylase system, D-citrulline formation **8A**,
268f
– exopolysaccharide **1**, 172
– succinoglycan formation **6**, 647
Agrobacterium radiobacter strain K84, secretion of
agrocin **10**, 487f
Agrobacterium rhizogenes **2**, 549
– effect in plant organ culture **1**, 600
– host range **2**, 574
– initiation of plant hairy root cultures **7**, 619
– root-inducing plasmid (Ri) **2**, 549
Agrobacterium sp., curdlan formation **6**, 635
Agrobacterium-transformed plant cell **2**, 550
– opine production **2**, 557
Agrobacterium tumefaciens **2**, 549ff, 629ff
– agrobactin **7**, 204
– effect in plant organ culture **1**, 600
– host range **2**, 574, 630f
– tumor-inducing plasmid (Ti) **2**, 549
– – map **2**, 550
– vectors for plant transformation **2**, 630
agrobactin, catecholate siderophore of *Agrobacterium tumefaciens* **7**, 204

– structure **7**, 204
agrocin **10**, 487f
agroclavine production, by immobilized *Claviceps* cells **1**, 240
Agrocybe aegerita, life cycle **1**, 533
agroinfection **2**, 569, 574f
– host range **2**, 575
– potential **2**, 575
– strategy **2**, 574
– viruses **2**, 575
agroinoculation **2**, 573
AHL *see* N-acyl-L-homoserine lactones
AIDS *see also* HIV
– gene therapy with retroviruses **5A**, 405
– monoclonal antibody therapy **5A**, 317
– treatment with fusion proteins **5A**, 169
– use of interleukin-4 antagonist **5A**, 152
– use of recombinant hormones **5A**, 164
– vaccine **5A**, 138
air bubble *see* bubble
air filter **3**, 177ff
– cartridges **3**, 181
– critical velocity **3**, 179
– design **3**, 178ff
– fibrous materials **3**, 177ff
– filter depth **3**, 180
– filter diameter **3**, 180
– length effects **3**, 179
– materials **3**, 180
– single-fiber collector, efficiency **3**, 177f
– sterilization criteria **3**, 179
air filtration, depth filtration **3**, 177ff
– log penetration theory **3**, 178
– set-up **3**, 182
– validation **3**, 183
– velocity effects **3**, 178
– with cartridges **3**, 181
air flow, measurement, accuracy of **4**, 21
airlift bioreactor **3**, 28, 84, 115f, 120
– aeration **3**, 238, 259ff
– – draught tube velocity **3**, 259
– – efficiency **3**, 260
– – mass transfer coefficients **3**, 259f
– – rate **3**, 259
– – superficial gas velocity **3**, 259
– heterogeneity **3**, 14
– operation scale **3**, 14
– oxygen supply **3**, 116
– – effects **3**, 328
– oxygen transfer rates **3**, 14
air lift loop reactor, axial dispersion coefficient **4**, 364f
– liquid velocities **4**, 363
air lift reactor, distributed parameter gas model, single-cell protein production **4**, 376f
– draft tube **4**, 374
– simulation of baker's yeast process **4**, 402
– structured model **4**, 372

air lift reactor flow, models **4**, 359ff
– – drift flux model **4**, 359f
– – energy balance approach **4**, 360ff
– – momentum balance approach **4**, 362f
air lift reactor models, concentration profiles **4**, 377
– cyclic operation **4**, 378
– for oxygen transfer **4**, 374
– for phenol degradation **4**, 374
– gas phase circulation **4**, 374f
– mixed cells in series **4**, 374f
– two-section loop reactor **4**, 364
air sterilization **3**, 157ff, 176ff
– modes **3**, 159
ajmalicine **7**, 607ff
– from plant cell cultures **7**, 608f
– – yields **7**, 608
– structure **7**, 607
ajmaline **8A**, 352f
– biosynthetic pathway, in *Rauwolfia* **8A**, 353f
– structure **8A**, 353
– treatment of arrhythmias **8A**, 353
D-alanine, biosynthesis **7**, 330
– enzymatic synthesis **6**, 534
alanine dehydrogenase **6**, 540f
– enzymatic synthesis of D-amino acids **6**, 533
– oxidative deamination **6**, 540
– substrate specificity **6**, 540
alanine dehydrogenases, applications of **8A**, 447
– characteristics of **8A**, 447
– reductive amination of pyruvate **8A**, 446f
L-alanine production **1**, 210
L-alanine production **3**, 457, **6**, 487ff, 512f, 521
– with *Pseudomonas* cells **6**, 487
alanine racemase **6**, 535f
– characterization **7**, 580
– role in cyclosporin biosynthesis **7**, 580f
albomycins **7**, 211f
– hydroxamate siderophores of *Stretomyces*, ferioxamine type **7**, 211f
– stucture **7**, 211
alcalase *see* subtilisin
Alcaligenes bronchisepticus, decarboxylation of α-aryl-α-methylmalonates **8B**, 73
Alcaligeneses eutrophus 2050, surfactant resistant bioreporter **11B**, 456
Alcaligenes eutrophus **6**, 406, 411ff
– biotechnological production of poly(3HB) **6**, 437
– denitrification by **11A**, 59
– heterologous gene expression **1**, 26
– PAH synthase, catalytic mechanisms **6**, 419
– – gene structure **6**, 418
– pathways of PHA biosynthesis **6**, 415
– poly-β-hydroxybutyrate **7**, 178
– poly-β-hydroxybutyrate (PHB) production **1**, 212
– structure of cell inclusions **6**, 411

– substrate spectrum **1**, 196
Alcaligenes eutrophus H850Lr, bioreporter
   strains **11B**, 453
Alcaligenes faecalis, curdlan production **6**, 621
– nitrilase activity **8A**, 297f
Alcaligenes latus **4**, 42ff
– production of poly-β-hydroxybutyric acid
   (PHB) **4**, 42ff
– production of poly(3HB) **6**, 439
alcaligin, hydroxamate siderophore of *Bordetella
   pertussis* **7**, 210f
– structure **7**, 211
alcohol *see* ethanol
– extraction **3**, 560
alcohol dehydrogenase, ethanol oxidation **6**, 387
– gene *(ADH)* **1**, 483
– horse liver **8B**, 362f, 372f, 379f
– yeast, keto ester reduction **8B**, 359ff
– – ketone reduction **8B**, 359
alcohol dehydrogenases (ADH), applications of
   8A, 436ff
– *CBADH* **8A**, 456f
– *CFADH* **8A**, 453f
– classification of **8A**, 429
– cross inhibition **8A**, 430f
– emulsion enzyme membrane reactor **8A**, 424f
– enantioselectivity of **8A**, 427ff
– enzyme membrane ractor, with continuous
   product extraction **8A**, 423f
– from extreme thermophiles **10**, 91
– HLADH **8A**, 437ff
– introduction of two chiral centers **8A**, 435f
– kinetics of **8A**, 427ff
– *LKADH* **8A**, 454
– *MJADH* **8A**, 452f
– *PADH* **8A**, 453
– PLADH **8A**, 452
– product inhibition **8A**, 430
– reaction mechanisms, ping-pong mechanism
   8A, 429
– – sequential mechanism **8A**, 428f
– resolution of racemic hydroxy compounds **8A**,
   434ff
– – introduction of two chiral centers **8A**, 435f
– stereochemistry of **8A**, 427
– substrate inhibition **8A**, 429f
– *TBADH* **8A**, 441f
– use in cyclodextrins **8A**, 420ff
– use in organic solvents **8A**, 420ff
alcohol dehydrogenase (secondary), from *Thermoanaerobium brockii* **8A**, 403
alcohol dehydrogenases of yeast *see* yeast alcohol
   dehydrogenases (YADH)
alcohol dehygrogenase, activity, in cross-linked
   yeast cells **1**, 238
alcohol fermentation, in clostridia **1**, 300f
alcoholic fermentation **9**, 472ff
alcoholic fermentations *see* ethanol

alcoholometry **6**, 114
alcohol oxidase, in methylotrophic yeasts **1**, 271,
   273, 280
alcohol oxidation, enantioselective **8A**, 435
alcohol oxidation reaction, rate equation **8A**, 433
alcohol oxidoreductase, in methylotrophic bacteria **1**, 273
alcohols, acyclic, reactions with *Pseudomonas* lipases **8A**, 79ff
– acylation of, use of tetrapeptide catalysts **8B**,
   512f
– analogs of **8A**, 98f
– axially-disymmetric **8A**, 94ff
– chiral, formation by microbial hydroxylation
   8A, 486
– – formation by soybean lipoxygenase **8A**, 503
– conversion of **11B**, 200
– conversion to aldehyde, by *Pichia pastoris* **8B**,
   345
– enantioselective oxidation of **8A**, 435
– flavor-active –, production of **8B**, 345f
– primary –, lipase enantioselectivity **8A**, 87ff
– secondary –, as lipase substrates **8A**, 68ff
– spiro compounds **8A**, 94ff
– tertiary –, enantioselectivity of lipases **8A**, 94
– with non-carbon stereocenters **8A**, 98
– with remote stereocenters **8A**, 97
alcoholysis, of triacylglycerides **8A**, 141
ALCON continuous fermentation process **6**, 155
aldehyde dehydrogenase, ethanol oxidation **6**,
   387
aldehyde-lyases **8B**, 45, 89f
aldehydes, asymmetric hydrocyanation of, catalyzed by *cyclo*-[(S)-Phe-(S)-His)] **8B**, 496
– conversion of **11B**, 200
– enantioselective hydrocyanation of, use of titanium dipeptide complex catalysts **8B**, 508f
– flavor-active –, production of **8B**, 345f
Aldesleukin **5A**, 151
aldol addition **8B**, 7ff
aldolase, gene *(FBA1)* **1**, 481
aldolases **8B**, 6ff
– asymmetric carbon-carbon bond formation **8B**,
   365
– classification of **8B**, 7
– dihydroxy-acetone phosphate-dependet – **8B**,
   7ff
– glycine-dependent – **8B**, 19f
– phosphoenolpyruvate-dependet – **8B**, 15ff
– pyruvate-dependent – **8B**, 15ff
– 2-deoxyribose 5-phosphate aldolase (DERA)
   8B, 17ff
aldol reaction **8B**, 388f
– use of calatytic antibodies **8B**, 472f
aleurodiscal **7**, 502f
– antifungal activity **7**, 503
– structure **7**, 502
algae *see also* macroalgae, microalgae

– – *see* oleaginous microalgae
– Bacillariophyta **10**, 114
– biodiversity of **10**, 110
– biosorption, of metals **10**, 230f
– – of radionuclides **10**, 230f
– biotechnological use of **10**, 445
– Chlorophyta **10**, 111
– Chrysophyta **10**, 113
– composition of **9**, 171
– Conjugaphyta **10**, 112
– Cryptophyta **10**, 114
– culture collections **10**, 116
– culture maintenance **10**, 115
– Dinoflagellata **10**, 115
– ecological applications **10**, 132f
– essential amino acids **9**, 174
– Euglenida **10**, 112f
– Eustigmatophyta **10**, 114
– Glaucocystophyta **10**, 111
– glycerol production **6**, 210, 217ff
– in artificial groundwater recharge **11C**, 432
– in biodeterioration **10**, 271
– marine – **10**, 445
– – DHA-synthesizing – **10**, 464
– – EPA-synthesizing – **10**, 464
– – production of iodomethane **8B**, 190
– – source of halometabolites **8B**, 178
– metal uptake by **11A**, 443
– non-viable –, metal sorption by **10**, 133
– Phaeophyta **10**, 115
– production of, agar **10**, 130
– – alginates **10**, 130
– – antioxidants **10**, 130
– – carrageenans **10**, 130
– – coloring substances **10**, 131
– – polysaccharides **10**, 130
– Prymnesiophyta **10**, 113
– Raphidophyta **10**, 114
– Rhodophyta **10**, 111
– taxonomy **10**, 107
– use in ecotoxicological testing **11B**, 132, 137
– Xanthophyta **10**, 113
algal blooms, cytotoxic activity **10**, 131
algal production, flow sheet **10**, 118
alginate, for cell entrapment **3**, 110ff
– in pseudomonads **1**, 413
alginate bead *see* cell immobilization
alginate beads *see also* microcapsules
– Ba^{2+} cross-linked **10**, 549f
– Ca^{2+} cross-linked **10**, 549f
– drop formation **10**, 550
– for cell encapsulation **10**, 549ff
– formation of **10**, 549f
– generators **10**, 550f
alginate capsules, atomic force microscopy images of **10**, 563
alginate gels, analytical tests **10**, 553
– immunological reactions **10**, 553

– implantation of **10**, 554f
– medically approved –, biocompatibility assays **10**, 553ff
– mitogenic impurities **10**, 553ff
– mixed lymphocyte test **10**, 553
– swelling properties **10**, 561
alginate production, by *Azotobacter vinelandii* **6**, 620, 633
– by *Pseudomonas* sp. **6**, 620f
alginates **6**, 622
– as products of marine biotechnology **10**, 466f
– commercial –, purification of **10**, 551ff
– composition of **10**, 549
– food uses **6**, 628
– of bacterial – **6**, 633
– production, by algae **10**, 130
– – of transplantation-grade – **10**, 551ff
– purification of, flow chart **10**, 552
– structures, of algal – **6**, 634
algorithmic interpretation **5B**, 124
alicyclic biotransformation products **8B**, 367ff
Alicyclobacillus **1**, 371
Alicyclobacillus acidocaldarius **1**, 378
alignment, consensus methods **5B**, 311
– EST sequences **5B**, 306
– global – **5B**, 303
– Hidden Markov Model **5B**, 307
– local – **5B**, 303f
– multiple – **5B**, 307
– Needleman and Wunsch method **5B**, 303
– pairwise – **5B**, 303, 305f, 372
– score **5B**, 303
– Smith and Waterman method **5B**, 303
alignment methods **5B**, 302ff
aliphatic biotransformation products **8B**, 353ff
aliphatic ester, hydrolysis of, use of catalytic antibodies **8B**, 451ff
aliphatic hydrocarbons, anaerobic degradation of **11B**, 171f
– bacterial degradation of **11B**, 148ff, 193ff
– – reaction types **11B**, 196ff
– biodegradability of **11B**, 468ff
– biodegradation mechanisms **6**, 687
– catabolic pathways **11B**, 203
– degradation by fungi **11B**, 157f
– environmental relevance **11B**, 194ff
– metabolism of **9**, 176f
– oxo-derivatives, conversion of **11B**, 200
– oxygenated –, environmental relevance **11B**, 195
– sources **11B**, 194f
alkaline phosphatase, phosphorylation by **8B**, 226
alkaline protease, purification **3**, 610
alkalinity, effect on activated sludge process **11A**, 266
alkaliphiles, biotopes of **10**, 68f
– characteristics of **10**, 69
– mode of growth **10**, 69

alkaliphilic bacilli *see* bacilli
alkaloid production, by *Claviceps purpurea* **1**, 240
– by plant cell culture **1**, 596
– calcium alginate effect **1**, 240
alkaloids *see also* benzylisoquinoline alkaloids,
   monoterpenoid indole alkaloids, morphine alkaloids, tropane alkaloids
– as drugs, development of new – **8A**, 330f
– biosynthetic pathways **8A**, 330
– biotransformations **8A**, 327ff
– chlorinated – **8B**, 185
– classification **8A**, 328
– from animals **8A**, 328f
– from bacteria **8A**, 328f
– from fungi **8A**, 328f
– from plant cell cultures **7**, 603ff
– from plants **8A**, 330f
– – use of cell culture **8A**, 331
– – use of metabolic engineering **8A**, 331
– historical uses **8A**, 328
– hydroxylation of **8A**, 495f
– semisynthetic derivatives of, cytotoxic action
   8A, 354
– skeletal types **8A**, 330
– structures of **8A**, 329
alkane hydroxylase **11B**, 196ff
alkane oxidation system **11B**, 203f
alkanes, anaerobic degradation of **11B**, 171f
– bacterial degradation of, peripheral pathways
   11B, 150
– branched –, debranching of **11B**, 199
– degradation of, by immobilized cells **11B**, 420f
– oxidation of, oxygenase reactions **11B**, 196ff
– pathways of oxidation **11B**, 197
– utilization by *Pseudomonas oleovorans* **11B**,
   203f
n-alkanes, aerobic degradation of, stoichiometric
   and kinetic coefficients **11A**, 388
– microbial oxidation **9**, 177
– substrates for biomass production **9**, 214
alkanols, effect on microbial growth **1**, 135
– resolution by PPL **8A**, 75ff
alkenes, biodegradation mechanisms **6**, 687
– dihydroxylation of **8A**, 512
– epoxidation of, use of poly-L-leucine **8B**, 503
– halogenation of **8B**, 196f
alkylated aromatic compounds, pathways of degradation **11B**, 213ff
alkylbenzenes, microbial degradation, pathway
   11B, 214
S-alkylcysteine *α,β*-lyase, substrate specificity **6**,
   527
alkyl mercury-lyase **8B**, 143
N-alkyl compounds *see* organic compounds
alkylphenol, bacterial degradation of **11B**, 153
S-alkyl compounds *see* organic compounds
alkynes, halogenation of **8B**, 197
alldimycin, antitumor agent **7**, 659

– structure **7**, 659
allelic isoenzyme (allozyme) **2**, 145
allergenicity, of microbial biomass **9**, 232f
allergy, mAb products **5A**, 222
– medical applications of recombinant proteins
   5A, 140, 145
– monoclonal antibody therapy **5A**, 316
alliacols, cytotoxic activity **7**, 507
– structure **7**, 505
allosamidin, chitinase inhibitor **7**, 440
allosamidins, structure **7**, 472
allostery **9**, 47
allotransplantation **5A**, 145f
– medical applications of recombinant proteins
   5A, 145
– of parathyroid tissue **10**, 557
alpha 1-antitrypsin deficiency (alpha1AD) **5B**, 70
Alteplase **5A**, 129
alternative splicing **5B**, 366, 370ff
alterobactin, peptide siderophore of *Alteromonas*
   luteoviolacea **7**, 212
– structure **7**, 212
Alteromonas haloplanctis, bisucaberin **7**, 210f
Alteromonas luteoviolacea, alterobactin **7**, 212
Alzheimer disease **5B**, 54, 68, 71, 117, 126
Alzheimer's disease, model in mice **2**, 783
Amadori rearrangement, during bread baking **9**,
   307
Ames test, coal bioconversion products **10**, 178ff
– coal derived reference substances **10**, 178f
ametantron, antitumor agent, structure **7**, 651
Amflow process **9**, 279ff
– flow sheet **9**, 281
amidase activity, in brevibacteria **8A**, 292
– of agrobacteria **8A**, 294
– of penicillin acylase **8A**, 229
– of pseudomonads **8A**, 294
– stereoselectivity **8A**, 308
amidases, production of non-natural amino acids
   8B, 285f
– synthesis of pharmaceutical precursors, kilogram-scale routes **8B**, 285
amide bond cleavage, antibody-catalyzed – **8A**,
   270
amide bond formation **8A**, 243f
– antibody-catalyzed – **8A**, 270
– hydrolysis of side chain mimics **8A**, 254f
– kinetic control **8A**, 249
– larger peptides, non *α*-amino acid substrates
   8A, 256f
– optimum water levels **8A**, 252
– regioselectivity **8A**, 253f
– thermodynamic control **8A**, 249
– use of immobilized enzymes **8A**, 255
– with non-activated esters as acyl donors **8A**,
   262
– with unnatural substrates **8A**, 253ff
amides, C-terminal –, hydrolysis of **8A**, 245

– delicious octapeptide amide **8A**, 256f
– formation by papain **8A**, 244f
– hydrolysis of, use of catalytic antibodies **8B**, 438f, 461f
– hydroxylations of **8A**, 479
– N-terminal –, hydrolysis of **8A**, 246
– synthesis, strategies for **8A**, 250
amikacin, semisynthetic aminoglycosides **7**, 461
amination, use of catalytic antibodies **8B**, 476
amine oxidase **1**, 274
amines **8A**, 98ff
– large-scale resolution of **8B**, 279f
– – involving a *Pseudomonas* lipase **8B**, 279f
– lipase-catalyzed deacylation of **8A**, 131f
– lipase-catalyzed resolutions of **8A**, 101
amino acid, deamination **1**, 91f
– degradation **1**, 91
– – in clostridia **1**, 304f
– estimation methods **1**, 126
– random sequence selection **2**, 351
– regulation of biosynthesis **2**, 251ff
– transport in *Lactococcus* **1**, 358
amino acid amides **8A**, 244f
amino acid biosynthesis **1**, 92
– aromatic amino acids **1**, 93
– aspartate family **1**, 454f
– by corynebacteria **1**, 454ff
– – regulation **1**, 445, 457
– by methylotrophs **1**, 279
– feedback inhibition **1**, 101
– oxaloacetate family **1**, 93
– oxoglutarate family **1**, 93
– phosphoglycerate family **1**, 93
– pyruvate family **1**, 93
– strain improvement **1**, 457
amino acid composition, in raw beet juice **6**, 14f
amino acid metabolism **1**, 91f
– in lactic acid bacteria **1**, 354, 358
– in yeast **1**, 489
– transamination **1**, 91
amino acid production **1**, 164
– by coryneforms **1**, 454
– by methylotrophic microorganisms **1**, 279
– engineering novel biochemical pathways **8B**, 324ff
– enzymology **6**, 503ff
– – alanine **6**, 512f, 533f
– – amino acid dehydrogenases **6**, 537ff
– – amino acid racemases **6**, 535ff
– – ammonia lyases **6**, 519f
– – hydrolytic enzymes **6**, 514ff
– – L-arginine **6**, 513
– – L-aspartic acid **6**, 512
– – L-citrulline **6**, 513
– – L-glutamic acid **6**, 506f
– – L-glutamine **6**, 512
– – L-histidine **6**, 513
– – L-isoleucine **6**, 510

– – L-leucine **6**, 510
– – L-lysine **6**, 507f
– – L-ornithine **6**, 513
– – L-phenylalanine **6**, 510
– – L-proline **6**, 513
– – L-serine **6**, 514
– – L-threonine **6**, 508ff
– – L-tryptophan **6**, 512
– – L-tyrosine **6**, 510ff
– – L-valine **6**, 510
– – pyridoxal 5′-phosphate enzymes **6**, 520ff
– L-lysine **8B**, 321ff
– L-phenylalanine **8B**, 314ff
– L-threonine **8B**, 321ff
– microbial pathways **8B**, 313ff
– strain construction **1**, 206ff
amino acid racemases **6**, 535ff
– from *Pseudomonas putida*, substrate specificity **6**, 536
amino acids **3**, 576
– advantages of microbial production **6**, 505
– anaerobic degradation of **11A**, 33f
– analysis, separation methods **3**, 733
– as nutrients, for mammalian cell cultures **3**, 145
– biocatalytic resolution of *N*-acetyl cysteine **8A**, 18
– biosynthetic pathway, of aromatic – **6**, 511
– – of aspartate-related – **6**, 509
– chlorinated – **8B**, 185
– cleavage of C-terminal residues, by carboxypeptidases **8A**, 247
– commercial production **6**, 465ff
– – L-glutamic acid **6**, 469ff
– enzymatic resolution of racemic – **6**, 543ff
– – asymmetric derivatization **6**, 544
– – asymmetric hydrolysis **6**, 544ff
– enzymatic synthesis **6**, 514ff
– – α-amino- -caprolactam hydrolase **6**, 514f
– – alanine dehydrogenase **6**, 540f
– – alanine racemase **6**, 535f
– – arginine deiminase **6**, 520
– – aspartase **6**, 519
– – aspartate β-decarboxylase **6**, 520f
– – aspartate racemase **6**, 537
– – β-chloro-L-alanine hydrogenchloride lyase **6**, 527
– – β-tyrosine **6**, 522f
– – cystathionine γ-lyase **6**, 530f
– – D-amino acid aminotransferase **6**, 532ff
– – D-cysteine desulfhydrase **6**, 528
– – D-selenocystine α,β-lyase **6**, 528
– – glutamate racemase **6**, 536f
– – glutamine synthetase **6**, 542f
– – hydantoinase **6**, 515ff
– – L-cysteine desulfhydrase **6**, 525
– – leucine dehydrogenase **6**, 537ff
– – L-methionine γ-lyase **6**, 528f
– – L-threonine alcolase **6**, 532

– – *O*-acetylhomoserine sulfhydrylase **6**, 530
– – phenylalanine ammonia lyase **6**, 519f
– – phenylalanine dehydrogenase **6**, 541f
– – *S*-adenosyl-L-homocysteine hydrolase **6**, 518f
– – *S*-alkylcysteine α,β-lyase **6**, 527
– – serine hydroxymethyltransferase **6**, 531
– – tryptophanase **6**, 523ff
– – tryptophan synthase **6**, 525ff
– – 2-amino-Δ²-thiazoline-4-carboxylate
  hydrolase **6**, 515
– essential –, in algae **9**, 174
– extraction **3**, 576ff
– – by liquid membrane techniques **3**, 582
– – by reversed micelles **3**, 583
– – enantiomeric separation **3**, 582
– – equilibrium constants **3**, 576
– – equilibrium reaction **3**, 576
– feed uses **6**, 468
– flavor-enhancing properties **6**, 467f
– food uses **6**, 467f
– for cosmetic applications **6**, 487ff
– for pharmaceutical applications **6**, 468, 487ff
– – direct fermentation from carbohydrates **6**,
  491
– – L-alanine **6**, 487ff
– – L-arginine **6**, 489
– – L-glutamine **6**, 489
– – L-histidine **6**, 489
– – L-isoleucine **6**, 492
– – L-leucine **6**, 492
– – L-methionine **6**, 492
– – L-proline **6**, 492
– – L-serine **6**, 493
– – L-valine **6**, 492
– found in lantibiotics **7**, 327ff, 347
– – D-alanine **7**, 330
– – didehydroalanine **7**, 328
– – didehydrobutyrine **7**, 328
– – lanthionine **7**, 328
– – methyllantionine **7**, 328
– – structures **7**, 327
– in bacteria **9**, 224
– industrial production methods **6**, 469f
– in fungi **9**, 224
– in microbial biomass **9**, 174, 223f
– interspecies formate transfer **11A**, 465
– in the feed industry **6**, 475ff
– in the food industry **6**, 469ff
– in yeast **9**, 224
– isoelectric point (p*I*) **3**, 576
– L-threonine **6**, 482ff
– L-tryptophan **6**, 484ff
– manufacture of **8A**, 17ff
– market **6**, 466ff
– molecular structure **6**, 466
– natural –, production of **8B**, 285
– non-natural –, **8B**, 285f
– p*K* value **3**, 576

– price **6**, 491
– production, by recombinant organisms **3**, 287ff
– – rate estimation **3**, 369
– production capacity **6**, 491
– production steps **6**, 472
– production volumes **6**, 470
– protein-forming – **6**, 466f
– protein stabilization effects **3**, 705
– quantitation, in bioprocess control **3**, 17
– residue properties **5B**, 335
– resolution by acylase I **8A**, 246ff
– side chains **9**, 10, 26
– syntrophic degradation of, acetate as a thermodynamic regulator **11A**, 465
– uncommon –, glycopeptide antibiotics **7**, 384ff
– unnatural –, enantioselective hydrolysis **8A**,
  248
– uses **6**, 465ff
α-amino acids, coupling of single amino acids, solvent effects **8A**, 250
– – structural effects **8A**, 250
– enantioselective hydrolysis of the *N*-phenylace-
  tyl derivatives **8A**, 261
D-amino acids, formation from racemic hydantoins **8A**, 268
– production of **8B**, 326
L-amino acids, manufacture by *N*-acetyl hydrolysis **8A**, 19
– production by immobilized aminoacylase, flow
  diagram **8A**, 19
amino acid weight matrices **5B**, 131
aminoacylase **6**, 545ff
– immobilization **3**, 448
– in immobilized enzyme reactors **3**, 448
– in industrial biotransformations **3**, 448f
– L-amino acid production, non-proteinogenic
  products **3**, 450
– – proteinogenic products **3**, 450
– production of L-amino acids **8A**, 19
– stability **3**, 448
– substrate specificity **3**, 448, **6**, 546
Aminoacyl-tRNA Synthetase Database **5B**, 405
δ-(L-α-aminoadiphyl)-L-cysteinyl-D-valine synthetase (ACVS) **7**, 254ff
δ-(L-α-aminoadipyl)-L-cystein-D-valine synthetase
  8B, 385
p-aminobenzonitrile, conversion to *p*-aminobenzo-
  ic acid **8A**, 296f
L-2-aminobutyrate, novel biosynthetic pathway
  8B, 324f
7-aminocephalosporanic acid, biocatalytic route
  8B, 280
aminochelin, catecholate siderophore of *Azotobacter vinelandii* **7**, 203
– structure **7**, 203
aminocyclitol aminoglycosides, producing bacterial
  genera **7**, 401
D-amino acid aminotransferase **6**, 532ff

D-amino acids, production **3**, 450
7-amino desacetoxycephalosporanic acid, synthesis
    of cephalosporin antibiotics **8A**, 258
2-amino-Δ^2-thiazoline-4-carboxylate hydrolase **6**,
    515
α-amino- -caprolactam hydrolase **6**, 514
α-amino- -caprolactam racemase **6**, 514f
aminoglycoside-like compounds, structures **7**, 402
aminoglycoside pathways, overview **7**, 448f
aminoglycoside producers, resistance in – **7**, 440ff
– – genes **7**, 440ff
aminoglycoside production, common gene pool **7**,
    462
aminoglycosides **7**, 397ff
– acarbose **7**, 438f
– adverse effects in humans **7**, 451
– allosamidin **7**, 440
– as translation inhibitors **7**, 450
– bactericidal effects **7**, 450
– biochemical classification **7**, 400
– biosynthesis of – **7**, 409ff
– biosynthetic genes, in *Streptomyces* **7**, 447
– bluensomycin **7**, 415f
– carbohydrate precursors, from primary metabol-
    ism **7**, 459
– compilation of **7**, 400
– decoding site of bacterial ribosomes **7**, 450
– deoxystreptamine-containing – **7**, 432ff
– – apramycins **7**, 436
– – biosynthetic pathways **7**, 433
– – destomycins **7**, 436
– – 4,5-disubstituted – **7**, 434
– – 4,6-disubstituted – **7**, 434f
– ecology **7**, 401
– evolutionary aspects **7**, 457f
– fortimicins **7**, 429
– functions of – **7**, 409ff
– genetics of **7**, 409ff
– glycosidase inhibitors **7**, 436f
– hexitol components, structures **7**, 406
– istamycins **7**, 429ff
– oligosaccharidic – **7**, 438f
– pathway engineering **7**, 459f
– pathway formula **7**, 400
– pentitol components, structures **7**, 406
– producing bacteria **7**, 401
– regulation of production, in stretomycetes **7**,
    443ff
– resistance mechanisms in **7**, 440ff
– semisynthetic **7**, 461
– sorbistins **7**, 439f
– streptomycin-related **7**, 427ff
– – boholmycin **7**, 428
– – hygromycin A **7**, 428f
– – kasugamycin **7**, 427
– – minosaminomycin **7**, 428
– – myomycin **7**, 428
– – spectinomycins **7**, 427

– streptomycins **7**, 415ff
– target site interactions **7**, 449
– trehalosamines **7**, 437f
– trehazolin **7**, 439
– validamycins **7**, 438f
amino groups, protection of, by phenylacetyl **8A**,
    261
amino hexitol components, structures **7**, 406
L-amino acids, dehydrogenases, in industrial bio-
    transformations **3**, 454
– production **3**, 448f
– – downstream processing **3**, 448
aminolysis, lipase-catalyzed **8A**, 263ff
6-aminopenicillanic acid (6-APA), production of,
    using penicillin G acylase **8B**, 280
6-aminopenicillanic acid (6-APA) **1**, 527
– manufacture of **8A**, 15f
– synthesis of penicillin antibiotics **8A**, 258
amino pentitol components, structures **7**, 406
aminopeptidase, of *Pseudomonas putida*, asym-
    metric hydrolysis of lupinic acid amide **8A**,
    245f
– – resolution of racemic α-amino acids **8A**, 245
aminopeptidase inhibitors, from basidiomycetes
    7, 518ff
– – structure **7**, 519
aminosorbitol, regioselective oxidation of, with
    Gluconobacter oxydans **8B**, 295
1-amino-1-deoxy-D-sorbitol, oxidation of, by *Glu-
    conobacter oxydans* **8B**, 300ff
– – protection groups **8B**, 302
– – pyridine formation from 6-amino-6-deoxy-L-
    sorbose **8B**, 302
– – time course **8B**, 304
– – with *Gluconobacter oxydans* **8B**, 304
AML *see* acute myelogenous leukemia
ammonia **1**, 174f
– as nitrogen source, in fermentation media **3**,
    129
– degradation of, in biological waste gas purifica-
    tion **11C**, 287
– effect on microbial growth **1**, 133
– estimation methods **1**, 126
– futile cycle **1**, 178
– inhibitory effects **3**, 129
– membrane transport **1**, 175
ammonia/ammonium coupled assays **4**, 90f
– analysis of amino acids **4**, 91
– in fermentation technology **4**, 91
ammonia lyases **6**, 519f, **8B**, 47, 125
ammonia monooxygenase **11A**, 56
ammonia oxidation **11A**, 56ff
– anaerobic – **11A**, 40
– anoxic – **11A**, 56f
– effect of nitrogenous oxides **11A**, 57
– physiology of **11A**, 56
ammonia oxidizers **11A**, 56f
– denitrification by **11A**, 57

ammonification **11A**, 37f
ammonium assimilation **1**, 92f, 174f, 387
ammonium lactate process **6**, 300ff
– scheme **6**, 301
ammonium sulfate, enzyme precipitation **9**, 94
amonabactins, catecholate siderophore of *Aero-
    monas hydrophila* **7**, 206
– structure **7**, 206
AMP *see* adenosine monophosphate
amperometric detector, thin-layer **4**, 84
– tubular **4**, 84
– wall-jet **4**, 84
Amperometric sensor, for biomass determination
    4, 194f
amphibolic pathway **1**, 52
amphotropic murine leukemia virus, use of vector
    systems, in HIV gene therapy **5A**, 479f
ampicillin, effect in mouse septicemia **7**, 380
– formation by penicillin G acylase **8A**, 259
amplicon generation **5B**, 233f
amplified fragment length polymorphism (AFLP)
    5B, 168
amplified fragment length polymorphisms
    (AFLP), detection of plant pathogens **10**, 500
amplified ribosomal DNA restriction analysis
    (ARDRA), in microbial community analysis
    11A, 87
AmpliTaq FS ™ **5B**, 197
Amycolatopsis, vanillin production by **8B**, 339f
Amycolatopsis methanolica, amino acid produc-
    tion **1**, 279
Amycolatopsis sp., production of dalbaheptides **7**,
    380
amylase **1**, 27
– from pseudomonads **1**, 421
– secretion **1**, 27
– thermostable amylolytic enzyme **1**, 28
α-amylase **6**, 143, **10**, 74ff
– action of **10**, 75
– expressed by *Bacillus licheniformis* **6**, 68
– extracellular **1**, 384
– gene *(AMY1)*, *Schwanniomyces occidentalis* **1**,
    201
– gene regulation **1**, 386
– heat-stable – **10**, 76
– in clostridia **1**, 310
– mutants **1**, 386
– of *Aspergillus oryzae* **6**, 68
– of *Bacillus subtilis* **6**, 68
– operon **1**, 386
– purification **3**, 604, 610
– synthesis, regulation in *Bacillus subtilis* **1**, 384
β-amylase **6**, 38, 143, **10**, 74f
– action of **10**, 75
– of *Bacillus licheniformis* **6**, 67f
α-Amylase production, by *Bacillus subtilis* **4**, 540
– kinetic model **4**, 540f
– optimal control of carbon and nitrogen **4**, 540ff
– optimal feed rate profile, caseinate **4**, 542
– – starch **4**, 542
– optimization **4**, 541
amylases **10**, 74ff
– heat stable – **10**, 76f
– in silage **9**, 778
– use in baking **9**, 675
α-amylases **5A**, 192ff
– biology **5A**, 192
– commercial products **5A**, 193
– *de novo* formation in malt **9**, 427
– detergent applications **5A**, 196ff
– – automatic dishwashing **5A**, 196f, 199
– determination, in a coupled assay **9**, 149f
– development **5A**, 197f
– – for the detergent industry **5A**, 197
– – high-fructose corn sirup **5A**, 197
– industrial applications **5A**, 192ff
– in flour **9**, 675f
– in paper production **5A**, 195f
– in starch liquefaction **5A**, 193f
– in textile desizing **5A**, 194f, 198
– properties on technical – **5A**, 193
– reaction catalyzed by – **5A**, 192
– stability **5A**, 192ff
– – against oxidants **5A**, 197f
– starch liquefaction **9**, 668f, 744f
– thermostability **5A**, 198
– use in baked goods **9**, 264
– use in brewing **9**, 682ff
β-amylases **9**, 669, 747
– in flour **9**, 675f
– in malt **9**, 427
– plant sources **9**, 747
amyloglucosidase **9**, 670f, 745f, 750ff
amylolysis **6**, 66ff
amylostatin family, structure of – **7**, 472
Anabaena azollae **1**, 237
– hydrogen production **1**, 237
anabasine, biosynthesis **7**, 625
– structure **7**, 624
anabolism **1**, 50
anaerobes **3**, 298ff
– facultative – **3**, 298
– growth stimulation **3**, 298
– strict – **3**, 298
anaerobic baffle reactor **11A**, 471
anaerobic contact process **11A**, 481f
– in industrial wastewater treatment **11A**, 479ff
– liquid-solid separation **11A**, 482, 487f
– mass balance concept **11A**, 485f
– mixing design **11A**, 486
– one-stage vs. two-stage anaerobic systems **11A**,
    486f
– reactor design **11A**, 484f
– reactor scheme **11A**, 481f
– start-up phase **11A**, 489f

anaerobic degradation *see* bacterial degradation, biodegradation
– of amino acids **11A**, 33f
– of biopolymers, by a food chain **11A**, 27
– – role of acetate **11A**, 29
– – role of molecular hydrogen **11A**, 29
– of carbohydrates **11A**, 30ff
– of cellulose **11A**, 30f
– of glucose **11A**, 23f
– – mass and energy balance **11A**, 23f
– of glycolipids **11A**, 34f
– of ligno-cellulose **11A**, 30f
– of lipids **11A**, 34f
– of neutral fats **11A**, 34f
– of phospholipids **11A**, 34f
– of proteins **11A**, 32ff
– of starch **11A**, 31f
– of whey, concentration of volatile acids **11A**, 537
– of xenobiotics **11A**, 44ff
– pathway **11A**, 536
anaerobic digestion **11A**, 456ff
– acetate decarboxylation **11A**, 457
– by methanogens **11A**, 458ff
– fermentation step **11A**, 456f
– inhibition by heavy metals **11C**, 154
– in wastewater treatment **11A**, 13
– kinetic data **11A**, 484
– metabolic steps **11A**, 456f
– – scheme **11A**, 457
– methane formation, inhibition by hydrogen sulfide **11A**, 541
– – inhibition by propionic acid **11A**, 541
– of municipal solid waste, processes **11C**, 156ff
– of waster streams, with high sulfate levels **11A**, 461
– reductive methane formation **11A**, 457
anaerobic ecosystems **11A**, 456ff
– hydrogen-consuming reactions **11A**, 29
anaerobic fermentation **3**, 295ff
– and composting, comparison of **11C**, 155f
– ATP generation **3**, 299ff
– available electron units **3**, 300
– bacteria **11C**, 153
– biogas from, composition of **11C**, 155
– – yields **11C**, 155
– biomass yield **3**, 299ff
– feedstock for **11C**, 163ff
– – scheme **11C**, 163
– influence of milieu conditions **11C**, 154f
– instrumentation **3**, 313
– measurement **3**, 313
– of biowaste **11C**, 157ff
– – agitation **11C**, 162
– – continuous operation **11C**, 161f
– – discontinuous operation **11C**, 161f
– – dry – **11C**, 160f
– – mesophilic – **11C**, 162

– – one-stage – **11C**, 160
– – procedures **11C**, 157ff
– – process engineering of **11C**, 159ff
– – thermophilic – **11C**, 162
– – two-stage – **11C**, 160
– – wet – **11C**, 160f
– process operation, mesophilic – **11C**, 162
– – thermophilic – **11C**, 162
– product inhibition **3**, 304
– products, gaseous – **3**, 303f
– – liquid – **3**, 304ff
– – solid – **3**, 309
– – yield **3**, 299ff
– scheme **11C**, 153
– substrates, concentration **3**, 299
– – consumption **3**, 299
– – gaseous – **3**, 303f
– – liquid – **3**, 304ff
– – solid – **3**, 309
– types **3**, 303ff
– with solid phases **3**, 311
anaerobic fermentations, biochemical fundamentals of **11C**, 152ff
– steps **11C**, 152f
anaerobicity, gene regulation in facultative anaerobs **2**, 210f
Anaerobic reactors, wastewater process models **4**, 467
anaerobic respiration **1**, 382
Analysis techniques **4**, 159ff
– continuous air-segmented analyzer **4**, 156ff
– detectors **4**, 169
– on-line flow-injection analyzer (FIA) **4**, 159ff
– on-line FPLC **4**, 167
– on-line HPLC **4**, 165ff
– on-line IC **4**, 167ff
analyte concentrations, determination with kinetic methods, catalytic assays **9**, 145f
– – first-order reactions **9**, 143f
– – zero-order reactions **9**, 144f
analytical investigation program, for solid matter **11B**, 495ff
– methods **11B**, 499ff
– – chromatographic procedures **11B**, 502
– – detectors **11B**, 502f
– – field measurements **11B**, 503ff
– – physical instrumental methods **11B**, 500f
– – spectrometric procedures **11B**, 501
analytical ultracentrifugation, of protein complexes **5A**, 91
analytical uses, of enzymes **9**, 137ff
– – coupled reactions **9**, 141ff
– – detection limits **9**, 139
– – end-point reactions **9**, 139f
– – in diagnostics **9**, 155ff
– – in food analysis **9**, 155ff
– – quality requirements **9**, 153ff
– – test strips **9**, 150

analyzer system, air-segmented automatic analyzer channels **4**, 158
anaplerotic reaction **1**, 85f
Anchusa officinalis, accumulation of rosmarinic acid **7**, 599f
ancovenin, primary structure **7**, 336f
androgen receptor antagonists **7**, 116
androgens, inhibition of – **7**, 685
anemometer, hot film anemometry (HFA) **4**, 138f
– laser-Doppler anemometry (LDA) **4**, 138
aneuploidy **2**, 8, **5B**, 75
– in plants **2**, 158
Angelmann syndrome (AS) **5B**, 73f
angina, unstable –, treatment with ReoPro **5A**, 376
angiogenesis **5A**, 133f
– medical applications of recombinant proteins **5A**, 133
angiogenesis inhibitors **7**, 673ff
– structures **7**, 674
angiostatin **5A**, 134
angiotensin converting enzyme **5B**, 117
angiotensin II **8B**, 356
angiotensinogen **5B**, 117
angiotensin synthesis, use of thermolysin **8B**, 355f
ang-kak **9**, 525ff
– production of **9**, 526
– therapeutic properties **9**, 527
anguibactin, catecholate siderophore of *Vibrio anguillarum* **7**, 206f
– structure **7**, 207
anhydrides, regioselective ring opening, by lipases **8A**, 111
aniline, bacterial degradation of **11B**, 153, 177f
animal, moral status **12**, 120
animal breeding **2**, 746
– historical importance **2**, 746
animal breeding program **2**, 777
– cattle **2**, 777
– pig **2**, 777
– time schedule **2**, 778
– traditional selection breeding **2**, 777
– transgenic line establishment **2**, 777
animal cell **1**, 38f
– culture **1**, 40f
– envelope **1**, 38
– genetic engineering **2**, 663ff
– requirements **1**, 41
animal cell culture **3**, 23ff, 106ff
– – *see also* mammalian cell culture
– adherent cells, shear stress effects **3**, 228
– aeration **3**, 219ff
– – methods **3**, 237ff
– anchorage-dependent – **3**, 27
– bioreactors **3**, 106, **3**, 223
– bubble entrainment **3**, 243
– carbon dioxide removal **3**, 224ff
– carbon dioxide sensitivity **3**, 227
– cell death, shear stress effects **3**, 229
– cell entrapment **3**, 110
– cell types **3**, 225
– control by NMR **3**, 35
– current applications **3**, 26ff
– future prospects **3**, 35
– genetic engineering **3**, 31
– historical perspective **3**, 24ff
– human tissue repopulation **3**, 35f
– hybridoma culture **3**, 25
– hydrodynamic damage **3**, 112
– implants **3**, 36
– in biotechnological manufacturing **3**, 23ff
– limitations **3**, 30
– mechanical stress **3**, 272ff
– media formulation **3**, 31
– microcarrier culture **3**, 108
– oxygen supply **3**, 224
– physical properties **3**, 234
– primary – **3**, 25
– products **3**, 106
– respiratory quotient **3**, 226
– shear damage **3**, 33, 228ff
– turbulence effects **3**, 272
animal cell cultures **12**, 94ff
– biosafety **12**, 94ff
– contamination with viruses **12**, 97
– continuous cell lines **12**, 94
– hybridomas **12**, 95
– immortalized cells **12**, 94
– maintenance of **10**, 27ff
– primary cell cultures **12**, 94
– vectors, adeno-associated **12**, 100
– – derived from adenoviruses **12**, 99f
– – derived from BPV-1 **12**, 100f
– – derived from HSV-1 **12**, 100
– – derived from poxvirus **12**, 98
– – derived from retroviruses **12**, 98f
animal cells **3**, 106
– adherent – **3**, 107
– as host cells **3**, 285
– electrofusion of, hybrid yield **10**, 515f
– – under microgravity conditions **10**, 515
– recombinants, in fermentation processes **3**, 285
– recovery, from bioreactors **3**, 473
– suspended – **3**, 107
animal cell technology, future developments **10**, 38
animal fats, commercial markets **7**, 135
– fatty acid composition **7**, 136
animal fats processing, wastewater concentrations **11A**, 204
animal feed, role of microalgae and cyanobacteria **10**, 127f
animal manure, cofermentation with biowaste **11C**, 23
animal protein *see* protein

animal rights **12**, 120
animals, alkaloids from **8A**, 328
animal tissue **1**, 40
– cell-cell connection **1**, 40
animal viruses, inactivation times, effect of temperature **11C**, 229
ankylosing spondylitis **5B**, 117
annotated databases **5B**, 286
annotation, automated – **5B**, 361ff
– eukaryotic – **5B**, 375
– of the human genome DNA sequence **5B**, 49ff
– – cataloguing the genes **5B**, 51ff
anonymization, of DNA samples **5B**, 421
ansamitocin, inhibition of the microtubular system **7**, 668
– structure **7**, 668
anthocyanin, structure **7**, 610
anthocyanins, from plant cell cultures **7**, 609f
anthracene, bioremediation of **11B**, 428
– bound residues, half life times **11B**, 117
– – mobilization of **11B**, 117
– complete metabolization and mineralization **11B**, 222ff
– depletion in soil **11B**, 115
– microbial degradation of **11B**, 116f
– – pathway **11B**, 222f
– oxidation by *Phanerochaete chrysosporium* **11B**, 233
– physical and chemical properties of **11B**, 218
– structure of **11B**, 216
anthracene bound residues, degradation in soil **11B**, 116
anthracite, as carrier materials of biofilters **11C**, 466
anthracycline analogs **7**, 660
anthracyclines, antitumor agents **7**, 652, 657ff
– – structures **7**, 657ff
– formation of oxygen radicals **7**, 661f
– pigmented antibiotics, as antitumor agents **7**, 660
– structures **7**, 34
anthramycins, antitumor agents **7**, 656
anthraquinones, anticancer drugs **7**, 652
– biosynthetic pathway **7**, 602f
– from plant cell cultures **7**, 601ff
– – yields **7**, 601
– structure **7**, 601
antialgal compounds, as products of marine biotechnology **10**, 458
antibacterial agents, from basidiomycetes **7**, 501ff
antibacterial compounds, as products of marine biotechnology **10**, 458f
anti-biotechnology effort **12**, 475ff
– in the U.S. **12**, 476ff
antibiotic **2**, 35
– biosynthesis genes **2**, 35
– improvement **2**, 36f
– production yield **2**, 31

– transcription inhibitors **2**, 248f
– translation inhibitors **2**, 248f
antibiotic export, regulatory cassettes, in streptomycetes **7**, 92
antibiotic formation, in stretomycetes, factors determining onset of – **7**, 86
– pathway-specific activator genes **7**, 89f
antibiotic hybrids **2**, 475ff
– isovaleryl-spiramycin **2**, 477
– mederrhodin A **2**, 447
antibiotic producer **2**, 538
– biosynthetic pathway investigation **2**, 538f
– β-lactam **2**, 539
antibiotic production, effect of a two-component regulatory system **7**, 83
– effect of metabolic imbalances **7**, 74
– gene clusters **7**, 81
– genes affecting also morphological differention **7**, 84f
– genetics **7**, 79ff
– in streptomycetes, metabolite interference **7**, 72ff
– multicopy inhibition of – **7**, 84
– onset of –, role of protein phosphorylation **7**, 81f
– organization of antibiotic biosynthetic genes **7**, 79ff
– regulation in *Streptomyces*, physiological aspects **7**, 71ff
– regulation of – **7**, 57ff
– – β-lactams **7**, 65ff
– – effect of growth conditions **7**, 59f
– – in stretomycetes **7**, 70ff
– – in unicellular bacteria **7**, 61ff
– – peptide antibiotics **7**, 67ff
– role of small diffusible compounds **7**, 78f
– stimulation by ppGpp **7**, 61ff, 75ff
– three-compartment model, *Myxococcus fulvus* **4**, 288ff
antibiotic regulatory genes, pleiotropic –, interactions **7**, 85ff
antibiotic resistance **1**, 447
– as selectable marker **1**, 536
– genes, in pseudomonads **1**, 410
– – in streptomycetes **1**, 447
– in streptomycetes, induction of – **7**, 92ff
– types **1**, 447
antibiotics **3**, 63
– – *see also* β-lactam antibiotics, *see also* dalbaheptides, *see also* lantibiotics, *see also* peptide antibiotics
– aminoglycosides, target site interactions **7**, 449
– anthracycline type **7**, 658
– as feed additives **9**, 780
– binding non-covalently to the DNA, structures **7**, 649
– chloroamphenicol **8B**, 178
– chlorotetracycline **8B**, 178

– evolutionary significance **7**, 60
– extraction **3**, 562f, 577
– – by liquid membrane techniques **3**, 582
– – equilibrium reactions **3**, 577
– griseofulvin **8B**, 178
– hydroxylation of **8A**, 496
– hystory of discovery **1**, 448
– intercalating – **7**, 650ff
– β-lactam biosynthesis **7**, 247ff
– macromolecular antitumor – **7**, 663ff
– producers **1**, 445
– production **1**, 214, 441ff, **3**, 9, 473
– – facility design **3**, 14
– – novel synthetic antibiotics **1**, 214
– – productivity **3**, 14
– – product quality **3**, 14
– – regulation of **1**, 441
– production, by filamentous fungi **1**, 527
– resistance mechanisms against –, producing microbes **7**, 442
– solvents, for liquid-liquid extraction **3**, 563
– streptomycetes genes **1**, 439
antibiotics, gene cluster **1**, 445
antibiotic synthesis, carbon catabolite regulation **4**, 150f
– enzyme induction **4**, 151
– key components **4**, 151f
– nitrogen metabolite regulation **4**, 151
– phosphate regulation **4**, 151
antibodies *see also* monoclonal antibodies, therapeutic antibodies
– activity of **5A**, 295ff
– – preclinical determination **5A**, 295f
– affinity for tansition state analog, screening of **8B**, 409
– as protein folding aids **3**, 541
– binding specificity **5A**, 312
– catalytic – **8A**, 269f
– chimeric – **5A**, 248
– choice of starting antibody **5A**, 292f
– detection of plant pathogens **10**, 499
– for sepsis **5A**, 329ff
– function of **8B**, 404
– galactopyranosidase **8B**, 434
– glycosidase, generation by *in vitro* immunization **8B**, 434
– immunogenicity **5A**, 289ff, 299f
– intracellular –, in HIV gene therapy **5A**, 477f
– kinetics **5A**, 289ff
– monoclonal –, hapten affinity binding studies **8A**, 269
– pharmacokinetics **5A**, 297f
– pharmacology **5A**, 289ff
– preclinical testing of **5A**, 289ff
– – affinity **5A**, 292
– – animal models **5A**, 296ff
– – carcinogenicity studies **5A**, 309
– – cell-based assays **5A**, 291

– – conditions of administration **5A**, 308
– – corss-reactivity **5A**, 306f
– – cross-reactivity **5A**, 295
– – determination of activity **5A**, 295
– – fragments **5A**, 294f
– – genotoxicity studies **5A**, 309
– – guidance documents **5A**, 305
– – immunoassays **5A**, 291
– – limit of detection **5A**, 292
– – local tolerance **5A**, 309
– – oncogenicity studies **5A**, 309
– – pharmacokinetics **5A**, 297f
– – pharmacology **5A**, 295
– – regulatory requirements **5A**, 305
– – reproduction toxicity studies **5A**, 308f
– – safety aspects **5A**, 303ff
– – safety testing program **5A**, 305
– – selection of isotypes **5A**, 293f
– – species used **5A**, 308
– – timing in relation to clinical trials **5A**, 306
– – tissue specificity **5A**, 306ff
– – use of transgenic models **5A**, 297
– – use of xenografts **5A**, 297
– – validation of assays **5A**, 291ff
– – whole antibodies **5A**, 294f
– produced in plants, as plant protection products **10**, 493
– production stages **8B**, 408
– product selection **5A**, 292f
– recombinant – **5A**, 248f
– regulation **5A**, 495ff
– sequence-specific peptide cleavage by **8B**, 420
– structure of **8B**, 404ff
– – antibody 48G7 **8B**, 411f
– therapeutic –, manufacture of **5A**, 245ff
– use for immunopurification **5A**, 275ff
antibody-antigen complex **5A**, 35, 99f
antibody-based products, manipulation of halflife **5A**, 234f
antibody binding sites **5A**, 221ff
– binding strength **5A**, 230
– bispecific antibodies **5A**, 231
– complementary determining regions **5A**, 223
– conformation predictions **5A**, 223
– formats **5A**, 229f
– generation of **5A**, 221ff
– human anti-murine antibody response **5A**, 227
– humanized – **5A**, 226ff
– low molecular weight – **5A**, 229
– murine – **5A**, 226f
– number per product **5A**, 230
– primatized – **5A**, 228
– sources **5A**, 226ff
– structures **5A**, 221f
antibody catalysis, hydrolysis of **8B**, 421
– spontaneous covalent – **8B**, 421
– spontaneous features of **8B**, 421ff
– spontaneous metal ion **8B**, 422f

– transition state stabilization **8B**, 423f
antibody-catalyzed amide bond formation **8A**, 270
antibody-catalyzed amide bond hydrolysis **8A**, 270
antibody cross-reactivity **5A**, 306f
antibody-dependent abzyme prodrug therapy *see* ADAPT
antibody-directed enzyme prodrug therapy *see* ADEPT
antibody-effector conjugates, ADAPT **5A**, 233
– ADEPT **5A**, 233
– cytokines **5A**, 232
– DNA (antibody targeted gene) therapy **5A**, 234
– drugs **5A**, 233
– enzymes **5A**, 232
– immunotoxins **5A**, 232
– radioisotopes **5A**, 234
antibody-effector fusions **5A**, 232
antibody engineering **5A**, 219ff
– effector mechanisms **5A**, 230f
– – ADAPT **5A**, 233
– – ADEPT **5A**, 233
– – antibody-effector conjugates **5A**, 232
– – antibody-effector fusions **5A**, 232
– – antibody Fc functions **5A**, 231f
– – bispecific antibodies **5A**, 231
– – cytokines **5A**, 232
– – drugs **5A**, 233
– – *E. coli* fermentation **5A**, 236
– – enzymes **5A**, 232f
– – glycosylation **5A**, 236
– – GRAS organisms **5A**, 237
– – immunotoxins **5A**, 232f
– – mammalian expression **5A**, 236
– – use of insect cells **5A**, 237
– – use of transgenic animals **5A**, 237
– – use of transgenic plants **5A**, 237
– half-life of products **5A**, 234f
– production of recombinant antibodies **5A**, 235ff
antibody expression **5A**, 219ff
– effector mechanisms **5A**, 230f
– – ADAPT **5A**, 233
– – ADEPT **5A**, 233
– – antibody binding sites **5A**, 234
– – antibody-effector conjugates **5A**, 232
– – antibody-effector fusion **5A**, 232
– – antibody Fc functions **5A**, 231f
– – bispecific antibodies **5A**, 232
– – cytokines **5A**, 232
– – DNA (antibody targeted gene) therapy **5A**, 234
– – drugs **5A**, 233
– – enzymes **5A**, 232f
– – immunotoxins **5A**, 232f
– – radioisotopes **5A**, 234

– half-life of products **5A**, 234f
– production of recombinant antibodies **5A**, 235ff
antibody Fc functions **5A**, 321f
antibody fragments **5A**, 283f, 293ff
– human/mouse chimeric monoclonal – **5A**, 365ff
– therapeutic applications **5A**, 313
antibody humanization **5A**, 226ff
antibody light-chain loci **5A**, 224ff
antibody primatization **5A**, 228
antibody product development **5A**, 292
antibody production **3**, 25
– human-anti-mouse-antibody (HAMA) response **3**, 30
– humanization **3**, 30
– monoclonal – **3**, 25, 68f
antibody purification **5A**, 263ff
– DNA contamination **5A**, 265
– economic aspects **5A**, 266
– methods **5A**, 267f
– – affinity chromatography **5A**, 267
– – ion exchange chromatography **5A**, 267
– – SDS-PAGE **5A**, 267
– – size exclusion chromatography **5A**, 267f
– protein A matrices **5A**, 267
– protein contaminants **5A**, 264f
– purity specification **5A**, 264
– requirements **5A**, 264f
– virus contamination **5A**, 265
antibody selection **5A**, 292
– of isotypes **5A**, 293f
– surface plasmon resonance analysis **5A**, 293
antibody-specific marker **1**, 11
antibody targeted chemotherapy **5A**, 355ff
antibody tissue specificity **5A**, 306f
anticancer drugs *see also* antitumor agents
– angiogenesis inhibitors **7**, 673ff
– – structures **7**, 674
– antifolates **7**, 648
– antimetastatic inhibitors **7**, 673f
– – structures **7**, 674
– baccatins, from the European yew tree **7**, 669
– binding non-covalently to the DNA **7**, 649f
– – structures **7**, 649
– DNA topoisomerase inhibitors, camptothecins **7**, 655
– – podophyllotoxins **7**, 655
– – structures **7**, 653f
– immunomodulators **7**, 675f
– inhibitors of, cellular mitogenic signal transduction pathway **7**, 675f
– – glutathione S transferase **7**, 672
– – mitosis **7**, 665
– – protein kinases **7**, 676ff
– – protein phosphatases **7**, 676ff
– – *ras*-farnesyltransferase **7**, 683
– – sexual hormone production **7**, 665
– – the microtubular system **7**, 665

– – tyrosine protein kinases **7**, 679
– intercalating antibiotics, actinomycins **7**, 650ff
– – anthracyclines **7**, 652
– – ellipticin **7**, 650
– non-classical approaches **7**, 670
– nucleobase analogs **7**, 647f
– nucleoside antibiotics **7**, 647f
– *p*-quinones **7**, 652
– quinoxaline antibiotics **7**, 652
– – structures **7**, 651f
– radical generation **7**, 667
– screening assays, regulation of the cell cycle **7**, 647
– – use of transformed cells **7**, 647
– side effects, reduction of – **7**, 669
anticipation **5B**, 69
anticytokine therapy, in sepsis **5A**, 333f
antifolates **7**, 648f
– structures **7**, 648
antifouling compounds, from marine organisms **10**, 467
antifungal angents, from basidiomycestes **7**, 501ff
antifungal compounds, as products of marine biotechnology **10**, 458f
antigens, human –, expressed in transgenic mice **5A**, 297
antihelminthic compounds, from secondary metabolites **7**, 10
anti-HIV activity, of cyclosporins **7**, 540ff, 550f
antiinflammatory compounds, as products of marine biotechnology **10**, 455ff
antimetastasis drugs **7**, 673f
antimicrobial agents, compilation of **7**, 4ff
antimicrobial compounds, as products of marine biotechnology **10**, 455ff
antimutagenesis **2**, 28f
antimutator gene **2**, 29
antineoplastic agents *see* antitumor agents
antioxidants, for mammalian cell cultures **3**, 151
– from microalgae **10**, 109, 130
– production by algae **10**, 130
antiparasitic compounds, as products of marine biotechnology **10**, 458f
antiport, over membranes **1**, 74ff
antisense RNA, in HIV gene therapy **5A**, 474f
antistaling enzymes, in baking **9**, 676ff
– – maltogenic amylase preparations **9**, 676ff
anti-TNF antibody **5A**, 332f, 343ff
antitumor agents **7**, 641ff
– – *see also* anticancer drugs
– biotechnical drugs **7**, 643
– classical – **7**, 647ff
– compilation of – **7**, 6ff
– forming covalent bonds with DNA **7**, 656ff
– – anthracyclines **7**, 657
– – azirridines **7**, 655f
– – belomycin **7**, 663
– – bleomycin **7**, 661

– – cyclopropane **7**, 655
– – epoxide compunds **7**, 655f
– – streptonigrin **7**, 656
– – structures **7**, 657
– from basidiomycetes **7**, 503ff
– from plant cell cultures **7**, 614ff
– non-classical action on tumors **7**, 646
– of marine organisms **7**, 685
– plant-derived drugs **7**, 643
– potentiators of cytotoxicity **7**, 670ff
– – structures **7**, 671
– prescreening for – **7**, 645ff
– radical generation, by anthracyclines **7**, 661
– – by bleomycin **7**, 663
– – by enedyines **7**, 664
– showing selective toxicity **7**, 682ff
antitumor compounds, as products of marine biotechnology **10**, 450ff
anti-tumor drug, molecular action **2**, 203
antitumor polysaccharides **7**, 508
antiviral agents, from basidiomycetes **7**, 509ff, 520
– – structures **7**, 510f
antiviral compounds, as products of marine biotechnology **10**, 455ff, 460
antiviral constructs, for HIV gene therapy **5A**, 473ff
– – antisense RNA **5A**, 474f
– – decoy RNA **5A**, 476
– – delivery of – **5A**, 479ff
– – protein inhibitors **5A**, 476f
– – ribozymes **5A**, 474ff
– – vector design **5A**, 479ff
antramycin, antitumor agent **7**, 662
– structure **7**, 662
AnWT reactor technology *see* wastewater treatment
AOC method **11C**, 459f
– and BDOC methods, comparison of **11C**, 462
– and BRP methods, comparison of **11C**, 462
– scheme **11C**, 460
AOT *see* sodium bis(2-ethylhexyl)-sulfosuccinate (AOT)
AOX (adsorbable organic chlorinated compounds), in municipal wastewater **11A**, 177
– monitoring of **11A**, 121, 133f
6-APA *see* 6-aminopenicillanic acid
APC gene **5B**, 83
APHIS (Animal and Plant Health Inspection Service) **12**, 52, 199, 241, 461ff, 571ff
– biotechnology regulations **12**, 201ff
– – notification **12**, 205f, 250, 252
– – petition **12**, 205, 250, 253
Aphyllophorales *see* Poriales
aplidine *see* dehydrodidemnin B
Apocynum cannabinum, shoot culture **1**, 601
apoenzyme **1**, 70
apolipoprotein E **5B**, 68, 117

apolipoproteins **5A**, 15f
– in human plamsa, function of **5A**, 16
apoptosis (programmed cell death) **10**, 576
apple, cell wall polysaccharides **9**, 707
apple cider **1**, 334
apple juice, enzymatic treatment **9**, 713
– production **9**, 713f
approval, for GMO release, decision tree **12**, 183ff
– of rDNA products **12**, 173ff
– – decision trees **12**, 175
apramycin family, structure of – **7**, 469
apramycins **7**, 436
aptamer **2**, 350
aquaculture **10**, 132ff
– integrated fish production **10**, 132f
– microbial PUFA production **10**, 465
– of marine animals **10**, 448f
– of microalgae and cyanobacteria **10**, 128
– of sponges **10**, 449
– production of macroalgae **10**, 448
aquatic ecotoxicity test systems, assessment of contaminated soils **11B**, 130ff
– – evaluation of results **11B**, 132f
– – use of algae **11B**, 132
– – use of bacteria **11B**, 132
– – use of daphnids **11B**, 132
– interference in **11B**, 133f
– sensitivity of **11B**, 133ff
– soil extracts for **11B**, 131
aquatic plants, use in phytoremediation **11B**, 379
aquifers, confined – **11B**, 45f
– unconfined – **11B**, 45f
Arabidopsis Genome Initiative **5B**, 26
Arabidopsis Information Resource (TAIR) **5B**, 285
Arabidopsis thaliana **5B**, 147ff
– functional assignments **5B**, 27
– functional genomics **5B**, 28
– genetic map **5B**, 25
– genome project **5B**, 24ff
– mitochondrial genome **5B**, 27
– PHA biosynthesis **6**, 443
– plant-specific functions **5B**, 28
– plastic genome **5B**, 27
– postgenome era **5B**, 27f
– repetitive elements **5B**, 27
– transposable elements **5B**, 27
arabitol, formation by yeasts **6**, 220
arachidonic acid, biosynthetic pathways **10**, 335
– production of **10**, 335
arachidonic acid (ARA), formation in, algae **7**, 161
– – microalgae **7**, 167f
– – phycomycetes **7**, 161
arbekacin, semisynthetic aminoglycosides **7**, 461
arbutin, production by biotransformation **7**, 622
ARC (Australian Research Council) **12**, 425

archaea **1**, 252f
– aminoglycoside-like glycolipid structure **7**, 402
– halophilic – **10**, 69f
– in bioleaching **10**, 202
– – cellulases, xylanases and chitinase from **10**, 83
– in high-temperature environments **10**, 67f
– – proteolytic enzymes from **10**, 86
– – starch-hydrolyzing enzymes from **10**, 79f
Archaebacteria **4**, 449
– cell wall **1**, 20
– genome **5A**, 51
Archaeoglobus fulgidus, electron micrograph **5B**, 16
– flap endonucleases **5B**, 19
– genes **5B**, 18
– genome project **5B**, 16ff
– IS elements **5B**, 18
– paralogous gene families **5B**, 18
– postgenome era **5B**, 19
– repeats **5B**, 18
– sulfate metabolism **5B**, 19
– sulfate reduction **5B**, 16
Archer Report **5A**, 526
ARD *see* Acoustic resonance densitometry
ARDRA *see* amplified ribosomal DNA restriction analysis
ARDS *see* adult respiratory distress syndrome
arene dioxygenase-catalyzed reactions **8A**, 483
arenes, dihydroxylation of **8A**, 481
– formation of dihydrodiols **8A**, 506
– heterobicyclic –, dioxygenation of **8A**, 509
arginase, in industrial biotransformations **3**, 451
arginine decarboxylase **8B**, 80
arginine deiminase **6**, 520
arginine deiminase pathway, in pseudomonads **1**, 420
arginine dihydrolase pathway, in lactic acid bacteria **1**, 358
arginine kinase **8B**, 224ff
– phosphorylation by **8B**, 224ff
L-arginine, decarboxylation of, by arginine decarboxylase **8B**, 80
L-arginine production **6**, 513
– by microbial fermentation **6**, 489
arginine-vasopressin receptor antagonists **7**, 121
ARIPO (African Regional Industrial Property Organisation) **12**, 314
aristeromycin, phosphorylation of **8B**, 230f
aroma chemicals *see also* natural flavors
– natural –, carboxylic acids **8B**, 344f
– – falvor-active lactones **8B**, 340ff
– – fermentative production of **8B**, 333ff
– – flavor-active alcohols **8B**, 345f
– – flavor-active aldehydes **8B**, 345f
– – methyl ketones **8B**, 347
– – phenolic compounds **8B**, 346ff
– – vanillin **8B**, 334ff

aroma substances, in bread **9**, 306
aromatic biotransformation products **8B**, 390ff
aromatic compounds, aerobic degradation of
  11B, 175
– alkylated –, microbial degradation of **11B**,
  213ff
– bacterial degradation of **11B**, 149ff, 175ff, 211ff
– – benzoyl-CoA pathway **11B**, 176f
– – *m*-cleavage **11B**, 151, 154
– – *o*-cleavage **11B**, 151, 154
– bioremediation with the heap technology **11B**,
  326
– chlorinated – **8B**, 188
– cometabolic degradation of **11B**, 158ff
– degradation by microfungi **11B**, 158ff
– mononuclear –, anaerobic degradation of **11B**,
  176f
– – benzoyl-CoA pathway **11B**, 176f
– *o*-ring cleavage by microfungi **11B**, 158f
– oxygenolytic ring cleavage **11B**, 151
– selective hydroxylation of **8B**, 390
– sources of **11B**, 212
aromatic hydrocarbons, anaerobic degradation of
  11B, 183ff
– bacterial degradation of **11B**, 149ff
– benzylic hydroxylation **8A**, 479f
– biodegradation mechanisms **6**, 686ff
– bioremediation, use of sophorolipids **6**, 700
– bioreporter systems for **11B**, 454f
– branched –, degradation by genetically engineered microorganisms **11B**, 445
– hydroxylation by pseudomonads **1**, 422
– mixed –, biodegradation of **11B**, 215
– oxidation **9**, 178
aromatic L-amino acid decarboxylase (AAAD),
  deactivation of **8B**, 84
– decarboxylation of tryptophan L-tryptophan
  8B, 83
aromatic ring hydroxylation **8A**, 499ff
aromatics, degradation, by clostridia **1**, 300ff
– – by denitrifiers **1**, 419
– – by pseudomonads **1**, 414
– – of chloro-aromatics **1**, 202f
– – *ortho-/meta*-cleavage pathway **1**, 414
– – strain improvement **1**, 202f
– demethylation, in clostridia **1**, 302
array format **5B**, 385
arraying, glass microscope slides **5B**, 228
Arrhenius equation **1**, 128, **9**, 24
ARS (Agricultural Research Service) **12**, 205
arsenate reductase **10**, 240
arsenic, microbial reduction pathways **11B**, 366
– removal by phytoextraction **11B**, 389
Arthrobacter, nitrilase activity **8A**, 293
– nitrile hydratase activity **8A**, 293f
Arthrobacter calcoaceticus, emulsan formation **6**,
  696
Arthrobacter lipase **8B**, 368f

Arthrobacter sp., arthrobactin **7**, 222
– hydantoin racemase **6**, 517
arthrobactin, siderophore of *Arthrobacter* sp. **7**,
  222
arthropods, use in plant protection **10**, 495f
arthrospores, of streptomycetes **1**, 437
Article 21, European Commission **12**, 650
artificial groundwater recharge *see* groundwater
  recharge
artificial intelligence **5B**, 327
– definition of **4**, 626
– expert systems **4**, 626ff
artificial intelligence (AI), for bioprocess design
  3, 345
artificial neural network, biomass estimation **3**,
  336ff
– for bioprocess analysis **3**, 376ff
– – applications **3**, 378
– – error back-propagation **3**, 377
– – feedforward network **3**, 333, 376ff
– – generalized delta rule **3**, 377
– – input **3**, 334, 376
– – input-output relationship **3**, 377
– – nodes **3**, 376
– – output **3**, 334, 376
– – training **3**, 377
– for fermentation modeling **3**, 333ff
– – construction procedure **3**, 334
aryl esters, hydrolysis of, tetrahedral intermediate **8B**, 406f, 455ff
– – use of catalytic antibodies **8B**, 406f
ASB *see* anaerobic sludge bed
ASCEND 21 **12**, 628f
ascidians, as sources of nonribosomal peptides **7**,
  313
Ascomycetes **1**, 517, 525ff, **2**, 78, 530
– reproduction **1**, 525
L-ascorbate **1**, 211
ascorbate manufacture, by sorbitol oxidation **8A**,
  19
ascorbic acid **9**, 31
ascorbic acid (vitamin C), synthesis of, chemical
  steps **10**, 328f
– – microbial steps **10**, 328f
L-ascorbic acid *see* vitamin C
L-ascorbic acid 2-phosphate, production of **10**,
  329
ASEAN, member states **12**, 409f
asepsis **3**, 159
aseptic sampling **4**, 152ff
– Biopem device **4**, 153
– for on-line analysis of broth **4**, 152
– for on-line *in situ* analysis **4**, 170f
– *in situ* sampling device **4**, 154f
– Millipore device **4**, 154
Ashby report **12**, 516
Asian newly industrialized countries (NICs) **12**,
  344ff

Asian-Pacific nations, agriculture **12**, 374
– biotechnology **12**, 369f
– – patents **12**, 373
– – publications **12**, 373
– status of science and technology **12**, 372
Asilomar Conference **5B**, 9, **12**, 101, 158, 242, 461, 467, 511ff
ASM *see* activated sludge models
ASM (American Society of Microbiology) **12**, 514f
ASN.1 **5B**, 403
– architectural model **5B**, 401
asparaginase, as therapeutic product **3**, 63
L-asparaginase, freeze-drying, protection by excipients **3**, 704
asparate racemase **6**, 537
aspartame **6**, 468, 473, **8A**, 250f
– large-scale process for **8A**, 251
– manufacture of **8B**, 125
– – commercial process **8B**, 286f
– synthesis, phenylacetyl protection **8A**, 262
aspartame production **3**, 452f
aspartame synthesis, enzymatic – **8B**, 355
aspartase **6**, 519f, **8B**, 126ff
– biotechnology of **8B**, 128f
– covalent modification **8B**, 127
– immobilization **3**, 457
– in industrial biotransformations **3**, 457
– of *Bacillus* sp. **8B**, 126ff
– of *E. coli* **8B**, 126ff
aspartate α-decarboxylase (AαD) **8B**, 72ff
– structure **8B**, 72
α-aspartate decarboxylase, mechanism of decarboxylation **8B**, 75
L-aspartate, biosynthetic pathway **6**, 509
– use of **8B**, 128
L-aspartate β-decarboxylase, in industrial biotransformations **3**, 457
L-aspartate 4-carboxy-lyase (aspartate β-decarboxylase) **6**, 520f
aspartate production **6**, 512
– by *Escherichia coli* **6**, 519
aspartate 4-decarboxylase **8B**, 74ff
D-aspartic acid production **3**, 458
aspartic endoprotease, in cocoa fermentation **9**, 570
L-aspartic acid production **3**, 457
– enzymatic process **6**, 473
aspartokinase-homoserine dehydrogenase (AKHDH), refolding, temperature sensitivity **3**, 547
aspartyl proteases **5A**, 207
– characteristics **8A**, 248
Aspergilli, infection of wine grapes **9**, 500
aspergillic acid, iron-binding compounds **7**, 233
– structure **7**, 233
Aspergillus **1**, 525
– bread spoilage **9**, 312

– in ogi fermentation **9**, 545
– Koji-mold **2**, 119
– strain improvement **2**, 118f
Aspergillus aculeatus, cell wall degrading enzymes **9**, 711f, **9**, 720
Aspergillus acylase I (AA) **8A**, 246ff
– hydrolysis of *N*-terminal amides **8A**, 246ff
– resolution of α-amino acids **8A**, 246
Aspergillus amstelodami, specific growth rate **1**, 130
Aspergillus clavatus, epoxysuccinic acid production **6**, 371
Aspergillus flavus, kojic acid formation **6**, 372
– malic acid formation **6**, 371
Aspergillus fumigatus, in composting **11C**, 74ff
Aspergillus itaconicus, itaconic acid production **6**, 366f
Aspergillus japonicus, pectin methylesterase **9**, 718f
Aspergillus nidulans **2**, 530ff, **7**, 250ff
– chymosin formation **9**, 680
– clustering of β-lactum biosynthetic genes **7**, 259ff
– gene expression, regulation of – **7**, 266
– interspecies recombination **2**, 119
– penicillin biosynthetic cluster, regulation of **7**, 21
– penicillin biosynthetic enzymes **7**, 255ff
– production of 3'-deoxyadenosine **6**, 596
– strain improvement **1**, 536f
Aspergillus niger **2**, 531
– amyloglucosidase **9**, 669, 671, 745
– as biosuspension model **3**, 194
– bioconversion of linalool **10**, 390
– catalases **6**, 349
– citric acid production **1**, 526f, **6**, 311ff
– glucoamylase **6**, 71
– gluconic acid production **6**, 348ff, 354
– – by immobilized – **6**, 354
– – calcium gluconate process **6**, 352f
– – localization of enzymes **6**, 349
– – sodium gluconate process **6**, 353f
– glucose oxidase **4**, 87, **6**, 349
– – induction of **6**, 350
– – recombinant production of **6**, 350
– immobilization procedures **6**, 354
– intraspecies recombination **2**, 118f
– morphology **1**, 116, 233f
– morphology of citric acid production mycelium **6**, 325
– oxygen consumption **1**, 228
– oxygen uptake **4**, 332ff
– process kinetics **6**, 332ff
– pyruvate carbocylase **6**, 312
– stereospecific hydroxylation **8B**, 373f
– strain improvement **1**, 536f, **6**, 317
Aspergillus oryzae, α-amylase **6**, 68
– batch culture, glucose uptake **3**, 82

- chymosin formation **9**, 680
- fermentation extract, use in feeds **9**, 781
- immobilized aminoacylase, in industrial bio-transformations **3**, 448
- kojic acid formation **6**, 372
Aspergillus oryzae, fungal α-amylase **9**, 669, 676, 746f
- in miso production **9**, 519f
- in natto fermentation **9**, 522f
- in soy sauce fermentation **9**, 515f
- maltose formation **9**, 746
Aspergillus phytase **5A**, 210f
Aspergillus sojae **9**, 515f
Aspergillus terreus, aconitate decarboxylase **8B**, 64f
- itaconic acid production **6**, 366f
asperlicin **7**, 117
- activities **7**, 122
- structure **7**, 119, 122
aspirin **5A**, 374
- in bledding events **5A**, 369
- in bleeding events **5A**, 354
assembly coverage **5B**, 394
assembly engine **5B**, 384
assembly software **5B**, 215
assimilable organic carbon *see* AOC method
astasin, cytotoxic compound **10**, 141
astaxanthin **10**, 466
- production by algae **10**, 131
- production by protists **10**, 145
- structure **7**, 174, **10**, 463
- use as feed for pen-reared salmonids **7**, 175
ASTEC (Australian Science and Technology Council) **12**, 425
asteromycin **8B**, 230
asthma, mAb products **5A**, 222
- medical applications of recombinant proteins **5A**, 145
- monoclonal antibody therapy **5A**, 316
asymmetric sulfoxidations, by flavin dependent monooxygenase enzymes, bakers' yeast **8A**, 574
- - mammalian tissue **8A**, 575f
- - *Photobacterium phosphoreum* **8A**, 574
- - *Pseudomonas putida* **8A**, 574
- - *Pseudomonas* sp. **8A**, 574
- by monooxygenase enzymes, occuring in bacteria **8A**, 570ff
asymmetric syntheses, lipase-catalyzed – **8A**, 66f
AT *see* acyl-coenzyme A:isopenicillin N acyltransferase
ATF process **11C**, 163f
atherectomy **5A**, 375f
- therapy with ReoPro **5A**, 375
Atlas **5B**, 292
atom-and-bonds representation **5B**, 402
ATP *see* adenosine triphosphate
- assay by bioluminescent readouts **10**, 48

- cofactor regeneration, in enzyme reactors **3**, 458f
ATPase **1**, 65
- ABC-family **1**, 77
- export ATPase **1**, 76
- types **1**, 74
ATPase inhibitors **7**, 519f
ATPases, cation transport **5A**, 30
- functions **5A**, 30
ATP hydrolysis **1**, 54
ATP sulfurylase **5B**, 210
ATP yield **9**, 193f
atramycin, antitumor agent, structure **7**, 662
atrazine, biodegradation by plants **11B**, 378
- in hybrid poplar leaves **11B**, 375f
- monitoring with biosensors **11A**, 132f
atrial natriuretic peptide receptor antagonists **7**, 120f
atromentic acid, iron-binding compounds **7**, 232
- structure **7**, 232
atromentin, iron-binding pigments **7**, 231
- structure **7**, 231
Atropa belladonna, transgenic –, enhanced scopolamine production **7**, 625f
atropine **8A**, 332
- degradation of, pathway **8A**, 335f
- microbial metabolism of **8A**, 335ff
atropine esterase **8A**, 335
attenuation **2**, 251
attenuator sequence **2**, 253
Aureobasidium pullulans, gluconic acid production **6**, 355
- pullulan production **6**, 644
Australia, biotechnology, companies **12**, 422ff
- - cooperative research centers (CRCs) **12**, 427ff
- - industry **12**, 422ff
- - research centers **12**, 426ff
- - universities **12**, 426f
- biotechnology regulatory policy **12**, 424ff, 429
- contaminated site management, financing schemes **11B**, 15
- - legal development **11B**, 8
- - principles **11B**, 11
- - risk assesment **11B**, 19
- - technical development **11B**, 8
- public opinion survey, on biotechnology **12**, 126ff
- regulatory concerns, action thresholds **11B**, 21
- - enforcement **11B**, 26
- - in soil contamination **11B**, 7ff
- - liability rules **11B**, 13
- - organization **11B**, 26
- - priority setting **11B**, 23
- - remediation targets **11B**, 24
- - verification of clean-up **11B**, 24
Austria, contaminated site management, financing schemes **11B**, 15

– – legal development **11B**, 8
– – principles **11B**, 11
– – risk assessment **11B**, 19
– – technical development **11B**, 8
– identification of contaminated sites **11B**, 17
– registration of contaminated sites **11B**, 17
– regulatory concerns, action thresholds **11B**, 21
– – conflict management **11B**, 28
– – enforcement **11B**, 26
– – in soil contamination **11B**, 7ff
– – liability rules **11B**, 13
– – organization **11B**, 26
– – priority setting **11B**, 23
– – remediation targets **11B**, 24
– – verification of clean-up **11B**, 24
authentication of cell lines **10**, 33ff
– DNA fingerprinting **10**, 35
– isoenzyme analysis **10**, 35
– karyotyping **10**, 35
autoamylolytical quotient, calculation of **6**, 111
– definition **6**, 73f
– of rye **6**, 63
– of triticale **6**, 63
– of wheat **6**, 63
– relation to ethanol yields **6**, 112
autoanalyzer system, continuous air-segmented analyzer **4**, 156ff
– – advantage **4**, 159
– – calibration curves **4**, 159
– – cephalosporin production **4**, 157
– – disadvantaged **4**, 159
autoimmune diseases **5A**, 38
– mAb products **5A**, 222
– medical applications of recombinant proteins **5A**, 140
– monoclonal antibody therapy **5A**, 315, 318
– treatment with cyclosporin A **7**, 540
autolysis **3**, 517
– of yeast cells **9**, 703
automated DNA sequencing **5B**, 432f
automation in biotechnology **4**, 563ff
– – *see also* Computer integrated fermentation
– application of fuzzy control **4**, 595f
– computer aided software engineering **4**, 597f
– distributed control **4**, 568ff
– hardware design routes **4**, 568ff
– heuristic knowledge-based methods **4**, 594
– initial applications **4**, 564ff
– investment costs for computer control systems **4**, 572
– math-function processing components **4**, 582ff
– process management bus **4**, 599
– quantitative measures of reliability **4**, 567f
– redundancy approach **4**, 569ff
– requirements for industrial systems **4**, 572
– system integration **4**, 575
– use for expert systems **4**, 588ff
automation systems, architecture **4**, 576

– basic computer components **4**, 577
– database operation **4**, 584
– data transmission **4**, 575
– hierarchy **4**, 577
– incorporation of heuristic knowledge, use of expert systems **4**, 588ff
– – use of fuzzy methods **4**, 595f
– modularity of the components **4**, 568ff, 596f
– off-line data input **4**, 584
– optimization, by *a priori* knowledge **4**, 588ff
– special functionality **4**, 582ff
– user interfacing **4**, 578ff
– window-based **4**, 579
autonomously replicating sequence (ARS) **2**, 534
– AMA1 from *A. nidulans* **2**, 534
– ARS-like sequence **2**, 522
– bending **2**, 213
autonomously replicating sequence (*ARS*), in yeast chromosomes **1**, 475
Autonomous Service for the Environment Development of Amazon Federal Territories *see* SADA
autonomy, of individuals **12**, 117ff
Autoregressive Moving Average (ARMA) models, in fermentation modeling **3**, 332
autotrophs **3**, 128
autotrophy **1**, 50
– definition of **10**, 275
auxin, effect in plant cell culture **1**, 591
avermectin **3**, 9
avermectin derivatives, hydroxylation of **8A**, 484
axial dispersion coefficient, in air lift loop reactors **4**, 364f
– in bubble column reactors **4**, 366f
azasugars, structure of **8B**, 12
– synthesis of, use of rabbit muscle aldolase **8B**, 10f
aziridines, antitumor agents **7**, 656f
azo dyes, anaerobic reduction of, scheme **11A**, 393
– biodegradation of **11A**, 392ff
– – wastewater treatment **11A**, 394
Azomonas, phylogenetic status **1**, 403
Azotobacter, phylogenetic status **1**, 403
Azotobacteriaceae, azotobactin **7**, 216
Azotobacter sp., biopolymer formation **6**, 617
Azotobacter vinelandii, alginate production **6**, 620, 633
– biotechnological production of PHA **6**, 439f
– catecholate siderophores **7**, 203
– protein extraction **3**, 611
azotobactin, chromophore **7**, 216
– peptide siderophore of Azotobacteriaceae **7**, 216
azotochelin, catecholate siderophore of *Azotobacter vinelandii* **7**, 203
– structure **7**, 203

B

baby hamster kidney cell lines (BHK), in manufacturing of antibodies **5A**, 253
– – serum-free media **5A**, 257
BAC *see* bacterial artifical chromosomes
baccatin **7**, 615
baccatins, antitumor agents **7**, 667
Bacillariophyta, characteristics of **10**, 114
bacilli **1**, 367ff
– α-amylase **1**, 384f
– acidophilic **1**, 381
– – mesophilic **1**, 381
– – thermophilic **1**, 381
– alkaliphilic **1**, 379f
– – adaption **1**, 380
– – bioenergetics **1**, 380
– – biotechnological application **1**, 380
– – ecology **1**, 380
– – proteases **1**, 380
– bacteriophages **1**, 567
– carbohydrate uptake **1**, 385f
– carbon metabolism **1**, 383ff
– – genes **1**, 385f
– cellulase **1**, 28
– DegS/DegU system **1**, 385
– DNA homology **1**, 372
– extracellular enzymes **1**, 384ff
– food fermentation **1**, 368
– genetic diversity **1**, 369
– genetic engineering **1**, 394ff
– glutamine synthetase **1**, 387
– habitats **1**, 374f
– heterologous gene products **1**, 396
– hexose catabolism **1**, 385
– insect toxins **1**, 397
– insertional mutagenesis **1**, 395
– interleucin production **1**, 396
– morphological groups **1**, 368
– nitrate reduction **1**, 383
– nitrogen metabolism **1**, 387f
– oxygen requirements **1**, 381
– pentose catabolism **1**, 385
– phylogenetic analysis **1**, 372
– physiological diversity **1**, 369
– physiology **1**, 381ff
– plasmid vectors **1**, 395
– – stability **1**, 395
– polysacchharide-degrading enzymes **1**, 384f
– psychrophilic **1**, 379
– soil bacilli **1**, 374
– sporulation **1**, 390ff
– – *cot* (spore coat protein) genes **1**, 391
– – gene expression timing **1**, 393
– – genetic regulation **1**, 392
– – *ger* (germination) genes **1**, 391
– – initiation **1**, 391f
– – morphological stage **1**, 390f

– – mutation **1**, 391
– – *out* (outgrowth) genes **1**, 391
– – RNA polymerase regulation **1**, 393
– – *spo* (sporulation) genes **1**, 391
– – *ssp* (SASP) genes **1**, 391
– taxonomy **1**, 368ff
– thermophilic **1**, 378ff
– – ecology **1**, 378
– – enzyme production **1**, 379
– – genetic engineering **1**, 379
– – product characteristics **1**, 378
– transformation **1**, 394f
– – of competent cells **1**, 395
– – plasmid introduction **1**, 395
– – protoplast transformation **1**, 395
Bacillus **2**, 457ff
– anaerobic wastewater process **4**, 447
– biology **2**, 457
– chemotaxonomy **1**, 369f
– classification **1**, 368ff
– – thermophilic bacilli **1**, 370f
– ecology **1**, 373ff
– endospores **1**, 388ff, **2**, 457
– evolution **1**, 372
– extracellular protease production, deficiency
  mutants **2**, 464
– – types **2**, 465
– genetic engineering **2**, 457ff
– – DNA amplification **2**, 462
– genetic linkage map **2**, 459
– identification by **1**, 368, 372ff
– – computerized systems **1**, 373
– – DNA probes **1**, 373
– – fatty acid composition **1**, 373
– – pyrolysis mass spectrometry **1**, 373
– industrial application **2**, 457
– numerical phenetics **1**, 369f
– physical chromosome map **2**, 459
– protein extraction **3**, 610
– recombination **2**, 106, 109
– RNA polymerase **2**, 458ff
– – σ-factor **2**, 236, 460
– *sacB* gene regulation **2**, 463f
– spore photoproduct (SP) **2**, 206
– sporulation regulation **2**, 235
– transcription control **2**, 458
– xylose isomerase gene **2**, 464
Bacillus acidopollulyticus, pullulanase **9**, 669, 746
Bacillus alcalophilus **1**, 380
Bacillus-animal association **1**, 376ff
Bacillus anthracis **1**, 376
Bacillus brevis **2**, 457
– cell wall proteins **2**, 465
– expression systems **9**, 79
– relatives **1**, 369f
Bacillus cereus, food poisoning **1**, 375
– single-cell analysis **5B**, 275
Bacillus circulans **1**, 372

Bacillus coagulans **1**, 378
– esterase **8A**, 219
– lactic acid formation **6**, 303
Bacillus fastidiosus **1**, 374
Bacillus firmus **1**, 374
Bacillus lentimorbus **1**, 376
Bacillus lentus **1**, 374
Bacillus licheniformis **1**, 368, 374, 382
– α-amylase **6**, 67f, **9**, 669, 744
– – expression **6**, 68
Bacillus macerans **1**, 382f
Bacillus megaterium, α-amylase **9**, 746
– endospore germination **1**, 394
Bacillus methanolicus **1**, 270
– alcohol oxidoreductase **1**, 273
Bacillus natto **1**, 368
– in natto fermentation **9**, 522f
Bacillus-plant association **1**, 375f
Bacillus polymyxa **1**, 375, 382f
– cell recycle **6**, 286
– relatives **1**, 369f
– 2.3-butanediol formation **6**, 270f
– – batch culture **6**, 276
– – effect of oxygen **6**, 278f
– – fed-batch culture **6**, 283
– – isomeric forms **6**, 273
– – productivity **6**, 276, 284
– – yield **6**, 276
Bacillus popiliae **1**, 376
Bacillus sp., acetolactate decarboxylase **8B**, 63
– aspartase **8B**, 126ff
– cyclodextrin production, enzymes **6**, 636
– production of adenosine, mutant strains **6**, 591
Bacillus sphaericus **1**, 376
– relatives **1**, 369f
Bacillus spores, for metal detoxification **10**, 471
Bacillus spp., peptide antibiotic production **7**, 67ff
– – surfactin **7**, 67ff
Bacillus stearothermophilus **1**, 379, **10**, 65
– α-amylase **9**, 744
– recombinants, in fermentation processes **3**, 284
Bacillus subtilis **1**, 372, **2**, 457, **9**, 669, 744
– α-amylase **6**, 68
– – purification **3**, 604
– α-amylase production **4**, 540
– – use of expert systems **4**, 628, 631ff
– batch fermentation, fault analysis **4**, 632f
– – use of expert systems **4**, 632ff
– biotope **5B**, 13f
– bread spoilage **9**, 311
– computer integrated fermentation **4**, 592
– control elements **5B**, 14
– cytidine production, guanosine **6**, 586ff
– – mutant strains **6**, 593f
– electron micrograph **5B**, 13
– esterase **8A**, 219
– expression systems **9**, 79

– gene expresson regulation **2**, 458
– gene families **5B**, 13
– genome evolution **5B**, 15f
– genome project **5B**, 12ff
– industrial processes **5B**, 15
– insertion sequences **5B**, 14
– metabolism **5B**, 13
– metal sorption isotherms **11A**, 438
– molecular chaperones **5B**, 14
– order of the genes **5B**, 16
– phage φ29 **1**, 548
– production of, glycerol **6**, 217
– – inosinate (5′) **6**, 570
– – inosine **6**, 582f, 589
– – orotic acid **6**, 595
– – orotidine **6**, 595
– – uridine **6**, 592f
– – xanthosine **6**, 585f
– recombinants, in fermentation processes **3**, 284
– relatives **1**, 369f
– replication fork movement **5B**, 14
– spatial distribution of repeats **5B**, 14
– spores **5B**, 14
– stationary-phase functions **7**, 67
– subtilin production **7**, 331
– surfactin **6**, 664
– surfactin production **7**, 67ff
– surfactin synthesis, scheme **7**, 68
– temperate phages **1**, 567
– three classes of genes **5B**, 15
– transformation **5B**, 15
– truncated mRNA molecule **5B**, 15
– virulent phages **1**, 567
Bacillus thuringiensis **1**, 368, 377f, **2**, 457
– delta-endotoxin gene **1**, 377, 424
– – transfer to plants **1**, 424
– delta-endotoxin protein **2**, 644
– – action mechanism **2**, 644
– – resistance development **2**, 645f
– – specificity **2**, 644
– endotoxins **5B**, 154
– inclusion bodies **1**, 26
– in plant protection **10**, 496ff
– single-cell analysis **5B**, 275
– toxin characteristics **1**, 377
Bacillus thuringiensis toxin genes, integration into plant genome **10**, 498
Bacillus thuringiensis toxins **10**, 497f
– resistance towards **10**, 498
– safety **10**, 497
backcross generation **2**, 147ff
bacteria **12**, 73f
– – *see also* microorganisms
– accumulation of PHB **7**, 178
– acetogenic – **11A**, 457, **11C**, 153
– aerobic –, biodegradation of nitroaromatics **11B**, 278ff
– – biodegradation of TNT **11B**, 283f

– – degrading organic pollutants **11B**, 147f
– aerobic metabolism of, energy flows **11C**, 38f
– affection by dalbaheptides **7**, 377ff
– affection by lantibiotics **7**, 354ff
– alkaloids from **8A**, 328
– aminocyclitol aminoglycoside-producing – **7**, 401
– anaerobic –, biodegradation of nitroaromatics **11B**, 284ff
– – degradation of aliphatics **11B**, 171f
– – degradation of aromatics **11B**, 175ff, 183ff
– anti-phage defense **1**, 559
– as producers of lantibiotics **7**, 326
– as source of biosurfactants **6**, 662ff
– as sources of nonribosomal peptides **7**, 304ff, 308f
– as sources of ribosomal peptides **7**, 299
– ATP production by **10**, 330
– bio-de-emulsification properties **6**, 698
– β-lactam producing –, actinomycetes **7**, 250
– – eubacteria **7**, 250
– catechol metabolism **6**, 689
– cell surface layer **1**, 22
– cellulolytic – **11A**, 25f
– cell wall synthesis **1**, 113
– classification, risk groups **12**, 74
– colonizing in diesel oil **11B**, 66
– composition of **9**, 171
– de-emulsification by – **6**, 701f
– denitrifying – **11A**, 39, 59, 92
– *E. coli* **12**, 84ff
– effect on cell cultures **10**, 34
– effects of microgravity **10**, 509
– elemental composition **3**, 131
– essential amino acids in **9**, 174, 224
– ethanol formation **6**, 125ff
– – list of species **6**, 129
– fermentative – **11C**, 153
– fermenting –, biodegradation of nitrated compounds **11B**, 286f
– filamentous –, in activated sludge flocs **11A**, 73
– fimbriae **1**, 17
– fluorescence monitoring for biomass concentration **4**, 187
– for human consumption, clinical tolerance studies **9**, 231ff
– forming acetone **6**, 231ff
– forming butanol **6**, 231ff
– forming 2,3-butanediol **6**, 271
– functions of PHA **6**, 424f
– gliding bacteria **1**, 114f
– glycerol production **6**, 217
– growth **1**, 113
– – *see also* microbial growth
– growth in sour doughs **9**, 296
– growth of, mass balance **11C**, 39
– halophilic – **10**, 69f
– in artificial groundwater recharge **11C**, 433

– in biodeterioration **10**, 269ff
– in biofilters, sessile – **11C**, 466f
– – suspended – **11C**, 466
– in bioleaching **10**, 202ff
– in BOD monitoring **11A**, 134
– in composting **11C**, 59f
– infection by phages **1**, 546
– in high-temperature environments **10**, 67f
– – cellulases, xylanases and chitinase from **10**, 83
– – proteolytic enzymes from **10**, 86
– – starch-hydrolyzing enzymes from **10**, 79f
– in soil **11B**, 95
– involved in floc formation **11A**, 73, 94f
– involved in foaming **11A**, 94f
– leaf spot causing –, in weed control **10**, 490
– lignin degradation **6**, 147
– lysogenic conversion **1**, 558
– magnetotactic – **10**, 469, 576
– manganese-oxidizing –, metal corrosion by **10**, 298
– marine – **10**, 444
– – antibacterial compounds of **10**, 458
– – antifungal compounds of **10**, 458
– – antiinflammatory compounds of **10**, 456
– – antiviral compounds of **10**, 460
– – biosurfactant-producing – **10**, 470f
– – cytostatic/antitumir compounds of **10**, 451
– – DHA-synthesizing – **10**, 464
– – EPA-synthesizing – **10**, 464
– – magnetosomes of **10**, 468f
– – polysaccharides from **10**, 466
– metal-reducing –, metal precipitation by **10**, 239f
– metal removal by **11A**, 437ff
– methanogenic – **11A**, 18f, 35f, 94, 456ff, **11C**, 153f
– methanol oxidation **1**, 271
– methanotrophic –, biostimulation of **11B**, 407
– – classes **11B**, 203
– – cometabolic degradation of trichloroethylene **11B**, 152f
– methanotrophs **11A**, 93
– methylotrophic **1**, 269
– minimum water potential **10**, 273
– modification of hard coal **10**, 157
– morphology **1**, 114
– mycorrhizal – **11B**, 374
– nitrate reducing –, anaerobic degradation of cresols **11B**, 179
– – anaerobic degradation of ethylbenzene **11B**, 184f
– – anaerobic degradation of resorcine **11B**, 180f
– – anaerobic degradation of toluene **11B**, 183f
– – hydroxyhydroquinone degradation **11B**, 181ff
– nitrifiers **11A**, 38, 59, 92, 98, 262, 395ff, 422
– nitrifying – **10**, 278

– occurrence of PHA, in *Alcaligenes eutrophus* **6**, 413
– – in *E. coli* **6**, 413
– of the composting process **11C**, 54ff
– of the rhizosphere, biodegradation of pollutants **11B**, 377
– oxidizing iron(II) ions **10**, 280
– oxidizing sulfur compounds **10**, 278
– PAH inclusions, structure **6**, 411
– parasitic for nematodes **10**, 489
– patented solvent-producing – **6**, 234
– pathways for PHA synthesis **6**, 414ff
– – *Acaligenes eutrophus* **6**, 415
– – in recombinant – **6**, 421ff
– – *Pseudomonas aeruginosa* **6**, 415
– – *Pseudomonas oleovorans* **6**, 415
– – *Rhodospirillum rubrum* **6**, 415
– phage receptor **1**, 552f
– phage species **1**, 548ff
– PHA inclusions **6**, 410ff
– phototrophic –, hydrogen production **10**, 471f
– pili **1**, 17
– plant pathogenic –, in organic wastes **11C**, 234
– polyphosphate accumulating – **11A**, 41f
– poly(3HB) complexes **6**, 410f
– predominant in polluted soils **11B**, 147f
– production of PUFA **7**, 147ff
– – eicosapentaenoic acid **7**, 147f
– pseudolysogenic **1**, 559
– PUFA-producing – **10**, 463f
– recombinant –, ethanol production **6**, 167
– – PHA production **6**, 421ff
– regulation of antibiotic production **7**, 57ff
– – in unicellular – **7**, 61ff
– reproduction **1**, 113
– resistance mechanisms in – **7**, 440ff
– sessile –, in biofilms **11A**, 77, **11C**, 448
– signaling molecules, structures **7**, 14
– specific properties **12**, 74
– sulfate-reducing – **10**, 277
– – anaerobic cometabolism **11B**, 433
– – anaerobic degradation of catechol **11B**, 180
– – anaerobic degradation of cresols **11B**, 179
– – anaerobic degradation of ethylbenzene **11B**, 184f
– – anaerobic degradation of hydroquinone **11B**, 179f
– – anaerobic degradation of toluene **11B**, 183f
– – biodegradation of nitrated compunds **11B**, 286
– – degradation of nitroorganics **11B**, 186
– – hydroxyhydroquinone degradation **11B**, 182
– – metal corrosion by **10**, 294ff
– – metal ion precipitation **11A**, 44, 93
– – metal precipitation by **10**, 240ff
– – suspension characteristics **1**, 585
– syntrophic – **11A**, 457f
– thermophilic –, temperature optimum **11C**, 40

– toluene oxydizing –, biostimulation of **11B**, 407
– used for inoculation of biological waste gas reactors **11C**, 278
– use in aquatic ecotoxicity test systems **11B**, 132
– use in plant disease control **10**, 487f
– use in terrestrial ecotoxicity test systems **11B**, 136
bacteria harvesting **4**, 608
bacterial activity enhancement, and soil decontamination **11B**, 425ff
– bioaugmentation **11B**, 426ff, 433ff
– – chlorinated organics **11B**, 434
– – polycyclic aromatic hydrocarbons (PAH) **11B**, 434f
– biostimulation **11B**, 426ff
– cometabolism **11B**, 432f
– – by sulfate reducing bacteria **11B**, 433
– – of halogenated pollutants **11B**, 433
– – of trichloroethylene **11B**, 432f
– environmental factors **11B**, 429ff
– – oxygen concentration **11B**, 430
– – pH value **11B**, 430f
– – soil moisture **11B**, 430
– – temperature **11B**, 429f
– methods **11B**, 428ff
– nutrient supplementation **11B**, 431f
– – C-N ratios **11B**, 431f
– – nitrogen **11B**, 431f
– – phosphorus **11B**, 431f
bacterial alginates **6**, 633
bacterial artificial chromosomes (BAC) **5B**, 173, 195
– binary – **5B**, 173
bacterial attachment to mineral surfaces **10**, 200
bacterial-based cloning vectors **5B**, 45
bacterial capsule **1**, 16
– electron microscopy **1**, 17
– Gram-negative bacteria **1**, 17
– Gram-positive bacteria **1**, 17
bacterial cell *see* prokaryotic cell
bacterial cell wall, synthesis inhibitors **2**, 103f
bacterial degradation *see also* biodegradation
– by growth **11B**, 147
– cometabolic – **11B**, 147, 153ff
– – of trichloroethylene (TCE) **11B**, 152ff
– cooperation of anaerobic and aerobic bacteria **11B**, 154f
– dioxygenase reaction **11B**, 149
– growth-associated – **11B**, 148f
– monooxygenase reactions **11B**, 149
– of acetone **11B**, 175
– of aliphatic compounds **11B**, 148f
– of alkanes, peripheral pathways **11B**, 150, 171ff, 196ff, 420ff
– of alkylphenol **11B**, 153
– of aniline **11B**, 153, 177f
– of anthracene **11B**, 116f, 222ff

– of aromatic compounds **11B**, 149ff, 154, 175ff, 183ff, 211ff
– – key reactions **11B**, 175f
– of benzene **11B**, 150, 153, 378
– of benzoate **11B**, 153
– of chlorinated compounds **11B**, 241ff
– of cycloaliphatic compounds **11B**, 148ff
– – metabolic pathways **11B**, 151
– of cyclohexane **11B**, 148ff
– – metabolic pathways **11B**, 151
– of fluoranthene **11B**, 225ff
– of nitrated hydrocarbons **11B**, 275ff
– of nitrobenzoate **11B**, 153
– of organic pollutants **11B**, 153ff, 169f
– of phenanthrene **11B**, 153
– of phenol **11B**, 153
– of phthalate **11B**, 153
– of *p*-hydroxybenzoate **11B**, 153
– of TCE **11B**, 378, 409f, 449ff
– of TNT **11B**, 283f, 286f, 289f, 378f
– of toluene **11B**, 153
– of xenobiotics **11B**, 149, 156ff
– principles of **11B**, 147ff
bacterial emulsifiers **6**, 685f
bacterial L-form, cell wall **1**, 20
bacterial metabolism in wastewater treatment
  11A, 17ff
bacterial pathogens, in household wastes **11C**, 232
– in wastewater and sludge **11C**, 216ff
bacterial protein production, by *Saccharomyces cerevisiae* **1**, 498
bacterial siderophores, acinetobactin **7**, 219
– carboxylate siderophores **7**, 223ff
– – cepabactin **7**, 224
– – rhizobactin DM4 **7**, 223
– – staphyloferrin A **7**, 223
– – staphyloferrin B **7**, 224
– – vibrioferrin **7**, 224
– catecholate siderophores **7**, 201ff
– – agrobactin **7**, 204
– – aminochelin **7**, 203
– – amonabactins **7**, 206
– – anguibactin **7**, 206f
– – azotochelin **7**, 203
– – chrysobactin **7**, 202f
– – desferrithiocin **7**, 207
– – enterobactin **7**, 201f
– – ferrorosamine A **7**, 208
– – fulvibactin **7**, 205f
– – myxochelin **7**, 203
– – parabactin **7**, 204f
– – protochelin **7**, 203
– – pyochelin **7**, 207f
– – siderochelin A **7**, 208
– – vibriobactin **7**, 205f
– – vulnibactin **7**, 205f
– – yersiniabactin **7**, 207f

– citrate hydroxamates **7**, 220ff
– – actinoferrin **7**, 221f
– – aerobactin **7**, 222f
– – arthrobactin **7**, 222
– – rhizobactin 1021 **7**, 221
– – schizokinen **7**, 220f
– frankobactin **7**, 219
– hydroxamate siderophores **7**, 209ff
– – albomycins **7**, 211f
– – alcaligin **7**, 210f
– – bisucaberin **7**, 210f
– – ferrimycins **7**, 211
– – ferrioxamines **7**, 209f
– keto hydroxy bidentates **7**, 225
– maduraferrin **7**, 219f
– mycobactins **7**, 217ff
– peptide siderophores **7**, 212ff
– – alterobactin **7**, 212
– – ferribactin **7**, 214
– – ferrocins **7**, 212f
– – pseudobactins **7**, 213ff
– – pyoverdins **7**, 213ff
bacterial slime **1**, 16
– electron microscopy **1**, 17
– exopolysaccharide **1**, 16
bacteriocin **1**, 358ff, 546
– formation by *Lactococcus lactis* **9**, 652
bacteriocin/cytolysin, structural gene **7**, 344
bacteriophage **2**, 29, 50, 60f, 384ff, 440
– EMBL3 vector **2**, 385
– insertion vector **2**, 385
– M13mp vector **2**, 386
– promoter introduction, gene product analysis
  2, 405f
– replacement vector **2**, 385
– temperate **2**, 50
– transposing **2**, 29
– vector **2**, 384ff
– – *in vitro* packaging system **2**, 381f
bacteriophage fd **1**, 548
bacteriophage φ105 **1**, 567
bacteriophage φ29 **1**, 548, 567
bacteriophage genome **1**, 549ff
– chromosome integration **1**, 558
– concatamers **1**, 557
– evolution **1**, 560
– recombination **1**, 560
– replication **1**, 546, 551, 554
– – φ-form **1**, 554f
– – rolling circle mechanism **1**, 554
– – σ-form **1**, 554f
bacteriophage lambda **1**, 562f, **2**, 60f, 384ff
– gt10 vector **2**, 385
– integration into chromosome **2**, 214
– maps **2**, 60, 386
– vector, genomic library construction **2**, 390
– – hfl selection **2**, 385
– – size selection **2**, 385

– – spi selection **2**, 385
bacteriophage Mu **1**, 562, **2**, 29, 440
– derivative **2**, 29
bacteriophage M13 **1**, 564, **2**, 386, **5B**, 195
bacteriophage "Psi"M1 **1**, 565
bacteriophage P1 **1**, 563
bacteriophage p22, tailspike protein, aggregation
  3, 547
bacteriophage Qβ **1**, 548
bacteriophage(s) **1**, 543ff, **5B**, 8
– *Acetobacter*-specific – **6**, 386
– actinophage **1**, 440, 568
– adsorption **1**, 546, 552f
– bacterial defense mechanisms **1**, 559
– characteristics **1**, 546f, 561
– classification **1**, 547
– control **1**, 561
– detection methods **1**, 560
– DNA packaging **1**, 556f
– – headful mechanism **1**, 557
– DNA protection **1**, 559
– dsDNA phage, morphogenesis **1**, 556f
– families **1**, 550
– Ff-group **1**, 564
– gene expression **1**, 554f
– gene organization **1**, 555
– host range **1**, 548
– host receptor **1**, 552f
– impact on host cells **1**, 553
– infection efficiency **1**, 559
– infection mechanism **1**, 546
– influence on lactic cultures **9**, 372, 398
– lysogeny **1**, 558f
– main groups **1**, 550
– morphogenesis **1**, 556f
– morphology **1**, 547
– mutual exclusion **1**, 554
– nucleic acid injection **1**, 552f
– nucleic acid structure **1**, 547
– of bacilli **1**, 567
– of clostridia **1**, 566
– of coryneform bacteria **1**, 568
– of *Escherichia coli* **1**, 562ff
– of lactic acid bacteria **1**, 566
– of methanogenic bacteria **1**, 565
– of methylotrophic bacteria **1**, 566
– of *Pseudomonas* **1**, 568
– progeny release **1**, 557
– prophage **1**, 558f
– receptor **1**, 546
– replication cycle **1**, 546
– routine monitoring **1**, 560
– species diversity **1**, 548
– ssRNA phages **1**, 565
– SSV1 group **1**, 550
– superinfection **1**, 554, 558
– temperate **1**, 558, 562
– – life cycle **1**, 558, 565

– therapeutic applications **1**, 545
– toxin gene transfer **1**, 545
– virulent **1**, 558, 563
bacteriophage SPβ **1**, 567
bacteriophage SPO1 **1**, 567
bacteriophage SPO2 **1**, 567
bacteriophage SSV1 **1**, 548
bacteriophage T1 **1**, 545ff
bacteriophage T4 **1**, 547, 563f
– detection **1**, 561
bacteriotoxins **5A**, 34
baculovirus **2**, 729ff
– expression, in insect cells **9**, 85ff
– in plant protection **10**, 496f
– promotors for heterologous expression **9**, 87
– recombinant virus construction **9**, 86f
Baeyer-Villiger biotransformations, by *Acineto-*
  bacter junii **8A**, 560
– by *Curvularia intermedia* **8A**, 560
– by *Cylindrocarpon destructans* **8A**, 563
– by HLADH **8A**, 565
– by *Nocardia* **8A**, 566
– by PCL **8A**, 568
– by *Pseudomonas putida* **8A**, 555ff
– by *Rhodococcus* **8A**, 560
– by *TBADH* **8A**, 568
– by *Xanthomonas autotrophicus* **8A**, 561, 566
– cofactor recycling and multi-enzyme systems
  8A, 567f
– Criegge intermediates in – **8A**, 540
– mediated by *Corynebacterium* amidase **8A**, 291
– multi-enzyme systems for **8A**, 567
– of bicyclo[2.2.1]heptanones **8A**, 563ff
– of bicyclo[3.2.0]heptanones **8A**, 560ff
– of bicyclo[4.2.0]octanones **8A**, 562f
– of monocycles **8A**, 555ff
– of polycyclics **8A**, 566f
Baeyer-Villiger monooxygenases **11B**, 198f
Baeyer-Villiger oxidation, of cyclohexanone derivatives with a genetically modified yeast **8A**,
  383
Baeyer-Villiger reactions, abiotic – **8A**, 537ff
– biological –, fenchone **8A**, 545
– – in camphor metabolism **8A**, 543f
– – listing of **8A**, 555ff
– – questin **8A**, 548
– – raspberry ketone **8A**, 548
– – ring cleavage of questin **8A**, 543
– – steroids **8A**, 541ff
– – tridecanone **8A**, 542
– – unnatural substrates **8A**, 546
– – use of *Acinetobacter* **8A**, 546
– Criegee intermediates **8A**, 538, 540, 551
– enantiometric intermediates **8A**, 552
– flavoenzymes catalyzing – **8A**, 549
– gross mechanism **8A**, 538
– intramolecular **8A**, 539
– large scale – **8A**, 568f

– mechanisms for cyclohexanone monooxygenase catalyzed – **8A**, 554
– N-terminal sequences of flavoenzymes **8A**, 549
– of steroids **8A**, 541
– peracids in **8A**, 538f
– reagents for **8A**, 539f
Baeyer-Villiger ring expansion **8A**, 535ff
BAF see biological air filter
bagasse **1**, 150
– composition **6**, 15
bagoong, production of **9**, 537
bait and switch strategy, generation of catalytic antibodies **8B**, 412ff
BAK (Bioindustry Association of Korea) **12**, 394f
baked goods see bread, baking
bakers' yeast **8A**, 364f, **8B**, 31f
– – see also Saccharomyces cerevisiae
– – see also yeast
– activity tests **9**, 345f
– acyloin condensation **8B**, 23ff
– age distribution of cells **4**, 389
– budding cells, correlation with fermentation rate **4**, 389
– combined batch and fed-batch cultivation **4**, 399
– commercial forms **9**, 342ff
– – aktive dry yeast **9**, 344
– – bulk liquid yeast **9**, 342ff
– – compressed yeast **9**, 344
– – instant dry yeast **9**, 344
– commercial production **9**, 321ff
– commercial propagation **9**, 330ff
– – oxygen limitation **9**, 332f
– computer integrated fermentation **4**, 592
– dewatering **9**, 339f
– Diels-Alder reaction **8B**, 30
– esterase **8A**, 220, **8B**, 370
– expression of cyclohexane monooxygenase in **8A**, 546
– fluidized-bed dryer **9**, 340f
– frequent contaminants **9**, 328f
– growth conditions **4**, 294
– growth simulation **4**, 294ff
– market segments **9**, 335
– Michael addition **8B**, 31f
– multiple activities **8A**, 31
– nutritional requirements, biotin **9**, 331
– – nitrogen **9**, 331
– – phosphorus **9**, 331
– – trace elements **9**, 331
– packaging **9**, 341f
– production **9**, 208f, 254
– – fed-batch culture **9**, 208
– pyruvate decarboxylase of **8B**, 23ff
– – synthesis of solerone **8B**, 27
– recovery by centrifuging **9**, 339
– seed yeast propagation **9**, 329f

– strain maintenance **9**, 325
– strain modification **9**, 325f
– – methods **9**, 325f
– strain selection **9**, 323ff
– sulfoxidations by **8A**, 574
– synthesis of lanosterol analogs **8B**, 382
– transaldolase **8B**, 20
– transketolase **8B**, 20
– use in wine making **9**, 471
– vitamin fortification of molasses **6**, 24
– 2,3-oxidosqualene lanosterol cyclase **8B**, 29f
baker's yeast fermentation, application of EKF **4**, 234ff
– biomass concentrations, estimation by EKF **4**, 236
– ethanol concentrations, estimation by EKF **4**, 237
– glucose concentrations, estimation by EKF **4**, 236
– product yield, estimation by EKF **4**, 237
– specific growth rate, estimation by EKF **4**, 237
– substrate yield, estimation by EKF **4**, 237
baker's yeast growth **4**, 340ff
– at different feeding points **4**, 342
– at different mean circulation times **4**, 343
– at different scales of operation **4**, 342
– computer-controlled feeding **4**, 341ff
– fermentative metabolism **4**, 388
– oxidative metabolism **4**, 388
– regulatory effects **4**, 388
– – Crabtree effect **4**, 388
– – glucose effect **4**, 388
– – Pasteur effect **4**, 388
– substrate distribution **4**, 340
– sugar feeding, fixed schedule **4**, 340f
baker's yeast process, costs **4**, 398
– development **4**, 387
– downstream processing model **4**, 398
– economic balance **4**, 400
– fraction of budding cells **4**, 401
– optimization **4**, 386ff
– – application of process model **4**, 400ff
– – criteria **4**, 404f
– – of aeration rate **4**, 403
– – of economic profit **4**, 400
– – of fermentation time **4**, 403
– – of product quality **4**, 401
– optimum operation **4**, 402ff
– outline **4**, 386, 390
– productivity **4**, 404f
– simulation of air lift reactor run **4**, 402
– upstream processing model **4**, 398
– verification of the process model **4**, 399
– yield **4**, 404f
baker's yeast production **4**, 539
– cell cycle model **4**, 396f
– cell model **4**, 396ff
– – dynamic regulation model **4**, 395

– – metabolic model **4**, 394f
– model **4**, 539
– – kinetic model parameters **4**, 539
– – optimal feed rate profiles **4**, 540
– reactor model **4**, 391ff
– – gas phase model **4**, 392f
– – liquid phase model **4**, 391
– – mass transfer model **4**, 393f
– wastewater of **11A**, 538
baking *see also* bread, dough
– additives **9**, 259ff
– Brimec process **9**, 283f
– Chorleywood process **9**, 283
– enzymes for antistaling **9**, 676ff
– formation of aroma substances **9**, 305ff
– history **9**, 244, 322f
– outline of technology **9**, 248ff
– physico-chemical changes **9**, 304f
– processing of, sour dough **9**, 285ff
– – wheat doughs **9**, 272ff
– raw materials, eggs **9**, 252
– – fat **9**, 252
– – flour **9**, 250ff
– – leavening agents **9**, 253ff
– – salt **9**, 252
– – sugar **9**, 252
– – water **9**, 252
– – yeast **9**, 253f
– temperatures **9**, 303f
– times **9**, 303f
– use of enzymes **9**, 674ff
baking loss **9**, 304
baking ovens **9**, 302f
baking powder **9**, 258
balanced polymorphisms **5B**, 65
baloon angioplasty *see* PTCA
banana, dell wall polysaccharides **9**, 707
– enzymatic treatment **9**, 717
– juice production **9**, 717f
BandCheck **5B**, 233, 235
Bangemann Communication **12**, 637
bank filtration **11C**, 379ff, 425ff, 431
– change of environmental conditions **11C**, 434ff
– filter layers, algae in **11C**, 432
– – biology of **11C**, 431
– process scheme **11C**, 381
– purification mechanisms **11C**, 435f
– redox succession during **11C**, 434ff
– removal of, microorganisms **11C**, 438ff
– – organic matter **11C**, 438
– – pathogens **11C**, 438f
– – trace elements **11C**, 438
– – xenobiotics **11C**, 438
– scheme **11C**, 380
banku, production of **9**, 544f
BAP (Biotechnology Action Programme) **12**, 534f, 537ff
bardenpho process **11A**, 289

barley **9**, 683
– analysis of **6**, 66
– as brewing adjunct **9**, 429
– as raw material for ethanol production **6**, 65
– processing in ethanol production **6**, 93
– stillage, nutrient content **6**, 107
barley malt **6**, 70
barleyy, malt raw material **9**, 426f
base **2**, 319
base calling **5B**, 212ff
base composition **5B**, 391
base pairing **2**, 196, 259, 320, 325
– Hoogsteen **2**, 325
– hydrogen bond **2**, 196
– Watson-Crick **2**, 325
Basic Local Alignment and Search Tool
  (BLAST) **5B**, 130
– BLAST-type searches **5B**, 332
Basidiomycetes **1**, 517, 532ff, **2**, 530, **11C**, 69f
– aminopeptidase inhbitors **7**, 518ff
– antibacterial metabolites **7**, 501
– antitumor compounds **7**, 503ff
– antitumor polysaccharides **7**, 508
– antiviral compounds **7**, 509ff
– ATPase inhibiting metabolites **7**, 520f
– biotransformation of mycrene **10**, 385
– characteristics **1**, 532
– cultivation of – **7**, 490f
– cytotoxic compounds **7**, 503ff
– degradation of lignin **11C**, 69
– herbicidal compounds **7**, 513ff
– immunosuppressive compounds **7**, 509
– in composting **11C**, 67f
– inhibitors of cholesterol biosynthesis **7**, 521f
– inhibitors of leukotriene biosynthesis **7**, 522f
– insecticidal compounds **7**, 515f
– life cycle **1**, 533, **7**, 490
– lignin degradation **11B**, 161ff
– metabolic inhibitors of signal transduction **7**, 524
– metabolites from – **7**, 489ff, 496ff
– – history **7**, 490
– – oudemansis **7**, 496ff
– – pleuromutilin **7**, 496
– – screening of – **7**, 496
– – strobilurins **7**, 496ff
– metabolites inducing differentiation of leukemia cells **7**, 524
– morphology **1**, 532
– nematicidal compounds **7**, 515f
– phospholipase inhibiting metabolites **7**, 518f
– platelet aggregation inhibitors **7**, 512f
– primary and secondary metabolism, interrelationship **7**, 492
– producing, ondemansins **7**, 499
– – strobilurins **7**, 499
– production of halometabolites **8B**, 176
– reproduction **1**, 533

– reverse transcriptase inhibitors **7**, 509ff
– taxonomy **1**, 534
– uses **7**, 490
batch bioreactor, adaptive optimization **4**, 557f
– optimization **4**, 528ff
batch culture **1**, 150, **3**, 28, 97f
– biomass concentration **3**, 373
– circulation time distribution **3**, 95
– kinetics **3**, 97
– mass balances **3**, 97
– microbial growth, exponential – **3**, 380
– model simulation **1**, 153
– oxygen – **3**, 224
– oxygen uptake rate (*OUR*) **3**, 367
– phases **9**, 201f
– sterilization criteria **3**, 174
– substrate concentration **3**, 373
batch experiment, identification functional **4**, 257
– initial substrate concentration **4**, 257
– trajectories **4**, 257
batch fermentation, principal time course **4**, 270
– sequential control **4**, 514
– types **4**, 270
batch operation, of slurry bioreactors **11B**, 335ff
batch penicillin fermentation model, optimum
    temperature profile **4**, 528ff
batch sterilization **3**, 164ff
batch stirred reactor, recycle model **4**, 305
Bayes' theorem **4**, 229, 238
Bayh-Dole Act **5A**, 522
BBC (Biotechnology Coordination Committee)
    12, 636ff
BCC (Biotechnology Coordination Committee)
    12, 56
BDOC method **11C**, 461f
– and AOC methods, comparison of **11C**, 462
– with sessile bacteria **11C**, 461f
bead mill, cell disintegration, rate constants **3**,
    514
– – residence time **3**, 513
– high-speed agitator – **3**, 510ff
– – agitator disc arrangement **3**, 512
– – construction **3**, 511ff
– – energy density **3**, 511
– – function **3**, 511ff
– – stirrer blade effects **3**, 514
Beauveria bassiana, synthesis of (*R*)-2-(4-hydroxy-
    phenoxy)proprionic acid **8B**, 282
Beauveria sulfurescens, as the classic hydroxylating
    microorganism **8A**, 486
– stereochemical model for hydroxylation **8A**,
    479
Beckwith–Weidemann syndrome **5B**, 74
beef tallow, fatty acid composition **7**, 136
beer, alcohol-free – **9**, 455
– – production of **9**, 453
– chillproofing **9**, 450, 685
– draft beer **9**, 455f

– dry beer **9**, 455
– ice beer **9**, 456
– lagering **9**, 450
– light beer **9**, 455
– oxygen content, effect on flavor **9**, 451
– stability **9**, 450ff
beer production **1**, 334, **11A**, 210
– beer sarcina **1**, 339
– light beer **1**, 200
– process improvement **1**, 495
– *Saccharomyces* strains **1**, 496
beer spoilage **9**, 446ff
beet molasses **6**, 17ff
– – *see also* molasses
– amino acid composition **6**, 21f
– as raw material, in levan production **6**, 25
– composition **6**, 18f, 138
– – mineral content **6**, 19
– – nitrogenous compounds **6**, 20f
– – organic acid content **6**, 20
– – vitamin content **6**, 23
beet processing, molasses yield **6**, 13
beet pulp, composition **6**, 15
– in biogas production **6**, 25
beet raw juice, amino acid composition **6**, 14f
– analytical values **6**, 13
– applicability **6**, 16
beet sugar processing, sugar content of raw mate-
    rial **6**, 12
beet thick juice, applicability **6**, 16
beet thin juice, applicability **6**, 16
Belgium, biotechnology regulations **12**, 605
– contaminated site management, financing sche-
    mes **11B**, 15
– – legal development **11B**, 9
– – principles **11B**, 11
– – registration **11B**, 17
– – risk assessment **11B**, 19
– – technical development **11B**, 9
– indentification of contaminated sites **11B**, 17
– regulatory concerns, action thresholds **11B**, 21
– – conflict management **11B**, 28
– – enforcement **11B**, 26
– – in soil contamination **11B**, 7ff
– – liability rules **11B**, 13
– – organization **11B**, 26
– – priority setting **11B**, 23
– – remediation targets **11B**, 24
– – verfication of clean-up **11B**, 24
Bel process, microbial biomass production **9**, 210
benastatins, glutathione S transferase inhibitors
    7, 672
Bence-Jones proteins **9**, 15
Benda Report **12**, 582
beneficience **12**, 118, 137, 140
benzaldehyde, stereoselective hydrocyanation of,
    use of cyclic dipeptides **8B**, 493ff

benzaldehyde derivatives, stereoselective hydrocyanation of **8B**, 493
benzamide, by hydrolysis of benzonitrile **8A**, 287
benz(a)pyrene, oxidation **11B**, 231f
– – by lignin peroxidase **11B**, 231f
– structure of **11B**, 216
benzene, bacterial degradation of **11B**, 150, 153
– biodegradation by plants **11B**, 378
– extraction of **11B**, 499
– pathways of degradation **11B**, 212f
– production of derivatives **3**, 459
– turnover, by normal flora in biofiltration plants
11C, 282
benzenes, alkylated –, degradation of **11C**, 286
– disubstituted –, dioxygenation reaction of **8A**,
506
– monosubstituted –, dihydroxylation of **8A**, 508
– polyalkylated **11C**, 286
– polyhydroxylated –, lipase-catalyzed deacylation **8A**, 129
benzenoids, bicyclic –, dihydroxylation of **8A**, 509
– substituted –, dihydroxylation of **8A**, 507
benzo[a]pyrene, oxidation by *Phanerochaete chrysosporium* **11B**, 233
– physical and chemical properties of **11B**, 218
benzoate, anaerobic degradation of **11B**, 176f
– – by *Thauera aromatica* **11B**, 177
– bacterial degradation of **11B**, 153
benzoate degradation **1**, 202
benzofurans, chlorinated – **8B**, 187
benzoic acid, *o*-ring cleavage by microfungi **11B**,
159
benzophenanthridine alkaloids **8A**, 338f
– biosynthesis **8A**, 339
– from plant cell cultures **7**, 603ff
– – yields **7**, 605
– structures **7**, 604
benzopyran, dihydroxylation of **8A**, 515
benzoyl-CoA pathway, in anaerobic degradation
of aromatics **11B**, 176f
benzoylformate decarboxylase (BFD) **8B**, 65ff
– of *Pseudomonas putida* **8B**, 65ff
benzylformate decarboxylase (BFD), decarboxylation of [*p*-(bromomethyl)benzoyl]formate,
mechanism **8B**, 67
benzylic hydroxylation, by *Mortierella isabellina*
8A, 487f
benzylisoquinoline alkaloids **8A**, 337ff
BEP (Biomolecular Engineering Programme) **12**,
518, 528ff
berberine, structure **7**, 604
berberine production, by *Thalictrum minus* cells
1, 598f
Berg letter **12**, 511f
Bermuda agreement **5B**, 439
Bertrand-Hudson rule **8B**, 299ff
Best equation **11B**, 77
Betabacteria **1**, 329

betacyanin, structure **7**, 610
betacyanins, from plant cell cultures **7**, 611
betaine, in wastewater, monitoring of **11A**, 534f
betalains, from plant cell cultures **7**, 611
betapol, interesterification of **8A**, 136
– production of **8B**, 289
betaxanthins, from plant cell cultures **7**, 611
beverage industry, wastewater of, amounts **11A**,
210
– – concentrations **11A**, 210
– – treatment **11A**, 209f
BFD *see* benzoylformate decarboxylase
bialaphos production, in *Streptomyces hygroscopicus,* regulatory genes **7**, 91
bibliographic databases **5B**, 283
bicyclic β-lactams *see* β-lactam antibiotics
BIDEA (Taiwan Bioindustry Development Association) **12**, 406
Bifidobacterium **1**, 335, 346
– as starter culture **1**, 346
– habitats **1**, 346
bile acid conversion, by clostridia **1**, 311f
binary bacterial artificial chromosome (BIBAC)
5B, 173
binary fission **1**, 113f
BIND **5B**, 401, 403f
Binding assays, immunochemical **4**, 79f, 93ff
– receptor-based **4**, 79f, 96
binding protein (BiP) **3**, 532
binding specificity, of antibodies **5A**, 312
bioaccumulation, of heavy metals **11A**, 441
– of metal ions **11A**, 42
bioactive compounds, from protists **10**, 140f
– hydroxylation of **8A**, 484
– screening of **10**, 44ff
bioadsorbent, preparation of **11A**, 440
– recycling of **11A**, 446
bioadsorbents, metal capacity **11A**, 437
bioaerosols, emission in composting plants **11C**,
239ff
– – particle sizes **11C**, 240, 242
bioassays, for emission control in coal bioconversion **10**, 176f
– for medically approved alginate gels **10**, 553ff
– for soils **11B**, 473f
– use of protists **10**, 140f
bioaugmentation **11B**, 408ff, 426ff, 433ff, 473
– adhesion-deficient strains **11B**, 410f
– – *Pseudomonas fluorescens* **11B**, 411
– application of **11B**, 428
– approaches **11B**, 408f
– definition of **11B**, 433
– degradation of, chlorinated organics **11B**, 434
– – PAH **11B**, 434f
– degradative organisms, enhancing energy reserves in **11B**, 411f
– for xenobiotic degradation **11A**, 46ff
– in *in situ* remediation **11B**, 356

– in soil **6**, 693
– trichloroethylene (TCE) as model target **11B**, 409
bioavailability, definition of **11B**, 63f
– monitoring of **11B**, 82f
– of pollutants **11B**, 61ff, 469
– – direct microbial uptake mechanisms from the non-aqueous phase liquid **11B**, 66f
– – improvement of **11B**, 78ff
– – increase of the desorption process **11B**, 79ff
– – increase of the mass transfer rate **11B**, 79ff
– – in non-aqueous phase liquids (NAPL) **11B**, 66
– – in soil **11B**, 67ff
– – in the water phase **11B**, 64ff
– – lack of **11B**, 82f
– – partitioning model **11B**, 68
– – soprtion retarded diffusion **11B**, 68
bioavailability number **11B**, 77
biobeneficiation, definition of **10**, 192
BIO (Bio-Industry Organization) **12**, 535
Biocarta **5B**, 405
biocatalysis **3**, 430
– in asymmetric synthesis **3**, 430
– thermodynamic cycle **8B**, 424
– under extreme conditions **10**, 61ff
– use in chemical synthesis **8A**, 11
Biocatalysis/Biodegradation Database **5B**, 405
biocatalysts **3**, 430
– – *see also* enzymes
– industrial use of **8B**, 278ff
– retention **3**, 433ff
biocatalytic processes, current large-scale processes **8A**, 20
– production volumes **8A**, 20
biocatalytic reactor *see* bioreactor
biocenoses, complex mixed culture, ecological relationships in **11C**, 277
– in biofilms **11C**, 279
biochemical oxygen demand (BOD), measurement of **4**, 452
biochemical pathway alteration **2**, 798f
– glyoxylate pathway **2**, 798
– in transgenic sheep **2**, 798
– milk composition **2**, 799f
biochemical reaction, relaxation time **1**, 138
biocomponent, of biosensors **4**, 77ff
– – cells **4**, 78f
– – enzymes **4**, 77f
– – organelles **4**, 79
– pre-column **4**, 82f
bioconversion, of coal, brown coal **10**, 160ff
– – environmental aspects **10**, 172ff
– – hard coal **10**, 157ff
biocorrosion *see also* biodeterioration, corrosion, microbial corrosion
– by nitrogen compounds **10**, 278f
– by sulfur compounds **10**, 277f

– definition of **10**, 267
– of concrete **10**, 279
– of iron **10**, 279f
– of metals **10**, 293ff
– – involved microorganisms **10**, 294
– – model **10**, 296
– – protection against **10**, 298
– – reactions **10**, 294
biocytin **9**, 34
bio-de-emulsification **6**, 697ff
– by bacteria **6**, 701, 702f
– – *Corynebacterium petrophilum* **6**, 705ff
– – *Corynebacterium* sp. **6**, 702
– – effect of autoclaving **6**, 703
– – heat stability **6**, 703
– – *Nocardia amarae* **6**, 704f
– – *Nocardia* sp. **6**, 702
– – reuse **6**, 709
– – *Rhodococcus* sp. **6**, 702
– effect of sophorose lipids **6**, 699
– mechanism **6**, 697
– of complex O/W field emulsions **6**, 708ff
– of complex W/O field emulsions **6**, 703ff
– properties of bacterial cells **6**, 698
– synergism with a synthetic flocculating agent **6**, 709
– use of biomass **6**, 711
– use of sludges **6**, 711f
biodegradability, of contaminants **11B**, 468ff
– – evaluation of **11B**, 470
biodegradable dissolved organic carbon *see* BDOC method
biodegradable organic matter (BOM), biofilters for BOM removal, characterization of **11C**, 465ff
– – efficiency of **11C**, 468ff
– – modeling of **11C**, 471f
– determination by, AOC method **11C**, 459f
– – BDOC methods **11C**, 461f
– – BRP method **11C**, 460f
– – comparative investigations **11C**, 462f
– effects of treatment processes on **11C**, 463ff
– – chlorination **11C**, 464f
– – coagulation **11C**, 463
– – flocculation **11C**, 463
– – ground filtration **11C**, 463
– – ozonation **11C**, 463f
– – precipitation **11C**, 463
– – river bank filtration **11C**, 463
– – slow sand filtration **11C**, 463
biodegradable substrate, enhanced biological phosphorus elimination **11A**, 301f
biodegradation *see also* bacterial degradation
– definition of **10**, 267
– kinetic description of **11B**, 71ff
– of aromatic compounds **6**, 688ff
– of atrazine **11B**, 375f, 378
– of azo dyes **11A**, 392ff

- of catechol **11B**, 153, 179ff
- of chlorinated compounds **11B**, 241ff
- – aliphatics **11B**, 251ff
- – haloaromatics **11B**, 257ff
- of chlorinated *n*-alkanes **11A**, 377ff
- of chlorobenzenes **11A**, 380ff
- of chlorophenol **11B**, 418f
- of chlorophenols **11A**, 382f
- of coal **10**, 157ff
- of cresols **11B**, 178f
- of dichloroacetic acid **11B**, 419f
- of diesel fuel **11B**, 420f
- of ethylbenzene **11B**, 184f
- of hydrocarbons, alkenes **6**, 687
- – aromatics **6**, 686f
- – chlorinated aromatics **6**, 687f
- of hydrocarbons in oily sludge **11B**, 419f
- of hydroquinone **11B**, 179f
- of marine oil spills **10**, 470
- of mineral oils **11A**, 386ff
- of naphthalene sulfonic acids **11A**, 390ff
- of nitroaromatics **11A**, 383ff
- of nitro compounds **11B**, 273ff
- of organic carbon **11A**, 115f
- of organic pollutants **11B**, 145ff
- of pentachlorophenol **11B**, 420
- of phenol **11B**, 418f
- of polycyclic aromatic hydrocarbons **11A**, 386ff
- of soil pollutants **11B**, 71ff
- – model of the overall process **11B**, 75ff
- of toluene **11B**, 183f, 407
- of toxic compounds in soil **6**, 689ff
- reactor design **11A**, 500ff
- scale-up **11A**, 500f
- space loading **11A**, 353f
- – BOD elimination **11A**, 354
- use for post-denitrification **11A**, 345
Bio-Denitro process **11A**, 276
biodeterioration *see also* biocorrosion, microbial corrosion
- by microorganisms **10**, 269ff
- – algae **10**, 271
- – bacteria **10**, 269ff
- – cyanobacteria **10**, 269ff
- definition of **10**, 267
- history of **10**, 268f
- of adhesives/glues **10**, 311
- of concrete **10**, 300ff
- of hydrocarbons **10**, 311
- of mineral materials **10**, 299ff
- of natural organic materials **10**, 305ff
- – laminated wood **10**, 306f
- – leather **10**, 307
- – paper and cardboard **10**, 307
- – parchment paper **10**, 307
- – textiles **10**, 308
- of plasters **10**, 311
- of synthetic organic materials **10**, 308ff

- of varnishes/lacquers **10**, 311
- role of extracellular polymeric substances (EPS) **10**, 271
biodistribution studies **5A**, 361f
- hCTM01 **5A**, 361
- tumors **5A**, 361f
biodiversity **12**, 121, 644
- conservation **12**, 438ff
- – international cooperation **12**, 442f
- impacts on **12**, 294ff
- in the tropics **12**, 433ff
- of algae **10**, 110
- research **12**, 443
- utilization **12**, 438ff
Biodiversity Convention *see* Convention on Biological Diversity (CBD)
bioelectrochemical fuel cells **10**, 5ff
- application of redox mediators **10**, 13ff
- cathode systems in, use of oxygen cathodes **10**, 15
- classification of **10**, 11ff
- – direct – **10**, 12f
- – indirect – **10**, 11f
- current output **10**, 7
- electricity yields **10**, 15
- examples for **10**, 15ff
- – use of *Clostridium butyricum* **10**, 15ff
- – use of *Escherichia coli* **10**, 16ff
- – use of *Proteus vulgaris* **10**, 15ff
- – use of *Pseudomonas ovalis* **10**, 18f
- experimental set-up **10**, 7
- flow-through –, for sensory applications **10**, 20
- function of **10**, 6
- history **10**, 6f
- overall reaction **10**, 10f
- polarization effects **10**, 16
- principle **10**, 10ff, 16f
- principle of anodic reactions **10**, 12
- use for educational purposes **10**, 20
- use in bioelectronics **10**, 17ff
- use of biosensors **10**, 17ff
- – measurement parameters **10**, 19
bioelectronics, use of bioelectrochemical fuel cells **10**, 17ff
bioemulsifiers, exopolysaccharides **6**, 664f
bioenergetic reaction, carrier-bonded hydrogen **10**, 8f
bioenergetics, global dimensions of **10**, 7ff
- of lactic acid bacteria **1**, 350f
bioerosion, definition of **10**, 267
bioethics **12**, 115ff, 582, 639
- – *see also* ethics
- and rDNA products **12**, 189ff
- conflicts **12**, 135ff
- cross-cultural **12**, 122
- decision-making **12**, 121f
- European Convention **12**, 610
- limits for biotechnology **12**, 150f

biofertilizers, from microalgae **10**, 128f
biofilm, poison resistance **1**, 243
– substrate profile **1**, 227
biofilm kinetics **4**, 418ff
– concentration profiles **4**, 418, 421
– limiting nutrient **4**, 419ff
– measurement, concentric-cylinder device **4**, 425
– – rotating-disc device **4**, 425
– parameter estimation **4**, 425
– rate-controlling nutrient **4**, 419ff
biofilm modelling, mathematical simplification **4**, 421f
– nitrification **4**, 422f
– sloughing criteria **4**, 424
biofilm processes, effect of colloids **4**, 423
– in wastewater treatment models **4**, 418ff
biofilm reactors *see also* fixed bed reactors, fluidized bed reactors, rotating biological contactors, suspension membrane reactors
– categories of **11A**, 350
– modelling **4**, 425
– use of sulfate-reducing bacteria for metal removal **10**, 244
biofilms **10**, 275f, 290ff, **11A**, 65ff, 496ff
– – *see also* flocs
– as precursors of multicellular organisms **11A**, 79
– attachment phenotype **11A**, 79
– attachment process **10**, 291f
– binding forces **11A**, 74
– biofouling layer **11A**, 68
– cell-cell communication **11A**, 78f
– – quorum sensing **11A**, 78f
– cell detachment mechanisms **10**, 292
– cellular aggregation **11A**, 75
– characteristics of **11A**, 497ff
– – external mass transfer **11A**, 497f
– – internal mass transfer **11A**, 498f
– coaggregation of microorganisms **11A**, 73
– composition of **11A**, 66ff, 71
– – controlling factors **11A**, 71
– concentration gradient of pollutants **11C**, 268
– conventional transport within **11A**, 78
– definition of **11A**, 66
– deleterious effects **10**, 284f
– dental plaque **11A**, 71
– development of **11A**, 67ff
– – phases **11A**, 71f
– dispersion forces **11A**, 74
– engineered – **11A**, 67
– enzyme activities in **11A**, 76
– extracellular polymeric substances **11A**, 74ff
– – bridging agents **11A**, 75f
– extracellular polymeric substances (EPS) in **11C**, 446ff
– – as structural elements **11C**, 448
– – function of **11C**, 448
– fixed bed substratum **11A**, 353

– formation of **11A**, 496, **11C**, 277ff
– – biomass immobilization **11A**, 496
– – in fixed bed systems **11A**, 496ff
– – in fluidized bed systems **11A**, 496ff
– – methanogens **11A**, 496
– fur-type **11A**, 357
– heterogeneous spatial arrangement of microorganisms **11A**, 77
– immobilization of microbial cells **11A**, 76
– in biodeterioration **10**, 290ff
– in biofiltration **11C**, 445ff
– in drinking water sand filters, scaning electron micrographs **11C**, 447
– in submerged fixed bed reactors, composition of **11A**, 356f
– – development of **11A**, 356f
– – structure of **11A**, 356ff
– – thickness of **11A**, 356
– kinetics of **11C**, 341
– layer-type **11A**, 357
– life forms in **11A**, 72f
– metabolic versatility **11A**, 78
– microbial communities in **11C**, 277ff
– microbial growth in **10**, 275f
– microorganisms in **11C**, 448ff
– – horizontal gene transfer **11C**, 450
– – *Pseudomonas aeruginosa* **11C**, 448ff
– – quorum sensing **11C**, 450
– – resistance against biocides and antibiotics **11C**, 450
– – spatial arrangement of **11C**, 449f
– modeling of **11C**, 341
– models of **11A**, 70f
– – dense confluent – **11A**, 70f
– – heterogeneous mosaic **11A**, 70f
– – nushroom-like – **11A**, 70f, 353, 357
– nitrifying –, *Nitrosomonas* **11A**, 98
– oxygen concentration profiles **11C**, 449
– population densities **11A**, 72f
– role of extracellular polymeric substances (EPS) **10**, 292
– sessile bacteria in **11A**, 77
– size of **11A**, 67
– structure **11A**, 68f, 71
– – analysis methods **11A**, 68f
– – controlling factors **11A**, 71
– – heterogeneous distribution of microbial aggregates **11A**, 68
– structure in biological trickling filters, scheme **11C**, 279
– support materials **11A**, 502ff
– transport processes in **11C**, 449
– use of immobilized inoculum **11A**, 508
– *Zoogloea*-type **11A**, 357f
– – microscopic photograph **11A**, 358
biofilm supports **11A**, 258f
– in activated sludge aeration tanks **11A**, 257f
biofilm thickness **4**, 420ff, 426ff, 429

– optimum **4**, 429
– steady-state **4**, 423f
– – sloughing **4**, 424
biofilter procedure **11C**, 331
biofilters **11C**, 321ff, 334f, 346, 359ff
– advanced –, for controlling styrene emissions **11C**, 366
– applications **11A**, 366f
– applications of **11C**, 362ff
– as biological main stage **11A**, 367f
– Biofilter procedure **11C**, 331
– biological fundamentals **11C**, 323f
– BIOLOG substrates **11C**, 452f
– – impact of ozonation on **11C**, 453
– biomass profile of **11C**, 451
– carrier materials **11C**, 466
– combinations with, adsoroption plants **11C**, 331
– – bioscrubbers **11C**, 331
– container filters, scheme **11C**, 328
– degradation efficiency **11C**, 326
– denitrification rates **11A**, 370
– design **11A**, 368
– dimensioning of **11A**, 371, **11C**, 329
– dynamic behavior of **11A**, 370
– elimination capacity of **11C**, 270
– filter material, bonding of pollutants with **11C**, 323
– – flow channels **11C**, 323
– – microorganisms in **11C**, 323f
– – qualities of **11C**, 324f
– – ripening of **11C**, 446
– filter materials, disposal of **11C**, 325
– filter moisture **11C**, 325f
– flow patterns of **11A**, 370f
– for BOM removal, bacteria in **11C**, 466f
– – characterization of **11C**, 465ff
– – efficiency of **11C**, 468ff
– – filter media **11C**, 465f, 468f
– – filter types **11C**, 465f
– – metazoa in **11C**, 468
– – pathogens in **11C**, 467f
– – protozoa in **11C**, 468
– high performance filters **11C**, 328
– horizontal filters **11C**, 328f
– humidifiers **11C**, 361
– hydraulic residence time **11A**, 370
– influence of mass transport **11A**, 366
– inorganic –, for odor control at a wastewater treatment facility **11C**, 364
– investment costs **11C**, 329
– in waste gas purification **11C**, 260f, 321ff
– in wastewater treatment **11A**, 365ff
– irrigation systems **11C**, 361
– low-cost systems **11C**, 330
– measuring and control technology **11C**, 326
– microbial communities in **11C**, 450ff
– microbial degradation of pollutants in **11C**, 265
– mode of operation **11A**, 366ff, **11C**, 324ff

– new developments of **11C**, 330f
– new technologies **11C**, 346ff
– nitrification rates **11A**, 369
– nutrient supply **11C**, 326
– open bed compost, scheme **11C**, 360
– operating costs **11C**, 329
– operating parameters **11C**, 265
– operational conditions **11A**, 367f
– oxidation of hydrogen sulfide in **11C**, 287
– phospholipid fatty acid analysis **11C**, 451f
– physical fundamentals **11C**, 323
– pollutant concentrations **11C**, 265
– population analysis **11C**, 451f
– procedure variants **11C**, 326ff
– process modes **11A**, 368
– process parameters **11C**, 468ff
– – backwashing **11C**, 471
– – contact time **11C**, 470
– – deposition of flocs and particles **11C**, 470f
– – filter media **11C**, 468f
– – flow rate **11C**, 470
– – temperature **11C**, 470
– rotor biofilters **11C**, 331, 346f
– – scheme **11C**, 347
– schematic representation of **11C**, 262, 359
– scheme **11A**, 368
– single-level filters, closed – **11C**, 327f
– – open – **11C**, 327
– – scheme **11C**, 327
– soil **11C**, 360
– soil bed **11C**, 363
– specific filter load **11C**, 329
– stagnating areas **11A**, 366
– tower filters **11C**, 328
– treatment of, gasoline vapor at a soil vapor extraction site **11C**, 365
– – VOC emissions from an optical lens manufacturer **11C**, 365
– use of mixed or inorganic media **11C**, 361f
– volume flow **11C**, 326
– volumetric rates **11A**, 368f
biofilter system, construction of **11C**, 324
– with adsorber **11C**, 366
biofiltration, biological fundamentals of **11C**, 276ff
– following ozonation **11C**, 451
– use of, activated carbon absorbers **11C**, 458ff
– – biofilms **11C**, 445ff
– – deep bed rapid filters **11C**, 458ff
biofiltration plants, benzene turnover **11C**, 282
biofiltration processes, for organic matter removal **11C**, 457ff
– history of **11C**, 458
– Mülheim process **11C**, 458
Bioflocculation **4**, 436ff
– effect of PHB **4**, 436
biofouling **10**, 284, 290f, **11A**, 66ff
– definition of **10**, 275, 291

biofuel cell **4**, 84f
biogas **11A**, 187
– – *see* also methane
– anaerobic wastewater purification, composition **4**, 454
– – gas flow **4**, 453ff
– composition of **11C**, 155
– – as a function of hydrostatic pressure **11A**, 556
– formation **11A**, 231f
– – from agricultured waste **11A**, 231f
– – in sewage sludge **11A**, 23
– – specific gas yield **11A**, 231f
– from anaerobic treatment of wastewater, amounts **11A**, 36f
– – composition **11A**, 36f
– – energy content **11A**, 37
– from landfills, cleaning with biofilters **11C**, 330
– microbial kinetics **11A**, 536ff
– production, in biogas reactors **11A**, 554ff
– production of, anaerobic digestion of waste **11C**, 207
– yields **11C**, 155
biogas digestion, decimation times, for indicator bacteria **11C**, 229
– – for pathogenic bacteria **11C**, 229
biogas plants, agricultural – **11C**, 227
– fecal salmonellas in **11C**, 228f
– fecal streptococci in **11C**, 228ff
– inactivation of pathogens **11C**, 228f
– sanitation concepts based on **11C**, 254f
– sanitation criteria **11C**, 230
biogas production **1**, 263
– in sanitary landfills **11C**, 14
– in the BIOPERCOLAT® process **11C**, 209f
– overall quantity **1**, 263
– volumetric rates from sewage fractions **11C**, 21
biogas reactors *see also* stirred tank reactors, up-flow anaerobic sludge blanket (UASB) reactor
– anaerobic process, effect of hydrogen sulfide **11A**, 535
– biogas production, effect of hydrostatic pressure **11A**, 554ff
– composition of biogas **11A**, 556
– fermentation broth **11A**, 537f
– – physico-chemical state **11A**, 537f
– gas bubbles, diameter of **11A**, 555
– – volume of **11A**, 553
– kinetics **11A**, 536ff
– mathematical model **11A**, 529f
– – elements of **11A**, 529f
– measuring techniques **11A**, 532ff
– methane production rate **11A**, 538ff
– model calculations **11A**, 556f
– – fit with experimental data **11A**, 556f
– modeling of **11A**, 527ff
– modules **11A**, 529f

– monitoring of organic substances, betaine **11A**, 534f
– on-line measurements **11A**, 532ff
– – use of HPLC **11A**, 534f
– pH range **11A**, 533
– reactor control **11A**, 556ff
– – variables **11A**, 556
– silicon membrane probe, calibration of **11A**, 534
– – configuration of **11A**, 533
– specific gas production, as a function of the concentration of undissociated H_2S **11A**, 542
biogas tower reactor (BTR) **11A**, 529f
– active sludge retention **11A**, 548
– axial dispersion coefficient **11A**, 546f
– biomass distribution **11A**, 548f
– biomass pellets **11A**, 530
– circulation velocity **11A**, 546f
– controller design **11A**, 557
– controller implementation **11A**, 557f
– effect of gas recirculation **11A**, 551ff
– gas collecting devices **11A**, 543
– gas holdup **11A**, 547
– hydrodynamic behavior **11A**, 543ff
– interdependence of reactor elements **11A**, 557
– liquid mixing behavior **11A**, 543ff
– – models **11A**, 544ff
– mass transport, liquid phase to gas phase **11A**, 550ff
– mathematicl modeling **11A**, 549f
– – produced biomass **11A**, 549
– mixing mechanism **11A**, 546
– model calculations, fit with experimental data **11A**, 553
– model of the process **11A**, 557
– modules **11A**, 543f
– pilot scale – **11A**, 532
– reactor height **11A**, 550
– scale-up strategy **11A**, 531f
– specification of **11A**, 531
– superficial gas velocity **11A**, 547
– suspended solids concentration **11A**, 548f
biogenic material, heat treatment at 70°C **11C**, 230
biohalogenation, commercial application of **8B**, 201f
– enzymology of **8B**, 192ff
– mechanistic aspects of **8B**, 184ff
bio-heap treatment **11B**, 472
bioherbicides **10**, 489ff
– preservation of **10**, 490
biohydrometallurgical processes, metal-containing solids **10**, 212
biohydrometallurgy, history **10**, 193
– terminology **10**, 192
bioinformatics **12**, 324ff
– European activities **12**, 331
– pharmaceutical – **5B**, 123ff

– tools **5B**, 435
– U.S. activities **12**, 331
bioleaching *see also* microbial metal leaching
– applications **10**, 208ff
– archaea in **10**, 202
– at low temperatures **10**, 201
– bacteria in **10**, 202ff
– definition of **10**, 267
– dump – **10**, 209
– economics of **10**, 213f
– fungi in **10**, 205f
– heap –, scheme **10**, 209
– heterotrophic – **10**, 214f
– history **10**, 192f
– influencing factors **10**, 197ff
– inhibition of **10**, 198f
– *in situ* **10**, 208f
– leaching efficiencies **10**, 198f
– microbial diversity in **10**, 200ff, 215
– microbiological parameters **10**, 198
– of electronic scrap **10**, 212f, 216
– of fly ash **10**, 196f, 210ff, 216
– of metals **10**, 191ff, **11B**, 365
– – cobalt **10**, 208
– – copper ores **10**, 208ff
– – gold **10**, 208
– – nickel **10**, 208
– – principles **10**, 194ff
– of sewage sludge **10**, 216
– physicochemical parameters **10**, 198
– standard test methods **10**, 197
– tank – **10**, 209
– terminology **10**, 192
– under thermophilic conditions **10**, 215
– vat – **10**, 209
– world production of metals **10**, 214
– yeasts in **10**, 206
bioleaching environments, acidic –, microbial diversity **10**, 202ff
bioleaching model, mechanistic –, scheme **10**, 196
biological air filter (BAF) **11C**, 349ff
– scheme **11C**, 350
biological air treatment, concept of **11C**, 359f
biological insect control **2**, 644f
biological membrane **2**, 95
– electropermeabilization **2**, 98f
biological models **4**, 277ff
– types **4**, 270
biological oxygen demand (BOD), elimination, in biofilm reactors **11A**, 354
– in wastewater **11A**, 259ff
– monitoring device **11A**, 135
– monitoring with biosensors **11A**, 134f
– – sensors with pure cultures of bacteria **11A**, 134f
– – sensors with pure cultures of yeast **11A**, 134
– of municipal sewage **11A**, 173ff
– process **11A**, 10ff

biological oxygen demand *see* BOD
biological pest control, public opinion **12**, 125
biological regrowth potential *see* BRP method
biological revolution **12**, 511
biological test systems, for bioreactor characterization **4**, 125
biological value, of protein, in bread **9**, 245f
biological waste **3**, 769ff
– characterization **3**, 772
– decontamination conditions **3**, 779
– definition **3**, 771
– designation **3**, 777
– from research and development origins **3**, 774ff
– generation **3**, 771ff
– handling **3**, 778
– infectious – **3**, 772ff
– labeling **3**, 777
– landfill disposal **3**, 782
– management **3**, 769ff, 776ff
– movement **3**, 778
– non-infectious – **3**, 772
– packaging, liquid waste **3**, 778
– – solid waste **3**, 777
– public health impacts **3**, 784f
– regulations **3**, 772, 777
– sanitary sewer disposal **3**, 783
– storage **3**, 778
– US Federal agencies **3**, 772
– waste stream segregation **3**, 777
biological waste gas purification *see* waste gas purification
biological waste treatment **3**, 769ff, 778ff
– batch – **3**, 779
– chemical disinfection **3**, 781
– continuous – **3**, 780
– dry heat – **3**, 780
– gamma radiation **3**, 781
– gaseous decontamination **3**, 781
– hydropulping **3**, 782
– implications on public health **3**, 784f
– incineration **3**, 780
– liquid waste **3**, 779
– methods **3**, 778
– – validation **3**, 779
– microwave – **3**, 782
– off-site – **3**, 777
– on-site – **3**, 777
– reclamation **3**, 782
– recycling **3**, 782
– shredding **3**, 785
– solid waste **3**, 780
– steam sterilization **3**, 779
– techniques **3**, 777
– thermal inactivation **3**, 779
– wet grinding **3**, 782
biological wastewater treatment *see* wastewater treatment
biologics license applications, statistics **5A**, 503

BIOLOG system **11C**, 452f
bioluminescence reaction, bacterial – **10**, 469f
bioluminescence readouts, in drug screening **10**, 48
bioluminescent genes **11B**, 453
biomanufacturing trends **5A**, 540
biomass **3**, 380
– – *see* microbial biomass
– assumptions for modelling **4**, 409f
– carbon yield **3**, 303
– de-emulsifying activity **6**, 711
– degradation, by an anaerobic food chain **11A**, 27ff
– determination, in bioreactors **3**, 16
– – techniques **3**, 328
– distribution, in biogas tower reactors **11A**, 548f
– formation, during glucose respiration **11A**, 21f
– growth **3**, 331
– immobilization of **11A**, 439f
– intracellular components, formation **3**, 85
– maintenance **3**, 301
– metabolic activity, in bioreactors **3**, 384
– metal removal by **11A**, 435ff
– pellets, in biogas tower reactors **11A**, 530
– production of **11A**, 439
– production rate, estimation **3**, 373
– recording, in bioprocesses **3**, 360
– viable cell number **3**, 386
– yield **3**, 301
biomass concentration **4**, 181ff
– definition **4**, 181
– determination methods **4**, 182ff
– inferential control **3**, 341
– on-line calcuation, anaerobic wastewater process models **4**, 470
biomass determination **4**, 181ff
– acoustic methods, acoustic resonance densitometry (ARD) **4**, 195
– – piezoelectric sensor **4**, 195
– calorimetric methods **4**, 188ff
– electrochemical methods **4**, 192ff
– – amperometric sensors **4**, 194f
– – fuel cell type sensors **4**, 193f
– – impedimetric methods **4**, 192f
– – potentiometric sensors **4**, 193
– filtration methods **4**, 190ff
– in anaerobic wastewater processes **4**, 450
– optical sensors, culture fluorescence **4**, 185ff
– – nephelometric methods **4**, 182ff
– thermogram **4**, 190
– viscosity methods **4**, 192
biomass estimation, artificial neural network, Kalman filter **3**, 336f
– by secondary variables **3**, 334
– indirect – **3**, 370
– model-based –, linear black-box – **3**, 337ff
– – non-linear black-box **3**, 338
– on-line – **3**, 339

– – computing **3**, 380f
– oxygen balancing **4**, 8
– process variables **3**, 339
– with state observers **3**, 379
biomass estimation analyte, estimation methods **1**, 126
– – *see also* cell mass estimation
biomass production, by cyanobacteria **10**, 115ff
– by microalgae **10**, 109, 115ff
biomass sensor, amperometric **4**, 194f
– fluorescence sensor **4**, 209
– fuel cell type **4**, 193f
– impedimetric **4**, 192f
– on-line **4**, 181ff
– optical **4**, 182ff
– piezoelectric **4**, 195
– potentiometric **4**, 193
biomass yield, as a function of COD **4**, 414
Biomax® **10**, 432
Biomedical Institute of Taiwan **12**, 403
biomedicinal products *see also* pharmaceutical products
– development **12**, 213ff, 352f
– – phases **12**, 229ff
– – plan **12**, 228ff
– from tropical forests **12**, 437f
– market research **12**, 216ff
– patent applications **12**, 299ff
– project management **12**, 234f
– quality assurance schemes **12**, 226f
– registration requirements **12**, 221ff
– – efficacy **12**, 223f
– – quality **12**, 221ff
– – safety **12**, 222f
biomembrane *see* membrane
Biomembrat process, flow diagram **11A**, 258
biomineralization, definition **10**, 267
biomining, definition of **10**, 192
biomolecules, chemical graph description **5B**, 402
– chirality of **5B**, 402
– dehydration **3**, 703
– properties **3**, 696
Bion **10**, 495
Bionolle® **10**, 432f
biooxidation, terminology **10**, 192
BIOPERCOLAT® process **11C**, 208ff
– biogas production **11C**, 209f
– main unit **11C**, 209
– mass balance of **11C**, 210
– pilot plant studies **11C**, 209ff
– scheme **11C**, 208
– treatment of municipal solid waste **11C**, 209
biopesticide, definition of **10**, 487
bio-pesticides **12**, 563
biopharmaceutical development, long-term perspectives **5A**, 541
biopharmaceutical manufacturing costs, transgenic animals **5A**, 539

biopile process **11B**, 313
bioplasts from biotechnologically liquefied lignite,
  PHAs **10**, 170ff
biopolymer degradation, aerobic – **11A**, 20ff, 34
– anaerobic – **11A**, 20ff, 34
– by a food chain **11A**, 27
– hydrolysis of, kinetic aspects **11A**, 25
– – role of acetate **11A**, 29
– – role of molecular hydrogen **11A**, 29
– thermophilic conditions **11A**, 34
– to methane and CO_2 **11A**, 30ff
biopolymer-producing microbes, culture conditions **6**, 618f
– isolation **6**, 616ff
– media **6**, 617f
– strain development **6**, 619ff
– strain enhancement **6**, 622f
biopolymer-producing microorganisms, derepression of polysaccharide synthesis **6**, 620f
biopolymer production *see* polymer production
biopolymers **10**, 424ff
– – *see also* polysaccharides
– as products of marine biotechnology **10**, 466f
– degradation by actinomycetes **11C**, 69
– extracellular –, metal binding by **10**, 234ff
– formation by, *Azozobacter* sp. **6**, 617
– – *Haloferax* sp. **6**, 617
– – *Pseudomonas* sp. **6**, 617
– – *Rhizobium* sp. **6**, 617
– – *Sphingomonas paucimobilis* **6**, 617
– – *Xanthomonas campestris* **6**, 617
– use in protective vaccines **6**, 616
biopreservaties, nisin **7**, 358
bioprocess **3**, 80, 94, 319ff, 355ff
– – *see also* bioreactor, fermentation, fermentation process
– averaged variables **3**, 364
– capital investment **3**, 19
– circulation time distribution **3**, 95
– consumption rates, specific – **3**, 384
– culture parameters **3**, 80
– culture variables **3**, 80
– – sensors **3**, 81
– development, stages **3**, 18
– effectiveness factor **3**, 94
– effects of, aeration **3**, 18
– – agitation **3**, 18
– elemental balancing **3**, 371
– fed-batch operation **3**, 100
– flow of calculations **3**, 362f
– flow scheme **3**, 361ff
– industrial scale **3**, 83
– input-output representation **3**, 360f
– integral variables **3**, 364
– internal feedback **3**, 361
– intracellular components **3**, 85f
– mass balances **3**, 96
– observers **3**, 371, 382

– oxygen supply **3**, 190
– phase transitions, handling **3**, 395
– physiological state, control structures **3**, 387
– – vector **3**, 362
– product formation **3**, 86
– production rates **3**, 368ff
– – crystallizing products **3**, 370
– – estimation **3**, 368ff
– – production/production ratios **3**, 385
– – production/uptake ratios **3**, 385
– – rate ratios **3**, 385
– – specific – **3**, 384
– regulations **3**, 20
– repeated fed-batch operation **3**, 101
– residence time distribution **3**, 95
– scale-up **3**, 18
– soft sensors **3**, 370ff
– state estimation **3**, 379ff
– – corrector **3**, 380, 382
– – linear system **3**, 380
– – non-linear system **3**, 381
– – predictor **3**, 380, 381
– uptake rates, potential – **3**, 386
– – production/uptake ratios **3**, 385
– – rate ratios **3**, 385
– – uptake/uptake ratios **3**, 385
– viable cell number **3**, 386
bioprocess analysis **3**, 355ff
– – *see also* fermentation monitoring
– artificial neural networks **3**, 376ff
– direct reconstruction **3**, 372ff
– elemental balancing **3**, 372
– empirical stoichiometry **3**, 373f
– expert systems **3**, 392
– foaming detection **3**, 396f
– Fourier transformation **3**, 391
– pattern recognition **3**, 388ff
– shape analysis **3**, 392ff
– state observers **3**, 379ff
– temporal reasoning **3**, 392
– trend analysis **3**, 389
– – real-time – **3**, 392ff
bioprocess control **3**, 16, 319ff, 355ff
– – *see also* bioreactor control, fermentation control
– adaptive – **3**, 340ff
– predictor-corrector method **3**, 380
– process phase transitions **3**, 395
– temporal reasoning **3**, 392
bioprocess design, continuous operation, steady-
  state design **3**, 99
bioprocessing **3**, 771ff
bioprocess kinetics, gene product formation **4**,
  490f
bioprocess modeling **3**, 77ff, 84ff
– applied dimensions **3**, 375
– dynamic measurement **3**, 195
– empirical stoichiometric models **3**, 373ff

– multiple substrate models **3**, 87
– population balance equation (PBE) **3**, 88
– population models **3**, 87f
– product formation models **3**, 87
– simple models **3**, 87
– single cell models **3**, 87
– steady-state measurement **3**, 195
– structured models **3**, 86ff
– unstructured models **3**, 86
bioprocess modelling **4**, 488ff
– bioplant model **4**, 489
– bioreactor model **4**, 489
– molecular model **4**, 488f
– population model **4**, 489
– simulation **4**, 488ff
– single-cell model **4**, 488f
bioprocess monitoring **3**, 16
– flow injection analysis (FIA) **3**, 82
– new techniques **3**, 16
– – image analysis – **3**, 16
– product concentration **3**, 17
– sampling **3**, 16f
– temporal profiles, real-time trend analysis **3**, 392
– water monitoring **3**, 757
bioprocess variables **3**, 360ff
– average –, time profile **3**, 384
– degrees of limitation **3**, 386
– direct reconstruction **3**, 372, 382
– integral – **3**, 383
– manipulated – **3**, 360
– measurable – **3**, 374
– production rates **3**, 385
– unmeasurable – **3**, 374
– – estimation **3**, 370ff
– uptake rates **3**, 385
– – potential – **3**, 386
bioproduct purification, principles **4**, 607
bioproduct separation **4**, 609ff
– centrifugation **4**, 609
– chromatography **4**, 611, 614f
– control **4**, 621
– extraction **4**, 610f
– membrane filtration **4**, 609f
– on-line sensors **4**, 621
– principles **4**, 607
biopulping *see also* pulping
– technology **10**, 533f
bioreaction kinetics **3**, 85ff
bioreactor **3**, 8, 79f, 105ff, 321ff, 365, **4**, 109ff
– – *see also* bioprocess, fermentation
– addition ports **3**, 421f
– aerated stirred tank – *see* stirred tank bioreactor
– air exhaust line **3**, 422
– airlift – *see* airlift bioreactor
– anaerobic operation **3**, 297
– baffles **3**, 268
– – effects of – **3**, 92

– batch operation **3**, 28, 432
– – ultrafiltration reactor **3**, 432
– biocatalytic – **3**, 427ff, 432
– biological test systems **4**, 125
– bottom jacket **3**, 412
– bubble column – **3**, 84
– bubble entrainment **3**, 243
– – effects on mass transfer **3**, 244
– bulk mixing properties **4**, 114ff
– cavity formation **3**, 92f
– cell densities **3**, 120
– cell retention **3**, 309
– cell retention devices **3**, 113f
– – dialysis membranes **3**, 115
– – external – **3**, 114
– – inclined sedimenters **3**, 115
– – internal – **3**, 114
– – rotating wire cage **3**, 114
– ceramic-matrix – **3**, 118
– ceramic monoliths **3**, 28
– characteristic properties **4**, 110ff
– circulating-bed – **3**, 119
– cleanability **3**, 419
– compact loop –**3**, 80
– configurations **3**, 15, 28, 120f
– continuous culture, dilution rate **1**, 156
– continuous flow, for NMR **1**, 193
– continuous operation **3**, 26, 98f
– – biomass concentration **3**, 99
– – dilution rate effects **3**, 99
– – steady-state design **3**, 99
– – substate concentration effects **3**, 99
– cooling equipment **3**, 411, 420f
– Corning spinner, impeller depth effects **3**, 242
– – power input **3**, 240
– – surface mass transfer **3**, 242
– design **1**, 595
– diameter **3**, 418
– drain valves **3**, 423
– electropolishing **3**, 419
– energy transfer **3**, 89ff
– enzyme – **3**, 427ff
– fed-batch operation **3**, 100
– fixed-bed – *see* fixed-bed bioreactor
– fixed-bed reactor **4**, 450
– – wastewater treatment **4**, 479
– fluidized-bed – *see* fluidized-bed bioreactor
– fluidized-bed reactor, wastewater treatment **4**, 427ff, 450
– foam control **3**, 422f, **4**, 14f
– for animal cells **3**, 24ff
– for high cell densities **3**, 34
– – perfusion culture **3**, 34
– for mammalian cell cultures **3**, 110ff
– for plant cell cultures **1**, 586ff
– gas agitated – **3**, 256ff
– gas balance, parameters **3**, 368
– gas holdup **3**, 409

– global measuring techniques **4**, 119ff
– head space gas analyzers **3**, 80
– heat generation **3**, 411
– heat transfer **3**, 417, **4**, 119
– heterogeneous – **3**, 27, 34
– high-performance – **3**, 80ff
– history **3**, 433
– hollow-fiber – *see* hollow-fiber bioreactor
– homogenous – **3**, 27
– hydrodynamics **4**, 113
– impellers *see* impeller
– industrial – **3**, 80, 83ff
– *in situ* sensors **3**, 81
– jackets **3**, 420
– laboratory – **3**, 80f
– local measuring techniques **4**, 128ff
– – bubble parameters **4**, 128ff
– – liquid flow properties **4**, 138ff
– loop – **3**, 84
– mass transfer **3**, 89ff, 190ff, **4**, 110ff
– – coefficients **3**, 190ff
– microbial growth **1**, 129
– microcarrier – **3**, 27f, 107ff
– MIMO stirred tank –, input-output linearizing control **3**, 343
– mixing behavior **4**, 116ff
– modelling **4**, 269ff
– modified packed-bed – **3**, 119f
– non-mechanically agitated – **3**, 14
– on-line sensors **3**, 81
– oxygen supply **1**, 135
– – *see* bioreactor aeration
– oxygen transfer **1**, 226ff, **4**, 38ff
– parameter values **1**, 151
– performance **3**, 96
– perfusion operation **3**, 28
– pH-auxostatic operation **4**, 474f
– pH-chemostatic operation **4**, 474
– pneumatic – **3**, 261ff
– power input **3**, 91, 240, 405ff
– probes *see* bioreactor instrumentation, fermentation instrumentation
– production –, size **3**, 8
– repeated fed-batch operation **3**, 101
– Reynolds number **3**, 241
– scale-up **3**, 185ff
– – influence parameters **3**, 210
– – tools **3**, 210
– scale-up effects, on characteristic process times **3**, 212
– – on gas residence time **3**, 212
– – on mixing **3**, 211f
– semi-continuous operation **3**, 28
– SISO, input-output linearizing control **3**, 343
– specific interfacial area **3**, 84, 90f
– static seals **3**, 421
– sterilization **3**, 165
– – coll-down effects **3**, 167

– – cycle design **3**, 167ff
– – degree **3**, 167
– – heat-up effects **3**, 167
– – hold-up time constant **3**, 168
– – steam in the jacket **3**, 167
– stirred tank – *see* stirred tank bioreactor
– stirred tank models **4**, 301ff
– stirred tank reactor, continuous – **1**, 150
– – oxygen transfer rate **1**, 125
– stirrer types **1**, 587
– surface finishes **3**, 419
– total volume **3**, 418
– tower reactor models **4**, 352ff
– tubular – *see* tubular bioreactor
– two-compartment, three-zone – **3**, 119
– types **1**, 587f, **3**, 12ff, 27ff, 111f, 120, 321, 431f
– vessel, material **3**, 419
– – welds **3**, 420
– vibration **3**, 424
– wall thickness **3**, 418
bioreactor aeration **3**, 32, 120, 185ff, 210, 219ff, 237ff, 401ff
– – *see also* bubble column aeration, membrane aeration, sparged aeration, surface aeration
– air exhaust line **3**, 422
– bubble size **3**, 261
– effects on mass transfer coefficients **3**, 201
– *ex situ* – **3**, 254f
– external – **3**, 33, 254
– gas phases **3**, 366
– heat transfer **3**, 411f
– indirect sparging **3**, 265f
– mass transfer **3**, 249
– – coefficients **3**, 269
– mechanical stress factors **3**, 272
– number **3**, 92, 412
– – of impellers **3**, 406
– of animal cell cultures, methods **3**, 237ff
– performance **3**, 268f
– survey **3**, 237ff
– system comparsion **3**, 268ff
bioreactor agitation **3**, 33, 410
– – *see also* bioreactor mixing
– drive **3**, 423f
– – mounting **3**, 424
– – seals **3**, 424
– – sizing **3**, 424
– effects **3**, 228
– – on mass transfer coefficients **3**, 201
– hydrodynamic forces **3**, 33
– turbulent collision severity factors **3**, 229
bioreactor control **3**, 16, 121, 321ff, 355ff, **4**, 512ff
– – *see also* bioprocess control, fermentation control
– automation **3**, 82f
– design of a control system **4**, 512
– direct digital control (DDC) system **3**, 80
– mechanisms **3**, 323

– sensors **3**, 323
– traditional policies **3**, 324ff
– use of expert systems **4**, 628
bioreactor design **3**, 14, 401ff, 412f, 432
– basic calculations **3**, 404ff
– effects of, inlet oxygen concentration **3**, 415
– – pressure **3**, 415f
– – vessel diameter **3**, 415
– enzyme reactors **3**, 452, 458ff
– for peptide synthesis **3**, 462
– for protein production **3**, 412ff
– gas flow **3**, 417
– heat load **3**, 417
– mechanical details **3**, 418ff
– membrane reactors **3**, 452, 458ff
– power input **3**, 416
– sizing **3**, 414ff
– vessel geometry **3**, 410
– vessel wall thickness **3**, 409
bioreactor identification **4**, 226ff
– – *see also* System identification
– application of EKF **4**, 232ff
– definition **4**, 226
– mathematical formalism **4**, 227ff
– model formulation, model equations **4**, 228ff
– – state variables **4**, 228ff
– models **4**, 227ff
– on-line **4**, 226ff
– sequential state parameter estimation (SSPE)
  4, 238
– SSPE application **4**, 241ff
bioreactor instrumentation **3**, 321, 359
– – *see also* fermentation instrumentation
– gas analyzers **3**, 323
– *in situ* probes **3**, 323
– on-line – **3**, 321ff, 327
– – instruments **3**, 323
bioreactor mixing **3**, 94f, 206ff, 410f
– – *see also* bioreactor agitation
– "cavern" formation **3**, 210
– circulation time **3**, 207f
– effects of, broth properties **3**, 411
– – impellers **3**, 96, 410
– – vessel geometry **3**, 410
– estimation methods **3**, 208
– flow-follower measurements **3**, 208
– highly viscous fluids **3**, 193
– low viscous fluids **3**, 193
– macromixing **3**, 94f
– maximal mixedness **3**, 94
– mechanical stress factors **3**, 272
– micromixing **3**, 94
– models, compartment – **3**, 95
– – loop – **3**, 95
– – network-of-zone – **3**, 95
– – turbulence – **3**, 95
– partial segregation **3**, 94
– power input **3**, 96

– single-phase – **3**, 207
– terminal mixing time **3**, 207
– total segregation **3**, 94
– tracer measurements **3**, 209
– two-phase – **3**, 207
– zone formation **3**, 210
bioreactor modeling **3**, 77ff
bioreactor monitoring *see* bioprocess analysis, fermentation monitoring
bioreactor optimization **4**, 526ff
– adaptive **4**, 551ff
bioreactors **11B**, 329ff, 345f
– – *see also* fixed bed reactors, rotating drum reactors, slurry bioreactors, solid state bioreactors
– and recycling of solid waste streams **11B**, 332ff
– comparsion of, economics **11B**, 345
– – technology **11B**, 345
– comparison of **11B**, 344f
– configurations, rotating drum dry solid bioreactors **11B**, 332f
– – slurry bioreactors **11B**, 332ff
– – solid state fixed bed bioreactors **11B**, 332f
– diffusion of the contaminants out of the solid particles **11B**, 333f
– operation modes **11B**, 332f
bioreactor state estimation **4**, 226ff
– detection of contamination, by SSPE **4**, 244ff
– techniques **4**, 543ff
bioreactor types, in manufacture of therapeutic antibodies **5A**, 261f
bioremediation *see also in situ* remediation
– biostimulation, methanotrophic bacteria **11B**, 407
– – toluene oxydizing bacteria **11B**, 407
– by metal leaching **10**, 245ff
– by reduction of metalloid oxyanions **10**, 241
– hierarchy of *in situ* treatments **11B**, 400f
– metalloid transformations **10**, 249
– natural attenuation **11B**, 401ff
– of chlorinated solvents **11B**, 407f
– of contaminated soil, *ex situ* processes **11B**, 313f
– – *in situ* processes **11B**, 314f
– of metal-contaminated sites **10**, 215ff
– of nitrile pollutants **8A**, 307
– of soils **6**, 689ff
– – cost **6**, 692
– – effect of biosurfactants **6**, 694ff
– – *in-situ* treatment **6**, 692
– – on-site treatment **6**, 692
– of wastes **8A**, 312f
– reductive metal precipitation, by metal-reducing bacteria **10**, 239f
– subsurface attenuation **11B**, 401
– use of genetically engineered microorganisms **11B**, 441ff

– use of microbial metal sulfide precipitation **10**, 242ff
bioremediation techniques **11B**, 471ff
– bio-heap treatment **11B**, 472
– evaluation of **11B**, 470f
– *ex situ* processes **11B**, 471f, 426ff
– – with heap technique **11B**, 319ff
– *in situ* processes **11B**, 472f, 426ff
– reactor processes, slurry state – **11B**, 472
– – solid state – **11B**, 472
bioremedition, of contaminated soil **3**, 312
– – halogenated organic compounds **3**, 313
bioreporter systems **11B**, 453ff
– for aromatic hydrocarbons **11B**, 454f
– for chlorinated compunds **11B**, 453f
– for metals **11B**, 455f
biosafety **12**, 63ff, 349
– – *see* safety
– CBD **12**, 629ff
– Code of Conduct **12**, 626
– ISNAR report **12**, 611f
– of animal cell cultures **12**, 94ff
– – viral vectors **12**, 98ff
biosafety levels **12**, 73ff, 244ff
– at large-scale practices **12**, 247
biosafety regulations, European approach **12**, 532ff
bioscreens **11B**, 361ff, 473
– as denitrification zones **11B**, 363
– composition of **11B**, 362
– process scheme **11B**, 362
bioscrubber plants, types of **11C**, 316
bioscrubbers **11C**, 305ff, 334, 346, 359
– Biosolv procedure **11C**, 348
– biosuspensions in **11C**, 277
– BioWave procedure **11C**, 348
– column types **11C**, 307
– degradation rate of solvents **11C**, 309
– design parameters **11C**, 308
– efficiency of **11C**, 313
– equalizing filter with countercurrent flow **11C**, 314ff
– for odor mitigation in a production plant, flow diagram **11C**, 312
– for waste air cleaning, flow diagram **11C**, 306
– industrial applications of **11C**, 312ff
– in waste gas purification **11C**, 261
– maximum pollutant load **11C**, 309
– microbial degradation of pollutants in **11C**, 265
– operational parameters **11C**, 265, 308
– parameters for **11C**, 306
– pollutant concentrations **11C**, 265
– schematic representation **11C**, 262
– scrubbing liquid quantities for the absorption of organic substances **11C**, 312
– special types of **11C**, 314ff
– specific kinds of **11C**, 347f

– with equalizing filter with countercurrent flow, simulation calculation **11C**, 315f
biosensor **4**, 77ff
– ATP-sensitive **4**, 211
– binding assays, enzymatic **4**, 79
– – immunosensors **4**, 79, 93
– – receptor-based sensors **4**, 96
– biocomponent **4**, 77ff
– cell-based **4**, 78f
– cell-based assay **4**, 91ff
– – analysis of BOD **4**, 92
– – analysis of inhibiting substances **4**, 92
– – analysis of specific metabolites **4**, 92f
– – analysis of vitamins **4**, 91
– construction **4**, 82ff
– definition **4**, 77
– enzyme-based **4**, 77f
– enzyme-based assay **4**, 86ff
– – FIA system **4**, 90
– – glucose analyzer **4**, 88
– – thermistor application **4**, 89ff
– for NADH **4**, 209, 211
– for NADPH **4**, 211
– immobilization of biocomponent on transducer **4**, 83
– immobilization of enzyme component **4**, 81
– in extreme environments **4**, 97f
– – biological liquids **4**, 97
– – operation in air **4**, 98
– – organic solvents **4**, 97
– optical sensor configurations **4**, 86
– organelle-based **4**, 79
– pre-column of immobilized biocomponent **4**, 83
– transducer **4**, 77ff
– – integration **4**, 81ff
– – types **4**, 83ff
biosensors **3**, 34f, **9**, 150f, **11A**, 125ff
– based on luminous light emission **11A**, 135
– concept **9**, 151
– definition of **11A**, 126
– enzymatic signal amplification **11A**, 131
– immuno-flow injection analysis **11A**, 133
– miniaturization of **11A**, 136
– monitoring of, BOD **11A**, 134
– – cyanide **11A**, 128
– – environmental processes **11A**, 125ff
– – fungicides **11A**, 131f
– – halogenated organic compounds **11A**, 133f
– – heavy metals **11A**, 127f
– – herbicides **11A**, 131ff
– – insecticides **11A**, 131ff
– – naphthalene **11A**, 134
– – nitrate **11A**, 128f
– – nitrite **11A**, 128f
– – pesticides **11A**, 131ff
– – phenolic compounds **11A**, 130f
– – sulfide **11A**, 128f
– – sulfite **11A**, 128

– – sum parameters **11A**, 134ff
– principles of monitoring **11A**, 126f
– scheme of **11A**, 126
– selectivity **11A**, 126
– toxicity control **11A**, 135f
– transducer elements **11A**, 126
– use of bioelectrochemical fuel cells **10**, 17ff
BIOSIS **5B**, 283
bioslurping **11B**, 473
– in soil remediation **11B**, 359f
– treatment wells **11B**, 360
bioslurry *see* fermentation broth
bioslurry treatment process, microbial tranforma-
    tion of TNT **11B**, 291f
– microbial transformation of nitroaromatics
    11B, 291f
biosolve plant, flow diagram of **11C**, 319
biosolve process **11C**, 316ff
– high-boiling oil **11C**, 317
– high-boiling oil in **11C**, 316ff
– principle of **11C**, 317
Biosolv technique **11C**, 348
biosorption, of heavy metals **11A**, 441
– of metals **11A**, 42, 436f
– – by immobilized biomass **10**, 231ff
– – by microorganisms **10**, 228ff
– – distribution equilibrium **11A**, 441
– process **11A**, 255
– terminology **10**, 228
biosparging **11B**, 473
– in soil remediation **11B**, 359f
– process scheme **11B**, 360
biospecific binding, reactant pairs **4**, 93
biostimulation **11B**, 406ff, 426ff
– aerobic conditions **11B**, 406
– anaerobic conditions **11B**, 406
– cost **11B**, 408
– examples **11B**, 427
– nutrient addition **11B**, 406
– use of oxygen **11B**, 406
biosurfactant **1**, 415
biosurfactant classes **6**, 662ff
biosurfactant production **6**, 665ff
– batch culture **6**, 667
– by resting cells **6**, 667f
– conditions **6**, 667f
– continuous culture **6**, 667
– from hydrocarbons **6**, 667f
– genetic engineering of microbes **6**, 696
biosurfactants **6**, 659ff, **7**, 180f
– – *see also* surfactants
– applications **6**, 660f
– – in cosmetics **6**, 670ff
– bacterial – **6**, 662ff
– – glycolipids **6**, 662
– – rhamnolipids **6**, 662
– emulsifying polysaccharides **6**, 650
– formation by yeasts **6**, 664

– function in hydrocarbon uptake **11B**, 149ff, 152
– in bioleaching **10**, 199f
– in enhanced oil recovery **6**, 668f
– in foods **6**, 683ff
– in soil bioremediation **6**, 694ff
– list of producing microbes **6**, 662
– medical applications **6**, 677ff
– – blood substitutes **6**, 678ff
– – parenteral emulsions **6**, 677f
– *n*-alkane substrates, biosynthetic pathway **6**,
    666
– of marine origin **10**, 470f
– pharmaceutical applications **6**, 677ff
– – as adjuvants for vaccines **6**, 680f
– – in drug administration **6**, 680
– – oral emulsion systems **6**, 681f
– – topical emulsion systems **6**, 682f
– properties **6**, 660f
– secreted by microorganisms **11C**, 290
– solubilization of coal **10**, 164
– sophorose lipids **6**, 669f
– – production cost **6**, 669
– triglyceride precursors **6**, 670
– use in environmental control **6**, 686ff
biosuspensions, in bioscrubber systems **11C**, 277
biosynthetic pathways, of secondary metabolites
    7, 31ff
biosynthetic strategies, formation of nucleoside
    antibiotics **7**, 32
BIOTASK **12**, 626
BIOTECH (National Institutes of Biotechnology
    and Applied Microbiology) **12**, 416
biotechnical drugs, as antitumor agents **7**, 641ff
biotechnological facility *see* facility
biotechnological inventions, legal protection of
    5B, 419
biotechnological manufacturing **3**, 283ff, 739ff
– – *see also* bioprocess, fermentation process
– component flow **3**, 743
– flow of personnel **3**, 743
– license **3**, 746
– product flow **3**, 743
– r-DNA based –, plant **3**, 472
– regulatory process **3**, 746f
– scale-up **3**, 742
– using municipal water **3**, 756
– using well water **3**, 756
– water pretreatment **3**, 756
– water treatment steps **3**, 756
– with recombinant cells **3**, 283ff
biotechnological process, characteristics **3**, 473
– pre-license inspection **3**, 743
– product changes **3**, 765
– validation *see* process validation, facility valida-
    tion
– waste treatment, validation **3**, 763
biotechnological processes, product formation, in-
    fluence of stress on **10**, 574

biotechnological products **3**, 696, 741
– annual production **6**, 48
– annual $ sales **6**, 49
– licensing policy interpretation **3**, 746
– markets for **6**, 48f
– production estimation of, gaseous – **3**, 369
– – non-volatile **3**, 369
– – volatile – **3**, 369
– production sequence **3**, 741
– production variables **3**, 743
– regulatory process **3**, 746f
– upstream considerations **3**, 40f
biotechnology, bioethical conflicts **12**, 135ff
– – ecological risks **12**, 138
– – future **12**, 139ff
– – health concerns **12**, 138
– – human misuse **12**, 139
– – regulatory concerns **12**, 138f
– conservation of tropical forests **12**, 440ff
– databases **12**, 323ff
– – bibliographic – **12**, 327
– – factual – **12**, 325ff
– – full-text – **12**, 328
– – hosts **12**, 328ff
– – referral – **12**, 327f
– – use **12**, 334ff
– difinition **12**, 41, 65, 581
– ethical – **12**, 115ff, 189ff
– – perceptions **12**, 123ff
– ethical limits **12**, 150f
– ethicity assessment, criteria **12**, 151f
– history **12**, 341
– in Asian-Pacific nations **12**, 369ff
– in Australia **12**, 422ff
– in China **12**, 396ff
– in developing countries **12**, 339ff, 343ff
– – capital base **12**, 347
– – company strategy **12**, 346
– – horizontal technology transfer **12**, 345f
– – human resource base **12**, 347f
– – industry structure **12**, 346
– – marketing and distribution **12**, 345
– – R & D costs **12**, 344
– in developing countriesr, production **12**, 344f
– information services **12**, 330
– in Japan **12**, 375ff
– in space **10**, 507ff
– intellectual property **12**, 281ff, 350f
– in the ASEAN member states **12**, 409ff
– – Indonesia **12**, 409ff
– – Malaysia **12**, 417f
– – Philippines **12**, 413ff
– – Singapore **12**, 418ff
– – Thailand **12**, 420ff
– in the Replublic of Korea **12**, 386ff
– myths **12**, 41ff, 475
– public education programs **12**, 148f

– public opinion surveys **12**, 123ff, 146ff, 449ff, 658ff
– regulations, history **12**, 505ff
– – in Belgium **12**, 605
– – in China **12**, 605
– – in Denmark **12**, 603f
– – in Europe **12**, 516ff
– – in France **12**, 587ff
– – in Germany **12**, 580ff
– – in India **12**, 605
– – in Japan **12**, 592ff
– – in Russia **12**, 605
– – in Sweden **12**, 604f
– – in Switzerland **12**, 606
– – in The Netherlands **12**, 597ff
– – in the United Kingdom **12**, 573ff
– – in the United States **12**, 513ff, 566ff
– risk scenarios **12**, 463ff
– safe developemnt of products **12**, 199ff
– standards **12**, 641f
– strategic regulations, science-based **12**, 200f
– – transparency **12**, 201ff
– – triggers **12**, 199ff
– Taiwan **12**, 403ff
– U.S. policy objectives **12**, 199f
– utilization of tropical forests **12**, 438ff
biotechnology-derived medicines **5A**, 535ff
– approval times **5A**, 535
– clinical research organizations **5A**, 538
– commercialization **5A**, 534
– contract manufacturing organizations **5A**, 537
– contract research organizations **5A**, 537
– costs **5A**, 535f
– economic environment **5A**, 535
– economic significance **5A**, 534f, 537
– industry consolidation **5A**, 537
– market **5A**, 535
– single sourced – **5A**, 536f
– virtual pharmaceutical companies **5A**, 537
– woldwide sales **5A**, 533
biotechnology evolution **5A**, 519f
biotechnology information **12**, 324ff
– international organizations **12**, 330f
– pre-conditions **12**, 332f
– problems **12**, 332
biotechnology products **5A**, 531ff
– approvals **5A**, 533ff
– approval times **5A**, 535
– cell culture **5A**, 539f
– characterization methods **5A**, 497
– clinical research organizations **5A**, 538
– commercialization **5A**, 534
– contract manufacturing organizations **5A**, 537
– contract research organizations **5A**, 537
– control procedures **5A**, 497
– costs **5A**, 535
– crisis management in Europe **5A**, 513
– development, long-term perspectives **5A**, 541

– economic considerations **5A**, 531ff
– economic environment **5A**, 532, 535
– economic significance **5A**, 534f, 537
– European Council regulation **5A**, 509
– genetic engineering **5A**, 540
– industry background **5A**, 532f
– industry consolidation **5A**, 537
– manufacturing costs **5A**, 539f
– manufacturing steps **5A**, 539
– manufacturing technology **5A**, 540
– market **5A**, 535
– – capitalization **5A**, 532
– marketing authorization holder, responsibilities **5A**, 512
– marketing in Europe **5A**, 510ff
– – organization **5A**, 510f
– registration of **5A**, 509
– – advisory committees **5A**, 499, 501ff
– – European Medicines Evaluation Agency **5A**, 506f
– – filing process **5A**, 500
– – Food and Drug Administration **5A**, 499ff
– – Food and Drug Administration Moderniza-tion Act **5A**, 503
– – in Europe **5A**, 506ff, 511
– – in the UK **5A**, 506
– – orphan drug legislation **5A**, 505f, 514
– – review process **5A**, 501ff
– – well-characterized product **5A**, 502f
– sales **5A**, 534
– single sourced – **5A**, 536f
– specific guidelines **5A**, 498f
– timetable for review of application, in Europe **5A**, 511
– trademarks in Europe **5A**, 513
– transgenic animals **5A**, 539
– use of **5A**, 534
– virtual pharmaceutical companies **5A**, 537
– worldwide regulation **5A**, 514
Biotex procedure **11C**, 347
– cleaning of the textile filter material **11C**, 348
biotin **1**, 72
– as DNA label **2**, 397
– biosynthesis in microorganisms **10**, 328
– chemical synthesis **10**, 328
– enzymic synthesis of **8B**, 384f
biotransformation, of terpene precursors **10**, 374f
– of terpenoids with flavor and fragrance impact **10**, 376ff
biotransformation products, alicyclic – **8B**, 367ff
– aliphatic – **8B**, 353ff
– aromatic – **8B**, 390ff
– heterocyclic – **8B**, 384ff
– polycyclic – **8B**, 376ff
– – use of alcohol dehydrogenases **8B**, 382f
biotransformations **3**, 427ff
– advantages over abiotic asymmetric catalysts **8A**, 26

– bioreactor types **3**, 431
– by enzymes, multi-enzyme systems **8A**, 33
– – practical aspects **8A**, 26, 32f
– by whole cells **8A**, 13, 26ff, 31f, 396ff, 483ff
– experimental procedures for **8A**, 27ff
– – hydrolase-catalyzed reactions **8A**, 27f
– – oxidoreductase-catalyzed reactions **8A**, 29
– industrial-scale **3**, 431
– in enzyme reactors **3**, 445ff
– in soil, kinetic data **11B**, 73ff
– literature **8A**, 27ff
– of dalbaheptides **7**, 389f
– – deacylation **7**, 389
– – deglycosylation **7**, 389
– – glycosylation **7**, 389
– on-line services **8A**, 27
– perspectives **8A**, 1ff
– – 1895–1935 **8A**, 6ff
– – 1935–1955 **8A**, 10ff
– – 1955–1975 **8A**, 11ff
– practical aspects **8A**, 25ff
– process options **8A**, 6
– products **3**, 431
– screening of microorganisms for **8A**, 31
– used enzymes **3**, 431
– with cultured plant cells **7**, 621ff
– with isolated enzymes **8A**, 502ff
– with yeast **8A**, 363ff
biotrickling filters *see* trickling filters
bioventing **11B**, 473
– in soil remediation **11B**, 356f
– process scheme **11B**, 357
biowaste *see also* waste
– aerobic/anaerobic treatment of **11C**, 213ff
– – hygienic considerations **11C**, 213ff
– amounts of **11C**, 104
– anaerobic digestion of **11C**, 19ff
– anaerobic fermentation of, agitation **11C**, 162
– – continuous – **11C**, 161f
– – discontinuous – **11C**, 161f
– – dry – **11C**, 160f
– – mesophilic – **11C**, 162
– – one-stage – **11C**, 160
– – procedures **11C**, 157ff
– – process engineering of **11C**, 159ff
– – thermophilic – **11C**, 162
– – two-stage – **11C**, 160
– – wet – **11C**, 160f
– anaerobic processes, full scale – **11C**, 20
– and sewage sludge, joint fermentation of **11C**, 224
– cofermentation, with animal manures **11C**, 23
– – with sewage sludge **11C**, 23
– composting of **11C**, 17ff, 128ff
– fermentation of, treatment steps **11C**, 158
– from households, sanitation of **11C**, 232
– hygienic considerations **11C**, 231
– methanation of **11C**, 19ff

– occupational health risks **11C**, 231
– opening of bins, fungal count in the air **11C**, 239
– volumetric biogas production rates **11C**, 21
– water content of, during the year **11C**, 109
– wet anaerobic digestion, wet processes **11C**, 22ff
BioWave technique **11C**, 348
biphenyl, hydroxylation of **8A**, 500
biphenyl dioxygenase, hybrid –, in *E. coli* **11B**, 448
biphenyls, degradation of, by genetically engineered microorganisms **11B**, 448
biphytanyl tetraether lipid, of methanogens **1**, 259
bipolar disorder **5B**, 80
bird embryo transformation **2**, 810
– frequency **2**, 810
β-bisabolene, biotransformation of **10**, 408f
bisanthrene, antitumor agent, structure **7**, 651
bispecific antibodies **5A**, 321
bisucaberin, hydroxamate siderophore of *Alteromonas haloplanctis* **7**, 210f
– structure **7**, 211
bitterness, of olive fruits **9**, 597
– of protein hydrolysates **9**, 693f
blackening, of ripe olives **9**, 617
Blackman kinetics **4**, 274f, 277
bladder cancer **5B**, 245
Blakeslea **1**, 522
blanket consent **5B**, 424
BLAST *see* Basic Local Alignment and Search Tool
blast-furnace cement, corrosion tests **10**, 302f
bleaching, laccase-mediator concept **10**, 537
– xylanase-aided – **10**, 535ff
blennin, leukotriene biosynthesis inhibitor, structure **7**, 523
bleomycin **7**, 661
– mode of antitumor action **7**, 663
– structure **7**, 663
blepharismin **10**, 140ff
– structures **10**, 142
– therapeutic potential **10**, 140f
β-blockers, lipase-catalyzed reactions **8A**, 84
– synthesis of **8B**, 385
BLOCKS **5B**, 289
blood cell hydrolysate **9**, 700
blood clotting cascade **5A**, 34
– intrinsic pathway **5A**, 32
blood glucose level **4**, 89
blood group antigens **5A**, 10
blood group makers **7**, 403
blood monitoring **4**, 88f
– glucose analyzer **4**, 88f
blood substitutes **6**, 678ff
– perfluorochemical emulsions, composition **6**, 680

bluensidine, biosynthetic pathway **7**, 417, 421
bluensomycin **7**, 415f
– biosynthetic pathway **7**, 416
BMFT (Federal Ministry for Research and Technology) **12**, 581ff
BOD *see* biological oxygen demand
BOD (Biological oxygen demand), cell-based biosensor assay **4**, 92
– in biofilms **4**, 419, 426ff
– in industrial wastewaters **4**, 92
Bodenstein number **4**, 364f, 369, 372f
boholmycin, streptomycin-related aminoglycosides **7**, 428
– structure **7**, 464
BOM *see* biodegradable organic matter
bone cleaning, enzymatic – **9**, 697f
bone marrow transplantation, monoclonal antibody therapy **5A**, 318
bone morphogenetic proteins **5A**, 162f
– function **5A**, 163
– in therapy **5A**, 162f
bone substitutes, as products of marine biotechnology **10**, 457
Bordetella pertussis, alcaligin **7**, 210f
boron, phytoextraction of **11B**, 391
Borrelia burgdorferi **5A**, 140
botryococcene **7**, 174ff
– structure **7**, 174
Botryococcus braunii, hydrocarbon production **7**, 175f
Botrytis cinerea **9**, 727
– changes in must components **9**, 498ff
– noble rot of wines **9**, 497ff, 727
botulinum toxin **1**, 311
boundary condition iteration **4**, 527
boundary predictions **5B**, 369
bound residues **11B**, 109ff
– age of **11B**, 116
– formation of **11B**, 109f
– – covalent binding **11B**, 112
– – from herbicides **11B**, 109f
– – from pesticides **11B**, 109f
– – from xenobiotic compounds **11B**, 110ff
– – physical entrapment **11B**, 112
– – sorption processes **11B**, 111f
– from anthracene, mibilization of **11B**, 117
– – microbial degradation of **11B**, 116f
– in the organic soil matrix **11B**, 109ff
– mibilization of **11B**, 114ff
– mineralization of **11B**, 115f
– – in soils contaminated by tar oil **11B**, 116f
– stability of **11B**, 115
– turnover of **11B**, 114ff
BovGBASE **5B**, 285
bovine growth hormone, hydrophobicity **3**, 549
– refolding, denaturant effects **3**, 548
bovine papilloma virus (BPV) **2**, 787
bovine spongiform encephalopathy (BSE) **10**, 29

Box-Behnken design, in medium optimization **3**, 136
box composting **11C**, 143f
Box-Wilson design, in medium optimization **3**, 136
BPI (Federal Pharmaceutical Industry Association) **12**, 586
BPM (Bioprocessing-Prototype Manufacturing Facility of Malaysia) **12**, 417
BPPT (Agency for Technology Assessment and Application) **12**, 412
BRAIN (Bio-oriented Technology Research Advancement Institution) **12**, 383f
brain-derived neurotrophic factor **5A**, 159
brain natriuretic peptide **5A**, 165
brandy **9**, 463ff
– definition **9**, 464
– higher alcohols in – **9**, 482
brazzein, proteinaceous sweetener **10**, 577f
BRCA1 **5B**, 71
BRCA2 **5B**, 71
bread *see also* baking
– aroma **9**, 305ff
– baking temperatures **9**, 303f
– baking times **9**, 303f
– caloric value **9**, 246
– defects **9**, 308f
– microbial spoilage **9**, 311f
– – by *Aspergillus* **9**, 312
– – by *Bacillus subtilis* **9**, 311
– – by *Monilia sitophila* **9**, 312
– – by *Penicillium* **9**, 312
– mineral content **9**, 247
– nutrient content, fat **9**, 246f
– – protein **9**, 245f
– packaging **9**, 313
– preservation, chemical – **9**, 315
– – pyhsical – **9**, 314
– production steps **9**, 249f
– quality **9**, 308ff
– – criteria **9**, 305
– shelf life **9**, 310
– staling **9**, 309ff, 676ff
– storage **9**, 313
– vitamin content **9**, 247
– yield **9**, 304
bread fermentation **3**, 312
bread production, history **1**, 327
break-even sugar price **6**, 50ff
– for citric acid production **6**, 52
– for ethanol production **6**, 51f
– for lactic acid production **6**, 52
– for single-cell protein production **6**, 52
breast cancer **5B**, 71
– familial – **5B**, 82
– monoclonal antibody therapy **5A**, 321
breeders' privilege **12**, 284f
breeding system **2**, 76ff

BRENDA **5B**, 291, 405
Brettanomyces, in wine spoilage **9**, 495
brevibacteria, amidase activity **8A**, 292
– nitrile hydratase activity **8A**, 292
Brevibacterium, nitrile hydratase defective mutants **8A**, 292
Brevibacterium ammoniagenes **2**, 34f
– cell homogenization **3**, 514
– mutagen treatment **2**, 35
– mutant genealogy **2**, 35
– nucleotide synthesis, inosinate (5′) **6**, 571ff
– – xanthylate (5′) **6**, 576f
– production of inosine **6**, 583ff
– – productivity of mutants **6**, 585
– purine nucleotide synthesis, by mutant strains **6**, 574
– – regulation **6**, 574
– – transformed strains **6**, 575f
– uridine production **6**, 593
– 5′-IMP production **2**, 35
Brevibacterium flavum, immobilized whole cells, fumarase activity in **8B**, 107ff
– L-glutamic acid production **6**, 506
– L-malic acid production, in bioreactors **3**, 457
– recombinanat L-threonine producing strains **6**, 486
Brevibacterium lactofermentum **2**, 479
– amino acid production by genetically constructed – **6**, 508
– glutamate production **6**, 507
– L-lysine biosynthetic genes **6**, 481f
– lysine production **6**, 507
– phenylalanine production **6**, 510
– threonine production **6**, 508ff
– tryptophan production **6**, 512
– tyrosine production **6**, 512
Brevibacterium linens, in cheese production **9**, 378
brewer's wort **9**, 433ff
– boiling **9**, 434
– cooling **9**, 434
– filtration **9**, 684
– – mash filter 2001 **9**, 433f
– mashing **9**, 433
– nutrients, carbohydrates **9**, 441f
– – nitrogen **9**, 442, 684
brewers' yeast **8A**, 364
– – *see also* yeast
– cell viability **9**, 448f
– co-flocculation **9**, 445
– cytoduction **9**, 437
– excretion products **9**, 442ff
– – alcohols **9**, 442f
– – carbonyl compounds **9**, 443
– – diacetyl **9**, 443
– – dimethylsulfide **9**, 444
– – esters **9**, 443
– – pentane-2,3-dione **9**, 443f

– – sulfur compounds **9**, 443
– – wort nitrogen **9**, 442
– flocculation **9**, 444f
– – genetic control **9**, 445
– – mechanisms **9**, 445
– genetically modified – **9**, 454
– hybridization **9**, 436
– immobilized cell technology **9**, 453f
– killer character **9**, 437f
– maltose permease **9**, 441
– microbial contaminants **9**, 446ff
– mutations **9**, 438
– nutritional requirements **9**, 440ff
– – ergosterol **9**, 440
– – oxygen **9**, 440
– pitching **9**, 448f
– propagation **9**, 446
– pyruvate decarboxylase of **8B**, 25
– rare mating **9**, 437
– storage **9**, 449f
– strain improvement **9**, 435ff
– – recombinant DNA **9**, 439f
– – spheroplast fusion **9**, 439
– taxonomy **9**, 435
– uptake of, wort carbohydrates **9**, 441f
– – wort nitrogen **9**, 442
– washing **9**, 448
brewery wastewater, amounts **11A**, 211
– composition of **11A**, 434
– concentrations **11A**, 211
– flow fractions **11A**, 210
– treatment **11A**, 211
brewing **9**, 419ff
– contamination of cultures **9**, 446ff
– – by bacteria **9**, 446f
– – by wild yeast **9**, 447
– continuous secondary fermentation **9**, 454
– fermentation by-product feeds **9**, 771f
– flavor maturation **9**, 450f
– high gravity brewing **9**, 452f
– history **9**, 421
– immobilized yeast systems **9**, 453ff
– raw materials **9**, 421ff
– – hops **9**, 423ff
– – malt **9**, 426ff
– – water **9**, 422
– use of enzymes **9**, 682ff
– yeast collection **9**, 449
– yeast pitching **9**, 448f
– yeast washing **9**, 448
brewing adjuncts **9**, 428ff, 682f
– barley **9**, 429f
– corn grits **9**, 429
– corn starch **9**, 430
– glucose syrup **9**, 431
– oat malt **9**, 432
– rice **9**, 429
– sorghum **9**, 430

– sorghum malt **9**, 432
– torrified cereals **9**, 431
– wheat malt **9**, 432
– wheat starch **9**, 430f
brewing process, scheme **9**, 422, 682
– water quality **9**, 422f
– yeast performance **9**, 440ff
BRIC (Biotechnology Regulation Inter-service Committee) **12**, 541ff, 641ff
BRIDGE **12**, 663f
brikollari composting **11C**, 141f
Brimec process **9**, 283f
BRITE **5B**, 405
brome mosaic virus (BMV) **2**, 566
– as plant transformation vector **2**, 566
bromine **8B**, 184
– content of the earth **8B**, 178
bromoperoxidases, isolated enzymes **8B**, 194
– producing cells **8B**, 194
– sequences from *Streptomyces venezuelae* **8B**, 182
– turnover number **8B**, 179
– X-ray crystal structure of **8B**, 195
5-bromouracil (BU), tautomerization **2**, 26f
bromoxynil, metabolism of **8A**, 298
brompperoxidases, of *Streptomyces aureofaciens* **8B**, 195
– X-ray crystal structure of **8B**, 195
Brookhaven Protein Database **5A**, 66, 92
broth, on-line analysis **4**, 150ff
– – immuno-techniques **4**, 172
– – spectroscopic techniques **4**, 171
– – supercritical fluid chromatography **4**, 172f
– – techniques **4**, 156ff
broth sampling system, *in situ* **4**, 154ff
– – discus-shaped body and flat membrane **4**, 154f
– – rod-shaped body and tubular membrane **4**, 155f
– integrated into a recirculation loop **4**, 152ff
– – flat membrane **4**, 152f
– – tubular membrane **4**, 153f
broth viscosity **4**, 13f
– measurement **4**, 13f
broth volume **4**, 10
brown algae **10**, 115
brown coal *see also* coal, lignite
– as growth substrate for microoranisms **10**, 163
– bioconversion of **10**, 160
– – ABCDE system **10**, 161f
– depolymerizing microorganisms **10**, 163
– solubilizing microorganisms **10**, 163
brown-rot fungi **1**, 534
BRP method **11C**, 460f
– and AOC methods, comparison of **11C**, 462
– scheme **11C**, 461
bryostatin, antitumor compound of marine origin **10**, 455

– aquaculture for production of   **10**, 449
bryostatin I, structure   **10**, 454
bryostatins, inhibition of protein kinase C   **7**, 677
Bryozoa   **10**, 445f
– cytostatic/antitumor compounds   **10**, 453
bryozoans   **10**, 446
BSC (Biotechnology Steering Committee)   **12**, 534, 543ff
BSCC (Biotechnology Science Coordinating Committee)   **12**, 570ff
BSE *see* bovine spongiform encephalopathy
BTA process   **11C**, 163f
BTEX, biodegradation of   **11B**, 315
– bioremediation of   **11B**, 427
– bioreporter systems   **11B**, 454
BTEX aromatics, biodegradability of   **11B**, 468f
– biodegradation by plants   **11B**, 378
– biomediation with the heap technology   **11B**, 326
BTR *see* biogas tower reactor
BTSF (European Biotechnology Stratetic Forum)   **12**, 330
bubble coalescence   **4**, 113f
– stirred tank reactors   **4**, 113
bubble column, axial dispersion coefficient   **4**, 366f
– back flow cells   **4**, 371
– carbon dioxide transfer   **4**, 370f
– flow models, circulation cell model   **4**, 354ff, 365f
– – model of cylindrical eddies   **4**, 356f, 366
– modelling of aerobic growth   **4**, 259ff
– models for hydrodynamics   **4**, 355ff
– – circulation cell model   **4**, 355f
– multiple circulation cells   **4**, 356
– profile of gas holdup   **4**, 357f
– recirculation patterns   **4**, 355
– shear stress   **4**, 357f
– structured models for mass transfer   **4**, 370f
– two-phase flow   **4**, 354ff, 367
bubble column aeration   **3**, 256ff
– coalescence rate   **3**, 256
– mass transfer coefficients   **3**, 256
– – column diameter effects   **3**, 257
– – effects of polymers   **3**, 257
– – medium effects   **3**, 257
– – sparger effects   **3**, 257
– sorption characteristics   **3**, 257
bubble column reactor, model for chlorination of ethylene   **4**, 375f
– process modelling   **4**, 372ff
bubble diameter   **4**, 131ff
bubble probe, conductivity probe   **4**, 128ff
– – multi-point probe   **4**, 130f
– – single-point probe   **4**, 129f
– – two-point probe   **4**, 129f
– isokinetic sampling   **4**, 132f
– optical fiber probe   **4**, 131f

– restrictions   **4**, 133
bubble recirculation   **4**, 113f
– stirred tank reactors   **4**, 113
bubble residence time   **4**, 111, 120
bubble rise   **4**, 116
– velocity   **4**, 111, 120, 130
bubble size, distribution   **4**, 111ff, 120, 131
bubble velocity   **4**, 112f
– distribution   **4**, 117, 129f, 134ff
– – by two-point bubble probes   **4**, 129f
– – measurement by ultrasound Doppler technique   **4**, 134f
– profile, within a stirred tank reactor   **4**, 112
bubble volume   **4**, 133
Budapest Treaty on the International Recognition of the Deposit of Microorganisms for the Purposes of Patent Procedure   **12**, 283, 385, 395, 429
– depository authorities   **12**, 310f
budding, in yeast   **1**, 115f
buffers, for fermentation media, bicarbonate/carbon dioxide system   **3**, 151
– – HEPES   **3**, 151
Bulgaria neritina   **10**, 446, 449
bulk-data-processing   **5B**, 124
bulking   **11A**, 73, 267
– bacteria involved in   **11A**, 94
bulk mixing properties   **4**, 114ff
– convective large-scale flows   **4**, 114
– disperson by rising bubbles   **4**, 116
– molecular diffusion   **4**, 115f
– regional dispersion mechanisms, turbulence   **4**, 114f
– viscous shear   **4**, 115f
Bunsen coefficient   **4**, 18
Burkholderia cepacia see also Pseudomonas cepacia
Burkholderia cepacia, degradation of 2,4-dichlorophenoxyacetic acid, influence of soil moisture   **11B**, 430
– trichloroethylene (TCE) degradation   **11B**, 410
Burkholderia cepacia BRI6001L, bioreporter strains   **11B**, 453
Burkholderia cepacia G4 5223-PR1, metabolization of trichloroethylene   **11B**, 450f
Buswell equation   **11A**, 36
butanediol, gas analysis in bioreactors   **4**, 60
2.3 butanediol fermentation, batch –, production rate   **6**, 282
– – productivities   **6**, 275ff
– – substrates   **6**, 275ff
– – yields   **6**, 275ff, 282
– continuous culture, productivity   **6**, 284
– effect of, acetate   **6**, 280
– – initial sugar concentration   **6**, 280
– – oxygen   **6**, 278ff
– – pH   **6**, 278
– – temperature   **6**, 277f

– fed-batch – **6**, 281ff
– modeling **6**, 281
– with cell recycle, productivity **6**, 284
– with immobilized cells, productivity **6**, 284
2.3-butandediol production **6**, 269ff
– biochemistry **6**, 271ff
– by Enterobacteriaceae **6**, 274f
– metabolic pathway **6**, 271
– microorganisms **6**, 270f
– regulation **6**, 274f
2,3-butanediol, formation in wine making **9**, 477
– isomeric forms **6**, 270f
– – formations mechanism **6**, 273
– – producing bacteria **6**, 271
– mixed acid-butanediol pathway **6**, 274
– product recovery, membrane processes **6**, 287
– – solvent extraction **6**, 287
– uses **6**, 288
butanol, bacterial formation of **6**, 231ff
– formation by clostridia **6**, 236
– – pathway **6**, 238
– – gas analysis in bioreactor **4**, 60
– physical properties **6**, 230
– production by *Clostridium acetobutylicum* **8B**, 61f
– production volumes 1910–1929 **8A**, 9
butanolides, of *Streptomyces virginiae*, structure **7**, 80
n-butanol, partition coefficients, in liquid-liquid extraction **3**, 560
butanol-producing clostridia, properties **6**, 235
butanol production **6**, 229ff
– by clostridia **6**, 236
butirosin family, structure of – **7**, 467
butirosins **7**, 434
butterfat, fatty acid composition **7**, 136
buttermilk, defects **9**, 410, 412f
– manufacture of **9**, 410, 413
– quality problems **9**, 417
butyl dinitrophenol (Dinoseb), catabolism of **11B**, 288
butyl-methyl threonine, biosynthetic pathway **7**, 581
– polyketide synthase **7**, 581
– role of transaminases **7**, 581
butyrate in syntrophic metabolism **11A**, 463f
γ-butyrolactone autoregulators, synthesis by *Streptomyces* sp. **7**, 446
γ-butyrolactones, as regulatory signals **7**, 27ff
– role in antibiotic production **7**, 78ff
– structures **7**, 80f

C

C1 compounds, metabolism **9**, 178ff
– – ribulose monophosphate pathway **9**, 179f

– – serine pathway **9**, 178f
C_1 metabolism, of methylotrophic microorganisms **1**, 266ff
C_4-dicarboxylic acids, manufacture of **8B**, 108
c7E3 Fab *see* ReoPro
C-1027 chromophore, antitumor agent, structure **7**, 666
CA *see* carbonic anhydrase
cabbage, composition of raw – **9**, 643
– for sauerkraut **9**, 639ff
– microorganisms on raw vegetables **9**, 634
– varieties **9**, 639
cabbage fermentation, controlled – **9**, 642
– natural – **9**, 641f
– process steps **9**, 639ff
CAB International **5B**, 283
cadeguomycin, antitumor activity **7**, 647
cadmium, bioprecipitation of **11B**, 365
– biosensors for depletion of **11B**, 456
– removal by phytoextraction **11B**, 389
– removal by rhizofiltration **11B**, 392f
CAD systems **5B**, 403
Caenorhabditis elegans **5B**, 148
– dispersed repeat families **5B**, 31
– ferritin **5B**, 34
– gene regulation **5B**, 32
– genome **5A**, 51
– genome project **5B**, 28ff
– introns **5B**, 30
– inverted repeats **5B**, 31
– maintenance **5B**, 31
– mitochondrial genome **5B**, 30
– of the predicted genes **5B**, 31
– physical map **5B**, 28
– possible function **5B**, 31
– postgenome era **5B**, 32f, 35
– proteome of **5B**, 34
– replication **5B**, 31
– sequencing **5B**, 29
– shotgun sequencing strategy **5B**, 29
– tandem repeats **5B**, 30, 32
caesium, removal by rhizofiltration **11B**, 392f
caffeic acid, from plant cell cultures **7**, 598ff
caffeic acid production **3**, 456
caffeine, determination by SFC **4**, 172
– in cocoa cotyledons **9**, 563
caffeoyl putrescine, from plant cell cultures, yields **7**, 600
– structure **7**, 599
CAL *see Candida antarctica* lipase
CAL-B *see Candida antarctica* B lipase
calcineurin **7**, 558f, 561f
– phosphatase activity **7**, 549f
– role in T cell activation **7**, 549
calcitonin **3**, 65, **5A**, 165
calcium gluconate, precipitation **6**, 356
calcium gluconate process, with *Aspergillus niger* **6**, 352f

– – production medium **6**, 353
– – product recovery **6**, 356
calcium hydroxide, reaction with sulfuric acid **10**, 281
calcium lactate process **6**, 298ff
– product recovery **6**, 299f
– purification procedures **6**, 299f
calculated sequence characteristics **5B**, 387
Caldariomyces fumago, haloperoxidase of **8B**, 197
– heme chloroperoxidase of **8B**, 193
caldariomycin, chlorination by a chloroperoxidases **8B**, 181
calicheamicin **5A**, 356ff
– binding to DNA **5A**, 356
– bioreduction **5A**, 357
– conjugation with monoclonal antibodies **5A**, 358
– disulfide reactions **5A**, 357f
– functionalization **5A**, 357
– structure **5A**, 356
– triggering **5A**, 357
calicheamicin β_1Br, antitumor agent, structure **7**, 666
Calin **5A**, 130
callus culture **1**, 582f
– media **1**, 582f
– sterilization **1**, 583
– texture **1**, 583
– tissue sources **1**, 583
caloporoside **7**, 518f
– inhibition of, phosphoinositol metabolism **7**, 680
– phospholipase, inhibition of – **7**, 518
– structure **7**, 520
caloric value, of bread **9**, 246
calorimeter, external-flow microcalorimeter **4**, 189
– heat-flux **4**, 189f
– twin-type **4**, 188
Calorimetric methods, for biomass determination **4**, 188ff
Calorimetry **4**, 188ff
– continuous **4**, 189
– dynamic **4**, 189
calorimetry of protein complexes **5A**, 90f
calphostin, antitumor activity **7**, 679
Calvin cycle **1**, 84, 276, **11A**, 38
calyculin, inhibition of protein kinase C **7**, 677
Cambridge Structural Database **5A**, 92
Camembert cheese **9**, 377
cAMP *see* cyclic adenosine 3′,5′-monophosphate
– in yeast **1**, 487f
camphor, bioconversion of **10**, 406ff
– catabolism by *Pseudomonas putida* **8A**, 544
– pathway for the degradation of **8A**, 543f
Camptotheca acuminata **8A**, 330

camptothecin, from *Campthotheca acuminata* **8A**, 330
camptothecin-type compounds, as inhibitors of topoisomerase II **7**, 655
Canada, contaminated site management, financing schemes **11B**, 15
– – legal development **11B**, 9
– – principles **11B**, 11
– – registration **11B**, 17
– – risk assessment **11B**, 19
– – technical development **11B**, 9
– identification of contaminated sites **11B**, 17
– regulatory concerns, action thresholds **11B**, 21
– – conflict management **11B**, 28
– – enforcement **11B**, 26
– – in soil contamination **11B**, 7ff
– – liability rules **11B**, 13
– – organization **11B**, 26
– – priority setting **11B**, 23
– – remediation targets **11B**, 24
– – verification of clean-up **11B**, 24
Canadian Bioinformations Resource **5B**, 229
cancer **5B**, 71
– – *see also* tumor
– chemotherapy **5A**, 321
– mAb products **5A**, 222
– monoclonal antibody therapy **5A**, 321
– multiple forms of **5B**, 88
– sporadic cases **5B**, 88
– treatment with fusion proteins **5A**, 169
– use of recombinant enzymes **5A**, 168
cancer cells *see also* tumor cells
– cell cycle **7**, 644
cancer genes *see* oncogenes
cancer vaccines **5A**, 135, 138
candicidin production, in *Streptomyces griseus* **7**, 71f
– – growth-phase dependency **7**, 71
– – phosphate repression **7**, 74
Candida **1**, 269
– in pozol fermentation **9**, 549
– in waries production **9**, 540
Candida albicans **1**, 474
– DNA microarrays **5B**, 229
– genome **5B**, 229
Candida antarctica B lipase (CAL-B) **8A**, 42f
– enantioselective reactions of carboxylic acids **8A**, 103f
– enantioselectivity toward secondary alcohols **8A**, 71f
– general features **8A**, 44
– resolution of acrylic secondary alcohols **8A**, 72
Candida antarctica lipase (CAL) **8A**, 263f
– amidation by **8A**, 263
– aminolysis by **8A**, 263ff
– ammonolysis by **8A**, 264ff
Candida boidinii, ATP formation **6**, 599

Candida boidinii alcohol dehydrogenase
  (*CBADH*), applications of **8A**, 457
– characteristics of **8A**, 456f
– formate dehydrogenase **8A**, 403
Candida bombicola, sophorolipid **7**, 180
Candida cylindracea lipase (CCL) **8A**, 263f, **10**,
  402
– aminolysis by **8A**, 265
Candida diddensii, in olive fermentation **9**, 612
Candida guilliermondii, in injera production **9**,
  546
– in kanji fermentation **9**, 541
Candida krusei, in olive fermentation **9**, 607
– in sour doughs **9**, 255ff, 297
Candida mycoderma, in ogi fermentation **9**, 545
Candida parapsilosis carbonyl reductases, applications **8A**, 445
– characteristics of **8A**, 454f
Candida parasilopsis, in olive fermentation **9**, 607
Candida rugosa lipase (CRL) **8A**, 41ff
– catalytic machinery **8A**, 70
– enantioselective reactions of carboxylic acids
  8A, 104
– enantioselectivity toward secondary alcohols
  8A, 72ff
– general features **8A**, 44
– resolution of acyclic secondary alcohols **8A**, 73
– resolution of cyclic secondary alcohols **8A**, 74
Candida sp., citric acid production **6**, 331
candidate gene approach **5B**, 109, 112
Candida tropicalis **1**, 32
– in kanji fermentation **9**, 541
Candida vini, in poi production **9**, 548
cane molasses **6**, 17ff
– – *see also* molasses
– composition **6**, 18f, 138
– – mineral content **6**, 19
– – nitrogenous compunds **6**, 20f
– – organic acid content **6**, 20
– – vitamin content **6**, 23
– xanthan gum production **6**, 26
canning, of ripe olives **9**, 617
canola *see* rape seed
cantharanthine **7**, 607f
– structure **7**, 607
capillary electrophoresis **5B**, 203, 233, 256
– analyte injection **5B**, 271
– capillary coatings **5B**, 271
– equipment **5B**, 270
– proteome analysis by **5B**, 269ff
– separation buffer **5B**, 271
– separation time **5B**, 272
– size-based protein separations **5B**, 274
– two-dimensional – **5B**, 276
capillary-free solution electrophoresis **5B**, 274
capillary gels **5B**, 200, 207
capillary isoelectric focusing **5B**, 273

capillary modules, in membrane processes **11C**,
  408
capillary zone elelctrophoresis **5B**, 258ff
-caprolactam, induction of nitrile hydratase **8A**,
  289f
-caprolactone, lipase-catalyzed polymerization
  8A, 145
capsduction **2**, 61
capsid **1**, 546
– morphogenesis **1**, 556f
CAPTURE trial **5A**, 376ff
carbamatase, antibody **8B**, 444f
– medical potential **8B**, 444f
carbamate ester, hydrolysis of, use of catalytic antibodies **8B**, 461
carbamate insecticides, monitoring with biosensors **11A**, 132
carbamoylase **3**, 449, **8A**, 267ff
carbapenem antibiotic **8B**, 350f
– synthesis of **8B**, 353f
carbapenem synthesis, by *Erwinia carotovora*, regulation of – **7**, 65f
– quorum regulation of – **7**, 66
carbazoles, chlorinated – **8B**, 187
carbocation cyclizations, use of catalytic antibodies **8B**, 430f
carbohydrate analogs, synthesis of **8B**, 17f
carbohydrate chemistry, selective protection of hydroxyl groups **8A**, 118
carbohydrate hydro-lyases **8B**, 115ff
carbohydrate lyases **8B**, 115ff
carbohydrate metabolism, biochemistry **1**, 352
– energetics **1**, 351
– enzyme coding genes **1**, 480ff
– in microorganisms **1**, 81
– mutants **1**, 480ff
– of bacilli **1**, 383ff
– of clostridia **1**, 297ff
– of lactic acid bacteria **1**, 350ff
– of methylotrophs **1**, 266, 271ff
– of *Saccharomyces cerevisiae* **1**, 478ff
– – regulation **1**, 486ff
carbohydrate production modification **2**, 651
carbohydrates, anaerobic degradation of **11A**,
  30ff
– anaerobic digestion of **11C**, 19ff
– as energy source in yeast fermentation **8A**,
  366f
– estimation methods **1**, 126
– fermentation of **11A**, 36f
– – biogas amounts and composition **11A**, 36f
– kinetic glycosylation **8B**, 246f
– regioselective esterification, by substitution **8A**,
  224
– thermodynamic glycosylation **8B**, 246f
carbohydrates, transport *see* saccharide transport
carbohydrate synthesis, use of rabbit muscle aldolase **8B**, 10f

carbohyrate-based sweeteners *see* sweeteners
carbon, degradation in wastewater purification **4**, 446
– determination of **4**, 452
carbon assimilation **1**, 276
carbonates, hydrolysis of, tetrahedral intermediates **8B**, 406f
– – use of catalytic antibodies **8B**, 406f, 459f
carbonatization **10**, 282
carbon balance, in bioprocesses **3**, 372
carbon-carbon bond formation reactions **8B**, 5ff, 365ff, 382, 388, 393
carbon-carbon lyases **8B**, 44f, 47ff
carbon catabolite inactivation **1**, 489
carbon catabolite regulation **4**, 150f
– need for on-line process analysis **4**, 150
carbon catabolite repression **3**, 129
– global regulation **7**, 59
– in cephalosporin C formation **3**, 129
– in cephamycin C biosynthesis **3**, 129
– in yeasts **1**, 489
carbon compunds, decomposition of **11A**, 19ff
carbon cycle **10**, 279
carbon cyle **11B**, 91
carbon dioxide *see also* CO$_2$
– as by-product in ethanol fermentation **6**, 171f
– as wine preservative **9**, 488
– concentration **3**, 225
– – pressure effects **3**, 226
– dissociation of, as a function of pH **11A**, 538f
– dissociaton equilibrium **3**, 225
– effect on microbial growth **1**, 135
– effects on microorganisms **3**, 304
– evolution rate (*CER*) **3**, 226, 369
– – as biomass sensor **3**, 370
– – specific – **3**, 385
– formation in wine making **9**, 472ff
– – inhibitory effects **9**, 483
– hydration of, by carbonic anhydrase **8B**, 100
– impact on concrete **10**, 279
– in biogas **4**, 453ff
– influence on pH, anaerobic waste water process **4**, 464f
– in water, dissociation **3**, 369f
– – solubility **3**, 370
– mass transfer, in bioreactors **3**, 225
– mass transfer resistance **4**, 393
– metabolic effects **3**, 227
– pH effects **3**, 225
– production **3**, 332
– – anaerrobic – **3**, 304
– – kinetics **3**, 303
– – rate estimation **3**, 369
– removal, from animal cell culture **3**, 225ff
– removal in activated sludge process, mass balance **11A**, 266
– substrates for biomass production **9**, 214
– transfer rate (*CTR*) **3**, 369

– uptake **3**, 304
– ventilation dependence **3**, 225
carbon dioxide analyzer, infrared **4**, 56f, 62f
– – characteristics **4**, 63
– paramagnetic **4**, 56f
carbon dioxide balance **4**, 37
carbon dioxide evolution rate (*CER*) **4**, 545f
– measurement **4**, 235, 545f
carbon dioxide fixation, in methanogens **1**, 261
– in plant cell cultures **1**, 580
carbon distribution, humic substances **11B**, 103
carbon flow **1**, 51
– in anaerobic wastewater processes **4**, 461
carbon-halide lyases **8B**, 142f
carbonic acid, microbial deterioration by **10**, 282
carbonic anhydrase B (CAB), aggregation pathway **3**, 548
– aggregation regime **3**, 546
– protein folding, PEG effects **3**, 542
– refolding pathway **3**, 548
– refolding regime **3**, 546
carbonic anhydrase (CA) **8B**, 99f
– hydration of carbon dioxide, mechanism **8B**, 100
carbon limitation, metabolic effect **1**, 172
carbon metabolism, of bacilli **1**, 383
– of methanogens **1**, 261
carbon monoxide, corrosion by **10**, 284
carbon-nitrogen lyases **8B**, 47, 125ff
carbon-nitrogen ratio, in composting **11C**, 44f
carbon-oxygen lyases **8B**, 46, 98ff
carbon pools, in soil **11B**, 114
carbon removal, activated sludge process, plant configurations **11A**, 269
– in a CSTR **11A**, 402
– in the activated sludge process **11A**, 259ff
– organic carbon **11A**, 115ff
carbon source **1**, 80
– – *see also* substrates
– effect on microbial growth **1**, 132
– for growth media **1**, 143
– microbial utilization **1**, 80
carbon sources, in fermentation media **3**, 128
– of microorganisms **9**, 174ff, 193
carbonyl compounds, asymmetric reduction, Prelog's rule **8A**, 427f
– enantioselective chemical reductions of, reagents **8A**, 394
– in beer **9**, 443
– microbial reductions of **8A**, 396f
carbonyl reductases, enantioselectivity of **8A**, 455
– novel – **8A**, 454f
– of *Candida parapsilosis* **8A**, 454f
– of *Rhodococcus erythropolis* **8A**, 456
– substrate specifity of **8A**, 455
– substrate spectrum of **8A**, 456
carbonyl reduction, Prelog's rule **8A**, 380
carboxykinases **8B**, 84f

– listing of **8B**, 85
carboxylate siderophores **7**, 223ff
– cepabactin **7**, 224
– rhizobactin DM4 **7**, 223
– staphyloferrin A **7**, 223f
– staphyloferrin B **7**, 224
– vibrioferrin **7**, 224
carboxylesterase NP, synthesis of (*S*)-naproxen **8B**, 281
carboxyl groups, lipase-catalyzed protection in peptide synthesis **8A**, 132
carboxylic acid degradation, in clostridia **1**, 303
carboxylic acids, aromatic –, by methyl hydroxylation **8A**, 488
– carrier extraction **3**, 566ff
– co-extraction degree **3**, 571f
– – temperature effects **3**, 572ff
– – with Cyanex **3**, 923, 573
– – with Hostarex A 327 + isodecanol **3**, 572ff
– conversion of **11B**, 200
– dissociation constants **3**, 567
– extraction **3**, 560f
– – by liquid membrane techniques **3**, 582
– – dissociation constants **3**, 567
– – equilibrium constants **3**, 566f
– – esterification **3**, 586
– – *in situ* – **3**, 586
– fermentative production of **8B**, 344f
– hydroxylation of, by *Pseudomonas putida* **8B**, 363f
– lipase-catalyzed enantioselective reactions of **8A**, 103ff
– precipitation **3**, 562
– re-extraction selectivity **3**, 573
– – Cyanex 923 system **3**, 573
– – Hostarex A 327 system **3**, 573
– used in flavor compositions, sensoric description of **8B**, 344
– with a stereocenter at the α-carbon **8A**, 103ff
– – lipase resolution **8A**, 107
– with a stereocenter at the β-position, lipase resolution **8A**, 108
– with quaternary stereocenters, lipase resolution **8A**, 109
– with remote stereocenters **8A**, 109f
– – lipase resolution **8A**, 110
– with sulfur stereocenters **8A**, 109f
– – lipase resolution **8A**, 110
carboxyl-lyases **8B**, 44, 47ff
carboxypeptidase, in cocoa fermentation **9**, 570
carboxypeptidase A (CP-A) **8A**, 247f
– preferred cleavage sites **8A**, 249
carboxypeptidases, cleavage of C-terminal L-amino acid residues **8A**, 247
carboxypeptidase Y, in industrial biotransformations **3**, 461
cardboard production, microbial problems **10**, 307

cardenolide **1**, 597
cardiac glycosides **7**, 617
– production by biotransformation **7**, 622
cardiopulmonary bypass, monoclonal antibody therapy **5A**, 316
cardiovascular disorders **5A**, 129f, 315f
– mAb products **5A**, 222
– therapeutic applications of recombinant proteins **5A**, 129f
– use of ReoPro **5A**, 366ff
carene, bioconversion of **10**, 406ff
L-carnitine, biocatalytic route to **8B**, 288f
– synthesis by *PADH* **8A**, 453
Carnobacterium **1**, 335
carnocin UI49, structure **7**, 336
β-carotene, biosynthetic pathway **10**, 332f
– formation in halophilic algae **7**, 175
– in fungi **7**, 174f
– in microalgae **7**, 172ff
– production by algae **10**, 131
– production by protists **10**, 145
– structure **7**, 173
– structure of **10**, 332
β-carotene production **1**, 211
carotenoids, biosynthesis genes, gentically engineered microorganisms **10**, 333
– biosynthetic pathway **10**, 332f
– determination by SFC **4**, 172
– occurrence in microalgae **7**, 172ff
– production by algae **10**, 131
– structures **7**, 173f
carrageenans, production by algae **10**, 130
carrier extraction **3**, 565ff
carrier transport *see* membrane transport
carrot, cell wall polysaccharides **9**, 707
carrot juice, enzymatic treatment **9**, 716
– production **9**, 716
carvone, biotransformation of **10**, 395
– product of limonene bioconversion **10**, 393f
caryophyllanes, structure **7**, 506
CAS (Chinese Academy of Science) **12**, 398
casein **5A**, 209
β-casein digestion **1**, 354ff
casein micelles **9**, 365f
– aggregation of **9**, 373f
– model **9**, 366
caseins **9**, 364ff, 680f
– fractions **9**, 365
– genetic polymorphism, influence on cheese yield **9**, 366f
– – influence on milk clotting **9**, 366f
cassava **6**, 156ff
– composition **6**, 141
– in industrial ethanol production **6**, 140, 156f
– – energetics **6**, 185f
– – flow diagram **6**, 158
cassava (*Manihot esculenta*) **1**, 149
catabolic genes *see* degratative genes

catabolism **1**, 50
catabolite activator protein (cap) **2**, 250f
catabolite inhibition, definition of **4**, 150
catabolite regulation, in microbial drug synthesis
  7, 11
catabolite repression **1**, 386, **7**, 59
– carbon catabolites in yeast **1**, 489
– definition of **4**, 150
catalase **5A**, 209
catalases, of *Aspergillus niger* **6**, 349
catalysis *see* enzyme catalysis
– free energy of activation **8B**, 406
catalytic activity, of lipases, increase in organic
  solvents **8A**, 51ff
– of yeast **8A**, 366
catalytic antibodies **8A**, 269f, **8B**, 403ff
– – *see* abzymes
– – *see also* abzymes
– acyl-transfer reactions **8B**, 412f, 474f
– – hapten design **8B**, 412f
– aldol reactions **8B**, 472f
– animation reactions **8B**, 476
– β-elimination reactions, hapten design **8B**, 412f
– bimolecular associative and substitution processes **8B**, 472ff
– carbamatase **8B**, 444f
– carbocation cyclizations **8B**, 430f
– cationic cyclization **8B**, 471f
– – of polyenes **8B**, 430f
– changing regio- and stereochemistry of reactions **8B**, 426ff
– cleavage of, carbamate prodrugs **8B**, 444f
– – glycosides **8B**, 432ff
– – phosphate esters **8B**, 435
– cleavage of, acetals **8B**, 432ff
– conjugate addition **8B**, 477
– conversion chorismic acid into prephenic acid
  8B, 416
– cycloreversions **8B**, 467
– decarboxylations **8B**, 466f
– detoxification by **8B**, 441f
– "Diels-Alderase" **8B**, 417ff
– Diels-Alder cycloadditions **8B**, 418, 426ff, 473
– – regioselectivity **8B**, 426f
– – stereoselectivity **8B**, 426
– disfavored regio- and stereoselectivity **8B**,
  427ff
– eliminations **8B**, 465f
– epoxide opening **8B**, 471
– esterolytic –, hapten design **8B**, 414
– formation of isomeric decalins **8B**, 431
– generation of, biochemistry **8B**, 409f
– – chemistry **8B**, 407ff
– – hapten for phosphodiester cleavage **8B**, 436f
– – immunonology **8B**, 409
– – methods for **8B**, 407ff
– – molecular biology **8B**, 410
– – suicide substrate **8B**, 434f

– hapten design **8B**, 410ff
– – bait and switch **8B**, 412ff
– – control of translational entropy **8B**, 417ff
– – entropic trapping **8B**, 415ff
– hydrolysis of, acetal **8B**, 465
– – aliphatic esters **8B**, 451ff
– – amides **8B**, 438f, 461f
– – aryl esters **8B**, 406f, 455ff
– – carbamate esters **8B**, 461
– – carbonates **8B**, 406f, 459f
– – cocaine **8B**, 441f
– – enol ether **8B**, 465
– – epoxide **8B**, 465
– – ether **8B**, 464
– – glycoside **8B**, 464
– – ketal **8B**, 465
– – phosphate esters **8B**, 463f
– – phosphotriesters **8B**, 437
– – trifluoramphenicol **8B**, 439f
– hyrolytic and dissociative processes **8B**, 451ff
– induction of catalysys, by reactive immunization **8B**, 439ff
– intramolecular processes **8B**, 468ff
– isomerizations **8B**, 468f
– Kemp decarboxylation **8B**, 419f
– ketalization in water **8B**, 465
– medical applications **8B**, 441ff
– nucleophilic substitution **8B**, 477
– oxidations **8B**, 477f
– peptide bond formation **8B**, 418f
– peroxidase-like – **8B**, 421
– porphyrin metalation **8B**, 477
– prodrug activation **8B**, 442ff
– rearrangements **8B**, 469f
– redox reactions **8B**, 477ff
– reductions **8B**, 478f
– resolution of diastereoisomers **8B**, 432f
– retro-aldol reactions **8B**, 468
– retro-Henry reactions **8B**, 468
– ring closure reaction, reversal of kinetiv control **8B**, 427ff
– side chains, interactions with transition state
  analog **8B**, 417
– substrate desolvation **8B**, 419f
– supplementation of chemical functionality **8B**,
  420ff
– *syn*-elimination, of β-fluoroketones **8B**, 429
– – of selenoxide **8B**, 419ff
catalytic mechanisms of enzymes **5A**, 25f
catechol, anaerobic degradation **6**, 690
– anaerobic degradation of **11B**, 179f
– central intermediates, in xenobiotic degradation **11B**, 153
– depletion in soil **11B**, 115
– *o*-ring cleavage, by microfungi **11B**, 159
catecholate siderophores, agrobactin, of *Agrobacterium tumefaciens* **7**, 204
– aminochelin **7**, 203

– anguibactin **7**, 206f
– azotochelin **7**, 203
– chrysobactin **7**, 202f
– desferrithiocin **7**, 207
– enterobactin **7**, 201f
– ferrosamin A **7**, 208
– fluvibactin **7**, 205f
– myxochelin A **7**, 203
– of *Azotobacter vinelandii* **7**, 203
– parabactin, of *Micrococcus denitrificans* **7**, 204f
– protochelin **7**, 203
– pyochelin **7**, 207f
– siderochelin A **7**, 208
– vibriobactin **7**, 205f
– vulnibactin **7**, 205f
– yersiniabactin **7**, 207f
catechol degradation, by pseudomonads **1**, 414
catechol metabolism, bacterial – **6**, 689
catechols, formation of **8A**, 509f
– metabolism of **11C**, 282
catenates **5A**, 446
Catenhusen Commission **12**, 582ff
Catharanthus roseus, accumulation of monoterpene indole alkaloids **7**, 607ff
– – vinblastine **7**, 614f
– – vincristine **7**, 614
– biosynthetic pathways for vindoline formation **8A**, 354f
– callus cultures **7**, 607
– indole alkaloids **7**, 665
– serpentine production **1**, 596
– strictosidine synthase **8A**, 352
– suspension cultures **7**, 608
cationic cyclization, use of catalytic antibodies **8B**, 471f
cauliflower mosaic virus (CaMV) **2**, 564, 570
– codet proteins **2**, 570
– DNA uptake capacity **2**, 571
– genome organization **2**, 570f
– RNA promoter **2**, 570
– transferred foreign genes **2**, 571
– vector design **2**, 570
caulimovirus **2**, 564, 570f
CBADH see Candida boidinii alcohol dehydrogenase
CBE *see* cocoa butter equivalent
CBO (Congressional Budget Office) **12**, 24
C–C bound formation, by yeast sterol cyclase **8A**, 378
C=C double bond reduction, by yeast, mechanism **8A**, 372
– prochiral –, by yeasts **8A**, 370f
– selective –, by yeasts **8A**, 377
C=C double bonds, asymmetric reduction by yeast enzymes **8A**, 375ff
– non-aromatic –, dihydroxylation of **8A**, 511
– preparation of chiral compounds by yeast reduction **8A**, 376

CCGM (Committee in Charge of the Control over Genetic Manipulation) **12**, 598ff
CCL *see Candida cylindracea* lipase
CC-1065 **5A**, 321
cDNA hybridization, application **2**, 399
cDNA sequences **5A**, 50f
– of human genes **12**, 142f, 291, 305
cDNA synthesis **2**, 390
CDP-ascarylose **8B**, 118ff
– biosynthesis of, by *Yersinia pseudotuberculosis* **8B**, 118
CDP-choline, biosynthetic pathway **6**, 603
CDP571, cross-reacting antibodies **5A**, 351
– elimination profiles **5A**, 348, 351
– generation of **5A**, 345
– immunogenicity **5A**, 345, 348
– in chronic therapy **5A**, 343ff
– in rheumatoid arthritis **5A**, 347ff
– – immunogenicity **5A**, 348ff
– – pharmacokinetics **5A**, 348ff
– – repeated-dose study **5A**, 347f
– – results of treatment **5A**, 349f
– preclinical studies, in non-human primates **5A**, 345f
– – in patients **5A**, 346ff
CDR *see* complementary determining regions
CD4 receptor **5A**, 472ff
CE *see* cholesterol esterase
CEBC (China-EC Biotechnology Center) **12**, 398
cedrol, bioconversion of **10**, 410
– preferred sites of oxygen attack on **10**, 410
CEFIC (Council of the European Chemical Industry) **12**, 331, 536f, 634
cell adhesion **3**, 107
cell aggregates **3**, 107, 109
– shear sensitivity **3**, 230
cell banking, in manufacturing of antibodies **5A**, 257f
– – documentation **5A**, 258
– – function of **5A**, 258
– – good manufacturing practice **5A**, 257
– – origin of **5A**, 257
– – preparation of **5A**, 257f
– – storage **5A**, 258
– principles **10**, 30f
– procedure **10**, 31
cell banking systems, regulations **5A**, 525
cell-based assays, testing of antibodies **5A**, 291
cell breakage **4**, 609f
cell-cell communication, in flocs and biofilms, quorum sensing **11A**, 78f
cell characterization **4**, 195ff
cell compartment **1**, 30f
cell component labeling **1**, 11
– nucleic acid preparation **1**, 11
– protein preparation **1**, 11
cell components, determination of **4**, 195ff
– – ATP **4**, 211

– – cellular protein content **4**, 205
– – DNA **4**, 204
– – intracellular lipid content **4**, 212
– – NADH **4**, 208ff
– – NADPH **4**, 208ff
– – RNA **4**, 204
cell composition **9**, 193
cell composting **11C**, 140
cell concentration *see also* Biomass concentration
cell counts, Coulter counter technique **4**, 200
cell culture **1**, 581, **3**, 23ff, **5A**, 539f
– – *see also* animal cell culture, mammalian cell
  culture
– – *see also* plant cell culture
– aeration *see* bioreactor aeration
– aggregate – **3**, 109
– batch – **3**, 113
– bioreactors **3**, 105ff
– cell retention **3**, 113f
– cell size **3**, 474
– continuous – **3**, 113
– eukaryotic cells **1**, 40f
– mechanical stress, impeller effects **3**, 407
– media *see* fermentation media, growth media
– microcarrier – **3**, 108
– microscale – **3**, 107
– mixing *see* bioreactor mixing
– nutrients **3**, 145f
– oxygen supply *see* bioreactor aeration
– perfusion – **3**, 113
– population modeling **3**, 87
– requirements **1**, 41
– semi-batch – **3**, 113
– substrates **1**, 195ff
– substrate uptake **3**, 365ff
– viscosity **3**, 235f
cell cultures *see also* animal cell cultures, plant cell
  cultures
– – *see* plant cell cultures
– authentication **10**, 33ff
– contamination by **10**, 34f
– – bacteria **10**, 34
– – fungi **10**, 34
– – mycoplasmas **10**, 34
– continuous cell line **10**, 29
– cryopreservation **10**, 31f
– – cell damage **10**, 31f
– – freezing rates **10**, 32
– – programmable controlled-rate freezer **10**, 32
– cryoprotectants **10**, 32
– culture collections **10**, 36f
– DNA fingerprinting **10**, 35
– effects of microgravity on **10**, 509ff
– finite cell line **10**, 28
– general banking principles **10**, 29ff
– general practice **10**, 29ff
– in proteomics **5B**, 245
– isoenzyme analysis of **10**, 35

– karyotype analysis of **10**, 35
– maintenance of eukaryotic – **10**, 27ff
– Master Cell Bank **10**, 30
– of macroalgae **10**, 448
– primary **10**, 28
– quality control **10**, 33ff
– – test for adventitious agents **10**, 33f
– regulatory considerations **10**, 35f
– safety aspects **10**, 35f
– serum-free media **10**, 29
– terminology **10**, 28f
– use of antibiotics **10**, 33f
– use of serum in **10**, 29f
– Working Cel Bank **10**, 30
cell culture safety **2**, 844
cell damage, agitation-induced – **3**, 228ff
– bubble induced – **3**, 231ff
– – bubble coalescence effects **3**, 233
– – bubble diameter effects **3**, 232f
– – in microcarrier cultures **3**, 231
– – in surface aerated bioreactors **3**, 233
– – liquid entrainment rate **3**, 232
– – mechanisms **3**, 232
– – Pluronic F68 effects **3**, 234
– – sparger location **3**, 233
– – surfactant effects **3**, 233f
cell death rate, effects of, interparticle collision **3**,
  229
– – shear stress **3**, 229
cell death ratem, bubble-induced – **3**, 232
cell debris, separation, comparison of methods **4**,
  611
– solid-liquid separation **4**, 609
cell debris removal *see* debris removal
cell differentiation, effect in plant cell culture **1**,
  600
cell disintegration *see* cell disruption
cell disruption **3**, 505ff
– analysis **3**, 506f
– automation **3**, 520ff
– by chemical treatment **3**, 515ff
– containment **3**, 520ff
– control **3**, 522
– efficiency **3**, 510
– enzymatic cell lysis **3**, 517f
– *Escherichia coli* **3**, 483f
– fragment size, analysis **3**, 507
– – distribution **3**, 507
– homogenizing pressure **3**, 509
– in high-pressure homogenizers **3**, 508ff
– in high-speed agitator bead mills **3**, 510ff
– kinetics **3**, 508
– – continuous disintegration **3**, 513
– mechanical methods **3**, 508ff, 518, 523
– non-mechanical methods **3**, 515ff
– quantitation **3**, 506
– – protein assays **3**, 507
– safety **3**, 520ff

– sedimentation analysis **3**, 506
– with alkali **3**, 517
– with detergents **3**, 516
– with guanidine-HCl **3**, 516f
– with organic solvents **3**, 516
– with toluene **3**, 516
– with Triton X-100 **3**, 516
– with urea **3**, 517
– with water-miscible solvents **3**, 516
cell disruptionn with chaotropic agents **3**, 517
cell division, phases **7**, 644
cell encapsulation **3**, 110
cell energy status, measurement by NMR **1**, 193
cell entrapment **3**, 110
– in calcium alginate **3**, 110
cell envelope, of microorganisms **3**, 506
cell flocculation **1**, 129
cell flotation **1**, 129
cell fusion **2**, 95f, 156
– chemically induced **2**, 93ff
– – Ca^{2-} effect **2**, 97
– laser-induced **2**, 101
– microcell fusion **2**, 156
– of mammalian cells **2**, 128ff
– – application **2**, 130
– – gene expression in products **2**, 130
– – intraspecific somatic hybrid **2**, 129
– – nuclear hybrid **2**, 129f
– – nucleus fusion **2**, 129
– – reconstituted cell **2**, 129
– somatic hybridization **2**, 131
– under microgravity conditions **10**, 515f
– virus-induced **2**, 96
cell growth, influence of stress on **10**, 574
cell immobilization **1**, 223ff, 598, **3**, 584
– – *see also* immobilized microorganisms
– by adsorption **1**, 224ff
– – materials **1**, 224
– by aggregate formation **1**, 225
– by covalent coupling **1**, 225
– – reaction mechanism **1**, 225
– by microorganism entrapment **1**, 225
– *Comamonas acidovorans* **1**, 129
– for process improvement **1**, 226
– in alginate beads **1**, 598
– methods **1**, 224ff
– plant cells **1**, 598
– suitable polymers **1**, 225
cell killing **5B**, 119
cell line **3**, 25ff
– Chinese hamster ovary (CHO) **3**, 27ff
– continuous –, VERO – **3**, 27
– development **3**, 30
– growth requirements **3**, 142
– immortal – **3**, 68
– mammalian –, anchorage-independent **3**, 25
– murine-human hybridoma **3**, 27
– murine-murine hybridoma **3**, 27

– murine myeloma **3**, 27ff
– productivity **3**, 30
– virus infection **3**, 31
cell lines, continuous – **10**, 29
– DNA fingerprinting **10**, 35
– finite – **10**, 28
– in therapeutic antibody manufacture **5A**, 248ff, 258f
– – stability parameters **5A**, 259
– – use of recombinant DNA technology **5A**, 259
– isoenzyme analysis of **10**, 35
– karyotyping analysis of **10**, 35
– quality control **10**, 33ff
– source materials **10**, 29
cell lysing **5B**, 244
cell lysis *see* cell disruption
cell mass estimation **1**, 121ff
– calorimetric estimation **1**, 126
– capacitance **1**, 127
– conductivity **1**, 127
– dry cell weight **1**, 121f
– elemental balance **1**, 124
– image analysis **1**, 123
– kinetic model **1**, 126
– mass balance **1**, 124
– model-based estimation **1**, 124
– off-line **1**, 122
– on-line **1**, 122
– oxygen consumption **1**, 125
– packed cell volume **1**, 123
– real-life measurement **1**, 122
– turbidity **1**, 123
– viscosity **1**, 126
cell membranes, chemical permeabilization **3**, 515ff
cell metabolism *see* metabolism
cell microcarriers *see* microcarriers
cell models **4**, 269ff
– baker's yeast process **4**, 394ff
– cybernetic **4**, 290ff
– – diauxic growth of *Klebsiella terrigena* **4**, 288
– growth kinetics **4**, 274f
– inhibition kinetics **4**, 275f
– – for a single inhibitor **4**, 276
– metabolic regulator model **4**, 292ff
– model equations **4**, 271ff
– structured **4**, 284ff
– – definition **4**, 271, 284
– – diauxic growth of *Klebsiella terrigena* **4**, 286ff
– – general formulation **4**, 284
– – secondary metabolite formation **4**, 288ff
– – thee-compartment model **4**, 288ff
– – two-compartment model, for diauxic growth **4**, 285ff
– substrate uptake kinetics **4**, 274f
– unstructured **4**, 270f, 273ff
– – alternative substrates **4**, 278

– – black-box models **4**, 282f
– – definition **4**, 270
– – diauxic growth of *Klebsiella terrigena* **4**, 279
– – essential substrates **4**, 277
– – interacting **4**, 277f
– – multi-substrate kinetics **4**, 279f
– – non-interacting **4**, 277f
– – primary metabolite formation **4**, 283
– – temperature effects **4**, 282
cell morphology, determination of **4**, 200f
cell number estimation **1**, 119ff
– Coulter counter **1**, 121
– direct microscopic count **1**, 119
– flow cytometer **1**, 121
– membrane filtration **1**, 120
– viable plate count **1**, 120
cellobiase **9**, 710
– – *see* β-glucosidase
cellobiohydrolases **5A**, 203ff, **9**, 710
cellobiose, determination with biosensors **4**, 78
– hydrolysis of **11A**, 26
cell organelle **1**, 31
cell pellet **1**, 116
– physical structure **1**, 117f
cell population **3**, 380
– – *see also* biomass
– in bioprocesses, recording **3**, 360
cell property distribution, modeling **3**, 88
cell recovery **3**, 8
cell retention, methods **3**, 305
cell rupture **9**, 91ff
– by chemical treatment **9**, 92
– by lysate clarification **9**, 92ff
– by pressure **9**, 92
– by sonication **9**, 91
– in bead mills **9**, 91f
cells, construction of biosensors **4**, 78f
cell separation **4**, 606ff
– by centrifugation, breakthrough **3**, 485
– – throughput **3**, 487
– centrifugation **4**, 608
– costs **3**, 501
– cross-flow membrane filtration **4**, 608
– efficiency **3**, 483
– from fermentation broth **3**, 469ff
– – *Escherichia coli* cells **3**, 482
– hollow-fiber filtration **4**, 608
– model system **3**, 482ff
cell-size determination **4**, 200f
– Coulter counter technique **4**, 200
– flow cytometry **4**, 201
– nephelometric methods **4**, 201
cell sorting **5B**, 246
cell spheroids **3**, 107f
– effects of, discharging **3**, 483
– – particle size distribution **3**, 483
– formation **3**, 109
– loss of solids **3**, 483

– model system **3**, 482ff
– of fermentation products **3**, 42
– – bioprocessing aids **3**, 43
– – flocculating agents **3**, 43
– of polydisperse systems **3**, 477
– process variables **3**, 478
– properties **3**, 109
– separation efficiency **3**, 483
– separation enhancement **3**, 486
– solids breakthrough **3**, 485
– three-dimensional structure **3**, 109
cell structure **1**, 5ff
cell surface, de-emulsifying activity **6**, 702
cell surface layer, bacterial **1**, 22
– in plants **1**, 34
cell suspension, ^{31}P NMR spectrum **1**, 192f
cell suspensions **3**, 107
cell therapy, microencapsulation-based – **10**, 547ff
– permanent cell lines **10**, 566
– xenogeneic donor cells **10**, 565f
– – porcine islets **10**, 565
cell-to-cell communication, in biofilms **11C**, 450
cell topology **1**, 9
cellular assays, testing of antibodies, relative potencies **5A**, 293
cellular circuitry **5B**, 403
cellular growth regulation, signal transduction pathway **7**, 644
cellulase **1**, 28
– carboxymethylcellulase **1**, 28
– *Clostridium thermocellum* **1**, 28
– – action mechanism **1**, 29
cellulases **6**, 149f, **9**, 710ff, **10**, 80ff, **11A**, 26
– activity **5A**, 203f
– applications of **10**, 82
– classification **6**, 149
– classification of **10**, 526f
– cloning **5A**, 204ff
– effects **5A**, 204
– from hyperthermophiles **10**, 83
– from thermophiles **10**, 83
– industrial applications **5A**, 202ff
– in silage **9**, 778
– in textile industry **5A**, 205f
– microbial – **5A**, 202ff
– properties **5A**, 202f
– recombinant **5A**, 204
– – in detergent industry **5A**, 204
– – in textile industry **5A**, 204f
– structure of **10**, 527
– thermostable – **10**, 81ff
cellulolytic organisms **11A**, 25
cellulose **6**, 620, **11A**, 25ff
– acid hydrolysis **6**, 147f, 160ff
– – dilute sulfuric acid process **6**, 147f, 160ff
– – kinetics **6**, 148
– – strong acid hydrolysis process **6**, 163

– action of enzymes on **10**, 532
– anaerobic degradation of **11A**, 30f
– bacterial –, applications **6**, 634
– comparsion of enzymatic and acid hydrolysis **6**, 150
– degradation of **10**, 285
– – by actinomycetes **11C**, 69
– enzymatic hydrolysis **6**, 149f
– – mechanism **6**, 149
– hydrolysis of **11A**, 25f
– in wood **10**, 526
– microbial metabolism **9**, 185
– models **6**, 144
– organization of **10**, 531
cellulose degradation, by clostridia **1**, 297
cellulose hydrolysis, application of EKF **4**, 232ff
– β-glucosidase activity, estimation by EKF **4**, 234
– cellobiose concentrations, estimation by EKF **4**, 234
– glucose concentrations, estimation by EKF **4**, 234
cellulose hydrolyzing enzymes **10**, 80ff
cellulosome **1**, 297
– function **1**, 29
– structure **1**, 29
cellulosomes **5A**, 203, **11A**, 26
cell viability, determination of **4**, 201f
– – flow cytometry **4**, 202
– – sample-flow analysis **4**, 202
cell wall **1**, 578, **2**, 109f
– bacterial **1**, 19ff
– cellulose synthase complex **1**, 35
– effects of microgravity on **10**, 510f
– lignin **1**, 36
– lytic enzymes **2**, 109f
– murein **1**, 19f
– of archaebacteria **1**, 21
– of bacterial L-forms **1**, 20
– of halobacteria **1**, 22
– of methanogens **1**, 21, 257
– of mycoplasms **1**, 20
– of plants **1**, 35, 578
– structure in fungi **2**, 109f
– synthesis in bacteria **1**, 113
cell wall degradation **9**, 710ff
cell yield coefficient, of fermentation broths **3**, 131
CEN (Comité Européen de Normalisation) **12**, 640f
Center for Biologics Evaluation and Research (CBER) **5A**, 499ff
– clinical studies **5A**, 504f
– investigational new drugs **5A**, 504f
– plasmid DNA, regulations **5A**, 459
– plasmid DNA therapeutics, guidelines **5A**, 445
– structure of **5A**, 500

Center for Drugs Evaluation and Research (CDER) **5A**, 499ff
– review process **5A**, 501
centrifugal filtration **1**, 191
centrifugal separation, modelling **4**, 612
centrifugation **3**, 469ff, 475ff
– bioproduct separation **4**, 609
– cell separation **4**, 608
– costs **3**, 501
– large-scale **4**, 612
– sedimentation coefficient **4**, 612
centrifuges **3**, 474ff
– area equivalents **3**, 477
– as closed systems **3**, 475
– cylinder bowl – **3**, 481
– decanter – **3**, 481f
– disc stack – *see* disc stack centrifuges
– for continuous processing **3**, 474
– grade efficiency curve **3**, 478
– materials **3**, 475
– nozzle – **3**, 481f
– retention time **3**, 475
– scale-up **3**, 480
– selection, for downstream processing **3**, 479ff
– self-ejection – **3**, 482
– separation efficiency **3**, 475
– Sharples decanter – **3**, 482
– sigma values **3**, 477
– size determination **3**, 480
– solids-ejecting – **3**, 481
– solids-retaining – **3**, 480f
– space requirement **3**, 475
– throughput **3**, 477, 480, 483
– types **3**, 480ff
– – applications **3**, 480ff, 486
– – handled sludge concentrations **3**, 479
– Westphalia's extraction – **3**, 482
cepabactin, carboxylate-type siderophore of *Pseudomonas cepacia* **7**, 224f
– structure **7**, 224
cephaloridine, effect in mouse septicemia **7**, 380
cephalosporin biosynthesis **7**, 251f
– genes **7**, 251
– – cloning of – **7**, 259
– – clustering of – **7**, 259ff
– pathway **7**, 252
cephalosporin C, production **1**, 214
– synthesis, biosynthetic pathway **1**, 528
cephalosporin C, two-step enzymatic side chain removal **8A**, 16
cephalosporin derivatives, conversion from penicillins **8A**, 16f
cephalosporin production **4**, 150f, 157
cephalosporins, semi-synthetic – **7**, 264
cephalosporin synthesis **8A**, 258ff
Cephalosporium **1**, 525ff
– in ogi fermentation **9**, 545
Cephalosporium acremonium **4**, 157f, **7**, 249ff

– cephalosporin production **1**, 214, 527
– gene expression, regulation of – **7**, 267
– on-line analysis of cultivation broth **4**, 157f
– penicillin biosynthetic enzymes **7**, 254ff
– use of recombinant DNA technology, in β-lactam biosynthesis **7**, 264f
Cephalosproium acremonium **2**, 538
– recombinant recovery **2**, 118
– strain improvement **2**, 118
cephamycin biosynthesis **7**, 252f
– genes **7**, 252
– – cloning of – **7**, 259
– – clustering of – **7**, 259ff
cephamycin C biosynthesis, chemically defined,
  medium **3**, 131
ceramic-matrix bioreactor **3**, 118, 121
– single-pass operation **3**, 118
ceramides, use in skin care formulations **7**, 184
cereals, grain fat **9**, 246f
– grain protein, nutritional value **9**, 245f
– production of **9**, 243
cerebral ischemia, medical applications of recombinant proteins **5A**, 130f
cerebrovascular thromboembolic disorders **5A**,
  129f
– therapeutic uses of recombinant proteins **5A**,
  129
cervical cancer, vaccines **5A**, 138
cetyltrimethylammonium bromide (CTAB), as reversed micellar medium **3**, 606
CFADH see Curvularia falcata alcohol dehydrogenase
CFF *see* crossflow filtration
CFSTR *see* continuous flow stirred tank reactor
  chemical oxygen demand (COD), COD removal, by the sequencing batch reactor
– in wastewater **11A**, 259ff
CFU *see* colony forming units
CGIAR (Consultative Group on International
  Agricultural Research) **12**, 364f, 440, 626
– centers **12**, 364
CGI programs **5B**, 381
C-glycosides, synthesis of, by rabbit muscle aldolase **8B**, 13
cGMP *see* cyclic guanosine 3´,5´-monophosphate
CGSC collection **5B**, 285
Chabrol model, application to drinking water
  treatment **11C**, 473
chaebols **12**, 387
Chaetomium test **11C**, 90
chalciporon, antimicrobial activity **7**, 495
chalcone, epoxidation of, use of poly-L-alanine
  8B, 499f
chalcone biosynthesis, mechanism **10**, 353f
chalcone synthases **10**, 352ff
– three-dimensional crystal structure **10**, 353
channel composting **11C**, 140

chanoclavine production, by immobilized *Claviceps* cells **1**, 240
chaperones *see* molecular chaperones
chaperonins **3**, 537, **9**, 22f
– in *Escherichia coli* **3**, 537
Charophyceae, characteristics of **10**, 112
chartreusin, structure **7**, 649
Cheddar cheese **9**, 375f, 681f
– curd handling **9**, 376
cheddaring **9**, 376
cheese, classification **9**, 356f
– maturation **9**, 681f
– – acceleration of **9**, 681
– – glycolysis **9**, 380
– – lipolysis **9**, 380
– – proteolysis **9**, 380f
cheese classes, Cheddar types **9**, 375f
– cheeses with eyes **9**, 376f
– Italian cheeses, grana types **9**, 378
– – pasta-filata types **9**, 378
– lowfat types **9**, 379
– mold-ripened types **9**, 377
– soft unripened types **9**, 379
– surface-ripened types **9**, 377f
cheese flavor **9**, 381
cheese-making, use of lipases **8A**, 40f
cheese production **9**, 353ff, 679f
– – *see* milk products
– clotting mechanism **9**, 366
– culture propagation **9**, 371ff
– curd handling **9**, 363
– milk culturing **9**, 360f
– milk enzymes **9**, 368f
– milk lactose **9**, 368
– milk lipids **9**, 367f
– milk minerals **9**, 368
– milk pretreatment **9**, 358ff
– milk proteins **9**, 364ff
– moisture control **9**, 362
– pH control **9**, 362
– process cheese products **9**, 363f
– use of cloned calf chymosin **5A**, 209
– use of enzymes **9**, 679ff
– use of lactic acid bacteria **9**, 360f, 369ff
– use of milk-clotting enzymes **9**, 373ff
– whey expulsion **9**, 361f
cheese types **9**, 356f
cheese whey **1**, 149
chemial pulps, composition of accessible carbohydrates **10**, 531
chemical compound libraries **10**, 52ff
– formats **10**, 53f
– use in drug discovery **10**, 52ff
– – combinatorial libraries **10**, 52ff
– – sreening libraries **10**, 55ff
chemical graph **5B**, 402
Chemical Industry Association of Germany *see*
  VCI

chemical intermediates, enantiomerically pure –, use of lipases **8A**, 146
chemical mutagenesis **5B**, 151
chemical oxygen demand (COD) **4**, 452f
– biomass yield **4**, 411
– measurement of **4**, 452
chemical reaction energetics **1**, 67
– transition state **1**, 66
chemicals control **12**, 623ff
– chronology **12**, 623f
chemical softening decarbonization, in drinking water treatment **11C**, 402ff
– process scheme **11C**, 403
chemiluminescence **5B**, 186
chemiosmotic theory **1**, 63
chemokines **5A**, 165f
– classes **5A**, 166
– function **5A**, 166
– in therapy **5A**, 165f
– receptors **5A**, 166
chemolithotrophs **3**, 128
chemoorganotrophs **3**, 128
chemoprevention, of colon cancer **5B**, 88
chemostat **1**, 156, **4**, 277f, 281
– – *see* bioreactor, continuous culture
– dilution rate **1**, 156
– glycerol-limited culture of *Klebsiella aerogenes* **4**, 281
– simulation of a glycerol- and oxygen-limited culture **4**, 272f
chemostat culture **9**, 203
– pulse and shift method **9**, 200f
chemostat pulse method **1**, 146
chemotaxis, in biofilm formation **10**, 291f
chemotaxonomical markers **1**, 404ff
– of euactinomycetic genera **1**, 435
chemotaxonomy *see* taxonomy
chemotherapy, antibody-targeted – **5A**, 355ff
chemotrophs **3**, 128
chemotrophy **1**, 50
– definition of **10**, 275
chicha, production of **9**, 548
ChickBASE **5B**, 285
chicken liver esterase **8A**, 211f
chillproofing, of beer **9**, 450
– – enzymes **9**, 685
chimera **2**, 761
– appearance frequency **2**, 762, 764f
– characterization **2**, 762
– Ec cell/embryo **2**, 761f
– – yield **2**, 764ff
– germ-line **2**, 762ff
– of farm animals **2**, 762
– production **2**, 761
– sex **2**, 762, 764f
– transgenic chicken **2**, 811
chimeric antibodies *see* antibodies
chimeric gene *see* heterologous gene

chimerism **5B**, 75
China, bio-products **12**, 397
– biotechnology, companies **12**, 403
– – history **12**, 397
– biotechnology regulatory policy **12**, 399ff, 605
– High-Tech Science Parks **12**, 399, 402
– joint venture projects **12**, 402
– National State Key Laboratories **12**, 399, 401
– public opinion survey, on biotechnology **12**, 127
– Science Economic Zones **12**, 399, 402
– structure of S&T **12**, 400
Chinese hamster ovary cell line (CHO) **5A**, 248f, 251f
– in manufacturing of antibodies, serum-free media **5A**, 257
Chinese Hamster Ovary (CHO) cells, as protein production host **3**, 530
– carbon dioxide sensitivity **3**, 227
Chinese Patent Office **12**, 399
chip technology **5B**, 24
chiral alcohols, formation by microbial hydroxylations **8A**, 486
chiral compounds, preparation by yeast reduction of activated C=C bonds **8A**, 375f
chirality **5B**, 402
chiral technology, biotransformations in **8A**, 26
chitin, biosorption of metals **10**, 229
– degradation of **10**, 84f
– degrading enzymes **10**, 84f
– from crustaceae **10**, 467
chitinase, from hyperthermophiles **10**, 83
– from thermophiles **10**, 83
chitinases **10**, 84ff
– thermostable – **10**, 85
chitin phosphate **10**, 236
chitooligosaccharides, synthesis by transglycosylation **8B**, 251
Chlamydomucor **1**, 522
Chlamydomucor oryzae, in lao-chao fermentation **9**, 525
– in tapé fermentation **9**, 529
Chlorella, in health food markets **10**, 129
– in human diet **10**, 126f
– mass cultivation of **10**, 144
Chlorella minutissima, formation of eicosapentaenoic acid **7**, 169
chlorides in municipal wastewater **11A**, 178f
chlorinated aromatics, from basidiomycetes **8B**, 176
chlorinated compounds *see also* halogenated compounds
– biodegradation of **11B**, 241ff
– – aliphatic – **11B**, 251ff
– – use of bioaugmentation **11B**, 434
– bioreporter systems for **11B**, 453f
– degradation by genetically engineered microorganisms **11B**, 445ff

chlorinated hydrocarbons *see also* halogenated hydrocarbons
– biodegradation of **11B**, 245ff
– halorespiration **11B**, 246f
– redox potential **11B**, 246
– reductive dehalogenation **11B**, 245ff
– – principles of **11B**, 245f
chlorinated phenols, degradation mechanism **11B**, 155
chlorinated solvents, bioremediation of **11B**, 407f
chlorination **11C**, 464f
– effect on BOM **11C**, 464f
– of drinking water **11C**, 402, 410
chlorination of municipal wastewater **11A**, 186
chlorine **8B**, 184
– content of the earth **8B**, 178
chloroalkanes, long-chain –, oxidative dehalogenation **11B**, 255f
– redox potential **11B**, 246
chloroalkenes, redox potentials **11B**, 246
chloroamphenicol **8B**, 178ff
chloroanilines, depletion in soil **11B**, 115
chloro-aromatics *see* aromatics
– dehalogenation reactions **11B**, 155
chlorobenzenes, aerobic biodegradation of **11B**, 257f
– annual production **11A**, 381
– biodegradation of **11A**, 381f
– – wastewater treatment **11A**, 382
– properties **11A**, 380
– reductive dehalogenation of **11B**, 258
– use **11A**, 380
chlorobenzoate, dehalogenation mechanisms **11B**, 261f
chlorobenzoates, degradation by genetically engineered microorganisms **11B**, 445f
chlorobiphenyls, degradation by genetically engieered microorganisms **11B**, 446ff
chloromethane, anthropogenic inputs **8B**, 184
– biodegradation of **11B**, 251f
– – reaction mechanisms **11B**, 252
– output of volcanoes **8B**, 184
chloromethyl anilines, aerobic biodegradation of **11B**, 258f
chloronitrobenzenes, degradation to chloroanilines **11B**, 259
chloropeptin I, inhibition of HIV replication **7**, 125
chloropeptins **7**, 124
– structures **7**, 126
chloroperoxidase, commercial availability **8A**, 482
– hydroxylation by **8A**, 504
chloroperoxidases, cloning from human eosinophils **8B**, 201
– halide-dependent reactions **8B**, 180
– isolated enzymes **8B**, 194
– producing cells **8B**, 194

– sequenced from microbes **8B**, 182
– turnover number **8B**, 179
chlorophenols, biodegradation in, wastewater treatment **11A**, 382f
– biodegradation of **11B**, 259
– bioremediation of **11B**, 427
– degradation by immobilized cells **11B**, 418f
– depletion in soil **11B**, 115
– properties **11A**, 383
– use **11A**, 383
Chlorophyceae, characteristics of **10**, 111f
chlorophyll **1**, 59
Chlorophyta, characteristics of **10**, 111f
– systematics of **10**, 111f
chloroplast **1**, 578, **2**, 169
– genome **2**, 169ff
chloropropanols, safety concerns **9**, 702
chloroquine, transfection level enhancement **5A**, 436f
chlorotetracycline **8B**, 178f
4-chloro-2-nitrophenol degradation **1**, 229, 235
CHMO *see* cyclohexanone monooxygenase
Choanephora **1**, 522
chocolate flavor **9**, 564, 574
cholecalciferol, structure **7**, 171
cholecystokinin receptor antagonists **7**, 116
cholera, as waterborne disease, outbreaks in the 19th and 20th century **11C**, 415f
cholesterol, determination **9**, 142
– structure **7**, 173
cholesterol biosynthesis, inhibitors of –, from basidiomycetes **7**, 518f
cholesterol esterase (CE) **8A**, 214ff
– fastest reacting enantiomers of racemic secondary alcohols with **8A**, 216f
– from bovine pancreas **8A**, 214
– kinetic resolution of phosphines and phosphine oxides **8A**, 217
– required presence of bile salts **8A**, 214
cholesteroloxidase, reaction catalyzed by – **5A**, 99
choline esterase inhibitors, monitoring with biosensors **11A**, 132
chorionic gonadotropin **5A**, 164
chorismate, common aromatic pathway **8B**, 316
chorismate mutase **8B**, 415f
– deregulation of **8B**, 317ff
– transition state **8B**, 416
– transition state analog **8B**, 416
Chorleywood process **9**, 283
chromatin domain **2**, 674f
Chromatium vinosum, PAH synthase, gene structure **6**, 418
chromatographic separation, affecting factors **4**, 618
– affinity adsorption, mass transfer **4**, 617f
– diffusion model **4**, 617f
– equlibrium model **4**, 614

– ideal stage model **4**, 615f
– liquid chromatography **4**, 614
chromatography **3**, 48, 718, 727ff
– affinity – (AFF) **3**, 727
– – in protein purification **3**, 48
– – ligand leakage **3**, 49
– antibody affinity methods **9**, 107
– bioproduct separation **4**, 611
– – affinity chromatography **4**, 615
– – chromatofocusing **4**, 615
– – gel filtration **4**, 615
– – hydrophobic **4**, 615
– – ion exchange **4**, 615
– columns for **9**, 99
– displacement – (DC) **3**, 50
– – in protein purification **3**, 48
– dye ligand method **9**, 107f
– FPLC **4**, 167
– HPLC **4**, 165ff
– hydrophobic interaction – (HIC) **3**, 48, 727
– – in protein purification **3**, 48
– hydroxyapatite – **9**, 55
– IC **4**, 167ff
– in enzyme purification **9**, 52ff, 99ff
– ion exchange – **9**, 53ff, 100ff
– – optimal pH **9**, 54
– ion exchange – (IEC) **3**, 728
– – matrices **3**, 728
– – supports **3**, 729
– medium properties **3**, 48
– SFC **4**, 172
– size exclusion – (SEC) **3**, 727f
– – in protein purification **3**, 48
– – matrices **3**, 727f
chromium, as electron acceptor for anaerobic respiration **11A**, 43
– microbial reduction pathways **11B**, 366
– removal by rhizofiltration **11B**, 392f
Chromobacterium violaceum **6**, 440f
– biotechnological production of PHA **6**, 440
Chromobacterium viscosum lipase (CVL) **8A**, 41ff
chromosomal aberration **2**, 159
chromosome, *in situ* hybridization **2**, 760
– size in eukaryotes **2**, 155
– transgene localization **2**, 760
chromosome abnormalities **5B**, 74ff
– numerical – **5B**, 75
– structural – **5B**, 76
chromosome banding, band size in eukaryotes **2**, 155
– gene localization **2**, 156
– techniques **2**, 156
chromosome dissection **2**, 161, 751
chromosome engineering **2**, 513ff
– allele rescue **2**, 513
– extragenic suppressor **2**, 514
– gapped plasmid integration **2**, 514

– targeted integration **2**, 513
chromosome homology **2**, 171f
– sex chromosome differentiation **2**, 177f
chromosome hybridization **2**, 159
– *in situ* **2**, 161
– *in vitro* **2**, 159f
– labeling techniques **2**, 160
– metaphase chromosomes **2**, 159ff
chromosome landing **5B**, 171
chromosome map **2**, 63
– genetical **2**, 154
– physical **2**, 154
chromosome maps **5B**, 42ff
chromosome mutation **2**, 7f
– deficiency **2**, 8
– duplication **2**, 8
– inversion **2**, 8
– translocation **2**, 8
chromosome organization **2**, 673
– linker **2**, 674
– locus control region (LCR) **2**, 675
chromosomes, animalies of, under spaceflight conditions **10**, 509
chromosome segregation **2**, 116ff, 127
– fused mammalian cells **2**, 129
– in fungal hybrids **2**, 116
– in plant hybrids **2**, 127
– of parental genome **2**, 116f
chromosome sorting **2**, 161
chromosome structure, matrix-associated region (MAR) **2**, 674f
– scaffold-associated region (SAR) **2**, 674f
– – transcription regulation **2**, 674
chromosome walking **2**, 162f, 398
chronic inflammatory disorders, monoclonal antibody therapy **5A**, 315
chronic therapy, with CDP571 **5A**, 343ff
chrorismic acid, conversion into prephenic acid, kinetic parameters **8B**, 416
– – thermodynamic parameters **8B**, 416
(+)-*trans* chrysanthemic acid, synthesis of, using an esterase from *Arthrobacter globiformis* **8B**, 283
chrysene, physical and chemical properties of **11B**, 218
– structure of **11B**, 216
chrysobactin **7**, 202f
– catechol-type siderophore **7**, 202f
– structure **7**, 202
Chrysophyta, characteristics of **10**, 113
chylomicrons **5A**, 15
chymosin **9**, 373ff, 680
– cloning of **5A**, 209
– industrial applications **5A**, 209
– manufacturing of cheese **5A**, 209
– production of recombinant **9**, 375
chymotrypsin **8A**, 226ff, **8B**, 356
– hydrolysis of ester bonds **8A**, 226ff

– increased stability of **8A**, 229
– kinetic resolution of racemic substrates **8A**, 227f
– preferred cleavage sites **8A**, 249
– stereospecificity **8A**, 228
α-chymotrypsin, dihydrocinnamate ester hydrolysis **8B**, 376f
– stereoselectivity of **8B**, 354
chymotrypsin catalyzed hydrolysis **8A**, 228
CIAA (Confederation of the Agro-Food Industry) **12**, 536
CIEC (China New-Investment Enterprise Company) **12**, 398
CIF *see* Computer integrated fermentation
cilia, mobility system **5A**, 32
ciliary neurothrophic factor **5A**, 159
CIMYTT **12**, 626
cineol, bioconcersion of **10**, 406ff
– biotransformation of **10**, 381f
cinnamaldehydes, biotransformation products of **8B**, 26f
cinnamic acid derivatives, from plant cell cultures **7**, 598ff
cinnamycin, antibiotic activity **7**, 357f
– prepeptide structure **7**, 345
– primary structure **7**, 336f
– solution structure **7**, 340
– structural gene **7**, 344
CIP (International Potato Center) **12**, 439f
circulation time **4**, 112ff
– estimates from simple flow methods **4**, 310f
– histogram **4**, 309
– influence of aeration **4**, 312
– in stirred tanks **4**, 304ff
– measurement **4**, 123f
– – flow-follower techniques **4**, 307ff
– – stimulus-response methods **4**, 304ff
– – tracer techniques **4**, 123f
– of gas bubbles **4**, 112
circulation-time distribution, in stirred tanks **4**, 304ff, 308ff
– single impeller systems **4**, 314
– tracer responses **4**, 313ff
– two-impeller systems, flow-follower techniques **4**, 315
– – tracer measurement methods **4**, 319f
circulin, proteinaceous sweetener **10**, 577f
citral, biotransformation of **10**, 377, 388
– hydroxylation of **8A**, 491
citrate dehydratase **8B**, 109
citrate hydroxamates, acetinoferrin **7**, 221f
– aerobactin **7**, 222f
– arthrobactin **7**, 222
– rhizobactin 1021 **7**, 221
– schizokinen **7**, 220f
– structure **7**, 220
citrate metabolism, in clostridia **1**, 306
citric acid, biosynthetic pathway **6**, 364

– break-even sugar price **6**, 52
– conversion cost **6**, 50
– distribution coefficients **3**, 571
– – in liquid-liquid extraction **3**, 561
– extraction **3**, 561f, 571
– growth of production volumes **8A**, 10f
– market prices **6**, 51
– producers **6**, 335
– production figures, present world production capacity **6**, 335
– production volumes 1910–1929 **8A**, 9
– specifications **6**, 334
– uses **6**, 334
citric acid accumulation, in *Aspergillus niger*, bypass of glycolytic regulation **6**, 311
– mechanism of regulation **6**, 312
– tricarboxylic acid cycle activity **6**, 313
citric acid cycle, lyases in **8B**, 48
citric acid fermentation **6**, 307ff
– current industrial technology **6**, 310
– history **6**, 308ff
– product recovery **6**, 327ff
– – by ion-exchange procedures **6**, 330
– – by precipitation **6**, 328f
– – by solvent extraction **6**, 329f
– – recovery of mycelial mats **6**, 327
– purification **6**, 327ff
– time-course **6**, 326
citric acid production, by *Aspergillus niger* **1**, 526f
– by fungi **6**, 330
– by-products, gypsum **6**, 333
– – mycelium **6**, 333
– by yeasts, *Candida* sp. **6**, 331
– classical surface process **6**, 320ff
– – flow-sheet **6**, 321
– conditions **1**, 527
– effect of, aeration **6**, 314
– – dissolved oxygen tension **6**, 314
– – nitrogen **6**, 315
– – pH **6**, 314f
– – phosphate **6**, 315
– – trace metal ions **6**, 313f
– Koji process **6**, 319f
– mixed-type fermentations **6**, 326
– molasses batch-fermentations **6**, 323
– molasses fed-batch fermentations **6**, 323
– process ecology **6**, 333
– process kinetics **6**, 332f
– pure sugar batch fermentations **6**, 324ff
– raw materials **6**, 318f
– – molasses **6**, 318f
– – starch hydrolyzates **6**, 318
– – unfiltered starch hydrolyzates **6**, 326f
– solid-state fermentations **6**, 319
– spore propagation **6**, 318
– strain improvement, mutagenic treatment **6**, 317
– – protoplast fusion **6**, 317

– strain maintenance **6**, 317
– strain selection **6**, 315ff
– – passage method **6**, 316ff
– – single-pore method **6**, 316
– submerged process **6**, 321ff, 322f
– – flow-sheet **6**, 321
– typical fermentation diagram **6**, 326
Citrobacter, in olive fermentation **9**, 605
Citrobacter freundii, tyrosine-phenol lyase **8B**, 95ff
citronellal, biotransformation of **10**, 377
– reduction of **10**, 384f
citronellol, biotransformation of **10**, 384ff
– hydroxylation of **8A**, 491
– microbial conversions of **10**, 387
citronellol acetate, biotechnological production of **10**, 397
D-citrulline formation, by *Agrobacterium radiobacter* hydantoinase **8A**, 268f
L-citrulline production **6**, 513
citrus fruits, enzymatic peeling **9**, 715f
citrus juice, enzymatic treatment **9**, 715f
– production **9**, 715f
cladistic classification **1**, 372
Cladosporium cladosporioides, in pozol fermentation **9**, 549
Cladosporium herbarum, in pozol fermentation **9**, 549
classification, of enzymes **9**, 7ff
classification of microorganisms, 16S rRNA sequence comparison **11A**, 86
clausenamide, preparation of, using poly-leucine catalyzed epoxidation **8B**, 501ff
Claviceps purpurea **1**, 525ff
– alkaloid formation **1**, 240
– genetic recombination **1**, 532
– immobilization effect **1**, 240
– life cycle **1**, 526
– morphology **1**, 234
– strain improvement **1**, 532
clavicoronic acid **7**, 509f
– structure **7**, 510
clay minerals **11B**, 93
clay organic complexes **11B**, 93f
CLEC lipases **8A**, 52
CLECs *see* cross-linked enzyme crystals
cleft lip **5B**, 66
Cleland's notation **9**, 36ff
client–server models **5B**, 434f
clinical tolerance studies, of microbial biomass for human consumption **9**, 229ff
clinical trials, adeno-associated virus vectors **5A**, 421
– adenovirus vectors **5A**, 413
– CMA-676 **5A**, 360f
– CMB-401 **5A**, 362
– macrophage stimulating factor **5A**, 157
– phases **12**, 223, 230f

– ReoPro **5A**, 365ff
– sepsis **5A**, 335ff
clinostat, study of microgravity effects **10**, 508
clitocine **7**, 509f
– structure **7**, 510
clogging, in biological waste gas reactors **11C**, 292
– in trickling filters **11C**, 337ff
– – avoidance of **11C**, 338f
– in water infiltration systems **11C**, 432
clone-by-clone approach **5B**, 46
cloned gene, molecular characterization **2**, 401ff
cloning **2**, 436
– human reproductive – **5B**, 420
– *in vivo* **2**, 437, 446
– therapeutic – **5B**, 420
cloning site **2**, 382
cloning strategies **2**, 389
– for artificial genes **2**, 392
– for gene library construction **2**, 389
– fragment subcloning **2**, 392
– role of *E. coli* **2**, 375
cloning techniques **2**, 376
cloning vector **2**, 268, 382ff
– direct selection vector **2**, 384
– M13 **2**, 345
– M13mp **2**, 268
– oligonucleotide single-strand integration **2**, 345
– "reporter" vector **2**, 384
cloning vectors **5B**, 195f
clostridia **1**, 285ff
– acetate production **1**, 309
– acetyl-CoA **1**, 299
– alcohol fermentation **1**, 300
– amino acid fermentation **1**, 304f
– – products **1**, 304f
– – reaction mechanisms **1**, 304f
– bacteriophages **1**, 566
– biochemistry of solvent-producing – **6**, 233ff
– butanol-producing –, properties **6**, 235
– citrate metabolization **1**, 303
– complex medium **1**, 295
– degradation of, aromatics **1**, 300
– – cellulose **1**, 297
– – flavonoids **1**, 303
– – nitrogenous compounds **1**, 304f
– – organic acids **1**, 300
– – pectin **1**, 297
– – purines **1**, 306f
– – pyrimidines **1**, 308
– – starch **1**, 297
– – xylan **1**, 297
– Embden-Meyerhof-Parnas pathway **6**, 237f
– enrichment culture **1**, 296
– Entner-Doudoroff pathway **6**, 239
– enzymes in pyruvate degradation **6**, 241
– enzymes involved in solvent formation **6**, 243ff
– extracellular enzymes **1**, 310

– G + C content **1**, 286ff
– genetics of solvent-producing –, strain improvement **6**, 246f
– glucose fermentation **1**, 299
– growth characteristics **1**, 293ff
– growth temperature **1**, 287ff
– handling **1**, 295
– homoacetogenic metabolism **1**, 300
– industrial application **1**, 308ff
– isolation **1**, 295
– media, complex – **1**, 295
– – group selective – **1**, 293f, 296
– – inoculation **1**, 296
– – redox potential **1**, 293f, 296
– – supplements **1**, 294
– – synthetic **1**, 295
– metabolism **1**, 297ff
– – reduction reactions **1**, 311f
– molecular genetics **1**, 293
– nitrogen fixation **1**, 308
– nutrition requirements **1**, 294ff
– oxygen sensitivity **1**, 293
– pathogenicity **1**, 286ff
– pathway of formation of, acetone **6**, 238
– – butanol **6**, 238
– – isopropanol **6**, 238
– pH requirements **1**, 294
– physiology **1**, 297ff
– products **1**, 308ff
– product stereospecificity **1**, 311
– product tolerance **1**, 309
– product yield **1**, 309
– properties **1**, 286ff
– protein production **1**, 310
– proteolysis **1**, 287ff
– saccharolysis **1**, 287ff
– species diversity **1**, 286
– sporulation **1**, 292f, 297
– starch-degrading enzymes **1**, 310
– storage **1**, 295f
– substrates **1**, 287ff
– – carbon dioxide **1**, 308
– – carbon monoxide **1**, 308
– – gaseous substrates **1**, 308
– – hydrogen **1**, 308
– – metabolization of **1**, 297ff
– succinate metabolism **1**, 303f
– temperature requirements **1**, 294
– transport systems **1**, 298
Clostridium **1**, 285ff, 368
– anaerobic wastewater process **4**, 447
– genetic engineering **2**, 489
Clostridium aceticum **1**, 300
Clostridium acetobutylicum **1**, 286, **6**, 236
– acetoacetate decarboxylase **8B**, 59ff
– acetone-butanol fermentation **1**, 309, **3**, 308
– biochemistry **6**, 233ff
– butyrate toxicity **1**, 181

– formation of solvents **6**, 243ff
– growth medium **1**, 295
– immobilization **6**, 252f
– lactate formation **6**, 239
– molecular genetics **1**, 293
– mono- and disaccharide uptake **6**, 237
– preservation by sporulation **6**, 233
– production of, acetone **8B**, 61f
– – butanol **8B**, 61f
– production of solvents **6**, 232ff
– – in chemostats **6**, 249ff
– pyruvate formation **6**, 239
– strain improvement **6**, 246f
Clostridium beijerincki, isopropanol production **8B**, 62
Clostridium beijerinckii, biochemistry **6**, 233ff
– formation of acids **6**, 239ff
– formation of solvents **6**, 243ff
Clostridium bifermentans, growth medium **1**, 295
Clostridium botulinum **1**, 286f
– toxin **1**, 311
– – medical application **1**, 311
Clostridium butyricum, BOD sensor **4**, 84
– growth kinetics **1**, 133
– use in bioelectrochemical fuel cells **10**, 15ff
Clostridium formicoaceticum **1**, 292f
– spore formation **1**, 292f
Clostridium histolyticum **1**, 293
Clostridium perfringens **1**, 293
– molecular genetics **1**, 293
Clostridium sphenoides **1**, 300f
Clostridium tetani **1**, 293
Clostridium tetanomorphum, β-methylaspartase **8B**, 129
Clostridium thermoaceticum, anaerobic formation of acetic acid **6**, 396f
Clostridium thermocellum **1**, 297, 310
– cellulase **1**, 28
– cellulosome **1**, 297
clostripain, preferred cleavage sites **8A**, 249
Clothier Committee on the Ethics of Gene Therapy **5A**, 527
clotting factors **5A**, 34
– in therapy **5A**, 168f
clozylacon, synthesis of **8B**, 108
Club of Rome **12**, 529f
clustering analysis **5B**, 149
clusters of orthologous groups of proteins (COGs) **5B**, 290
CluSTr **5B**, 288, 290
CMAS system **4**, 409f, 416, 435f
CMA-676 treatment **5A**, 356ff
– adverse events **5A**, 359f
– development **5A**, 358
– efficacy data **5A**, 361
– immune reaction to **5A**, 361
– of acute myeloid leukemia **5A**, 358ff
– patient enrollment **5A**, 359

– phase I study **5A**, 360ff
– – preliminary results **5A**, 360
– safety data **5A**, 359
CMB-401 **5A**, 356ff, 361f
– clinical studies **5A**, 362
– development **5A**, 358
– imaging studies **5A**, 361f
– in tumor treatment **5A**, 361
CMC solution, characteristic shear rates **3**, 204
– mass transfer coefficient **3**, 203
– – dimensionless correlation **3**, 204
– viscosity **3**, 203
CMP *see* cytidine
CNCBD (China National Center for Biotechnology) **12**, 398
Cnidaria **10**, 445f
– antibacterial compounds of **10**, 459
– antiinflammatory compounds of **10**, 456
– cytostatic/antitumor compounds of **10**, 452
– substances of **10**, 446
C-N ratios, and biodegradation in soil **11B**, 431f
coagulation, effect on BOM **11C**, 463
coal *see also* brown coal, hard coal, lignite
– bioconversion of, bioassays for emission control **10**, 176f
– – environmental aspects **10**, 172ff
– – perspectives **10**, 180f
– – to methane **10**, 169
– biodegradation of **10**, 157ff
– biotechnology of **10**, 153ff
– – developments **10**, 158
– – composition of **10**, 155
– conversion into polyhydroxyalkanoic acids **10**, 171
– depolymerization by laccases **10**, 168f
– modification of **10**, 157ff
– – bituminous – **10**, 159
– – sub-bituminous – **10**, 159
– organosulfur compounds in, classification of **10**, 174
– pyrite removal from **10**, 173f
– sulfur removal from **10**, 172ff
coal bioconversion products **10**, 176ff
– ecotoxicological testing of **10**, 177f
– mutagenicity testing of, Ames test **10**, 178ff
coal biodesulfurization, application potential **10**, 175
coal derived reference substances, ecotoxicological testing of **10**, 177f
– mutagenicity testing of, Ames test **10**, 178f
coal desulfurization **10**, 173ff
– biological mechanisms, use of model compounds **10**, 174f
– use of ethanol **6**, 198
coal humic substances, depolymerization, by laccases **10**, 168f
– – by manganese peroxidase **10**, 167
coalification process, stages **10**, 155

coal particles, binding of hydrophobic pollutants **11B**, 71
cobalt, bioleaching of **10**, 208, 213
cocaine **8A**, 332
– hydrolysis of, use of catalytic antibodies **8B**, 441f
– microbial metabolism of **8A**, 337f
cocaine esterase **8A**, 337
Cochliobolus lunatus **1**, 525, 529
cocoa beans, color changes **9**, 571
– flavor development **9**, 574f
– roasting **9**, 574f
– – phases **9**, 574
– storage **9**, 567
cocoa butter, fatty acid composition **7**, 136
cocoa butter equivalent (CBE) **7**, 150
– costs of yeast process **7**, 155
– formation, from *Cryptococcus* mutants **7**, 153
– growth of *Cryptococcus*, oxygen limitation **7**, 154
– price **7**, 148
cocoa butter equivalents, lipase catalyzed synthesis **8B**, 289
cocoa butter substitutes **8A**, 146
– use of lipase in manufacture of **8A**, 135f
cocoa fermentation **1**, 375, **9**, 561ff
– aerobic phase **9**, 572
– anaerobic phase **9**, 570ff
– aroma precursors **9**, 571
– bean curing **9**, 567
– compositional changes **9**, 569
– determination of quality **9**, 573
– drying **9**, 566, 572f
– enzymes **9**, 570f
– measurement, correlation coefficients **9**, 574
– – cut test **9**, 573f
– methods **9**, 565f
– – use of sweat-boxes **9**, 566
– microflora **9**, 568
– pulp fermentation **9**, 567ff
cocoa fruit, description **9**, 562f
– harvesting **9**, 564
– types, Criollo **9**, 562f
– – Forastero **9**, 562f
cocoa seed, components **9**, 563
– composition **9**, 563
– enzymes **9**, 569f
coconut kernel, cell wall polysaccharides **9**, 707
coconut milk, enzymatic treatment **9**, 717
coconut oil, enzymatic extraction **9**, 726
COD *see* Chemical oxygen demand
CODATA (Committee on Data for Science and Technology) **12**, 330
codeine, hydroxylation of **8A**, 496
– *N*-demethylation by *Cunninghamella* **8A**, 345f
– potent analgesic **8A**, 340
– redox reactions of **8A**, 350
codeine *N*-oxide, structure **8A**, 348

Code of Federal Regulations **5A**, 462
Code of Practice for Vinegar **9**, 588ff
coding regions **5B**, 362f
– DNA and protein analysis of **5B**, 366ff
– separation from non-coding sequences **5B**, 363ff
coding sequence **5B**, 392
codominance, of gene loci **2**, 147
codon preference statistics **5B**, 320f
codon usage analysis **5B**, 318
coenzyme A **9**, 31
– applications of **10**, 327
– determination **9**, 145
– microbial production of **10**, 325ff
coenzyme Q **9**, 31
– – *see* ubiquinone Q
coenzyme regeneration, characteristics of selected enzymes **8A**, 445
– dehydrogenases **8A**, 399f
– enzyme-coupled – **8A**, 402
– methods **8A**, 402ff
– nonenzymatic – **8A**, 400f
– substrate-coupled – **8A**, 401f
– use of LDHs **8A**, 442
– use of YADH **8A**, 440
coenzymes **1**, 70ff, 259, **5A**, 24, **9**, 28ff
– definition of **10**, 320
– distribution of **1**, 71
– function of **1**, 71
– F_{420} **1**, 259
– F_{430} **1**, 259
– methanofuran **1**, 259
– of methanogenic bacteria **1**, 72, 259
– production of **10**, 321
– – microbial and enzymatic processes **10**, 322f
cofactor recycling, Baeyer–Villiger biotransformations **8A**, 567f
cofactor regeneration **8A**, 368
– in enyzme reactors **3**, 454f
– novel schemes **3**, 458
cofactors, as protein folding aids **3**, 541
– of enzymes **9**, 28ff
– – amino acid derivatives **9**, 30
– – cosubstrates **9**, 29
– – metal ions **9**, 30
– – prosthetic groups **9**, 29
co-fermentation **11A**, 228f
– treatment of agricultural waste **11A**, 228f
– – plant scheme **11A**, 229
COGENE (Scientific Committee on Genetic Experimentation) **12**, 46, 527, 555
cohumulone **9**, 423
coiled-coils **5A**, 67f
Colby cheeses **9**, 376
colcemid **2**, 8
colchicine **2**, 8
– demethylation of **8A**, 497
cold mash process **6**, 85

Coleus blumei, accumulation of rosmarinic acid **7**, 598ff
– callus culture **1**, 582
– rosmarinic acid production **1**, 590ff
– – sucrose effect **1**, 594
– suspension culture **1**, 582
colinearity between genomes **5B**, 148
colitis, ulcerative – **5A**, 319
collagens **5A**, 4, 32f
– functions **5A**, 33f
collection systems for wastes **11C**, 9f
collision induced dissociation (CID) cells **5B**, 331
colloids, transport of contaminants, types of **11B**, 54f
collybial, antiviral activity **7**, 520
– structure **7**, 522
colon cancer **5B**, 81
– early diagnosis of **5B**, 86ff
– familial – **5B**, 81ff
– prevention of **5B**, 86ff
– sigmoidoscopic surveillance for polyps **5B**, 87
colony-forming unit (CFU) **1**, 120
colony forming units (CFU), in drinking water **10**, 287
– in the atmosphere **10**, 287
colony stimulating factors **5A**, 155ff
– genetic engineering of **5A**, 540
– in therapy **5A**, 155
colorectal carcinoma, monoclonal antibody therapy **5A**, 320
coloring, of food products, by microalgal pigments **10**, 131
coloring substances, from microalgae **10**, 109, 131
colupulone **9**, 423
COL1A1 gene **5B**, 69
COL1A2 gene **5B**, 69
Comamonas acidovorans, immobilized cell **1**, 129
Comamonas testosteroni, degradation of, chlorobenzoate **11B**, 448
– – chlorobiphenyls **11B**, 448
combinatorial chemistry, in drug discovery **10**, 52ff
combinatorial libraries, formats **10**, 53f
– use of solid-phase techniques **10**, 54f
– use of solution-phase techniques **10**, 53
comb types, rectangular – **5B**, 211
– shark – **5B**, 211
– tooth size – **5B**, 211
cometabolic degradation, of PAH, by bacteria **11B**, 226ff
cometabolism **11C**, 281
– and soil bioremediation **11B**, 432f
– – halogenated pollutants **11B**, 433
– – TCE **11B**, 432
commercialization **5B**, 427
– of biological and medical discoveries **5B**, 98
commitment building, interactive **12**, 22ff
Committee for Responsible Genetics **12**, 465

common mortality **12**, 123
comosting technologies *see* composting process
compactin, synthesis by yeast alcohol dehydrogenase **8B**, 360
comparative genomics **5B**, 127, 361ff
comparative mapping, in plants **5B**, 148
compatible solutes (osmoprotectants), from bacteria **10**, 468
– production of **10**, 468
COMPEL **5B**, 406
competence **12**, 162
– for DNA uptake *see* DNA uptake competence
complement, in therapy **5A**, 166
complementary determining regions (CDR) **5A**, 223
complement C5a receptor antagonists **7**, 121
complement receptors **5A**, 166
complement system **5A**, 36
– regulation of **5A**, 38
complete domains **5B**, 138
complete gene sequencing **5B**, 116
completely stirred tank reactor (CSTR) **11A**, 401f
– carbon removal in **11A**, 402
– design **11A**, 484
– in industrial wastewater treatment **11A**, 479ff
– laboratory-scale investigations **11A**, 489
– mass balance concept **11A**, 485f
– mixing design **11A**, 486
– one-stage vs. two-stage anaerobic systems **11A**, 486f
– pilot-scale investigations **11A**, 489
– start-up-phase **11A**, 489f
– treatment of agricultural waste and slurry **11A**, 226f
complex media *see* fermentation media, growth media
composition percentage **5B**, 392
compost, actinomycetes in **11C**, 60ff
– aerobic/anaerobic treatment of **11C**, 213ff
– – hygienic considerations **11C**, 213ff
– as an additive in soil remediation **11B**, 343
– as fertilizer **11C**, 17ff
– control of soilborne pathogens **11C**, 77f
– end uses of **11C**, 129f
– heavy metals in **11C**, 145
– hygienic safety, validation and supervision strategy **11C**, 236
– isolation of thermophilic actinomycetes **11C**, 61
– isolation of thermophilic fungi **11C**, 66
– maturity of **11C**, 89f
– – *Chaetomium* test **11C**, 90
– – determination of **11C**, 90
– microorganisms, as biocontrol agents **11C**, 78
– preparation to end product **11C**, 112f
– process control **11C**, 237
– process validation **11C**, 237
– product control **11C**, 237

– production of, hygienic requirements **11C**, 237
– quality **11C**, 123f
– quality criteria **11C**, 145ff
– screening of **11C**, 115
– stability classes **11C**, 135
– sythetic organics in **11C**, 145f
– thermophilic actinomycetes in, analysis of **11C**, 62f
– toxic compounds in, maximally allowed values **11C**, 18
– uses of **11C**, 145ff
– water extracts of, control of foliar diseases by **11C**, 78
compost biofilter **11C**, 360ff
– open – **11C**, 360
– open bed, for wastewater plant odor control **11C**, 364
compost heaps, microbial activity in **11C**, 109f
– – products **11C**, 110
– self-ignition of **11C**, 41
composting **11A**, 187, **11B**, 95, 343, 430
– – *see also* laboratory composting
– aerated pile, with automatic turning machines **11C**, 140f
– aeration **11C**, 132f
– – effect of **11C**, 132
– aerobic conversions **11C**, 110
– anaerobic conversions **11C**, 110
– and anaerobic fermentation, comparison of **11C**, 155
– and humic substances **11B**, 107f
– bacteria in **11C**, 59f
– basic steps of **11C**, 112f
– box **11C**, 143f
– bricollari process **11C**, 141f
– bulk density of wastes **11C**, 106
– cell **11C**, 140
– channel **11C**, 140
– C/N ratio **11C**, 110
– container **11C**, 143f
– definition of **11C**, 36, 102, 128
– disintegration **11C**, 113ff
– drug substance of wastes, heavy metal content **11C**, 105
– – nutrient content **11C**, 105
– environmental control **11C**, 129
– exhaust gases **11C**, 135
– feedstocks **11C**, 128f
– – biodegradability of **11C**, 130
– fungi in **11C**, 64f
– hygienic aspects **11C**, 70ff, 213ff, 231
– – inactivation of pathogens **11C**, 70ff
– inactivation of plant pathogens **11C**, 77
– influencing factors **11C**, 110f, 130ff
– in Herhof boxes, process control **11C**, 134
– isothermal – **11C**, 68f
– *Legionella* in **11C**, 71
– liquid emissions **11C**, 135

– main groups of **11C**, 56ff
– mat composting **11C**, 139
– material structure, effect of **11C**, 132
– matrix water potential **11C**, 46
– mesophilic phase **11C**, 59
– microbial successions in **11C**, 65ff
– – bacterial – **11C**, 68
– – fungal – **11C**, 67f
– microbial transformation, of nitroaromatics **11B**, 290ff
– – of trinitrotoluene **11B**, 290ff
– microbiology of **11C**, 35ff
– – methods **11C**, 55
– microorganisms in **11C**, 54ff
– moisture content **11C**, 131ff
– – effect of **11C**, 131
– nutrients effect of **11C**, 133
– occupational health risks **11C**, 231
– odors from, appearance of **11C**, 88
– of biowastes **11C**, 17f
– of plant material **11C**, 17
– – consequences for **11C**, 112
– of plant residues **11C**, 101ff
– of sewage sludge **11C**, 222ff
– – optimal temperature **11C**, 42
– of solid organic wastes **11A**, 233
– of urban garbage, temperature course **11C**, 41
– of wastes **11C**, 15ff
– optimal temperature of **11C**, 48
– oxygen content **11C**, 132
– phases of **11C**, 134
– pH value **11C**, 48
– – effect of **11C**, 133f
– phytopathological aspects of **11C**, 76ff
– plant materials for **11C**, 102
– post-processing **11C**, 137
– pre-processing **11C**, 137
– process control **11C**, 134
– processing steps **11C**, 134f
– properties of plant wastes **11C**, 106
– rate of biomass turnover **11C**, 44f
– raw material preparation and conditioning, basic steps **11C**, 112
– reactors, classification of **11C**, 122f
– rotating drum process **11C**, 144
– rotting **11C**, 15ff
– rotting systems **11C**, 207
– *Salmonella* in **11C**, 71f
– self-ignition of **11C**, 41
– socioeconomic factors **11C**, 129
– starting material, carbon-nitrogen ratio **11C**, 44f
– – water content **11C**, 45f
– static pile **11C**, 138f
– strategies of **11C**, 127ff
– technology of **11C**, 127ff
– temperature, effect of **11C**, 133
– thermophilic phase **11C**, 59
– tunnel reactors **11C**, 142
– use of, flow reactors **11C**, 123
– – rotating drum reactors **11C**, 123
– vermicomposting **11C**, 139f
– vertical flow **11C**, 144f
– windrow – **11C**, 138, 140
– yield coefficient **11C**, 44f
composting configurations, for agricultural wastes, windrows **11A**, 233
composting ecosystem, physical model **11C**, 50ff
composting plants, bioaerosols in **11C**, 239ff
– emission of aerosols **11C**, 72
– – actinomycetes in **11C**, 74
– – bacterial population of **11C**, 73f
– – predominating fungi **11C**, 75
– emission of germs **11C**, 73
– emission of microorganisms **11C**, 72ff
– – health risks **11C**, 72ff
– heat recovery from **11C**, 85f
– hygienic safety, validation and supervision strategy **11C**, 236
– microbial emission in **11C**, 239ff
– occupational health risks **11C**, 237ff
composting process **11C**, 15ff, 109ff, 112ff
– and products of metabolism **11C**, 109
– balancing of **11C**, 78ff
– basic flow sheet **11C**, 112f
– calculation of a hypothetical – **11C**, 81
– characteristics of **11C**, 111
– control of **11C**, 140
– energy balance **11C**, 82
– heat production **11C**, 82f
– – and air requirement **11C**, 84
– – and removal of water **11C**, 85
– – balance of **11C**, 82
– heat transfer **11C**, 82ff
– – routes of **11C**, 82f
– hygienic safety **11C**, 233f
– – indicating organisms **11C**, 233ff
– – process validation **11C**, 233f
– machines and equipment **11C**, 113ff
– main composting **11C**, 111f
– mass balance **11C**, 79ff
– mathematical modeling of **11C**, 87
– microbial processes **11C**, 109
– microorganisms **11C**, 109
– odor from, control of **11C**, 87ff
– – formation of **11C**, 87ff
– – relation to microbial population **11C**, 89
– phases **11C**, 40ff, 111f
– post-composting **11C**, 111f
– pre-composting **11C**, 111f
– temperature **11C**, 111
composting systems **11C**, 134ff
– classification of **11C**, 117, 136f
– enclosed reactor systems **11C**, 140ff
– field composting **11C**, 117f
– geometric shapes of windrows **11C**, 120

– in-vessel reactor systems **11C**, 142ff
– non-reactor systems **11C**, 138ff
– – classification of **11C**, 118
– on market **11C**, 136
– reactor composting **11C**, 122f
– reactors for **11C**, 119
– windrow composting **11C**, 118f
compost pile, aeration of **11C**, 46f
– as a microbial habitat **11C**, 37ff, 42ff
– free air space **11C**, 47
– oxygen supply **11C**, 46f
– pore volume **11C**, 47
– structure of **11C**, 46f
– temperature **11C**, 47f
– ventilative cooling (Rutgers strategy) **11C**, 47
– volume of solid matter **11C**, 47
– water availability in **11C**, 45f
compost quality, regulatory guidelines **11C**, 146
comprehensive automation sytem, general structure of **4**, 573
computer aided software engineering (CASE) **4**, 597f
computer control **4**, 521f
– configuration **4**, 522
– direct digital control (DDC), advantages **4**, 523
– – position algorithm **4**, 522
– – velocity algorithm **4**, 523
– of fermentations, analog control **4**, 566
– – digital control **4**, 566
– – history **4**, 564ff
– setpoint control (SPC) **4**, 522
– supervisory control **4**, 522
computer control systems, fault detection **4**, 571
– graphical visualization techniques **4**, 574
– industrial installations, hardware building-block systems **4**, 567ff
– – redundancy approach **4**, 567ff
– open systems **4**, 574
– reliability **4**, 567ff
– software developments **4**, 573
– software engineering **4**, 570f
– tandem processors **4**, 570
computer integrated fermentation (CIF) **4**, 574ff
– – *see also* Automation in biotechnology
– *Bacillus subtilis* **4**, 592
– baker's yeast **4**, 592
– central real-time database **4**, 576
– control profile generation, use of filter banks **4**, 585
– database operation **4**, 584
– data-storage techniques **4**, 585
– diagram of network nodes **4**, 578
– distributed process management software **4**, 575
– *Escherichia coli* **4**, 592
– implementation of *a priori* knowledge **4**, 588ff
– implementation of models **4**, 585ff
– industrial uses **4**, 574

– math-function modules **4**, 583f
– model attachment to process management systems **4**, 585
– nodes in the local area network **4**, 577f
– – PC-systems **4**, 577
– – workstations **4**, 577
– off-line data input **4**, 584
– operator interface capabilities **4**, 578
– operator interface to the model bank **4**, 587
– programming techniques **4**, 596
– software structure **4**, 576
– user interface **4**, 580
– X-windows, application **4**, 580ff
– – operator surface **4**, 581
concatemers **5A**, 446
concerted fungus breeding **2**, 82f
– edible mushrooms **2**, 82f
– hybrid breeding **2**, 83
– metabolite producers **2**, 83
– strain improvement **2**, 82
concrete **10**, 299ff
– biodeterioration of **10**, 300ff
– corrosion tests **10**, 302ff
– impact of CO_2 on **10**, 279
Condromyces crocatus, fruiting body **1**, 115
Conductivity meter, for biomass determination **4**, 193
conductivity probe technique **4**, 128ff
confidence level **5B**, 382
confidentiality **12**, 120
– agreements **12**, 301
confirmation pages **5B**, 390
confirmatory sequencing **5B**, 194
confocal scanning **5B**, 236
conformation of the bound state **5B**, 353
congestive heart failure, medical applications of recombinant proteins **5A**, 132
conidia, of *Aspergillus* **1**, 116
– of streptomycetes **1**, 437
coniine, hemlock alkaloid **8A**, 328f
– synthesis of **8A**, 6
Conjugaphyta, chatacteristics of **10**, 112
conjugate addition, use of catalytic antibodies **8B**, 477
conjugation **2**, 48ff, **12**, 80, 93f, 163
– chromosome mobilization **2**, 53, 57
– Gram positive bacteria **2**, 55ff
– high frequency of recombination **2**, 49, 51
– in bacilli **2**, 461
– in Corynebacteria **2**, 483
– – with *E. coli* **2**, 484
– in *E. coli* **2**, 49ff
– in *Lactococcus* **1**, 361
– in Streptomycetes **2**, 472
– unrelated bacteria **2**, 58f
conjugative plasmid **2**, 53ff
– braod host range plasmid, IncP1 host range **2**, 53

– broad host range plasmid **2**, 53
– Gram negative bacteria **2**, 53f
– pAMβ1 **2**, 56
– R68.45 **2**, 51f
– table of **2**, 54
connective tissue growth factor **5A**, 132
consensus methods **5B**, 311
conserved sequence patterns **5B**, 333
consistent display idioms **5B**, 382
consumer acceptance, of rDNA products **12**, 135, 157ff, 192f
Consumers' Union **12**, 465
contact leaching **10**, 195
contact reactor, treatment of agricultural wastewater and slurry **11A**, 226f
contact stabilization process **11A**, 255
contact stabilization system, model **4**, 434f
– treatment of municipal sewage **4**, 433ff
contained use, of GMOs **12**, 555
– – Council Directive **12**, 555ff
container composting **11C**, 143f
container filter, scheme **11C**, 328
containment levels **12**, 73, 77, 246f, 257ff, 261ff, 468
containment measures **12**, 72f
contaminant concentrations, and natural attenuation **11B**, 402f
contaminants *see also* pollutants
– bioavailability of **11B**, 61ff
– degradability of, laboratory test methods **11B**, 354f
– – limiting factors **11B**, 351
– diffusion processes, in bioreactors **11B**, 333f
– distribution in solids fractions of soil **11B**, 331, 485
contaminated areas, investigation strategy **11B**, 480ff
contaminated site management *see also* regulatory concerns
– action thresholds **11B**, 38f
– definition **11B**, 11f
– efficient system elements **11B**, 30ff, 40f
– financing models **11B**, 15f, 37
– identification of sites **11B**, 17f, 32, 37
– international approaches **11B**, 5ff
– legal development **11B**, 8ff
– legislative status **11B**, 36
– listing and delisting **11B**, 37f
– monitoring **11B**, 39f
– preferred treatment techniques **11B**, 35f
– principles **11B**, 11ff, 36
– registration of sites **11B**, 17f, 32, 37
– risk assessment **11B**, 19f, 32f, 38
– technical development **11B**, 8ff
contaminated sites, remediation methods, classification of **11B**, 306
contaminated soil, addition of compost **11B**, 343
– biological *ex situ* treatment **11B**, 313f

– – biopile process **11B**, 313
– – bioreactor process **11B**, 313
– biological *in situ* treatment **11B**, 314f
– – scheme **11B**, 314
– bioremediation of **11B**, 313ff
– – methods **11B**, 305ff
– – use of heap technique **11B**, 319ff
– chemical/physical *ex situ* treatment **11B**, 310ff
– – scheme **11B**, 311
– – scrubbing **11B**, 310f
– chemical/physical *in situ* treatment, pump and treat processes **11B**, 312
– – soil vapor extraction **11B**, 312
– disposal of **11B**, 315
– distribution of contaminants over the solids fractions **11B**, 331
– ecotoxicological assessment, use of aquatic test systems **11B**, 130ff
– – use of terrestrial test systems **11B**, 135ff
– ecotoxicological chracterization of **11B**, 137f
– elution test **11B**, 490ff
– thermal *ex situ* treatment **11B**, 307f
– – scheme **11B**, 309
– – technologies **11B**, 307f
– thermal *in situ* treatment **11B**, 308f
– treatment of **11B**, 330ff
– – bioreactors **11B**, 332ff
contaminated solids *see also* contaminated soil, sediments
– characteristics of **11B**, 331f
contamination, in HAVAC systems **3**, 753
– in multiple-use facilities **3**, 744
– of disused sites, parameters **11B**, 495f
contiguous sequences (contigs) **5B**, 384
continuous air-segmented analyzer, failures **4**, 173
continuous culture **1**, 155, **3**, 98f, **9**, 202ff
– chemostat **9**, 202f
– dilution rate **9**, 203
– effects of, dilution rate **3**, 99
– – substrate concentration **3**, 99
– in physiological engineering studies **3**, 100
– mass balances **3**, 98
– model simulation **1**, 157
– residence time distribution **3**, 95
– steady-state design **3**, 99
– sterilization criteria **3**, 174
continuous flow processes, applications **11A**, 329f
– in municipal wastewater treatment **11A**, 321ff
continuous flow reactors, application of experience to SBR systems **11A**, 331f
– comparison with SBR systems **11A**, 328ff
– mass balance equation **11A**, 329
– mass balances **11A**, 329
continuous flow stirred tank reactor (CFSTR), concentration profiles **11A**, 324
continuously stirred tank reactor (CSTR) *see also* bioreactors

– treatment of contaminated soils **11B**, 335
continuous (quantitative) traits **5B**, 65
continuous sterilizer **3**, 170ff
– cell balance **3**, 172
– continuous steam injecion **3**, 175
– Damköhler number **3**, 174
– designing strategy **3**, 174
– dispersion model **3**, 172
– effects of solids **3**, 175
– equipment **3**, 175
– flow type **3**, 174
– fluid residence time **3**, 171
– heat exchangers **3**, 175
– length **3**, 174
– sterility level **3**, 172
– temperature profile **3**, 176
contractile proteins **5A**, 31ff
control, types **4**, 513
Controlled Autoregressive Integrated Moving Average (CARIMA) models, in fermentation modeling **3**, 333, 342
Controlled Autoregressive Moving Average (CARMA) models, in fermentation modeling **3**, 332, 342
controlled evaporative drying **3**, 707
controlled pore glass **5B**, 231
controller modes **4**, 515
controllers *see* fermentation control
controller setting, calculation of **4**, 516ff
– use of computer control **4**, 521
controller tuning, automatic **4**, 518, 521
– process reaction curve method **4**, 517f
– ultimate gain method **4**, 516f
control of bioreactor systems **4**, 263
– – *see also* Process control
– high-level control **4**, 523ff
– – profile optimization **4**, 525ff
– low-level control **4**, 513ff
– – adaptive control **4**, 520ff
– – computer control **4**, 521f
– – constant setpoint optimization **4**, 524
– – controller selection **4**, 515
– – feedforward control **4**, 519ff
– – multiloop control **4**, 519f
– – sequence control **4**, 514
– – single-loop control **4**, 514ff
– – tuning methods **4**, 516ff
control variable iteration **4**, 529
control vector iteration **4**, 527
Convention on Biological Diversity (CBD) **12**, 143, 296, 442, 628ff
conversion cost, of monomeric sugar products **6**, 50
cooling coils, for bioreactors **3**, 420
cooling water, in industrial wastewaters **11A**, 195
Coon's medium, composition **3**, 143f, 154
cooperativity **9**, 47, 125f

Coordinated Framework for the Regulation of Biotechnology **5A**, 520
coordination office **5B**, 438
copper, biorecipitation of **11B**, 365
– biosensors for depletion of **11B**, 456
– effect on plant cell culture **1**, 594
– monitoring with biosensors **11A**, 127f
copper ores, bioleaching of **10**, 208ff
coprogens, structure **7**, 227
– trihydroxamate siderophores **7**, 227
Coptis japonica, accumulation of protoberberines **7**, 604f
copyright **12**, 301
coriolin, ATPase inhibiting metabolite **7**, 521
– structure **7**, 521
corn, as raw material for ethanol production **6**, 64f
– in industrial ethanol production **6**, 140, 156f
– – energetics **6**, 185f
– – flow diagram **6**, 157
– processing in ethanol production **6**, 92f
– stillage, composition **6**, 108
– – nutrient curent **6**, 107
corn based sweeteners *see* sweeteners
corn germ, cell wall polysaccharides **9**, 707
corn grain, analysis of **6**, 64f
corn grain silage **6**, 64f, 93
corn grits, as brewing adjunct **9**, 429
corning spinner *see* bioreactor
corn mash, saccharification of **6**, 72ff
cornstalks, acid hydrolysis **6**, 164
– in industrial ethanol production **6**, 163f
– – energetics **6**, 187f
– – flow diagram **6**, 164
corn starch, as brewing adjunct **9**, 430
– composition **6**, 32, 34f
– production **6**, 33
– wastewater from – **11A**, 202
corn steep liquor, as feed **9**, 774
corn steep liquor (CSL) **1**, 149
corn syrup *see also* syrups
– high-fructose – **5A**, 194, 197, **9**, 672ff
– – production costs **9**, 673
– production of **9**, 669ff
corn wet milling, by-products **9**, 774
corollosporin, structure **10**, 459
coronary diseases **5A**, 366ff
– treatment with ReoPro **5A**, 366
coronary heart disease **5B**, 68, 71
coronary interventions, use of ReoPro **5A**, 367ff, 371
coronary stenting, therapy with ReoPro **5A**, 375
coronary syndromes, treatment with ReoPro **5A**, 371
corrosion *see also* biocorrosion
– by biogenic sulfuric acid, mechanism **10**, 299f
– definition of **10**, 267
– flatspread – **10**, 294

– localized – **10**, 294
corrosion damage, definition of **10**, 267
corrosion tests, for sewage pipelines **10**, 302ff
– of cement **10**, 302f
– simulation of **10**, 302ff
Corterra® **10**, 433
corticosteroids, in sepsis therapy **5A**, 334
Corticoviridae **1**, 550
cortisol **8A**, 13
corynebacteria **1**, 433, 453ff, **2**, 478ff
– amino acid biosynthesis **1**, 454ff
– amino acid production **1**, 456ff, **2**, 479
– bacteriophages **1**, 568
– biochemical pathway engineering in **8B**, 321
– biology **2**, 478
– biosynthetic pathway of, L-homoserine **8B**, 323
– – L-isoleucine **8B**, 323
– – L-lysine **8B**, 323
– – L-methionine **8B**, 323
– – L-threonine **8B**, 323
– biotechnological application **1**, 434, 456ff
– chemotaxonomic marker **1**, 435
– chemotaxonomy **2**, 479
– culture **1**, 453
– ecology **1**, 453
– G + C content **1**, 435
– genetic engineering **1**, 453, **2**, 478ff
– genetics **1**, 453
– industrial application **2**, 479
– isolation **1**, 453
– morphology **1**, 435
– nitrile hydratase activity **8A**, 290ff
– nucleotide production **1**, 458
– overproduction of amino acids **8B**, 322
– primary metabolism **1**, 454ff
– product alteration **1**, 458
– protein secretion **2**, 485
– protoplast **2**, 483
– secretory pathways **1**, 454
– spheroplast **2**, 483
– strain improvement **1**, 457f
– substrates **1**, 454
– taxonomy **1**, 434, 453
Corynebacterium, amidase **8A**, 291
– Baeyer-Villiger reaction, mediated by **8A**, 545
– hydrolysis of amino nitriles **8A**, 290f
– hydrolysis of cyclohexane dinitriles **8A**, 291
– in ogi fermentation **9**, 545
– recombinants, in fermentation processes **3**, 284
Corynebacterium diphtheriae, glucose-6-monocory-
   nomycolate formation **6**, 663
Corynebacterium glucamicum, L-glutamic acid biosynthesis, regulation **6**, 471
Corynebacterium glutamicum **1**, 206, 434, **2**, 478f,
   485ff
– arginine production **6**, 513
– aromatic amino acid production **1**, 457ff
– – strain improvement **1**, 457f

– glutamic acid production **1**, 456
– – biotin effect **1**, 456
– histidine production **6**, 513
– isoleucine production **6**, 510
– lactose metabolization **1**, 195
– L-lysine biosynthesis, genes **6**, 481f
– – regulation **6**, 476
– L-phenylalanine biosynthesis, regulation **6**, 474
– L-tryptophan producing mutants **6**, 488
– lysine export **1**, 209
– lysine production **6**, 507
– mechanisms of amino acid secretion **6**, 471
– phenylalanine production **6**, 510
– production of, glutamate **8B**, 321ff
– – lysine **8B**, 321ff
– – threonine **8B**, 321ff
– recombinant L-tryptophan producing strains **6**,
   490
– threonine production **6**, 508ff
– tryptophan production **6**, 512
– tyrosine production **6**, 510
Corynebacterium manihot, in gari fermentation **9**,
   543
Corynebacterium petrophilum, bio-de-emulsifying
   acitivity **6**, 705ff
Corynebacterium sp., bio-de-emulsification **6**, 702
coryneform bacteria, genes of L-lysine biosynthesis **6**, 481f
– glutamate production **6**, 506f
– L-lysine producing mutants **6**, 478ff
– – auxotrophic **6**, 478
– – regulatory **6**, 478
– overproduction of L-glutamic acid **6**, 469ff
– production of phenylalanine **8B**, 314
COS *see* monkey kidney cell lines
cosmetic emulsions, constituents **6**, 671ff
– – acyl lactylates **6**, 674
– – amine oxides **6**, 674
– – glycerine esters **6**, 673
– – glycerol monostearate **6**, 676
– – lanolin **6**, 672
– – N-acylamino acids **6**, 676
– – nonionic ethoxylates **6**, 672f
– – phospholipids **6**, 676f
– – phosphoric acid esters **6**, 673
– – protein condensates **6**, 674
– – silicones **6**, 674f
– – sucrose esters **6**, 673
– – sugar esters **6**, 677
cosmetics, biosurfactant uses **6**, 670ff
– use of amino acids **6**, 487ff
cosmid **1**, 562, **2**, 387f
– library construction **2**, 387f
– pJB8 **2**, 387
– rescueing **2**, 535
– vector **2**, 387
cosmids **5B**, 195

cosmomycins, redifferentiation of cancer cells **7**, 682

cosolvents, as protein folding aids **3**, 542
– mechanisms **3**, 543
– physiochemical properties **3**, 543

cottage cheese **9**, 379

Coulter counter **1**, 121
– cell counts **4**, 200
– cell-size determination **4**, 200
– limitations **4**, 200

Council of Europe **12**, 525, 609
– organs **12**, 609f

countercurrent distribution method, partition **4**, 613

counting chamber **1**, 119

coupling methods, for dehydrogenases **8A**, 407

covalent chromatography, in enzyme purification **9**, 59

covered anaerobic lagoons, for industrial waste-water treatment, scheme **11A**, 483

CO₂ see also Carbon dioxide

CO₂ exit, measurement, accuracy of **4**, 21

CO₂ fixation see carbon dioxide fixation

CO₂ measurement, in penicillin fed-batch fermentation **4**, 70f

CO₂ partial pressure see p_{CO_2}

CO₂ production rate (*CPR*) **4**, 9
– determination of **4**, 34

CO₂ profile, in bubble columns **4**, 370f

CO₂ transfer rate (*CTR*) **4**, 34

CP-A see carboxypeptidase A

CpG islands **5B**, 175, 315ff

CPR see CO₂ production rate

Crabtree effect **4**, 340
– during yeast growth **4**, 388
– in yeast **1**, 133f, 154, 486

cracking mills, disintegration of organic wastes **11C**, 114f

craft pulp, treatment with xylanase **10**, 84

cream cheese **9**, 379

creatine kinase **8B**, 224ff
– phosphorylation by **8B**, 224ff

crenarchaeota **1**, 252

m-cresol, degradation, in biotrickling filters **11C**, 286

cresols, anaerobic degradation pathways **11B**, 178f

CREST (Scientific and Technical Research Committee) **12**, 518ff

Criegee intermediates **8A**, 538, 551ff
– as reagents for Baeyer-Villiger reactions **8A**, 540
– of cyclohexanone monooxygenase, of *Acinetobacter* **8A**, 553

CRIFC (Central Research Institute for Food Crops) **12**, 412

crinipellins, cytotoxic activity **7**, 508
– structure **7**, 506

CRITICA **5B**, 17

CRL see *Candida rugosa* lipase

Crohn's disease, monoclonal antibody therapy **5A**, 315, 319
– role of tumor necrosis factor **5A**, 346

crop domestication **5B**, 146

crop genomics **5B**, 145ff

crop improvement **5B**, 146f

crop plants, generation of mutations **5B**, 151ff
– genomic mapping **5B**, 165ff
– positional cloning **5B**, 165ff
– production of PHAs **10**, 430
– transgenic – **5B**, 152ff

crop protection agents, formation of bound residues **11B**, 109ff
– – in soil **11B**, 109

crop residue, nutritive value, enhancement **9**, 779f

crops, herbicide resistant – **10**, 493f

crossflow filtration (CFF) **3**, 43ff, 50, 440f, 469ff, 487ff, 493
– cake deposit, thickness **3**, 494
– diafiltration **3**, 50
– effects of, flow velocity **3**, 494
– – particles **3**, 491f
– – rheological properties **3**, 495
– – solids concentration **3**, 495
– – solutes **3**, 489ff
– – tangential flow velocity **3**, 493f
– – temperature **3**, 495
– – transmembrane pressure drop **3**, 493f
– – tube diameter **3**, 494
– enzyme transmission rate **3**, 498
– filtrate flux **3**, 493f, 497
– flow resistance **3**, 490
– fluidization theory **3**, 491f
– flux **3**, 492
– – decline minimization **3**, 497
– – time-dependent decline **3**, 496f
– gel polarization theory **3**, 489ff
– harware selection **3**, 498
– hollow-fiber systems **3**, 45f
– membranes **3**, 44
– – characteristics **3**, 495
– – see also filtration membranes
– – selection **3**, 498
– modeling **3**, 489ff
– module configurations **3**, 498f
– optimization of, flow conditions **3**, 497
– – tangential velocity **3**, 497
– – transmembrane pressure **3**, 497
– performance improvement **3**, 497
– permeate flux **3**, 493
– plate and frame systems **3**, 45f
– process parameters **3**, 492ff
– scale-up **3**, 499
– sedimentary pipe flow **3**, 491f
– – critical velocity **3**, 491

– settling force **3**, 491
– solids build-up **3**, 493
– spiral-wound systems **3**, 45f
cross-indexing **5B**, 291
crossing over **2**, 434ff
– double event **2**, 436ff
– mitotic **2**, 84
– single event **2**, 434ff
cross inhibition, of alcohol dehydrogenases **8A**, 429f
cross-linked enzyme crystals (CLECs) **8A**, 52
– in peptide synthesis **8A**, 256
cross-linked enzymes **8A**, 407f
cross protection **10**, 492
crown gall tumor **2**, 549
CRS (Congressional Research Service) **12**, 24
crude oil, elimination from tidal environments **6**, 697
crude preparations, of lipases **8A**, 43f
crustaceans **10**, 445f
cryophiles **10**, 273
cryopreservation, of cell cultures **10**, 31f
cryoprotectants **3**, 703ff
– formulation characteristics **3**, 706
– in cell cultures **10**, 32
Cryptococcus curvatus **7**, 152ff
– cocoa butter fatty acids **7**, 153
– fatty acyl composition **7**, 153
Cryptococcus elinovii, immobilization effect **1**, 242f
– phenol degradation **1**, 241f
– phenol resistance **1**, 243
Cryptococcus laurentii, α-amino- -caprolactam hydrolase **6**, 515
Cryptophyta, characteristics of **10**, 114
cryptosporidiosis, as waterborne disease, outbreaks in the 19th and 20th century **11C**, 415f
Cryptosporidium, in drinking water **11C**, 374, 377
– in river water **11C**, 381
Cryptosporidium parvum, elimination from drinking water **11C**, 420ff
– transmission via drinking water **11C**, 414ff
– – listing of outbreaks **11C**, 418f
CsA *see* cyclosporin A
CSIRO (Commonwealth Scientific and Industrial Research Organization) **12**, 425f
CSNDB **5B**, 406
CSTP (Committee for Scientific and Technological Policy) **12**, 612f, 617
CSTR *see* completely stirred tank reactor
– – *see* stirred tank bioreactor
CTM01, biodistribution studies **5A**, 361f
CTP *see* cytidine triphosphate
CTR see CO₂ transfer rate
CUBC (China United Biotechnology Corporation) **12**, 399
CUBE (Concertation Unit on Biotechnology in Europe) **12**, 450ff, 534, 544

cucumber fermentation **9**, 631ff
– controlled – **9**, 636f
– natural – **9**, 634f
– process schemes **9**, 633
– purging **9**, 637
cucumbers *see also* pickles
– bloater formation **9**, 636, 638f
– malolactic fermentation **9**, 636
– microorganisms on raw vegetables **9**, 634
– pasteurization **9**, 632f
– pickling varieties **9**, 631
– pure culture fermentation **9**, 637
– refrigeration **9**, 631ff
culture collections **10**, 36f
– addresses **8A**, 32
– listing **8A**, 32
– listing of **10**, 37, 116
– of algae **10**, 116
– of cyanobacteria **10**, 116
culture maintenance, algae **10**, 115
– animal cells **10**, 27ff
– plant cells **10**, 27ff
culture media *see* fermentation media
– – *see* media
cumenes, hydroxylation of **8B**, 390f, 392
cumulative energy expenditure **11A**, 573
Curagen Pathcalling **5B**, 406
curcumin, as substrate for vanillin production **8B**, 335
– hydrolysis of **8B**, 335
curdlan, consumption **6**, 627
– formation by *Agrobacterium* sp. **6**, 635
– price **6**, 627
– production, by *Alcaligenes faecalis* **6**, 621
– structure **6**, 635
– use in Japanese food industry **6**, 636
Curvularia falcata alcohol dehydrogenase (*CFADH*), characteristics of **8A**, 453
Curvularia intermedia, Baeyer-Villiger biotransformations **8A**, 560
Curvularia lunata see Cochliobolus lunatus
Curvularia lunata, Baeyer-Villiger biotransformations **8A**, 559
custers effect, in yeast **1**, 486
cut-and-feed culture **3**, 28
cutting mills, disintegration of organic wastes **11C**, 114ff
– mobile – **11C**, 116
CVI (Children's Vaccine Initiative) **12**, 358
CVL *see Chromobacterium viscosum* lipase
cyanamide, conversion to urea **8A**, 295
cyanide, biotransformation of **8A**, 299ff
– monitoring with biosensors **11A**, 128
– – use of immobilized yeast **11A**, 133
cyanide hydratases **8A**, 299ff
– fungal –, basic mechanisms **8A**, 299
– – use in bioremediation of wastes **8A**, 312f

cyanide wastes, fungal cyanide hydratase, use of **8A**, 312f

β-cyanoalanine synthase, properties **8A**, 280

β-cyanoalanine synthesis, by cyanogenic bacteria **8A**, 280

– – by cyanogenic plants **8A**, 280

cyanobacteria, aquaculture of **10**, 128

– as sources of nonribosomal proteins **7**, 307f

– biology of **10**, 108ff

– biotechnological use of **10**, 105ff

– blooms, therapeutic value of toxins **10**, 131

– culture collections **10**, 116

– exploitation of **10**, 126ff

– genetic engineering of **10**, 110f

– hydrogen metabolism of **10**, 472

– hydrogen production by, optimization of **10**, 472f

– in biodeterioration **10**, 269ff

– in human diet **10**, 126f

– mass cultivation **10**, 119ff

– – in tubulat photobioreactors **10**, 123ff

– – plate type reactors **10**, 125f

– – ponds **10**, 119ff

– production of biomass **10**, 115ff

– – human diet **10**, 126f

– taxonomy **10**, 107, 109f

– use as biofertilizer **10**, 128f

cyanobacteriae, production of biomass, animal feed **10**, 127f

cyanogenic glycosides **8A**, 280ff

– biosynthetic pathway in plants **8A**, 282

– general scheme for the release of cyanide **8A**, 283

– metabolic pathways in plants **8A**, 280

cyanohydrine production, enzymatic formation **3**, 460

– yield **3**, 460

cyanohydrin formation, enzymatic – **8B**, 366

cyanohydrins, biosynthesis **8A**, 301

– biotransformation of **8A**, 301f

– enzyme-catalyzed formation **8A**, 301

cybernetic models **4**, 290ff

– defintion **4**, 290

– formulation **4**, 291

cybrid **2**, 127, 129

– mammal cell **2**, 129

– plant **2**, 127

cycle sequencing, use of thermostable DNA polymerases **10**, 89

cycle-to-cycle optimization, definition **4**, 558

cyclic adenosine 3′,5′-monophosphate (cAMP), fermentative production **6**, 580

cyclic diesters, asymmetrization of, by acetylcholinesterase **8A**, 213

cyclic diesters (racemic –), kinetic resolution by pig liver esterase **8A**, 201

cyclic dipeptides, catalysis of Strecker reaction **8B**, 496f

– enantiodiscrimination mechanism **8B**, 497f

– hydrocyanation by **8B**, 492ff

– – aminol mechanism **8B**, 498f

– – autoinduction **8B**, 497

– – imidazole/cyanide salt mechanism **8B**, 498

– – mechanism **8B**, 497ff

– – optimal substrate **8B**, 492

– – production of natural products and drugs **8B**, 496

– hydroxyanation by, optimal conditions **8B**, 492

– stereoselective hydrocyanation by, of benzaldehyde **8B**, 493ff

– – of benzaldehyde derivatives **8B**, 493ff

– used for enantioselective hydrocyanation, intermolecular hydrogen bonding **8B**, 497

– use in stereoselective synthesis **8B**, 492ff

cyclic guanosine 3′,5′-monophosphate (cGMP), fermentative production **6**, 581

cyclitols **7**, 454f

– basic pathways **7**, 402ff

– benzoic acid derivatives, structure **7**, 407

– biogenesis **7**, 402ff

– components of nucleoside-type antibiotics **7**, 454

– dehydroquinate pathway **7**, 403ff

– *myo*-inositol pathway **7**, 403ff

– sugar components, attachment of – **7**, 405ff

– synthesis of **8B**, 13

cycloaliphatic compounds, bacterial degradation of **11B**, 148ff

– – metabolic pathway **11B**, 151

cycloalkanes, Baeyer-Villiger monooxygenase type reactions **11B**, 198f

– ring cleavage of **11B**, 198

cyclocymopole, structure **10**, 454

cyclodextrin **1**, 381

β-cyclodextrin **8B**, 142

cyclodextrin glucanotransferase (CGT) **1**, 381

cyclodextrin glycosyltransferases, bacterial – **8B**, 267f

– thermoactive – **10**, 77f

cyclodextrin glycotransferase **10**, 76ff

cyclodextrins **9**, 674, 748

– alcohol dehydrogenases in **8A**, 420ff

– applications **9**, 762

– as co-solvents **8A**, 426

– oxidoreductions in **8A**, 425ff

– product complexes with **8A**, 426

– production of – **9**, 761f

– properties **9**, 762

– properties of natural – **6**, 637

– structure of **8A**, 425

– substrate complexes with **8A**, 426

cyclohexane, bacterial degradation of **11B**, 148ff

– – metabolic pathway **11B**, 151

– hydroxylation of **11B**, 199

– ring cleavage of **11B**, 198

– scrubbing liquid quantities for the absorption of **11C**, 312
cyclohexane dinitriles, hydrolysis of, by *Corynebacterium* **8A**, 291
cyclohexanone monooxygenase (CHMO) **8A**, 570ff
– active site model **8A**, 551
– Baeyer-Villiger reaction, mechanisms for **8A**, 554
– Criegee intermediates of *Acinetobacter* **8A**, 553
– from *Acinetobacter* sp., mechanism of the Baeyer-Villiger reactions **8A**, 548
– of *Acinetobacter calcoaceticus* **8A**, 570ff
– oxidation of sulfides to sulfoxides **8A**, 572
– stereoselectivity of **8A**, 575
– thiolactone inhibitors of **8A**, 573
cyclohexanone monooxygenases, expression in yeast **8A**, 546
cyclohexan-1.2-diol, metabolism to **8A**, 547
cyclohexene derivatives, asymmetrization by pig liver esterase **8A**, 197
cycloleucomelone, iron-binding compounds **7**, 232
– structure **7**, 232
cyclooxygenase, hydroxylation by **8A**, 504
cyclopentane, ring cleavage of **11B**, 198
cyclopentane dicarboxylates, asymmetrization by pig liver esterase **8A**, 198
cyclophilin A, complexed with cyclosporin A, 3D-structure **7**, 566
– complexed with cyclosporin derivatives **7**, 568ff
– – X-ray structures **7**, 568
– peptidyl–prolyl isomerase active site **7**, 564ff
– target of cyclosporin A **7**, 560
– 3D-structure, complexed with a tetrapeptide **7**, 563f
cyclophilin B, complexed with a cyclosporin derivative **7**, 570f
– – X-ray structure **7**, 570
cyclophilin C, complexed with cyclosporin A, X-ray structure **7**, 571
cyclophilins **7**, 547f, 560ff
– amino acid residues **7**, 560f
cyclopropane, antitumor agent **7**, 655
cyclopropanes, halogenation of **8B**, 197
cycloreversions, use of catalytic antibodies **8B**, 467
cyclosporin **3**, 10
cyclosporin A **7**, 538ff
– effector site **7**, 548
– immunosuppressive properties **7**, 539ff
– indications **7**, 539
– mode of immunosuppressive action **7**, 546ff
– – dual domain concept **7**, 546, 548
– structure **7**, 538
– T cell selective immunosuppressant **7**, 539
– X-ray structure **7**, 553
– 3D-structure, in apolar environment **7**, 561
– – in polar environment **7**, 562
cyclosporin A analogs, non-immunosuppressive –, antiviral activity **7**, 541, 551
– – effect on multidrug resistant tumor cells **7**, 540, 551
– SDZ NIM 811, structure **7**, 541
– SDZ PSC 833 **7**, 541
– SDZ 214-103 **7**, 580
– – structure **7**, 538
– – X-ray structure **7**, 562
cyclosporin biosynthesis **7**, 572ff
– building units **7**, 573
– cofactors **7**, 573
– cyclosporin synthetase **7**, 572ff
– mechanistic aspects **7**, 577ff
– modified thiotemplate mechanism **7**, 578
cyclosporin C, cleavage of peptide bonds **7**, 558
cyclosporin–cyclophilin complexes **7**, 549f
– crystal structures **7**, 565ff
cyclosporin receptors **7**, 547f
cyclosporins **7**, 535ff
– activity against tumor multidrug resistance **7**, 540, 551
– analogs **7**, 294f
– anti-HIV activity **7**, 540ff, 550f
– antiinflammatory effects **7**, 550
– antimalarial activity **7**, 551
– binding to cyclophilins **7**, 547ff, 558f
– – effects of valine for leucine substitutions **7**, 560
– chemical synthesis **7**, 552ff
– constituent amino acids **7**, 551f
– cyclic peptide backbone, chemical transformations of – **7**, 555
– fermentation **7**, 552ff
– immunosuppressive activity **7**, 539ff
– – mechanism **7**, 546ff
– – relation to structure **7**, 556ff
– incorporating non-proteinogenic amino acids **7**, 556
– macrocycle, selective ring opening **7**, 555
– mode of immunosuppressive action **7**, 542ff
– resistance to systemic proteases **7**, 551
– structural investigations **7**, 561ff
– structure–activity relationships **7**, 557ff
– structures in polar and non-polar solutions **7**, 552
– treatment of autoimmune diseases **7**, 540
– use in transplant rejection therapy **7**, 539f
– variety of conformations **7**, 571
cyclosporin synthetase, characterization **7**, 573f
– enzymatic activities **7**, 572f
– gene **7**, 573f
– – manipulation by integrative transformation **7**, 574ff
– – structure **7**, 575
– theoretical reprogramming **7**, 576
cyclosporin variants, biosynthesis **7**, 580

cyclovariegatin, iron-binding compounds **7**, 232
– structure **7**, 232
Cylindrocarpon destructans, Baeyer-Villiger biotransformations **8A**, 563
D-cysteine desulfhydrase **6**, 528
L-cysteine, food uses **6**, 468
L-cysteine desulfhydrase **6**, 525
L-cysteine production, from ATC **6**, 515
– keratin hydrolysis **6**, 475
cysteine protease, from protists **10**, 144
cysteine proteases **5A**, 207
cysteine synthesis, alteration **2**, 798
cystic fibrosis **5A**, 472
– gene therapy with adenovirus vectors **5A**, 413, 421
– use of liposomes in gene therapy **5A**, 433
– use of recombinant enzymes **5A**, 167
cystic fibrosis (CF) **5B**, 64
– gene, linkage analysis **5B**, 77f
– – positional cloning **5B**, 77f
– gene therapy of **5B**, 84
– prevention of **5B**, 84
cystic fibrosis transmembrane conductance regulator (CFTR) **5B**, 64
– gene **5B**, 85
– knock-out mouse **5B**, 86
Cystoviridae **1**, 550
cytarabine ocfosfate, antimetabolic effects **6**, 565f
– metabolic pathway **6**, 566
– structure **6**, 566
cytidine, (CMP) fermentative production **6**, 579
– formation by *Bacillus subtilis*, mutant strains **6**, 593f
cytidine diphosphate choline *see* CDP-choline
cytidine triphosphate (CTP), and phosphodiester formation **8B**, 232
cytochrome bc$_1$ complex, in pseudomonads **1**, 417
cytochrome c **5A**, 20
cytochrome P-450-dependent enzymes **8A**, 505
cytochrome P-450-dependent monooxygenase (cyt. P-450) **8A**, 477f
– expression in *Escherichia coli* **8A**, 483
– expression in yeast **8A**, 483
cytochrome P-450 monooxygenases, metabolism of polycyclic aromatic hydrocarbons **11B**, 219
cytodifferentiation, in *Streptomyces*, A-factor **7**, 28
– – signal cascade **7**, 27
– role of secondary metabolism **7**, 16
cytoduction **9**, 437
cytokine antagonists, in therapy **5A**, 140
cytokine receptors **7**, 545
cytokines **5A**, 135ff, 146ff
– – *see also* interferons, interleukins
– definition **5A**, 146
– in IBD therapy **5A**, 144
– medical applications of recombinant proteins **5A**, 146

– release upon T cell activation **7**, 545f
cytolysin/bacteriocin **7**, 335f
– primary structure **7**, 335
cytomegalovirus **5A**, 472
– monoclonal antibody therapy **5A**, 317
Cytophaga, in activated sludge **11A**, 97f
cytoplasm **1**, 15
cytoplasmic membrane *see* membrane
cytosine **2**, 194, 320
cytosine methylation **2**, 596
cytoskeleton **1**, 30, 38
– effect of microgravity on **10**, 512
cytosol **1**, 14
cytostatic compounds, as products of marine biotechnology **10**, 450ff
cytotoxic agents, from basidiomycetes **7**, 503ff
cytotoxic lymphocytes, HIV reactive – **5A**, 481f
cytotoxic metabolites, marine organisms **7**, 687
Cy3 **5B**, 234
Cy5 **5B**, 234

D

DAGD *see* dialkylglycine decarboxylase
DAGs *see* diacylglycerides
DAHP synthase **8B**, 315
– activity, deregulation of **8B**, 317
dairy industry *see* milk products
dairy products, cultured – **9**, 385ff
– – composition of **9**, 392
– – consumption of **9**, 388f
– – fermentation **9**, 407
– – heat treatment **9**, 405f
– – homogenization **9**, 406f
– – ingredients **9**, 401ff
– – manufacturing procedures **9**, 407ff
– – mix preparation **9**, 405
– – packaging and storage **9**, 416
– – plant equipment **9**, 404f
– quality control **9**, 416f
– starter cultures **9**, 390f, 394ff
– – *Lactobacillus acidohilus* **9**, 397
– – *Lactobacillus casei* **9**, 397
– – *Lactobacillus delbrueckii* **9**, 397
– – *Lactobacillus lactis* **9**, 396
– – *Leuconostoc mesenteroides* **9**, 396
– – *Streptococcus thermophilus* **9**, 397
– use in baked goods **9**, 260
– use of enzymes **9**, 679ff
dairy wastewater **11A**, 221
– amounts **11A**, 208
– concentrations **11A**, 208
– treatment **11A**, 208f
dalbaheptide fermentation, feedback inhibition of – **7**, 382f
– media **7**, 381f

– – carbon sources **7**, 381f
– – effect of phosphates **7**, 382
– – nitrogen sources **7**, 382
– product recovery **7**, 383
– purification **7**, 383
dalbaheptides **7**, 369ff
– actinoidin type, naturally occurring – **7**, 375
– biological activity **7**, 377ff
– – *in vitro* antibacterial activity **7**, 377f
– – *in vivo* efficacy **7**, 379f
– – mechanism of action **7**, 378f
– – pharmacology **7**, 379f
– biosynthesis **7**, 384ff
– biotransformation of – **7**, 389f
– – deacylation **7**, 389
– – deglycosylation **7**, 389
– – glycosylation **7**, 389
– chemical modifications of – **7**, 387f
– definition **7**, 371
– descriptive chemistry **7**, 371ff
– effect in mouse septicemia **7**, 380
– effect on pathogens **7**, 378
– glycolysation of – **7**, 387
– glycopeptide skeleton **7**, 372
– lipoglycopeptides, naturally occuring – **7**, 375
– physicochemical properties **7**, 376f
– – isoelectric points **7**, 376
– – reverse-phase HPLC retention times **7**, 377
– producing organisms **7**, 372, 374f, 380
– resistance to – **7**, 379
– ristocetin type, naturally occurring – **7**, 372
– screening methods **7**, 381
– uncommon amino acids in – **7**, 384ff
– vancomycin type, naturally occurring – **7**, 374
Dalton's law **4**, 32
damascenone, biotransformation of **10**, 411
damascone, biotransformation of **10**, 382, 411
α-damascone, hydroxylation of **8A**, 493
Damköhler number **3**, 173f, **11A**, 569f
– calculation, for continuous sterilizer design **3**, 174
dangerous substances, in municipal wastewater **11A**, 163
Danish Distilleries Process **6**, 154
daphnids, use in ecotoxicological testing **11B**, 133, 137
Darcy's law **3**, 490ff
data, experimentally generated – **5B**, 136
– predicted – **5B**, 136
data abstraction **5B**, 386
data analysis, in bioprocesses **3**, 355ff
data assessment, in monitoring of wastewater treatment **11A**, 111ff
databanks **5B**, 124ff
– genetic epidemiology – **5B**, 140
– knowledge – **5B**, 126
– primary – **5B**, 126
– secondary – **5B**, 126

database management system **5B**, 124
Database of Ribosomal Cross-links (DRC) **5B**, 407
database parsers **5B**, 295
databases, further development **12**, 333f
– relevant to biotechnology **12**, 323ff
– – bibliographic – **12**, 327
– – factual – **12**, 325ff
– – full-text – **12**, 328
– – referral – **12**, 327f
– – use **12**, 334ff
database searches of protein sequences **5A**, 53ff
data integration systems **5B**, 126, 292
data organization, flat file **5B**, 125
data warehousing **5B**, 296
Daucus carota, suspension culture characteristics **1**, 586
daunomycin, antitumor agents **7**, 652
daunorubicin, antitumor agent, structure **7**, 658
daunorubicin production, in *Streptomyces peuceticus,* regulatory genes **7**, 91
dawadawa, nutritional characteristics **9**, 542
– production of **9**, 542f
DBA-WABIO process **11C**, 163f
DBGET **5B**, 291
DBT *see* dibenzothiophene
DCA *see* dichloroethane
– – *see* 1,2-dichloroethane
DCB (Development Center of Biotechnology of Taiwan) **12**, 403ff
DCM *see* dichloroethane
– – *see* dichloromethane
DDBJ **5B**, 283
DDM degradation **1**, 229
DDT, dehydrochlorination of **8B**, 206
– dehydrohalogenation of **8B**, 205f
DDT degradation **1**, 229, 235
deacetyllanatoside *C* **1**, 597
deacylation, of dalbaheptides **7**, 389
dead-end filtration **3**, 440f
dealkylation *see also* demethylation
dealkylation reactions **8A**, 496ff
death rate, specific –, temperature sensitivity **3**, 163
Debaryomyces hansenii, in olive fermentation **9**, 612
debris removal, by centrifugation, breakthrough **3**, 485
– costs **3**, 501
– effects of particle size distribution **3**, 477, 483
– from fermentation broths **3**, 469ff
– – of *Escherichia coli* cells **3**, 482
– model system **3**, 482ff
debromohymenialdisine, antiinflammatory compound of marine origin **10**, 455f
Debye-Hückel parameter **5A**, 94
γ-decalactone, formation of **8B**, 340ff

– – β-oxidation pathway of ricinoleic acid **8B**, 342
– – by *Yarrowia lipolytica* **8B**, 341f
– sensoric description of **8B**, 341
decalins, formation of isomeric –, use of catalytic antibodies **8B**, 431
decarbonization, in drinking water treatment **11C**, 402ff
– rapid – **11C**, 403
– "rapid slow" process **11C**, 404
– slow – **11C**, 403
decarbonization reactors, types of **11C**, 403
decarboxylation, of α-acetolactate **8B**, 63
– of acetoacetic acid **8B**, 59f
– of α-methylornithine **8B**, 79
– of *cis*-aconitate **8B**, 65
– of dimethylglycine **8B**, 87
– of fluoropyruvate **8B**, 54, 56
– of L-arginine **8B**, 80
– of L-histidine **8B**, 81
– of L-lysine **8B**, 80
– of L-ornithine **8B**, 79
– of L-tyrosine **8B**, 82
– of malonic acid **8B**, 69
– of malonyl-CoA **8B**, 69
– of orotidine-5′-phosphate **8B**, 82
– of oxaloacetate **8B**, 56
– of phenylpyruvate **8B**, 86
– pyruvate decarboxylase mediated – **8B**, 49
– – catalytic sequence **8B**, 52
– use of catalytic antibodies **8B**, 466f
DECHEMA **12**, 526ff, 580ff
dechlorination, aerobic –, cometabolism-based – **11C**, 285
– of polychlorinated biphenyls (PCB) **11A**, 46
– of xenobiotics **11A**, 45f
decision-making, ethical **12**, 121f
– in product development, delegation of responsibility **12**, 232f
– – project failure **12**, 233
– – routine schemes **12**, 232
– interactive rational – **12**, 32
decision tree, evaluation of rDNA products **12**, 189f
– for approval, of GMO release **12**, 183ff
– for food components **12**, 179
– for foods **12**, 175ff
– – containing rDNA-encoded materials **12**, 177ff
– – derived from GMOs **12**, 182f
– – derived from new plant varieties **12**, 176ff
– – derived from transgenic animals **12**, 181
– – derived from transgenic plants **12**, 180
– for rDNA products **12**, 175ff
decontaminated soil, utilization of **11B**, 315f, 473
decontamination, of biotechnological facilities, validation **3**, 763
de-emulsification *see* bio-de-emulsification

deep bed rapid filters, removal of organic matter **11C**, 458ff
defective mutants, glycolysis-defective yeasts **1**, 480
defence response to pathogens **5B**, 150
defense proteins **5A**, 34f
– – *see also* antibodies
– blood clotting **5A**, 34
– clotting factors **5A**, 34
– of the immune system **5A**, 35
– preventing penetration of disease causing factors **5A**, 34
deferrization, in drinking water treatment **11C**, 404f
deglycosylation, of dalbaheptides **7**, 389
degradability tests, anaerobic wastewater purification **4**, 453
degradative genes **11B**, 444ff
– for branched aromatic hydrocarbons **11B**, 445
– for chlorinated compounds **11B**, 445f
– for chlorobenzoates **11B**, 445
– for chlorobiphenyls **11B**, 446ff
– for nitroaromatic compounds **11B**, 452
– for polychlorinated biphenyls **11B**, 446ff
– for sulfur compounds **11B**, 452
– for trichloroethylene **11B**, 449ff
– for 2,4-dichlorophenoxyacetic acid **11B**, 451
– of *Burkholderia* sp. **11B**, 450
– of *Comamonas* sp. **11B**, 448
– of *Escherichia* sp. **11B**, 448
– of *Pseudomonas* sp. **11B**, 447
dehalogenases **8B**, 181ff
– classification of **8B**, 182ff
– commercial application of **8B**, 211f
– dehalogenation via epoxide formation **8B**, 183
– dehydrohalogenation **8B**, 183
– glutathione-dependent – **8B**, 205
– halidohydrolase-type **8B**, 204f
– haloalcohol **8B**, 205
– hydrolytic dehalogenation **8B**, 183ff
– – classification of **8B**, 183f
– oxygenolytic dehalogenation **8B**, 182f
– reductive dehalogenation **8B**, 182
dehalogenation, by clostridia **1**, 313
– – tetrachloromethane degradation **1**, 313
– by intramolecular substitution of vicinal haloalcohols **11B**, 249f
– dehydrodehalogenation **11B**, 249f
– hydrodehalogenation **11B**, 249f
– hydrolytic – **11B**, 248f, 256f
– oxygenolytic –, by dioxygenases **11B**, 248
– reductive – **11B**, 185, 247
– – enzymes **11B**, 247f
– – lignin peroxidase **11B**, 246
– – microorganisms **11B**, 247
– – principles of **11B**, 245f, 261f
– thiolytic –, of dichloromethane **11B**, 249
dehalogenations **8B**, 202ff

– aliphatic – **8B**, 203
– catalyzed by specific enzymes **8B**, 181
– dioxygenase-catalyzed – **8B**, 208ff
– hydrolytic – **8B**, 183, 204f, 210
– monooxygenase-catalyzed – **8B**, 210
– of chlorinated aromatic compounds **8B**, 207
– oxidative – **8B**, 203f, 208
– oxygenolytic – **8B**, 183
– reductive – **8B**, 182, 203, 207
– via epoxide formation **8B**, 183
dehydratases **8B**, 98f
dehydration, of biomolecules **3**, 703
dehydroalanine residues in peptides, formation
    of **8B**, 134
dehydro-amino acids **8A**, 253f
dehydrochlorination, of DDT **8B**, 206
– of lindane **8B**, 206
7-dehydrocholesterol, structure **7**, 171
dehydrodidemnin B (aplidine), antitumor com-
    pound of marine origin **10**, 455
dehydrogenases **8A**, 391ff
– – *see also* oxidoreductases
– coenzyme recycling **8A**, 402
– coenzyme regeneration sytem, requirements of
    8A, 399
– cofactor regeneration, in enzyme reactors **3**,
    454
– commercial – **8A**, 436ff
– – biochemical characteristics of **8A**, 438
– – manufacturers of **8A**, 449
– enantioselective reduction, of β-ketoacid esters
    8A, 412
– – of β-ketoacids **8A**, 412
– enzymatic transformations **8A**, 396ff
– enzyme-coupled regeneration **8A**, 402
– noncommercial **8A**, 448ff
– – biochemical characteristics of **8A**, 450
– purification **3**, 611
– substrate spectra **8A**, 437
– total turnover number **8A**, 399
– turnover number **8A**, 398f
– typical reactions **8A**, 394ff
– whole-cell transformations **8A**, 396ff
dehydrohalogenation **8B**, 183, 205f, 211
– of DDT **8B**, 205f
– of lindane **8B**, 205f
dehydroquinate pathway **7**, 403ff
dehydroquinate synthase, substrates **7**, 405
3-dehydroquinate dehydratase (DHQD) **8B**, 111f
de-inking, of recycled paper pulp **10**, 539f
deionization (DI), in production water treatment
    3, 757
delicious octapeptide amide, synthesis **8A**, 256f
Delors White Paper **12**, 648ff
– follow-up **12**, 650ff
demanganization, in drinking water treatment
    11C, 404f
Demeter's genomes **5B**, 285

demethylation, of colchicine **8A**, 497f
– of immunomycin **8A**, 497
– of magnolin **8A**, 498
denaturing gradient gel electrophoresis (DGGE),
    in microbial community analyis **11A**, 88ff
– – applications **11A**, 88
– – limitations **11A**, 88ff
– – method **11A**, 88f
denim processing, use of industrial cellulases **5A**,
    204
denitrification **10**, 279, **11A**, 55ff, 59ff, 118f, 256f,
    264f, 335f, 341f
– activated sludge process **11A**, 256f, 264f
– – pre-anoxic zone – **11A**, 256, 270f
– – step-feed – **11A**, 272ff
– aerobic – **11A**, 59ff
– by ammonia oxidizers **11A**, 57
– by nitrite oxidizers **11A**, 58
– capacity in non-aerated basins **11A**, 344
– degradation of aromatics **1**, 419
– denitrifying bacteria **11A**, 39
– enzyme properties **1**, 417ff
– following anaerobic wastewater treatment **11A**,
    407f
– in an activated sludge stage **11A**, 343f
– in fixed growth reactors **11A**, 335ff
– in pseudomonads **1**, 417, 419ff
– microbial fundamentals **11A**, 55ff
– mixing tanks for – **11A**, 268
– modeling of **11A**, 422f
– nitrate removal **11A**, 39
– of landfill leachates **11C**, 188
– physiology of **11A**, 59
– process scheme **11A**, 118
– use of rotating disc filters **11A**, 346
– use of trickling filters **11A**, 341ff
– – scheme **11A**, 342
denitrification rates, in biofilters **11A**, 370
denitrifiers, substrate requirements **11A**, 59
denitrifying bacteria, oligonucleotide probes **11A**,
    92
Denmark, biotechnology regulatory policy **12**,
    603f
– contaminated site management, financing sche-
    mes **11B**, 15
– – legal development **11B**, 9
– – principles **11B**, 11
– – registration **11B**, 17
– – risk assessment **11B**, 19
– – technical development **11B**, 9
– identification of contaminated sites **11B**, 17
– public opinion survey, on gene technology **12**,
    660
– regulatory concerns, action thresholds **11B**, 21
– – conflict management **11B**, 28
– – enforcement **11B**, 26
– – in soil contamination **11B**, 7ff
– – liability rules **11B**, 13

– – organization **11B**, 26
– – priority setting **11B**, 23
– – remediation targets **11B**, 24
– – verification of clean-up **11B**, 24
de novo mutation **5B**, 69
dense membranes, in biological waste gas purification **11C**, 349ff
– – scheme **11C**, 350
dental caries **1**, 344
dental plaque **11A**, 71, 73
3'-deoxyadenosine, production by *Aspergillus nidulans* **6**, 596
3-deoxy-D-arabino-heptulosonate 7-phosphate synthase *see* DAHP synthase
6-deoxyerythronolide B synthase **10**, 354f
– modular organization of **10**, 355
– structural studies of **10**, 357
6-deoxyhexoses **7**, 410ff
– as moieties of antibiotics **7**, 456
– biosynthesis, general pathway **7**, 408
– biosynthetic pathways, gene clusters **7**, 455
– structural variants, in secondary metabolites **7**, 410ff
– – occurrence in natural products **7**, 414
deoxyilludin M, antitumor activity **7**, 503
– structure **7**, 504
1-deoxynojirimycin **8B**, 296ff
– α-glucosidase inhibitor **8B**, 297
– industrial synthesis of **8B**, 295
– production of, by chemical synthesis **8B**, 299
– – by extraction from plants **8B**, 298
– – by fermentation **8B**, 298f
– – combined biotechnological-chemical synthesis **8B**, 299ff
– sources **8B**, 297f
– structure **8B**, 296ff
deoxynucleoside triphosphate, chemical structure **2**, 272
deoxynucleoside-3'-phosphorothioamidate **2**, 326
– synthesis of non-oxygen bridged – **2**, 328
deoxyribonucleotide **2**, 193
2-deoxyribose-5-phosphate aldolase **8B**, 17ff
– acetaldehyde condensations **8B**, 18
– from *Escherichia coli* **8B**, 17
– synthesis of carbohydrate analogs **8B**, 17f
deoxysugar synthesis **8B**, 23
deoxy-2'-methylidenecytidine, antitumor action **6**, 567
– structure **6**, 567
dependent patents **12**, 309
dephlegmators **6**, 100f
depositories, for microorganisms **12**, 210f
– – in Australia **12**, 429
– – in Japan **12**, 385
– – in Korea **12**, 395
depth filtration **3**, 440
derivatization, fluorometry **4**, 165
– reagents **4**, 165

desalination, by reverse osmosis **11C**, 405
desferrithiocin, orally available iron chelator **7**, 207
designed oligospecies culture systems **11C**, 280f
designer proteins **5B**, 355
desmosome **1**, 40
desoxyribonucleic acid *see* DNA
destomycin family, structure of – **7**, 469
destomycins **7**, 436
Desulfobacterium, anaerobic wastewater process **4**, 447
Desulfovibrio, anaerobic wastewater process **4**, 447
detectors **11B**, 502f
– used for GC **11B**, 502f
– used for HPLC **11B**, 502f
detector signal, in DNA sequencing **5B**, 204f
detergents, in wastewater **11A**, 12
– – municipal – **11A**, 175f
– use of, α-amylases **5A**, 196
– – cellulases **5A**, 204
– – lipases **5A**, 200f
– – recombinant proteases **5A**, 208
Deuteromycetes **2**, 530
Deutsche Gesellschaft für Technische Zusammenarbeit *see* GTZ
devazepide, activities **7**, 122
– structure **7**, 122
devazepide receptor antagonist **7**, 121
developing countries **12**, 339ff
– biotechnology regulatory approaches **12**, 56f, 350
– entry into biotechnology **12**, 343ff
– – agriculture **12**, 358ff
– – barriers **12**, 345ff
– – health care **12**, 352ff
– – innovations **12**, 345ff
– – threshold factors **12**, 344ff
– food supply, impacts of biotechnology **12**, 361ff
– health care problems, impacts of biotechnology **12**, 353ff
– immunization programs **12**, 357f
– national biotechnology policy **12**, 348f
– R & D capabilities **12**, 343f
– tropical – **12**, 438
dextran, consumption **6**, 627
– in two-phase extraction, as phase-forming polymer **3**, 596, 599
– price **6**, 627
dextran production, by *Leuconostoc mesenteroides* **6**, 638
– saleable by-products **6**, 638
– use of dextransucrase **6**, 637
dextrans, ion exchange uses **6**, 639
– molecular weight ranges **6**, 639
– pharmaceutical uses **6**, 639
dextransucrase, dextran formation **6**, 637
dextrose *see* glucose

– applications **9**, 754f
– commercial products **9**, 753f
– – crystaline – **9**, 753
– – liquid – **9**, 753
– – pharmaceutical – **9**, 753f
– properties **9**, 754
– refining **9**, 752f
dextrose equivalent **6**, 35ff
– definition **9**, 739
dextrose hydrolyzates *see* glucose syrups
dextrose production **9**, 669ff, 745, 750ff
DGGE **11A**, 377ff
– – *see* denaturing gradient gel electrophoresis
1,2-dichloroethane (DCA), biodegradation of
– denaturing gradient gel electrophoresis 1,2-dich-
loroethane (DCA), in wastewater treatment
11A, 378f
– properties **11A**, 377
– use **11A**, 377
DGRST (Délégation Générale à la Recherche
Scientifique et Technique) **12**, 587ff
DG XII *see* European Commission's Directorate
General for Science, Research and Development
DHA *see* docosahexaenoic acid
DHAD *see* α,β-dihydroxy acid dehydratase
DHAP *see* dihydroxyacetone phosphate
DHFR (dihydrofolate reductase) expression systems **5A**, 250ff
DHGLA *see* dihomo-γ-linolenic acid
dhokla, nutritional chracteristics **9**, 539
– production of **9**, 540
DHQD *see* 3-dehydroquinate dehydratase
diabetes **5B**, 71
diabetes mellitus **5A**, 134f, 161, 164
– antibody therapy **5A**, 314, 319
– cell implantation **10**, 548
– encapsulated-cell therapy **10**, 548
– implantation of pancreatic islets **10**, 555f
– insulin dependent – **5A**, 144
– medical applications of recombinant proteins
5A, 134, 144
– non-insulin dependent – **5A**, 134f
– treatment of, by Miglitol **8B**, 298
– treatment with CDP517 **5A**, 346f
diacetyl, avoidance in brewing **9**, 685
– in beer **9**, 443f, 684f
diacylglycerides (DAGs), acyl migration in **8A**,
139
– formation from glycerol **8A**, 142
– general features **8A**, 137ff
– lipase-catalyzed syntheses of **8A**, 138
dialkylbenzenes, microbial degradation of **11B**,
214
dialkylglycine decarboxylase (DAGD) **8B**, 86f
– decarboxylation catalyzed by **8B**, 88
– transamination catalyzed by **8B**, 88
dialysis bioreactors **3**, 441

dialysis fermentation, of *Sulfolobus solfataricus*
10, 71
cis-diaminedichloroplatinum (*cis*Pt) **2**, 203
diaminopimelate decarboxylase, decarboxylation
of L-arginine **8B**, 80
D,L-diaminopimelate biosynthetic pathway **1**, 208
L,L-diastereomers, protease-catalyzed synthesis
8A, 253
diastereotopic faces, of flavin adenine dinucleotide **8A**, 550
diauxic growth, *Klebsiella terrigena*, cybernetic
model **4**, 288
– – formal kinetic model **4**, 286
– – two-compartment model **4**, 287
– – unstructured model **4**, 279
– *Saccharomyces cerevisiae*, metabolic regulator
model **4**, 296
– two-compartment model **4**, 286
diazaqinomycin B, thymidylate synthase inhibition **7**, 648
dibenzodioxin, aerobic degradation of **11B**, 263
dibenzothiophene (DBT), degradation by genetically engineering microorganisms **11B**, 452
dichloroacetic acid, degradation by immobilized
cells **11B**, 419f
dichloroanilines, anaerobic biodegradation **11B**,
258
dichloroethane (DCA), biodegradation by *Xanthobacter autotrophicus* **11B**, 253
dichloroethane dehalogenase **11B**, 253f
– structure of the active site **11B**, 254
dichloromethane, anaerobic biodegradation **11B**,
252f
– scrubbing liquid quantities for the absorption
11C, 312
dichloromethane (DCM), biodegradation by dehalogenases **11B**, 251f
– biodegradation of **11A**, 377ff
– – in wastewater treatment **11A**, 378f
– degradation of, in biotrickling filters **11C**, 285
– properties **11A**, 377
– use **11A**, 377
dichlorophenol, degradation of **8B**, 207
2,4-dichlorophenoxyacetic acid, aerobic degradation of **11B**, 261
– degradation by genetically engineered microorganisms **11B**, 451
– – use of bioaugmentation **11B**, 434
– mineralization in soil **11B**, 110
dichloropropanes, biodegradation of **11B**, 255
dictyosome **1**, 40
2,3-didehydroalanine **7**, 328ff
2,3-didehydrobutyrine **7**, 328ff
didemnin B, structure **10**, 454
3,6-dideoxyhexoses, naturally occurring – **8B**, 120
3,6-dideoxyhexose pathway, of gram-negative bacteria **7**, 414
dideoxynucleotide **2**, 271f

– use in DNA sequencing **2**, 271f
3,6-dideoxysugars **8B**, 117ff
dielectrophoresis, of cells **2**, 99f
"Diels-Alderase", antibody **8B**, 417
Diels-Alder cycloadditons, control of translational entropy **8B**, 417
– of tetrachlorothiophene and *N*-ethylphthalimide **8B**, 418
– use of catalytic antibodies **8B**, 473f
Diels-Alder reaction, enantio- and diastereoselectivity of **8B**, 426
– – for *ortho*-approach **8B**, 426
– in biosynthesis **8B**, 30ff
– intramolecular – **8B**, 30f
– use of catalytic antibodies **8B**, 426ff
diepoxybutane (DEB) **2**, 24
diesel fuel, biodegradability **6**, 693
– degradation by immobilized cells **11B**, 420f
diesel oil, biodegradation of **11B**, 429
– bioremediation of **11B**, 427, 432
– colonizing bacteria **11B**, 66
diesters (*meso*), enantioselective hydrolysis by pig liver esterase **8A**, 196
diesters (prochiral –), enantioselective hydrolysis of **8A**, 195
diethylether, degradation of, in biotrickling filters **11C**, 286
– scrubbing liquid quantities for the absorption of **11C**, 312
diethylsulfate (DES) **2**, 24
diffuse large B cell lymphoma **5B**, 92
diffusion **1**, 72, **3**, 93
– effectiveness factor **3**, 435
– effects on biocatalysis **3**, 438
– facilitated – **1**, 77
– film – **3**, 435
– in immobilization beads **3**, 435
– of contaminants, in bioreactors **11B**, 333f
– over membranes **1**, 73
– pore – **3**, 435
– strong pore – **3**, 438
diffusion coefficient, shear-enhanced –, of filtration membranes **3**, 492
diffusion-collision model, of protein folding **3**, 544
diffusion membranes **3**, 245ff
– – *see also* aeration membranes
– bioreactor aeration **3**, 250
– mass transfer coefficients **3**, 247, 251
– permeabilities **3**, 248
– pressurized tubing **3**, 247
– reinforced tubing **3**, 247
– relative tubing area **3**, 252
– silicone – **3**, 245
– solution – **3**, 246
– – liquid supersaturation **3**, 250
– temperature effects **3**, 248
Digitalis, suspension culture **1**, 584

Digitalis lanata **7**, 621ff
– β-methyldigitoxin hydroxylation **1**, 596f
– cells in alginate bead **1**, 599
– tissue culture process **7**, 621
digitoxin **1**, 597
– production by biotransformation **7**, 622
digoxigenin, as DNA label **2**, 397
digoxigenin derivatives **8A**, 133
digoxin **1**, 597
– production by biotransformation **7**, 622
dihomo-γ-linolenic acid (DHGLA), as precursor of prostaglandins **7**, 160
– formation by deletion of desaturase **7**, 160
dihydroflavonols, preparation of, using poly-leucine catalyzed epoxidation **8B**, 501f
dihydrofolate reductase **8B**, 387
– folding **3**, 540
dihydrolipoamide **9**, 34
dihydrostreptose, biosynthetic pathway **7**, 422f
α,β-dihydroxy acid dehydratase (DHAD) **8B**, 111f
– of *Salmonalla typhimurium* **8B**, 112
dihydroxyacetone phosphate analogs, structure of **8B**, 8
dihydroxyacetone phosphate (DHAP), structure of **8B**, 8
– synthesis of **8B**, 8ff
dihydroxylation **8A**, 475ff
– definition of **8A**, 480f
– of alkenes **8A**, 512
– of alkenic monoterpenes **8A**, 512f
– of arenes **8A**, 481
– of benezenes **8A**, 508
– of benzenoids **8A**, 507ff
– of benzopyran **8A**, 515
– of fluoranthene **8A**, 511
– of geranylacetone **8A**, 513f
– of heterobicyclic arenes **8A**, 509f
– of naphthalenes **8A**, 509
– of nerolidol **8A**, 513f
– of non-aromatic C=C bonds **8A**, 511
– of precocene **8A**, 515
– of sesquiterpenes **8A**, 513f
– routes **8A**, 480f
– use of isolated enzymes **8A**, 482
– use of whole-cell catalysts **8A**, 481f
dihydroxylation reactions, dioxyenase-catalyzed –, non-aromatic C=C bonds **8A**, 511
– dioxygenase-catalyzed –, arenes **8A**, 506ff
– – heteroarenes **8A**, 509f
– monooxygenase-catalyzed – **8A**, 511ff
α-diketones, conversion to diols by yeast **8A**, 372
– yeast reduction of, use of selective enzyme inhibitors **8A**, 382
diltiazem, commercial synthesis, use of lipases **8A**, 147
– commercial synthesis of, lipase catalyzed – **8B**, 279

– preparation of, using poly-leucine catalyzed epoxidation **8B**, 501f

dilution rate, optimization, in a two-stage continuous culture system **4**, 497ff

dimethyldicarbonate, as wine preservative **9**, 489

dimethylglycine, oxidative decarboxylation of, by dialkylglycine decarboxylase **8B**, 87

dimethylsiloxane, aeration membranes **3**, 248

dimethylsulfide, in beer **9**, 444

dimethylsulfide (DMS) oxidation **1**, 274

dinitriles, aromatic –, selective hydrolysis of **8A**, 308

– prochiral –, selective amidase hydrolysis of **8A**, 310

– – selective nitrile hydratase hydrolysis of **8A**, 310

– regiospecific biotransformation of **8A**, 308

2,4-dinitrophenol (DNP), effect on microbial metabolism **1**, 172

– membrane proton transport **1**, 181

2,4-dinitrotoluene (2,4-DNT), biodegradation of **11A**, 383ff

– – in wastewater treatment **11A**, 385f

Dinoflagellata, characteristics of **10**, 115

Dinoseb *see* butyl dinitrophenol

dioecism **2**, 77

– morphological **2**, 77

– physiological **2**, 77

diol fermentation, in clostridia **1**, 301

dioxamycin, antitumor agent, structure **7**, 662

dioxane, bacterial degradation of **11B**, 205

dioxins, chlorinated – **8B**, 188

– degradation of **8B**, 207

dioxygenase **8B**, 208ff

– dihydroxylation of arenes **8A**, 506ff

– dihydroxylation of heteroarenes **8A**, 509f

dioxygenase enzymes **8A**, 481

dioxygenases, bacterial degradation of xenobiotics **11B**, 149

– biodegradation of nitrophenols **11B**, 281

– degradation of nitro-substituted aromatics **11B**, 279

– degradation of organic compounds **11B**, 171

– in alkane biodegradation **11B**, 198

– microbial degradation of benzene **11B**, 213

– oxygenolytic dehalogenation **11B**, 248

DIP – Database of Interacting Proteins **5B**, 406

dipeptide ligands, enantioselective hydrocyanation of aldehydes **8B**, 508

dipicolinic acid (DPA), in endospores **1**, 388

Dipodascopsis uninucleata, formation of prostanoid-type lipids **7**, 184f

direct cDNA selection **5B**, 175

directed evolution, of lipases **8A**, 149f

directory databases **12**, 327f

disaccharides, by immobilized glycosyltransferases **8B**, 258

– regioselective acylation by subtilisin **8A**, 224f

– synthesis of, by glycosylation **8B**, 253f

discodermolide, antitumor compound of marine origin **10**, 453

discontinuous (dichotomous) traits **5B**, 65

discrete-time controller **4**, 252f

– z-transform **4**, 252

disc stack centrifuges **3**, 476ff

– grade efficiency curve **3**, 478

– particle migration **3**, 476

– solids recovery **3**, 478

disease genes, identification of **5B**, 53ff, 108ff

disease phenotype, delayed onset **5B**, 71

disease resistance gene **2**, 806ff

– anti-sense RNA **2**, 809

– intracellular immunization **2**, 808

– Mx gene **2**, 806

– – foreign gene expression **2**, 807

– promotor efficiency **2**, 806

disease resistance genes, of crop plants **5B**, 170

diseases *see also* genetic diseases

– adult multifactorial – **5B**, 66

– caused by, *Cryptosporidium parvum* **11C**, 414ff

– – *Giardia lamblia* **11C**, 414ff

– – *Toxoplasma gondii* **11C**, 414ff

– congenital – **5B**, 66

disinfection **3**, 159, 781

– chemical, implications on public health **3**, 784

– of drinking water **11C**, 409ff

– of sewage sludge **11C**, 220ff

– of wastewater **11C**, 219

dispersing machine **6**, 79f

– scheme **6**, 80

dispersing mash process **6**, 88f

– scheme **6**, 89

dispersion mechanisms, in bioreactors **4**, 114ff

– large-scale convectional flows **4**, 114

– molecular diffusion **4**, 115

– rising bubbles **4**, 116

– turbulence **4**, 114f, 118

– viscous shear **4**, 115f

displacement chromatography (DC) *see* chromatography

dissolved organic carbon (DOC), removal efficiency in biofilters **11C**, 469

dissolved oxygen **3**, 196, 206

– – *see also* oxygen

– cascade control **3**, 16

– concentration, in bioreactors **3**, 13, 366ff

– control **4**, 515

– – in stirred tank reactors **3**, 113

– effect in activated sludge process **11A**, 265f

distamycin, antitumor agent **7**, 650

– structure **7**, 649

distillation *see* ethanol distillation

distillers' mashes, dispersing mash process **6**, 88f

– preparation at saccharification temperature **6**, 85

– pressure boiling processes **6**, 82ff

distillery wastewater **11A**, 211ff
– amounts **11A**, 212
– concentrations **11A**, 212
– treatment **11A**, 212
distilling industry, analytical methods, mash hydrosizing **6**, 111ff
– – raw materials **6**, 109ff
– – yeast mash analysis **6**, 114
– ethanol determination **6**, 113
– fermentation by-product feeds **9**, 772f
distributed control systems **4**, 568ff
– general layout **4**, 569
– hardware modularity **4**, 568ff
distribution coefficients, for acetic acid **3**, 566
– – in liquid-liquid extractions **3**, 561ff
– – polar diluent effect **3**, 570
– for citric acid **3**, 571
– – Alamine 336 **3**, 571
– – concentration dependence **3**, 571f
– – diluent effect **3**, 571f
– – in liquid-liquid extractions **3**, 561
– – temperature dependence **3**, 571f
– – with tri-dodecyl amine **3**, 571
– for penicillin G, in liquid-liquid extractions **3**, 563
distribution constants, for carboxylic acids, in liquid-liquid extractions **3**, 567
diterpenes, hydroxylation of **8A**, 494f
– labdane diterpene hydroxylation **8A**, 495
diterpenoids, occurrence in basidiomycetes **7**, 493
DITS reactor *see also* bioreactors
– – *see* dual injected turbulent separation (DITS) reactor
DMT-BIODYN process **11B**, 338f
– configuration of **11B**, 337
DNA **2**, 193ff, 234, 320
– – *see also* plasmid DNA
– adapters **2**, 379f
– attachment to nuclear scaffold **2**, 675
– banked samples **5B**, 420ff
– – anonymization of **5B**, 421
– – archivation of **5B**, 421
– binding with polylysine **5A**, 435
– calcium phosphate precipitation **5A**, 430f
– cassettes **12**, 165
– changes **12**, 161
– compaction of – **5A**, 435
– composition **2**, 193ff
– content of various cells **5A**, 4
– damage of, by solar UV radiation **10**, 516f
– direct transfer **5A**, 429f
– duplication **12**, 161
– fixation **2**, 396
– for transgene expression **5A**, 437
– G + C content **2**, 193
– immunization using particle bombardment **5A**, 432
– in eukaryotes **2**, 143

– intercalator **2**, 28
– international data base **2**, 305
– investigation, history **2**, 193
– *in vitro* modification **2**, 377
– *in vivo* targeted delivery **5A**, 435
– linker **2**, 379f
– linker DNA **2**, 143
– microinjection **5A**, 430
– modifying enzymes **2**, 378f
– particle bombardment for transfection **5A**, 432
– plasmid, gene delivery **5A**, 435
– premutational lesion **2**, 7, 12
– – AP site **2**, 16
– protection **2**, 581f
– – by liposomes **2**, 581f
– – by proteins **2**, 581f
– receptor-mediated endocytosis **5A**, 434f
– recombinant –, environmental risk avoidance **3**, 775
– single-strand generation **2**, 341
– – for PCR **2**, 287
– single-strand preparation **2**, 275
– targeting ligands **5A**, 436
– terminal restriction fragment length polymorphisms **11A**, 90f
– thermal resistance of **10**, 72
– use of, anionic liposomes for transfection **5A**, 434
– – cationic liposomes for transfection **5A**, 432
– – cationic polymers for transfection **5A**, 431
– – electroporation for transfection **5A**, 431
– – neutral liposomes for transfection **5A**, 434
– viscosity **3**, 236
DNA alkylation **2**, 23ff
– mutagenic effect **2**, 25
– product **2**, 24f
DNA amplification **2**, 267
– directional overlapping subcloning **2**, 267
– *in vitro* **2**, 269, 333, 393
– *in vitro see also* polymerase chain reaction
– *in vivo* **2**, 267
DNA banking, access to archived samples **5B**, 420f
– ethical aspects **5B**, 417
– legal norms **5B**, 420
DNA bending **2**, 196, 212ff
– biological significance **2**, 213
– by protein binding **2**, 213
– detection **2**, 214
– effect on transcription initiation **2**, 214
– examples for bent loci **2**, 212
– intrinsic bending **2**, 212
DNA binding protein **1**, 106, **2**, 671
– activator **1**, 106
– amphipathic helix **2**, 672
– characteristics **2**, 672
– effector **1**, 106
– helix-loop-helix **2**, 672

– helix-turn-helix **2**, 672
– in yeast **1**, 492f
– leucine zipper **2**, 672
– repressor **1**, 106
– zinc finger **2**, 672
DNA/chemical interaction, Z-DNA **2**, 203
DNA cloning **2**, 267ff
– for sequencing **2**, 278ff
– impact on sequencing **2**, 267f
– shotgun cloning **2**, 267
DNA conformation **2**, 200
– A-DNA **2**, 194
– – biological effects **2**, 206
– B-DNA **2**, 194ff
– – D-DNA **2**, 195
– – T-DNA **2**, 195
– B-to-A transition **2**, 197ff
– B-to-Z transition **2**, 201
– effects of synthetic inserts **2**, 203ff
– flexibility **2**, 221
– function effects of synthetic inserts **2**, 203ff
– G-DNA **2**, 220
– glycosyl-bond **2**, 196, 200
– groove proportions **2**, 196
– H-DNA **2**, 218
– influencing parameters, NaCl **2**, 202
– in solution **2**, 201ff
– – detection **2**, 201
– microheterogeneity **2**, 221
– non-B-DNA **2**, 200
– *osmZ* gene effects **2**, 210ff
– physical distance **2**, 196
– proportions **2**, 196ff
– rotation sense **2**, 196
– synthetic oligonucleotides **2**, 202
– Z-DNA **2**, 194ff
– Z-to-A transition **2**, 202
DNA cruciform structure **2**, 215
– d(AT)$_n$ **2**, 217
– formation **2**, 215f
– gene regulation **2**, 217f
– induction by environmental stress **2**, 218
– interaction with proteins **2**, 218
– investigations **2**, 215f
– *in vivo* existence **2**, 218
– junction **2**, 215ff
– – indentification **2**, 215f
– loop **2**, 215ff
– – identification **2**, 215f
– probing **2**, 215f
– telestability **2**, 217
DNA–DNA hybridization **5B**, 228
DNA double helix **2**, 195ff, 320, 325
– bending **5B**, 314
– hedydration effects **2**, 202
– left handed **2**, 194
– structure **2**, 195ff
– structure-influencing parameters **2**, 197ff

DNA/drug interaction, Z-DNA **2**, 203
DNA electrophoresis **2**, 264f
– agarose gel of PCR products **2**, 448
– direct transfer to cellulose nitrate **2**, 280
– post-electrophoresis visualization **2**, 279f
– temperature problem **2**, 279
DNA fragment **2**, 381
– annealing **2**, 379ff
– – blunt end ligation **2**, 379
– – cohesive end generation **2**, 379f
– – – *in vitro* joining **2**, 379
– homopolymer tailing **2**, 381
– – *see also* DNA nested set fragment
– modification **2**, 392
– separation of large fragments **2**, 403
– subcloning **2**, 392
DNA fragmentation **2**, 376ff
– base specific **2**, 276
– by endonuclease cleavage **2**, 376
– by mechanical shearing **2**, 376
DNA glycosylase **2**, 17
DNA homology, in bacilli **1**, 372
DNA hybridization **2**, 396ff
– applications **2**, 397
– heterologous **2**, 398
– hybrid arrested translation **2**, 399
– hybrid selected translation **2**, 399
– probe **2**, 332ff
– – allele-specific oligonucleotide (ASO) **2**, 340
– subtractive **2**, 398ff
– with reporter molecules **2**, 284
DNA integration **2**, 434
– stable genome integration **2**, 436ff
DNA labeling **2**, 332, 396f
– autoradiography **2**, 279
– biotinylation **2**, 333
– – biotin detection **2**, 333
– chemiluminescent detection **2**, 282f
– colorimetric detection **2**, 282
– digoxigenin binding **2**, 333
– end-labeling **2**, 266, 277f
– enzyme **2**, 333, 397
– fluorescent **2**, 281, 333
– immunological **2**, 396
– indirect, by hybridization **2**, 282
– labeling reactions **2**, 397
– nick translation **2**, 397
– non-isotopic **2**, 332
– radioactive **2**, 280, 332, 396
– single-end labeling **2**, 278
– stable isotope labeling **2**, 282
– tagging **2**, 281f
DNA ligase **2**, 340, 378
– chain reaction **2**, 340
– of *Thermus aquaticus* **2**, 340
– thermostable –, use for detection of single base
  mutations **10**, 90f
– – use in ligase chain reaction **10**, 89ff

DNA manufacturing **5A**, 443ff
DNA measurement, flow cytometry **4**, 204f
– sample-flow analysis **4**, 205
DNA methylase **2**, 377
DNA methylation **2**, 586f, 673
– DNA recognition mechanism **2**, 673
– effect on transcription activity **2**, 596
– influencing factors **2**, 597
– maintenance methylation **2**, 673
DNA methyltansferase **2**, 19
– Ada methyltransferase **2**, 19
DNA microarrays **5B**, 90
– – *see also* gene chips
– *Candida albicans* **5B**, 229
– fabrication of **5B**, 227ff
– microarraying robots **5B**, 234
– probing and scanning **5B**, 234f
DNA micro-injection *see* micro-injection of DNA
DNA multiplex sequencing **2**, 263, 285ff
– dideoxynucleotide chain termination **2**, 286
– method **2**, 285
– multiplex walking **2**, 290
DNA nested set fragment **2**, 261ff
– as PCR primer **2**, 338
– automated enzymatic generation **2**, 304
– automated generation of fluorescently labeled **2**, 299
– denaturation **2**, 265
– detection **2**, 279ff
– fractionation **2**, 279ff
– generation **2**, 261f
– – by PCR **2**, 289
– labeling **2**, 289
– preparation **2**, 270ff
– sequencing pattern **2**, 300ff
– – ABI 373A **2**, 300f
– – dependence on nested set generation **2**, 300ff
DNA photolyase **2**, 14
DNA polymerase **2**, 270ff, 378
– characteristics **2**, 270ff
– from *Thermus thermophilus* **2**, 337
– manganese effect **2**, 274
– phage T7 **2**, 273
– reverse transcriptase **2**, 272
– thermostable, for PCR **2**, 274
– utility for PCR **2**, 337
DNA polymerase I, fill-in reaction **2**, 342
– Klenow fragment **2**, 272, 342
DNA polymerases **5B**, 196f
– Taq **5B**, 196f
– thermostable –, applications of **10**, 90
– – proofreading activity **10**, 87f
– – use in DNA sequencing **10**, 89
– – use in high temperature reverse transcription **10**, 88f
– – use in polymerase chain reaction **10**, 87f
DNA-processing enzymes **10**, 87ff
DNA/protein interaction **2**, 197, 207f

DNA recombination **2**, 147ff
– allele selection **2**, 149
– frequency **2**, 149f
– mapping mathematics **2**, 147ff
– progeny type **2**, 147
– rate **2**, 147
– statistics **2**, 147ff
– – LOD score **2**, 150f
– – maximum likelihood **2**, 147
– – mixture distribution **2**, 151
DNA repetitive sequence, in yeasts **1**, 475
– Ty (transposon yeast) **1**, 475
DNA replication **1**, 95ff, **2**, 9
– DNA bending effect on initiation **2**, 213
– errors **2**, 9f
– inhibition of topoisomerases **7**, 653
– post-replicative mismatch repair **2**, 9
– primer **2**, 270
– – biotinylated **2**, 275
– – biotinylated for sequencing **2**, 282
– – tagged for sequencing **2**, 282
– – universal **2**, 274
DNA/RNA hybrid **2**, 197
– structure **2**, 197
DNA-rRNA hybridization **1**, 403
– $T_{m(e)}$ dendrogram **1**, 403
DNA samples, coded – **5B**, 424
– commercial exploitation of **5B**, 423
DNAse I **2**, 674
DNA separation technology **5B**, 433
DNA sequence **2**, 215
– inverted repeat **2**, 215
– palindrome **2**, 215ff
– scaffold attachment region (SAR) **2**, 595
DNA sequence analysis **5B**, 210ff
– confidence values **5B**, 215f
DNA sequencers **2**, 298ff
– commercial systems **2**, 298f
– comparison of currently commercialized – **5B**, 208
– detection optics **5B**, 206
– electrophoresis **2**, 298
– fluorescence detection **2**, 298
– lane finder results **5B**, 212
– laser excitation and fluorescence emission **5B**, 206
– scheme **5B**, 204
DNA sequencing **2**, 259ff, 402
– – *see also* DNA multiplex sequencing, genomic sequencing
– automated – **5B**, 185, 432f
– automation **2**, 297ff, 333
– biochemistry of **5B**, 192ff
– chain termination –, primers **2**, 274f
– – template **2**, 2274
– chemical degradation – **2**, 262f, 266, 276ff
– – autoradiograph **2**, 277
– – base-specific reactions **2**, 266f, 276f

– – chemical agents **2**, 276f
– contig **2**, 291
– data acquisition routes **2**, 267
– data interpreatation **2**, 305
– *de novo* **5B**, 193
– dideoxy chain termination – **2**, 261f, 265, 270ff
– – autoradiograph **2**, 273
– directed approaches, enzymatic nested deletion method **5B**, 194
– – primer walking **5B**, 194
– – transposon insertion **5B**, 194
– electrophoresis **5B**, 186
– ExoMeth method **2**, 290
– fast atom bombardement mass spectrometry (FAB) **2**, 263
– fluorescence detection **5B**, 186ff, 197
– – instrumentation **5B**, 198ff
– gel-based mass dependent **2**, 261ff
– history **2**, 259ff
– information throughput **5B**, 201ff
– in pGV451 plasmids **2**, 289
– large target regions **2**, 290ff
– limitations **2**, 278
– manual **5B**, 185
– Maxam and Gilbert method **5B**, 185
– methodology **2**, 264ff
– nested set ladder **2**, 264
– of the human genome **5B**, 45ff
– – cDNA sequencing **5B**, 47
– – systematic mapping **5B**, 47
– oligomer walking **2**, 290
– pattern **2**, 300ff
– PCR-based **2**, 287ff
– – cycle sequencing **2**, 289
– "plus and minus" technique **2**, 261
– primer extension analysis **2**, 405
– primer library **2**, 290
– primer walking sequencing **2**, 274
– real-time detection **2**, 279ff
– resequencing of segments **5B**, 193
– robotics in **5B**, 202
– robotization, commercial systems **2**, 304
– sample preparation **5B**, 202
– Sanger dideoxy sequencing **5B**, 185f
– scanning tunneling electron microscopy **2**, 263
– semi-automated **2**, 289
– single-molecule detection **2**, 263
– solid-phase **2**, 275
– steps amenable to automation **2**, 297
– targeted gene walking with PCR **2**, 295
– technology **2**, 261f
– template preparation by PCR **2**, 293f
– transposon-based techniques **2**, 285f
– use of thermostable DNA polymerases **10**, 89
– using DNA polymerase I **2**, 261
– using *E. coli* RNA polymerase **2**, 261
DNA sequencing, automation, template purification **2**, 304

DNA sequencing laboratories **5B**, 436
DNA sequencing technology **5B**, 183ff
DNA structure *see also* DNA cruciform structure
– $(CA)_n$ tract **2**, 200
– – biological role **2**, 201
– extragenic palindromic sequence (REP) **2**, 208
– Hoogsteen base pairing **2**, 218
– primary **2**, 193, 259, 320
– – depiction **2**, 259, 319f
– X-ray crystal structure **2**, 194ff
DNA subcloning **2**, 268
– concept **2**, 268
– deletion strategies **2**, 290ff
– overlapping clone engineering **2**, 290f
– shotgun cloning **2**, 290f
– – flow chart **2**, 291
– – limitations **2**, 292
– unidirectional deletion **2**, 291f
– – in M13mp clones **2**, 292f
DNA supercoiling **2**, 207ff
– energetics **2**, 208
– function **2**, 207ff
– gene regulation **2**, 211
– – *tonB* promoter **2**, 211
– influencing genes **2**, 210f
– negative **2**, 207
– osmolarity effect **2**, 210
– structure **2**, 207ff
– topA mutant **2**, 208ff
DNA superhelix **2**, 215
DNA synthesis **2**, 271, 319ff
– – *see also* oligonucleotide synthesis
– chemical reaction **2**, 271
– high-throughput – **5B**, 230ff
– inhibition by antitumor agents **7**, 655
DNA synthesizers, high-throughput –, suppliers **5B**, 232f
DNA tagging **2**, 333, 395f
– *in vitro*, of eukaryotes **2**, 396
– *in vivo*, of Gram negative bacteria **2**, 396
– tag detection strategies **2**, 395
DNA technologies **5B**, 301ff
DNA template, double-stranded – **5B**, 195
– preparation of **5B**, 195
– single-stranded –, bacteriophage M13 **5B**, 195
DNA topoisomerase inhibitors **7**, 654ff
– camptothecins **7**, 655
– podophyllotoxins **7**, 655
– structures **7**, 654f
DNA topoisomerases, as targets of antitumor drugs **7**, 653ff
DNA transcript analysis, S1 nuclease mapping **2**, 404
DNA transcription regulation **2**, 718ff
– modification **2**, 720
DNA transduction **2**, 429
– Gram negative bacteria **2**, 429
DNA transfer, biolistic approach **2**, 579, 632

– by *Agrobacterium* **2**, 549
– DNA construct design **2**, 586
– DNA integrity **2**, 586
– DNA stability **2**, 586
– DNA topology **2**, 586
– target cell physiology **2**, 586
– targets **2**, 579
– to monocotyledonous plants **2**, 574f, 579
– to plant cells **2**, 555ff, 629ff
– – by *Agrobacterium tumefaciens* **2**, 629
– – cereal transformation **2**, 632
– – direct transfer to protoplasts **2**, 631
– – methods **2**, 629ff
– – micro-injection **2**, 632
– – to organelles **2**, 632
– – vectorless **2**, 575ff
– transformated plant species **2**, 579
DNA transfer technique **2**, 381
– protectifer **2**, 582
DNA transformation **2**, 429
– Gram negative bacteria **2**, 429
DNA triplex **2**, 218ff, 325
– formation **2**, 356
– formation conditions **2**, 219
– occurence **2**, 219
– poly d(A) **2**, 220
– poly d(G) **2**, 219
– poly d(T) **2**, 220
– probing **2**, 220
– structure **2**, 218
DNA uptake **2**, 460
– by sperms **2**, 772
– – defence mechanism **2**, 774
– – DNA binding regions **2**, 772
– – electroporation **2**, 772
– – frequency **2**, 772
– – insect sperms **2**, 772
– – liposome-mediated **2**, 772
– induced **2**, 461
– natural **2**, 460
DNA uptake competence **2**, 62f, 460, 508
– induction **2**, 62
– linear DNA **2**, 62
– of bacilli **1**, 395
– plasmid **2**, 62
– preparation of competent cells **2**, 381
– stability **2**, 62
DNA vaccination **5A**, 430ff
2,4-DNT *see* 2,4-dinitrotoluene
docetaxel, inhibition of the microtubular system **7**, 668
– structure **7**, 668
docosahexaenoic acid (DHA) **8A**, 137
– from protists **10**, 142
– in bacteria **7**, 147f
– in microalgae **7**, 169f
– in phytoplankton species **7**, 169f
– nutritional role **7**, 162

– occurrence in marine fungi **7**, 163
– producing organisms **10**, 462ff
γ-dodecalactone, formation of, from oleic acid **8B**, 343
n-dodecane emulsions **11A**, 387f
– concentration of protein **11A**, 388
– oil droplet distribution **11A**, 387
dolastatin, antitumor compound of marine origin **10**, 454
– structure **10**, 454
domains, oligonucleotides specific for **11A**, 96
Do-Maker process **9**, 279ff
– flow sheet **9**, 281
domestication of animals, historical importance **2**, 746
Domestic Policy Council Working Group on Biotechnology **5A**, 520
domestic wastes, pathogens in **11C**, 232ff
domestic wastewater *see* municipal wastewater, municipal sewage
Donnan exclusion **4**, 168
L-DOPA, synthesis of **8B**, 95
– – hybrid pathway **8B**, 98
L-DOPA decarboxylase *see* aromatic L-amino acid decarboxylase (AAAD)
dopamine, multi-enzyme synthesis of **8B**, 98
dopamine receptor antagonists **7**, 114
dosa, production of **9**, 538
DOST (Department of Science and Technology of Philippines) **12**, 416f
dot plot **5B**, 130, 302
doubled haploids **5B**, 166
doubling time, of cells **1**, 118, 139, 585
dough, characteristics **9**, 266
– fermentation **9**, 301
– formulations **9**, 279
– frozen yeast leavened – **9**, 284f
– leavening **9**, 253ff, 272ff
– liquid pre-ferment processes **9**, 274ff
– make-up **9**, 301f
– no time doughs, Brimec process **9**, 283f
– – Chorleywood process **9**, 283
– proofing **9**, 302
– sponge dough process **9**, 273ff
– straight dough process **9**, 273ff
– temperature **9**, 266, 270f
dough conditioners **9**, 259ff
– α-amylases **9**, 264
– acidifying agents **9**, 264f
– dairy products **9**, 260
– emulsifying agents **9**, 261f
– formulated additives **9**, 265
– in malt preparations **9**, 263f
– lecithin **9**, 261
– oxidants **9**, 262
– proteases **9**, 263
– reducing agents **9**, 262f
– regulation of water absorption **9**, 259f

– soy products **9**, 260
dough mixing **9**, 266ff
– continuous – **9**, 267ff, 278ff
– – Amflow process **9**, 279ff
– – Do-Maker process **9**, 279ff
– mixers **9**, 267ff
– stages **9**, 271
– time **9**, 270
Down snydrome **5B**, 75
downstream processing **3**, 39ff
– aerosols **3**, 520
– by reversed micellar extractions, equipment **3**, 611
– cell disruption **3**, 505ff
– environmental impacts **3**, 520
– flowsheet **3**, 594
– flow-sheeting **4**, 621
– *in situ* – **3**, 604
– integration with upstream processing **4**, 622
– liquid-liquid extraction **3**, 594ff
– – product purification **3**, 557ff
– modelling **4**, 606ff
– optimization, multilevel technique **4**, 621
– particle removal **3**, 469ff
– principle of unit operation **4**, 606ff
– – modelling **4**, 612ff
– product separation, nonsecreted products **3**, 505ff
– synthesis of separation processes **4**, 619ff
– with two-phase systems, equipment **3**, 605
doxorubicin **7**, 651f
– antitumor agent **7**, 651f, 658f
– structure **7**, 651, 658
DPInteract **5B**, 406
DRANCO process **11C**, 163ff
drinking water, AOC concentrations in **11C**, 460
– CFU conductivity of **10**, 287
– disinfection of **11C**, 409ff, 421
– – by chlorination **11C**, 410
– – by ozonation **11C**, 411
– – by-products **11C**, 410
– – limits **11C**, 410
– – thermal – **11C**, 411
– – ultrafiltration **11C**, 411
– – UV light **11C**, 411
– filtration of **11C**, 421
– hygiene aspects of **11C**, 413ff
– limit of nitrate concentration **11A**, 178
– parasites in **11C**, 417ff
– – dissemination of **11C**, 417ff
– – regulations **11C**, 373f
– – transmission of **11C**, 417ff
– production of, bank filtration **11C**, 431
– – groundwater recharge **11C**, 427
– role of **11C**, 372f
– treatment of **11C**, 371ff, 375ff
– – adsorption processes **11C**, 395ff
– – by-products **11C**, 376f

– – chemical softening **11C**, 402ff
– – deferrization **11C**, 404f
– – demanganization **11C**, 404f
– – flocculation in **11C**, 385f
– – flotation **11C**, 386ff
– – fluxes of substances **11C**, 376
– – gas exchange **11C**, 399f
– – membrane filtration **11C**, 405ff
– – organic matter removal, by biofiltration **11C**, 458ff
– – oxidation **11C**, 400ff
– – precipitation in **11C**, 385f
– – process alternatives **11C**, 384
– – processes **11C**, 377ff
– – rapid filtration **11C**, 389ff
– – reactions of important oxidants **11C**, 401
– – removal of BOM by biofiltration **11C**, 457ff
– – sedimentation **11C**, 386ff
– – types of raw water **11C**, 377ff
– – use of biofilms **11C**, 445ff
– turbidity in **11C**, 375, 422
– use of reverse osmosis **10**, 289
– waterborne diseases **11C**, 414ff, 418
– – listing of outbreaks **11C**, 415f
– – measures to eliminate causative agents of **11C**, 420ff
drinking water quality, demands on **11C**, 373ff
– – in Germany **11C**, 374
– – regulations **11C**, 373ff
Drosophila melanogaster **5B**, 148
– genome project **5B**, 32ff, 48
– maps **5B**, 32
– number of genes **5B**, 34
– re-annotating **5B**, 375f
– telomeres **5B**, 34
– X chromosome **2**, 160
– – cytological map **2**, 160
– – genetic map **2**, 160
drug design, use of databases **12**, 335
drug discovery, and bioinformatics **5B**, 123ff
– gene-centric – **5B**, 125
– lead structure optimization **10**, 54
– miniaturization of detection systems **10**, 50f
– screening methods **10**, 41ff
– screening of bioactive compounds **10**, 44ff
– target-centric – **5B**, 125
– use of combinatorial libraries **10**, 53ff
– use of screening libraries **10**, 54ff
drug master file, plasmid DNA manufacturing **5A**, 457f
drug production, in plant, leuenkephalin **2**, 650
– in plants **2**, 650f
– – human proteins **2**, 650
drugs *see* microbial drugs
– – *see* therapeutic products, peptide drugs, poly-peptide drugs, protein drugs
– of plant origin, camptothecin **8A**, 330
– – taxol **8A**, 330

– phenotypic response to **5B**, 93
– plant-derived **1**, 605
– preparation of, using poly-leucine catalyzed epoxidation **8B**, 501f
– production of, using cyclic dipeptide catalyzed hydrocyanation **8B**, 496
– synthesis of, enzymic hydroxylation in **8B**, 392
drug screening, biochemical assays **10**, 44ff
– – enzyme assays **10**, 48f
– – fluorescent readouts **10**, 45ff
– – microvolume techniques **10**, 47
– – radiolabeled readouts **10**, 44f
– – single molecule detection **10**, 47
– functional assays **10**, 49f
– – cell-based – **10**, 49f
– – reporter gene assays **10**, 49f
– – second messenger assays **10**, 49
– – two-hybrid – **10**, 50
– ligand binding assays **10**, 47f
– miniaturization of, detection systems **10**, 51f
– – flow technologies **10**, 51
– – liquid handling **10**, 51
– technologies **10**, 43ff
– use of chemical compound libraries **10**, 53ff
drug screening assays, applications of fluorescence polarization **10**, 46
drug targets, identification of **5B**, 118ff, 129ff
– – profile-based methods **5B**, 133ff
– – sequence-based search methods **5B**, 130ff
– validation of **5B**, 228
– virtual screening of **5B**, 353f
drug therapy **5B**, 118ff
– response to **5B**, 119
drying, of baker's yeast **9**, 340f
dual injected turbulent separation (DITS) reactor **11B**, 339ff
– – *see also* bioreactors
– configuration of **11B**, 336, 341
dual-media filters, design of **11C**, 391
Duchenne muscular dystrophy **5B**, 54, 64, 72
Dulbecco's modified Eagle's medium (DMEM), composition **3**, 143f, 154
dumps, environmental impacts **11C**, 195f
Dunaliella **10**, 66
– glycerol production **6**, 210f, 217ff
– in human diet **10**, 126f
– mass cultivation of **10**, 144
Dunaliella salina, β-carotene content **7**, 173
duocarmycins, antitumor agents **7**, 657f
duramycins, antibiotic activity **7**, 357f
– primary structure **7**, 336f
– solution structure, stereorepresentation of – **7**, 340
duroplasts **10**, 308f
Dutch Institute for Consumer Research **12**, 192
dye-labeled primer sequencing **5B**, 197
dye-labeled terminator sequencing **5B**, 197

dyes, aerobic degradation of metabolites **11A**, 393f
– biodegradation of **11A**, 392ff
dynamic programming method **5B**, 304
dynein **5A**, 4
dynemicin A **7**, 664ff
– antitumor activity **7**, 664
– antitumor agent **7**, 665
– structure **7**, 666
dystrophin gene **5B**, 65
2D gel databases **5B**, 328
2D gel electrophoresis **5B**, 150, 327ff
– for proteomics **5B**, 241ff
– principles **5B**, 247f
2D gels, intergel variability **5B**, 328
3D folds, prediction of **5B**, 338f
3D homology model **5B**, 367

E

Eadie-Hofstee kinetics **1**, 68
Eadie-Hofstee plot **9**, 35
earthworms, use in ecotoxicological testing **11B**, 136, 138
– vermicomposting **11C**, 139f
EASC process **11A**, 301
– in enhanced biological phosphorus removal **11A**, 298
Eastar Bio® **10**, 433
EBCG (European Biotechnology Co-ordination Group) **12**, 536f
EBI (European Bioinformatics Institute) **12**, 331, 333
EBI SRS server **5B**, 294f
– available databases **5B**, 295
EBV *see* Epstein Barr virus
EC Commission, Community Plant Variety Rights **12**, 289
– Council Directive, on the Legal Protection of Biotechnological Inventions **12**, 287ff
ECDC **5B**, 285
ECE (UN Economic Commission for Europe) **12**, 610f
echinomycin **7**, 561f
– antitumor agent **7**, 651f
– structure **7**, 651
EcoCyc **5B**, 285, 407
Ecoflex® **10**, 433
E. coli see Escherichia coli
E. coli, aspartase **8B**, 126f
– expression of, human insulin **5A**, 112
– – recombinant antibodies **5A**, 236
– – somatostatin **5A**, 112
– fumarases **8B**, 101ff
– genome **5A**, 3
– in plasmid DNA manufacturing **5A**, 457f

– in plasmid production **5A**, 453
– plasmid vectors, guidelines **5A**, 452
– removal of endotoxins **5A**, 455, 465
E. coli K12 **12**, 162
– biological features **12**, 84f
– classification **12**, 85f
– in plasmid DNA manufacturing **5A**, 447ff
– properties **12**, 168
economic analysis, ethanol production, effect of
  plant size **6**, 181
ecotoxicity test systems **11B**, 130ff
ecotoxicological assessment, of contaminated soils,
  contact assays **11B**, 135
– – methodologies **11B**, 129ff
– – test procedures **11B**, 129ff
– – with aquatic test systems **11B**, 130ff
– – with terrestrial test systems **11B**, 135ff
ecotoxicological characterization of contaminated
  soils **11B**, 137f
ecotoxicological testing, of coal derived reference
  substances **10**, 177f
ECRAB (European Committee on Regulatory
  Aspects of Biotechnology) **12**, 537, 550ff
EC specifications, for crystal sugars **6**, 10
– for liquid sugars **6**, 11
ecteinascidin, antitumor compound of marine origin **10**, 453f
ectoines **10**, 70
– production of **10**, 468
ectomycorrhizal fungi **10**, 579
ectotoxicological assessment of contaminated
  soils **11B**, 127
EDB (Economic Development Board) **12**, 418
Edman degradation **5B**, 251, 263
Edward syndrome **5B**, 76
EEC, chemical control **12**, 623ff
– Council Directives **12**, 547ff
– – Sixth Amendment **12**, 547ff
EEC Directive, contained use of GMOs **12**, 266ff
– protection of workers **12**, 271
– release of GMOs **12**, 269
EFB (European Federation of Biotechnology)
  12, 526, 608f
effectiveness factor, of bioprocesses **3**, 94
EFPIA (European Federation of Pharmaceutical
  Industry Associations) **12**, 536
EGF *see* endothelial growth factor
EGSB *see* expanded granular sludge bed
eicosapentaenoic acid (EPA) **8A**, 137
– from protists **10**, 142
– in bacteria **7**, 147f
– in microalgae **7**, 168f
– physiological effects **7**, 162
– precursor of prostaglandins **7**, 162
– producing organsims **10**, 462ff
eicosatrienoic acid (ETA), occurrence in marine
  fungi **7**, 163
EKF *see* Extended Kalman filter

elastin fibers **5A**, 33
elective culture, definition of **8A**, 546
– isolation of *Acinetobacter* **8A**, 547
electrical neutral dyes **5B**, 248
electrochemical analyzer, fuel cell **4**, 69
– gas analysis **4**, 69
– potentiometric **4**, 69
Electrochemical methods, for biomass determination **4**, 192
electrochemical potential **1**, 55, 62
– standard electrode potential **1**, 55
electrodes, ion-selective carrier membrane, types
  4, 22
electrofusion **2**, 98f
– cell agglutination **2**, 99
– electric field **2**, 99
– membrane permeability **2**, 99
– of plant and animal cells, hybrid yield **10**, 515f
– – under microgravity conditions **10**, 515
– used species **2**, 99
electrokinetic injection **5B**, 271
electromagnetic radiation techniques, gamma den-
  sitometer **4**, 137
– X-ray radiation **4**, 137
electromagnetic spectrum **5B**, 187
electron density map **5B**, 351
electronic scrap, bioleaching of **10**, 212f, 216
electron impact ionization **4**, 63
electron microscopy **1**, 9ff
– electron spectroscopic imaging (ESI) **1**, 13
– element analysis **1**, 13
– sample visualization **1**, 12f
electron transfer proteins **5A**, 20
electron transport, in pseudomonads **1**, 442
– photosynthetic **1**, 60
– respiratory **1**, 58f
electron transport phosphorylation **1**, 57ff
– chemo-lithotrophic **1**, 59
electroosmosis **5B**, 271
electrophoresis **3**, 718
– capillary – (CE) **3**, 726
– embodiments for fluorescence DNA sequencing **5B**, 207f
– genetic change detection **2**, 145
– pulsed field system for DNA separation **2**, 403
electrophoretic cell transfection (ETR) **2**, 578
electroporation **2**, 62, 377, 578, **5A**, 431, **12**, 93
– bacterial genera **2**, 62
– embryonic stem cells **2**, 762
– in Corynebacteria **2**, 483
– – transformation frequency **2**, 483f
electrospray ionization **5B**, 329
– of analytes **5B**, 254
elemental balancing **4**, 41ff
– in bioprocesses **3**, 372
– synthesis of poly-β-hydroxybutyric acid (PHB)
  4, 42ff
– – instantaneous error analysis **4**, 51ff

– systematics **4**, 41f
elemental composition matrix **4**, 42
eleutherobin, antitumor compound of marine origin **10**, 454
elicitation, of plant secondary compound synthesis **1**, 593f
– signal transduction **1**, 593
eliminations, use of catalytic antibodies **8B**, 465f
ELISA *see* enzyme-linked immunosorbant assay
– – *see* immunoassay
ellipticin **7**, 650
ELSI **12**, 151
eluate investigation **11B**, 490ff
– aim of **11B**, 490
– bottle tests **11B**, 492
– column tests **11B**, 492f
– elution agents **11B**, 491f
– factors affecting – **11B**, 490f
– liquid-solid ratio **11B**, 492f
– pH value **11B**, 492
– scheme **11B**, 494
elution agents **11B**, 491f
elution liquid **11B**, 491f
elymoclavine production, by immobilized *Claviceps* cells **1**, 240
EMBASE **5B**, 283
EMBC (European Molecular Biology Conference) **12**, 522
Embden-Meyerhof-Parnas (EMP) pathway **1**, 82
Embden-Meyerhof-Parnas pathway **9**, 182f
– in clostridia **6**, 237f
– scheme **6**, 125
EMBL Data Library **2**, 163
EMBL-EBI **5B**, 283
EMBL (EBI) Data Library **12**, 326
EMBO (European Molecular Biology Organization) **12**, 520ff, 598, 608
embryonic carcinoma cell (EC) **2**, 760f
– isolation **2**, 760
embryonic stem cell (ES) **2**, 748, 760ff
– chimera production **2**, 761
– colony **2**, 761
– directed mutagenesis **2**, 785
– homologous recombination **2**, 763
– – recombinant enrichment **2**, 763
– – screening strategies **2**, 763
– introduction of targeted mutations **2**, 764f
– *in vitro* cultivation **2**, 760
– isolation **2**, 760f
– lines **2**, 764
– transformation **2**, 762
embryo preparation, for micro-injection **2**, 752ff
– – cattle **2**, 754ff
– – donor/recipient synchronization **2**, 755
– – goat **2**, 754
– – mouse **2**, 752ff
– – oocyte isolation **2**, 752ff
– – pig **2**, 753ff

– – rabbit **2**, 752ff
– – sheep **2**, 753ff
– – treatment of donor animals **2**, 752ff
embryo research **5B**, 419
embryo transfer, to recipients **2**, 754f
– – directly after micro-injection **2**, 756
– – intermediate cultivation **2**, 755
– – reproductive data of farm animals **2**, 756
– – techniques **2**, 755f
EMEA *see* European Medicines Evaluation Agency
emirical stoichiometric models, of bioprocess **3**, 373ff
EMP **5B**, 291, 407
emulsan, formation, by *Acinetobacter calcoaceticus* **6**, 650
– – by *Arthrobacter calcoaceticus* **6**, 696
– structure **6**, 650
emulsifiers *see* bioemulsifiers, bio-de-emulsification
emulsifying agents, use in baked goods **9**, 261f
emulsion enzyme membrane reactor, scheme **8A**, 424f
emulsions *see* biosurfactants, surfactants
emusions, water-in-oil **3**, 606
enantiomerically pure pharmaceutical intermediates, synthesis of **8B**, 278ff
– – amines **8B**, 279f
– – chiral C_3 building blocks **8B**, 283
– – diltiazem **8B**, 279
– – naproxen **8B**, 281f
– – pyrethroid precursors **8B**, 282f
– – 6-aminopenicillanic acid **8B**, 280
– – 7-aminocephalosporanic acid **8B**, 280f
encapsulated-cell therapy **10**, 548ff
– macrocapsules **10**, 548
– microcapsules **10**, 548ff
encapsulated tissue, clinical trials **10**, 555ff
– pancreatic islets **10**, 555f
– parathyroid tissue **10**, 556f
enceophalomyelitis, monoclonal antibody therapy **5A**, 319
endergonic processs **1**, 53
endo-β-glucanases, *de novo* formation in malt **9**, 427
endogenous metabolism, effect on growth rate **4**, 280f
endoglucanase **5A**, 203ff
endoglucanases, cell wall degradation **9**, 710
endoglycosidases **8B**, 255
Endomycopsis, in lao-chao fermentation **9**, 525
Endomycopsis fibuliger, in tapé fermentation **9**, 529
endonuclease **2**, 16
– AP endonuclease **2**, 16
– UvrABC system **2**, 16
endopeptidases **5A**, 206
– nomenclature **8A**, 244

endoplasmatic reticulum (ER) **1**, 39
endosome release, transfection level enhancement **5A**, 436f
endospore(s) **1**, 368, 388ff
– – *see also* sporulation
– activation **1**, 394
– coat protein **1**, 391
– DNA conformation **1**, 388
– formation, in clostridia **1**, 292
– germinants **1**, 394
– germination **1**, 394f
– – genes **1**, 394
– – mutants **1**, 394
– internal structure **1**, 389
– of bacilli **1**, 388ff
– resistance properties **1**, 388ff
– small acid-soluble spore protein (SASP) **1**, 388f
– spore cortex **1**, 391
– sporulation **1**, 390ff
– structure **1**, 388ff
– thermoresistance **1**, 388
– UV resistance **1**, 388
– water content **1**, 388
endostatin **5A**, 134
endothelial growth factor (EGF), in therapy **5A**, 141
endothelin receptor antagonists **7**, 116ff
endothermic reaction **1**, 52
Endothia parasitica, milk-clotting emzyme **9**, 375
endotoxin, bacterial, composition of **5A**, 331
end walking **5B**, 177
enedyines, cytotoxic antitumor agents **7**, 664ff
– macromolelcular antitumor antibiotics **7**, 663ff
– mode of action **7**, 664ff
– structures **7**, 666
energy balance, for ethanol production **6**, 117
– in microorganisms **1**, 142
energy balancing, in fermentation control **3**, 329
energy charge **1**, 55
energy consumption, in ethanol production **6**, 115ff
– – from corn **6**, 116
– – from potatoes **6**, 116
– – from wheat **6**, 116
energy coupling **1**, 62ff
– chemical – **1**, 62
– electrochemical – **1**, 62
– – in halobacteria **1**, 64
– proton gradient **1**, 64
– uncouplers **1**, 64
energy distribution, energy dissipation rate **4**, 334f
– in stirred tanks **4**, 304ff, 331ff
energy flow **1**, 51
energy ligands, Scatchard plot **9**, 48f
energy metabolism **1**, 50ff
– energy yielding reactions **1**, 260

– of lactic acid bacteria **1**, 350ff
– of methanogens **1**, 260
– of microorganisms **1**, 170f
energy of activation **9**, 24
energy status *see* cell energy status
energy surplus **1**, 174
energy transfer, in bioreactors **3**, 89ff
engineering of microbial pathways, for amino acid production **8B**, 303ff
engines, alcohol burning – **6**, 188ff
– – emissions **6**, 190
– – ignition **6**, 190f
– – lubrication **6**, 191f
– – wear **6**, 192
– conversion of conventional – **6**, 192ff
– modifications **6**, 192ff
enhanced oil recovery **6**, 629, 632
– use of biosurfactants **6**, 668f
– with bio-de-emulsifyers **6**, 706
enhancer trap lines **5B**, 152
enlightenment view, certainty-based **12**, 19, 25
– completely-objectivistic **12**, 19
enoate reductase, of yeast **8A**, 372
– – reduction of triple substituted double bonds **8A**, 375
– – substrates for **8A**, 375
enolase **8B**, 112f
enolase gene (*ENO*) **1**, 482
enol esters, in lipase-catalyzed acylations **8A**, 56ff
enol ether, hydrolysis of, use of catalytic antibodies **8B**, 465
EnsEMBL **5B**, 286
ensilage, yield of fermentation juice **11A**, 223
ensiling, improvement with bacterial inoculants **9**, 777
– stages **9**, 775f
Enterobacter, in olive fermentation **9**, 605
Enterobacter aerogenes **6**, 596f
– production of ribavirin **6**, 596ff
– transarabinosylation **6**, 596
Enterobacteriaceae, in cabbage fermentation **9**, 642
– 2,3 butanediol production **6**, 274f
Enterobacter XM6 exopolysaccharide, structure **6**, 651
enterobactin **7**, 201f
– catecholate siderophore **7**, 201f
– of enteric bacteria **7**, 202
– structure **7**, 201
Enterococcus **1**, 346
– habitats **1**, 346
– isolation media **1**, 349
– pathogenicity **1**, 346
Enterococcus faecalis **2**, 489
– conjugative plasmids **2**, 55
– pyruvate fermentation **1**, 182
Enterococcus faecium **1**, 346
Enterococcus feacalis **1**, 171, 346

Enterovirus, in medical wate **3**, 783
– in residential waste **3**, 783
enthalpy **1**, 52
Entner-Doudoroff (KDPG) pathway **1**, 82f
Entner-Doudoroff pathway **9**, 183
– in clostridia **6**, 239
– in *Gluconobacter* **6**, 387
Entomophtera **1**, 522
entrapment *see* cell immobilization
– of lipases **8A**, 51f
entrapment methods **8A**, 409f
entrapment of microorganisms **11B**, 417
Entrez **5B**, 291
entropic trapping, rotational entropy **8B**, 415ff
– translational entropy **8B**, 417ff
entropy **1**, 52
environmental certificates **6**, 55f
environmental control, use of biosurfactants **6**, 686ff
Environmental Defense Fund **12**, 465, 476ff
environmental impact index **11A**, 572
Environmental Impact Statement (EIS) **12**, 204
environmental load units (ELU) **11A**, 572
environmental monitoring **3**, 763
– aseptic processing, FDA guideline **3**, 763
– of biotechnological processes **3**, 763
environmental processes **11A**, 125ff
– monitoring of, with biosensors **11A**, 125ff
environmental protection **2**, 838
– in chemical production processes **11A**, 563
– production-integrated – **11A**, 199, 562ff
environmental taxes **6**, 55f
environmental variables, in bioprocesses **3**, 360, 371
– – direct reconstruction **3**, 372ff
– unmeasurable – **3**, 379
enzymatic activities, of yeast **8A**, 370ff
enzymatically catalyzed reaction, effects of freeze-concentration **3**, 699
enzymatic peeling, of citrus fruits **9**, 715
ENZYME **5B**, 291, 407
enzyme activity, determination of, initial rate **9**, 148
– – pH effect **9**, 148
– – substrate concentration **9**, 148
– – temperature effect **9**, 146f
– in immobilized microorganisms **1**, 236
– international unit (IU) **3**, 730
– measurement **3**, 730
– optimization of, by site selective mutagenesis **8A**, 26
– optimization of, by heterologous expression **8A**, 26
– regulation **1**, 70, 101
– – allosteric control **1**, 101
– – binding site **1**, 101
– – covalent modification **1**, 101
– – effector **1**, 101

– – feedback inhibition **1**, 101
– solvation effects on **8A**, 252
enzyme assays **9**, 60ff
– coupled enzymes **9**, 60, 149f
– determination of protein concentration **9**, 60ff
– immunoassays **9**, 151ff
– – heterogeneous **9**, 152
– – homogeneous **9**, 152ff
enzyme catalysis **1**, 65ff
– active site residues **9**, 26
– concept of **8B**, 404ff
– covalent – **9**, 27f
– energetic **1**, 66
– energy of activation **9**, 24
– entropic trapping, rotational entropy **8B**, 415ff
– enzyme-catalyzed reactions, synthetic applications of **8B**, 351ff
– general acid – **9**, 25f
– general base – **9**, 25f
– pH dependence **9**, 49f
– substrate binding **9**, 24f
– transition state stabilization **8B**, 423f
enzyme-catalyzed reaction **1**, 67
– kinetics **1**, 67
– ping-pong mechanism **1**, 69
– sequential mechanism **1**, 69
enzyme classes **5B**, 137
enzyme cofactor, of methanogens **1**, 259
– – *see also* coenzymes
enzyme cofactors, metal ions **5A**, 19
– trace elements **5A**, 19
Enzyme Commission system, classification **9**, 7ff
enzyme construction **1**, 216
enzyme electrode **4**, 81
enzyme engineering **8A**, 17f
enzyme immobilization **3**, 432
– covalent bonding **3**, 436f
– methods **3**, 436f
– surface charge density **3**, 435
enzyme inactivation, mechanisms **9**, 110
enzyme induction **4**, 151
enzyme inhibition, competitive **1**, 70
– non-competitive **1**, 70
– selective –, improvement of yeast biotransformations **8A**, 382f
enzyme-inhibitor complexes **5A**, 96f
enzyme inhibitors, biological roles **9**, 46f
– irreversible – **9**, 41ff, **9**, 41ff
– – affinity reagents **9**, 41f
– – photoaffinity labels **9**, 41f
– – side-chain selective reagents **9**, 41
– – suicide reagents **9**, 41f
– medical applications **5A**, 167
– of the growth-regulating pathway **7**, 646
– reversible – **9**, 43ff
– – competitive **9**, 43ff
– – non-competitive – **9**, 45
– – uncompetitive – **9**, 45f

– substrate inhibition **9**, 46
enzyme kinetics **5A**, 22f, **9**, 30ff
– multi-enzyme systems **9**, 121ff
– multi-substrate reactions, Cleland's notation **9**, 36ff
– – King-Altman method **9**, 38
– – ping-pong reaction **9**, 37f
– simple cases **9**, 31ff
– turnover number **8A**, 31
enzyme-linked immunosorbant assay (ELISA) **5A**, 290
enzyme-linked immunosorbent assay (ELISA), detection of plant pathogens **10**, 499
enzyme membrane reactors, modeling of **8A**, 432f
– technology **8A**, 409
– with continuous product extraction **8A**, 423f
enzyme nomenclature **8A**, 29f
– classification system **8A**, 30
enzyme production **3**, 10, 473
– – *see also* protein production, extracellular enzymes
– *Bacillus* protease **1**, 380
– by clostridia **1**, 310
– by filamentous fungi **1**, 522
– by immobilized microorganisms **1**, 239
– by plant cell cultures **1**, 589
– by pseudomonads **1**, 421
– by streptomycetes **1**, 451
– by white-rot fungi **1**, 534f
– kinetics **1**, 239
– methane monooxygenase **1**, 280
– thermostable amylolytic – **1**, 28
enzyme purification **9**, 88ff
– – *see* protein purification
– affinity chromatography **9**, 56ff, 106ff
– buffers **9**, 51
– cell rupture, by chemical treatment **9**, 92
– – by pressure **9**, 92
– – by sonication **9**, 91
– – in bread units **9**, 91f
– centrifugation **9**, 96
– choice of the affinity ligand **9**, 57f
– covalent chromatography **9**, 59
– desalting **9**, 52f, 109
– – dialysis **9**, 52
– – size exclusion chromatography **9**, 52f
– extraction **9**, 50f
– – stability of crude extracts **9**, 50f
– filter press **9**, 97
– fluidized bed technology **9**, 96
– heparin agarose chromatography **9**, 108
– high pressure liquid chromatography **9**, 56
– hydrophobic interaction chromatography **9**, 58f, 103f
– hydroxyapatite chromatography **9**, 55, 109
– ion exchange chromatography **9**, 53ff, 100ff
– – optimal pH **9**, 54

– methodologies **9**, 97ff
– microfiltration **9**, 95
– perfusion chromatography **9**, 109f
– precipitation steps **9**, 51f
– pseudoaffinity chromatography **9**, 58
– size exclusion chromatography **9**, 56, 104ff
– ultrafiltration, tangential flow filtration **9**, 95f
enzyme purity assessment, by high pressure liquid chromatography **9**, 65
– by polyacrylamide gel electrophoresis **9**, 62ff
– specific activity **9**, 59ff
enzyme reaction, cofactor regeneration, in industrial biotransformations **3**, 454f, 458
– engineering **3**, 442
– in bioreactors, conversion **3**, 446f
– – dual-enzyme reaction **3**, 446f
– – enzyme consumption **3**, 446f
– – reactor volume **3**, 446f
– – residence time **3**, 446f
– – single-enzyme reaction **3**, 446
– – substrate concentration **3**, 447
– – yield **3**, 446f
– in membrane bioreactors **3**, 446f
– kinetic modeling **3**, 442f
– kinetics **3**, 443
– lipase-catalyzed – **3**, 453
– parameter estimation **3**, 444
– rate **3**, 444
– systems **3**, 443
– thermodynamics **3**, 444
enzyme reaction rate, initial velocity data **9**, 39
– pH effect **9**, 49f
enzyme reactions, multiple – **8B**, 366
enzyme reactors **3**, 427ff
– batch stirred tank reactor **8A**, 431f
– biotransformations **3**, 445ff
– characteristics of **8A**, 432
– continuous stirred tank reactor **8A**, 431f
– design **3**, 432
– immobilization reactors **3**, 434
– membrane bioreactors **3**, 431
– – new design **3**, 458
– – process design **3**, 452
– – two-stage – **3**, 461
– modeling of **8A**, 431
– multiphase bioreactors **3**, 458
– plug flow reactor **8A**, 431f
enzyme(s) **1**, 66, 165f, 492, **3**, 433ff, 441, **5A**, 22ff, **9**, 5ff
– – *see also* biocatalysts
– – *see also* extracellular enzymes, proteins
– – *see also* proteins
– – *see also* synthetic enzymes
– α-amylases **5A**, 192ff
– active site **1**, 67
– amplification system **9**, 161
– amylolytic – **10**, 74ff
– – action of **10**, 75

– – thermoactive – **10**, 76ff
– analytical uses **9**, 137ff
– – coupled reactions **9**, 141ff
– – detection limits **9**, 139
– – end-point reactions **9**, 139f
– – in diagnostics **9**, 155ff
– – in food analysis **9**, 155ff
– – quality requirements **9**, 153ff
– – test strips **9**, 150
– and carbon-carbon bond formation **8B**, 5
– and food protein **9**, 685ff
– as catalysts **9**, 7ff
– as products of marine biotechnology **10**, 457ff
– as proteins, excretion **9**, 21
– – misfolding **9**, 21
– – primary structure **9**, 9ff
– – secondary structure **9**, 12f
– – tertiary structure **9**, 13ff
– as silage additives **9**, 778f
– biotransformations of unnatural substrates **8A**, 32
– catalytic effect **5A**, 23
– catalytic mechanisms **5A**, 25f
– cellulose-hydrolyzing – **10**, 80ff
– cell wall degrading – **9**, 706ff
– – cellulases **9**, 710ff
– – hemicellulases **9**, 709
– – multienzyme complexes **9**, 711f
– – pectinases **9**, 707f
– chitin-degrading – **10**, 84f
– classification **1**, 66
– coenzymes **5A**, 24, **9**, 28
– cofactor-dependent –, commercially available **8A**, 393
– cofactors **9**, 28ff
– cold-adapted – **10**, 459ff
– – applications of **10**, 64
– – structural properties of **10**, 459f
– conformational stability **9**, 11f
– construction of biosensors **4**, 77f
– control coefficient **1**, 165ff
– – negative **1**, 166
– DNA ligases **10**, 89ff
– DNA polymerases **10**, 87ff
– enantioselectivity **3**, 445
– encapsulation **3**, 441
– entrapment, in hydrophilic membranes **3**, 442
– expression, in *Bacillus brevis* **9**, 79
– – in *Bacillus subtilis* **9**, 79
– – in *E. coli* **9**, 75ff
– – in *Hansenula polymorpha* **9**, 85
– – in *Pichia pastoris* **9**, 85
– – in *Yarrowia lipolytica* **9**, 85
– – in yeast **9**, 80ff
– formation by protozoa **10**, 142ff
– for pulp and paper processing **10**, 525ff
– for starch modification **9**, 667ff
– – production of sweeteners **9**, 742ff

– – scheme of action **9**, 670
– glycolytic **1**, 480
– – *GCR1* (glycolysis regulation) gene **1**, 492
– – gene promoters **1**, 492
– – gene transcription **1**, 492
– – regulation **1**, 486
– housekeeping **1**, 105
– hydrolases **5A**, 25
– immobilization of, in peptide synthesis **8A**, 255
– in carbohydrate chemistry **8B**, 243ff
– induced fit **1**, 67
– induced fit hypothesis **5A**, 22
– industrial **9**, 663ff
– – quality assurance **9**, 667
– – regulatory considerations **9**, 666f
– – world consumption **9**, 666f
– industrial applications **5A**, 189ff
– in feed products **9**, 782
– in food processing **9**, 663ff
– in milk **9**, 368f
– – lipase **9**, 369
– – plasmin **9**, 368f
– *in situ* extraction **3**, 604
– in soil **11B**, 98f
– isomerases **5A**, 25
– ligases **5A**, 25
– lignin-modifying – **10**, 529ff
– lock and key hypothesis **5A**, 22
– lyases **5A**, 25
– market **5A**, 191
– medical applications **5A**, 167
– Michaelis-Menten equation **5A**, 23f
– mitochondrial **1**, 485
– multi-enzyme systems **8A**, 33
– multisubunited – **9**, 18
– nomenclature **5A**, 24ff, **9**, 7ff
– of psychrophilic microorganisms **10**, 459ff
– – applications of **10**, 460ff
– – structural properties **10**, 459
– of *Saccharomyces cerevisiae* **1**, 480
– overproduction **9**, 75ff
– oxygen detoxifying **1**, 352
– production of, as fine chemicals **9**, 73ff
– – fermentation **9**, 89ff
– prosthetic group **5A**, 24
– prosthetic groups **9**, 28f
– proteolytic – **10**, 85f
– pullulytic –, action of **10**, 75
– reactions catalyzed by – **5A**, 22
– recombinant, use for analytical purposes **9**, 161
– recombinant technology **5A**, 191
– release, from periplasmic space **3**, 516
– – of intracellular – **3**, 508ff, 513
– screening in extremophiles **10**, 73f
– solubilization **3**, 513ff
– stability, determination methods **3**, 444
– – half-life **3**, 444f
– – residence time **3**, 444

- starch-degrading **1**, 310
- structural motifs **9**, 13ff
- substrate inhibition, kinetics **3**, 331
- synthetases **5A**, 25
- therapeutic recombinant – **5A**, 167f
- thermoactive – **10**, 72f
- tolerance to organic solvents **8A**, 32f
- transferases **5A**, 25
- use in, baking **9**, 674ff
- – brewing **9**, 682ff
- – dairy products **9**, 679ff
- – fruit juice production **9**, 712ff
- – oil extraction **9**, 719ff
- – wine making **9**, 727f
- use in paper manufacture **10**, 538ff
- xylan-degrading – **10**, 82ff
enzyme specificity, exploiting combinations **8B**, 382
enzyme stabilization **9**, 110ff
- addition of, polyhydroxy compounds **9**, 111
- – salts **9**, 111f
- – substrate **9**, 111
- by disulfide bonds **9**, 17
- by freeze-drying **9**, 113f
- by hydrogen bonding **9**, 16
- by hydrophobic effects **9**, 16ff
- by undercooling **9**, 113
- low temperature storage **9**, 112f
- permazyme technology **9**, 114
enzyme-substrate binding, allostery **9**, 47
- cooperativity **9**, 47
enzyme-substrate complex **5A**, 22
enzyme synthesis regulation **1**, 105
- – see also translation regulation, gene expression
- translation regulation, glycolytic enzymes **1**, 480ff
- – two-component system **1**, 106
enzyme thermistor **4**, 90
EPA *see* eicosapentaenoic acid
EPA's Biotechnology Science Advisory Committee (BSAC) **12**, 470
EPA (U.S. Environmental Protection Agency) **12**, 41, 53, 198, 207, 241, 461ff, 570ff
EPC (European Patent Convention) **12**, 142, 283, 287ff, 302ff, 532
EPD **5B**, 284
epelmycin A, antitumor agent **7**, 659
- structure **7**, 659
ephedrine, formation of **8A**, 9f
- industrial synthesis of **8B**, 24
- production using the Knoll process **8B**, 53f
- synthesis of **8B**, 393
D-ephedrine manufacture, by aldol condensation **8A**, 19
EPIC trial **5A**, 367ff
epidermal growth factor, in therapy **5A**, 162
- receptor **5A**, 162

epidermin, C-terminus, aminovinyl cysteine residue **7**, 349
- medical uses **7**, 359
- prepeptide structure **7**, 345
- primary structure **7**, 332ff
epidermin biosynthesis, from pre-epidermin, gene cluster **7**, 342
- – scheme **7**, 342
- gene cluster **7**, 348
- regulation of – **7**, 351
epidermis, of plants **1**, 579
EPI (Expanded Programme on Immunization) **12**, 357
epilancin K7, primary structure **7**, 332ff
EPILOG trials **5A**, 374ff
epirubicin, antitumor agent, structure **7**, 658
epistasis **5B**, 70
episuccinic acid, production by *Aspergillus clavatus* **6**, 371
EPO *see* erythropoietin
EPO (European Patent Office) **12**, 314
- Examination Guidelines **12**, 291
- practice **12**, 291ff
- Technical Board of Appeal **12**, 292f
epolactaene, structure **10**, 459
epoxidation, by cyt. P-450 monooxygenase **8A**, 481
- of "chalcone-type" substrates, using poly-L-leucine **8B**, 502
- of digeminal and trisubstituted alkenes, use of poly-L-leucine **8B**, 503
- poly-leucine catalyzed – **8B**, 501f
- – preparation of natural products and prodrugs **8B**, 501f
- regio and enantioselective, of farnesamide-*N*-oligopeptides **8B**, 507
- stereoselective, biphasic – **8B**, 499f
- – triphasic – **8B**, 499f
epoxide carboxylase **11B**, 202
epoxide compounds, antitumor agents **7**, 656f
epoxide hydrolases **8A**, 481, **11B**, 201f
epoxides, conversion of **11B**, 200ff
- formation of, by degradation of vicinal haloalcohols **8B**, 206f
- hydrolysis of, use of catalytic antibodies **8B**, 465
- metabolism of **11B**, 201f
- opening of, use of catalytic antibodies **8B**, 471
epoxysuccinic acid, production **6**, 370f
EPS *see* extracellular polymeric substances
Epstein Barr virus (EBV) **2**, 788
- vaccines **5A**, 139
equilizing filter **11C**, 314ff
ER *see* endoplasmatic reticulum
Eremothecium ashbyii **1**, 525
ergocalciferol, structure **7**, 173
α-ergocryptine **1**, 531
ergometrine **1**, 531

– production, by immobilized *Claviceps* cells **1**, 240
ergosterol **7**, 171f
– effect on brewer's yeast metabolism **9**, 440
– production of **10**, 333f
– structure **7**, 171
ergot alkaloids **1**, 530f
– biological effect **1**, 530
– biosynthesis **1**, 531
ergotamine **1**, 531
– as receptor antagonist **7**, 112
Error analysis, for gas balancing **4**, 44ff
– – measurement errors **4**, 49ff
– – simplifications **4**, 46ff
– statistical error propagation **4**, 49f
Erwinia, vitamin precursor production **1**, 211
Erwinia carotovora **6**, 596f
– ATP production **6**, 600f
– carbapenem production, regulation of – **7**, 65
– cell-free phosphorylase **6**, 601
– immobilized phosphorylase **6**, 601
– production of ribavirin **6**, 596ff
Erwinia herbicola, tyrosine-phenol lyase **8B**, 95ff
Erwinia rhapontici, ferrorosamine A **7**, 208
erythritol, formation by yeasts **6**, 220
erythroid differentiation factor **5A**, 156
erythromycin, branched chain analogs **10**, 361
– chemical structure **10**, 342
– effect in mouse septicemia **7**, 380
– enzymatic synthesis of **8B**, 358
erythromycin A **7**, 122ff
– activities **7**, 124
– structure **7**, 123
erythromycin polyketide synthase, structural studies of **10**, 357
erythropoietin (EPO) **5A**, 155f
– genetic engineering of **5A**, 540
– in therapy **5A**, 155
– market **5A**, 533
– sales **5A**, 533
ESC (Economic and Social Committee) **12**, 521ff
Escherichia coli **1**, 168, **2**, 18, 50ff, 208, 375
– – *see* E. Coli
– as host cell **3**, 284, 530
– autolysis **2**, 415
– bacteriophage λ promoters **9**, 77
– bacteriophages **1**, 545ff, 562ff
– bacteriophage T7 promoters **9**, 77f
– basal media supplements **1**, 143
– β-galactosidase activity **3**, 484
– carbon catabolite-repressible operons **7**, 60
– cell disruption **3**, 483f
– – protein release **3**, 516
– cell homogenization **3**, 484
– chaperonins **3**, 539
– chymosin formation **9**, 680
– computer integrated fermentation **4**, 592
– contamination of yeast cultures **4**, 247

– deoxyribose-5-phosphate aldolase **8B**, 17
– ethanol production **1**, 205
– export of heterologous proteins **9**, 78
– expression of cyt. P-450 **8A**, 483
– expression systems **9**, 75ff
– F pilus **1**, 18
– fusion proteins **9**, 78f
– gene duplications **5B**, 12
– gene expression system **2**, 405
– gene-product yield **4**, 500
– genetic engineering **2**, 375ff
– genome project **5B**, 8ff
– genome sequencing **5B**, 10
– grey holes **5B**, 12
– growth kinetics **1**, 133
– heat shock proteins, functions in protein folding **3**, 539
– – groEL **3**, 539
– – groES **3**, 539
– in olive fermentation **9**, 605
– K-12 strain **4**, 495, 500
– lac promoter **9**, 76
– lactose operon transfer **1**, 195f
– L-aspartate production **6**, 519
– – use of immobilized cells **6**, 473
– L-aspartic acid production, in bioreactors **3**, 457
– *lexA* gene **2**, 19
– L-lysine biosynthesis, regulation **6**, 485
– lpp promoter **9**, 77
– L-threonine producing mutants **6**, 483f
– L-tryptophan biosynthesis **6**, 489
– membrane lipid-protein ratio **1**, 243
– microcin C7 synthesis **7**, 61ff
– multifunctional genes **5B**, 12
– mutagenization **3**, 289
– nitrogen requirements **3**, 131
– nutrient limitation effect **1**, 173
– origin or replication **5B**, 12
– overexpression of heroin esterase **8A**, 351
– pBR322 ampicillin resistance gene, transcription control **2**, 209
– penicillin acylase **8A**, 15
– penicillin G acylase **8A**, 258
– phage fd **1**, 564
– phage f1 **1**, 564
– phage-infected cell **1**, 547
– phage lambda **1**, 562f
– phage MS2 **1**, 565
– phage Mu **1**, 562f
– phage M13 **1**, 564
– phage P1 **1**, 563
– phage Qβ **1**, 565
– phage SP **1**, 565
– phage T4 **1**, 564
– phenylalanine hyper-secreting phenotype **8B**, 321
– physical genetic map **5B**, 10

– plasmids, stabilization **3**, 291
– postgenome era **5B**, 12f
– potassium efflux systems **1**, 177
– potassium uptake systems **1**, 176
– ppGpp **7**, 61ff
– – synthesis **7**, 63
– protein, export **3**, 534
– – *in vitro* folding **3**, 549
– – separation **3**, 603
– re-annotating **5B**, 373ff
– *recA* gene **2**, 19
– recombinant –, biotechnological production of poly(3HB) **6**, 442f
– – ethanol production **6**, 167f
– – growth **3**, 290, 292
– – in fermentation processes **3**, 284
– – phenotypes **3**, 287
– – tryptophan hyperproduction **3**, 292f
– – tryptophan production **3**, 287f
– recombinant L-threonine producing strains **6**, 486f
– recombinant L-tryptophan producing strains **6**, 490
– recombinant strains, production of PHAs **10**, 430
– restriction map **5B**, 10
– RNA polymerase σ-factors **2**, 235f
– RNA polymerase sigma factor **7**, 61ff
– – regulation of – **7**, 64
– scanning electron micrograph **5B**, 8
– SecB export **3**, 534
– separation **3**, 482ff
– – from fermentation broth **3**, 479, 501
– single-cell analysis **5B**, 275
– specific growth rate **1**, 128
– statistical features of the genome **5B**, 12
– strains, nontoxic – **5B**, 12
– – toxic – **5B**, 12
– strain 6F484 **3**, 289
– Tc^r cells **3**, 290
– termination of replication **5B**, 12
– trp promoter **9**, 77
– tryptophan uptake **3**, 291
– use in bioelectrical fuel cells **10**, 16ff
– 1,3-propanediol production **1**, 206
– 2D gel electropherogram profile **5B**, 249
Escherichia coli DPD2511, biosensor, for oxidative stress **11B**, 457
Escherichia coli FM5/pKY287, oxidation of, toluene **11B**, 449
– – trichloroethylene **11B**, 449
Escherichia coli HB101/pMY402, oxidation of, toluene **11B**, 449
– – trichloroethylene **11B**, 449
Escherichia coli HMS174(pOS14/pOS15), biosensor for detection of Hg(II) **11B**, 455
Escherichia coli HMS174(pOS25), bioreporter strains **11B**, 455

Escherichia coli JM109(pDTG601), degradation of trichloroethylene **11B**, 450
Escherichia coli JM109 (pSHF1003)/(pSHF1007) **11B**, 448f
– biphenyl dioxygenase **11B**, 448
Escherichia coli WM1202(pRY002)/TV1061 (pGrpElux.3 or pGrpElux.5), biosensors for environmental stress **11B**, 457
ESF (European Science Foundation) **12**, 520ff, 526, 598, 608, 653
– Liaison Committee on Recombinant DNA **12**, 523f
ESI-MS/MS, definition of **5B**, 254
ESNBA (Europpean Secretariat of National Bioindustry Associations) **12**, 635
esperamicin A$_1$ **7**, 664ff
– antitumor agent **7**, 666
– enrichment pattern **7**, 667
– structure **7**, 666
ESPRIT (European Strategic Programme of R & D in Information Technology) **12**, 529
esterase activity, of penicillin acylase **8A**, 229
esterases **8A**, 28, 193ff, 250
– by aminolysis **8A**, 263ff
– differences from lipases **8A**, 194
– formation of amides **8A**, 263ff
– in coal solubilization **10**, 164
– in peptide synthesis **8A**, 262
– microbial – **8B**, 357f
– – *Bacillus coagulans* **8A**, 219
– – *Bacillus subtilis* **8A**, 219
– – bakers' yeast **8A**, 220
– – *Penicillium frequentans* **8A**, 221
– – *Pichia miso* **8A**, 219f
– – *Rhizopus nigricans* **8A**, 219
– solvent-stable – **10**, 92
– stability in anhydrous solvents **8A**, 262
– synthesis of (+)-*trans* chrysanthemic acid **8B**, 283
– thermostable – **10**, 92
esters, formation in wine making **9**, 479
ester synthesis, with lipases, solvent-free – **10**, 400ff
estimation *see* process estimation
estimation techniques **4**, 542ff
– balancing methods **4**, 544ff
– deterministic techniques **4**, 546f
– – exponential filter **4**, 546
– – moving average method **4**, 546
– – recursive least-square estimation technique **4**, 547
– indirect measurements **4**, 543f
– stochastic, Kalman filter **4**, 547ff
estrogen receptor agonists **7**, 112, 123
estrogen receptor antagonists **7**, 115f, 686
estrogens, hydroxylation of **8A**, 489
– inhibition of –, treatment of hormone-dependent tumors **7**, 685

estrone, combined synthesis **8A**, 14
EST searches **5B**, 367
EST sequences, alignment of **5B**, 306
ETA *see* eicosatrienoic acid
ethanol, as a liquid fuel **6**, 131f, 188ff
– – applications **6**, 194f
– – comparison with other motor fuels **6**, 196f
– – costs **6**, 197
– – emissions **6**, 190
– – engine performance **6**, 190
– – safety **6**, 192
– as energy source in yeast fermentation **8A**, 367f
– as raw material, in acetic acid production **6**, 383
– as wine preservative **9**, 488
– break-even sugar price **6**, 51f
– by-products **6**, 169ff
– – stillage **6**, 169ff
– cell-based biosensor assay **4**, 92
– classical methods **6**, 59ff
– conversion cost **6**, 50
– conversion to lipid **7**, 142
– degradation rate in bioscrubbers **11C**, 309
– determination **9**, 143
– effect, on membranes **1**, 136
– – on microbial growth **1**, 136
– estimation methods **1**, 126
– extraction **3**, 559ff
– – costs **3**, 561, 585
– – *in situ* – **3**, 584
– – solvent selection **3**, 560
– fermentation in clostridia **1**, 300
– formation in wine making **9**, 472ff
– – inhibitory effects **9**, 483
– from biomass, maximum yield **6**, 166
– from pentoses **1**, 204
– from xylose **1**, 197
– gas analysis in bioreactors **4**, 60
– gas balancing **4**, 70
– industrial – **6**, 121ff
– – as cattle feed **6**, 198
– – from cassava **6**, 140f
– – from cassava root **6**, 156
– – from cellulose **6**, 160
– – from corn **6**, 140, 156
– – from cornstalks **6**, 163f
– – from food industry wastes **6**, 139f
– – from fruit crops **6**, 138
– – from Jerusalem artichoke **6**, 141f, 159f, 178ff
– – from municipal solid waste **6**, 166
– – from newspaper **6**, 165
– – from potatoes **6**, 158f
– – from sorghum **6**, 141
– – from starch **6**, 140
– – from sugar beet **6**, 137f
– – from sugarcance **6**, 136f
– – from sweet potato **6**, 141

– – from waste sulfite liquor **6**, 138f, 168
– – from whey **6**, 139, 169
– – from wood **6**, 161, 167f
– – from xylose **6**, 167
– – in coal desulfurization **6**, 198
– – in the production of biodegradable plastics **6**, 198
– – production cost **6**, 173ff
– inhibition of microbial growth **6**, 124f
– – kinetic constants **6**, 128
– in syntrophic metabolism **11A**, 463ff
– in vinegar **9**, 586f
– market prices **6**, 51
– mass transfer resistance **4**, 393
– partition coeefficients, in liquid-liquid extraction **3**, 560
– production **3**, 305f, 473
– – by yeasts **3**, 329
– – influencing factors **3**, 305
– – rate estimation **3**, 369
– – self-inhibition **3**, 305
– – yield **3**, 305
– production in yeast cultures **4**, 70, 388
– production volumes 1910–1929 **8A**, 9
– removal by trickling filters **11C**, 336
– uptake, in bioprocesses **3**, 366
ethanol distillation, apparatus **6**, 100f
– batch rectifying apparatus, scheme **6**, 102
– combined distillation and rectification, apparatus **6**, 103f
– raw spirit **6**, 100ff
– recification of product spirit **6**, 192f
ethanol distillery, diagram **6**, 152
ethanol fermentation, batch – **6**, 97ff
– by-products, carbon dioxide **6**, 171f
– – fusel oils **6**, 172f
– contaminants, acidification of mashes **6**, 99
– – acrolein **6**, 99
– energy aspects **6**, 179
– – from cassava **6**, 185ff
– – from corn **6**, 185
– – from cornstalks **6**, 187f
– – from sugarcane **6**, 185ff
– – from wood **6**, 186f
– modes of industrial interest **6**, 151ff
– – batch process **6**, 151
– – continuous processes **6**, 152ff
– – semicontinuous processes **6**, 151f
– – simultaneous saccharification and fermentation process **6**, 155f
– – solid-phase fermentation **6**, 155
– – vacuum fermentation **6**, 155
– productivity, of continuous systems **6**, 133
– – of immobilized cell systems **6**, 135
– productivity factors **6**, 151
– reactions **6**, 126f
– thermodynamics **6**, 126f
ethanol formation, by bacteria **6**, 125ff

– – kinetic parameters **6**, 130ff
– – list of species **6**, 129
– – *Zymomonas mobilis* **6**, 128ff
– by thermophilic, *Thermoanaerobacter ethanolicus* **6**, 132
– by thermophilic organisms **6**, 131f
– by yeast **6**, 123ff
– comparsion of bacteria and yeasts **6**, 132f
– use of immobilized cell systems **6**, 133ff
– – effect of dilution rate **6**, 134
– – effect of pH **6**, 134f
– – effect of temperature **6**, 134f
ethanol-gasoline blends, applications **6**, 194f
– as motor fuel **6**, 189ff
– water tolerance **6**, 191
ethanol markets, cattle feed **6**, 198
– coal desulfurization **6**, 198
– motor fuel **6**, 197f
– plastics **6**, 198
ethanol metabolism, in yeasts **9**, 181
ethanol oxidation, by alcohol dehyydrogenase **6**,
387
– by aldehyde dehydrogenase **6**, 387
ethanol production **1**, 204f
– artificial operon construction **1**, 205
– autoamylolytic processes, corn **6**, 96
– – potatoes **6**, 96
– – rye **6**, 96
– – triticale **6**, 95
– – wheat **6**, 95
– cellulosic material, pretreatment **6**, 145f
– economic analysis **6**, 173ff
– – capital costs **6**, 176
– – distribution of costs **6**, 174
– – effect of plant size **6**, 179
– – ethanol from Jerusalem artichoke **6**, 178ff
– – production costs **6**, 177
– energy balance **6**, 117
– energy consumption **6**, 115ff
– genetic improvement **1**, 204f
– in immobilized *Saccharomyces cerevisiae* **1**, 236
– mashing equipment **6**, 77ff
– – dispersing machines **6**, 79f
– – heat exchangers **6**, 81
– – mills **6**, 78
– – mash tubs **6**, 80f
– processing of, barley **6**, 93
– – corn **6**, 92f
– – grain **6**, 90ff
– – manioc **6**, 94
– – potatoes **6**, 90
– – rye **6**, 92
– – sorghum **6**, 94
– – triticale **6**, 92, 95
– – wheat **6**, 91f, 95
– raw materials, analysis of **6**, 109ff
– – barley **6**, 65
– – corn **6**, 64f

– – manioc **6**, 66
– – potaoes **6**, 62
– – rye **6**, 63
– – sorghum **6**, 65f
– – triticale **6**, 63f
– – wheat **6**, 62f
– spoilage by lactobacilli **1**, 334
– substrates **6**, 136ff
– technical parameters **6**, 175
– yeast mash preparation **6**, 96f
ether, degradation of **11B**, 172ff, 200f
ether bond cleavage, catalysis of **11B**, 200f
ether glycoside, hydrolysis of, use of catalytic antibodies **8B**, 464
ether lipids, archaebacterial – **7**, 181ff
– – structures **7**, 182
ethical aspects, mutation screening **5B**, 117
– of genome research **5B**, 97f, 417ff
ethical biotechnology **12**, 115ff, 189ff
– perceptions **12**, 123ff
– – of benefit and risk **12**, 124ff
ethical framework, of biomedical research **5B**,
418
ethical norms **5B**, 418
ethics, key principles **12**, 117ff
– – animal rights **12**, 120
– – autonomy **12**, 117f
– – beneficence **12**, 118f
– – confidentiality **12**, 120
– – environmental ethics **12**, 121
– – justice **12**, 119f
– – non-maleficence **12**, 119
– – rights **12**, 118
ethics review **5B**, 424
ethoxylates, as nonionic surfactants **6**, 672f
ethylacetate, degradation rate in bioscrubbers
11C, 309
– scrubbing liquid quantities for the absorption
of **11C**, 312
ethylbenzene, anaerobic degradation of **11B**, 184f
4-ethylbenzoate degradation **1**, 202
ethylene, effect on plant cell cultures **1**, 587
ethylene formation, inhibition in transgenic
plants **2**, 649
– in plants **2**, 649
ethyleneimine **2**, 24
ethyl methanesulfonate (EMS) **2**, 24
euactinomycetes **1**, 435
euakryotes, function of PHA **6**, 424
eubacteria, genome **5A**, 51
eugenics, public opinion **12**, 133f
eugenol, degradation of, by *Pseudomonas* sp. **8B**,
337f
– for ferulic acid production **8B**, 337
Euglena gracilis, production of paramylum **10**,
145f
Euglenida, characteristics of **10**, 112f
eukaryotes, DNA transfer to prokaryotes **7**, 263

– effects of microgravity on **10**, 509ff
– genome **5A**, 51
– occurrence of PHA **6**, 412f
– phylogenetic tree **10**, 139
eukaryotic annotation **5B**, 375
eukaryotic cell **1**, 30ff
– structure organization **1**, 30ff
eukaryotic cells, DNA content **5A**, 4
– organelles **5A**, 6ff
– protein compartmentalization **5A**, 6
– protein content **5A**, 4
– secreted proteins **5A**, 9
eukaryotic genome **2**, 143
eukaryotic protein production, by *Saccharomyces cerevisiae* **1**, 498
Eumycota **1**, 517ff
Euphorbia milli, anthocyanin production **7**, 610
Eurobarometer surveys **12**, 450ff, 602, 660, 664ff
EUROFAN **5B**, 24
Europe, biotechnology regulatory policy **12**, 55ff, 349, 486ff
– – Chemical Legislation **12**, 546ff
– – Concertation Action **12**, 538ff
– – history **12**, 516ff
– Common Agricultural Policy **12**, 535
– national bioindustry associations **12**, 635
– public opinion survey, on biotechnology **12**, 133f, 658
– recent conflicts **12**, 633ff
European Agency of Medicinal Products, good manufacturing practice **5A**, 460ff
– – definition **5A**, 460
European Biotechnology Information Strategic Forum *see* BTSF
European Commission, Article 21 **12**, 650
– DG III **12**, 535ff, 543ff, 635ff
– DG VI **12**, 543ff, 638ff
– DG XI **12**, 544ff, 638ff, 642ff
– DG XII **12**, 518ff, 543ff, 638ff, 644f
– number of approvals of biotechnology products **5A**, 512
European Commission's Directorate General for Science, Research and Development (DG XII) **12**, 518ff, 543ff, 638ff, 644f
European Community, Council Recommendations **12**, 518ff
– regulations of gene therapy **5A**, 461
European Medicines Evaluation Agency (EMEA) **5A**, 506ff, 513, 527
– clinical studies **5A**, 504f
– function of **5A**, 507f
– investigational new drugs **5A**, 504f
– organization of **5A**, 507
– structure of **5A**, 508
European Union **5A**, 506ff
– biotechnology regulatory policy **12**, 241ff
– – EEC directives **12**, 264ff
– pharmaceutical legislation **5A**, 506ff

"Europe +30" report **12**, 529
euryarchaeota **1**, 252
Eustigmatophyta, characteristics of **10**, 114
eutectic point, of solutions **3**, 698ff
eutrophication **11A**, 288
evaluation by ecopoints **11A**, 572
evidence information **5B**, 389
evolution, of antibodies **5A**, 3
– of bacilli **1**, 372
– of DNA **5A**, 2f
– of enzymes **5A**, 3
– of metabolic pathways **5A**, 3
– of multicellular organisms **5A**, 3
– of proteins **5A**, 2f, **5B**, 349f
– of RNA **5A**, 2f
– of self-replicating systems **5A**, 2f
– of the genetic code **5A**, 3
evolution tree, *Clostridium* branch **1**, 328f
exergonic process **1**, 53
exobiology, in space **10**, 516f
exochelins **7**, 217f
– structure **7**, 218
exoglucanases **10**, 81
exoglycosidases **8B**, 255
exon **1**, 97, **2**, 143, 676
exon trapping **5B**, 175
exonuclease **2**, 393
– for DNA fragment subcloning **2**, 393
exopeptidases **5A**, 206
– nomenclature **8A**, 244
exopolysaccharides *see also* polysaccharides
– – *see* polysaccharides, biopolymers
– in bacteria **1**, 16
– in pseudomonads **1**, 413
exothermic reaction **1**, 52
expanded bed reactor **11A**, 470
expanded granular sludge bed (EGSB) **11A**, 519ff
– reactor scheme **11A**, 521
– reactor technology **11A**, 519ff
expert control **4**, 591
expert systems **4**, 626ff
– combination with control algorithms **4**, 591
– completeness problem, fuzzy reasoning **4**, 594
– construction of a knowledge base **4**, 627
– definition of **4**, 626
– fault diagnosis **4**, 591, 630ff
– – fermentation processes **4**, 632f
– – pH control **4**, 630f, 633ff
– for bioreactors, prototype **4**, 628ff
– – specific tasks **4**, 628
– for error diagnosis **4**, 173
– for failure correction **4**, 173
– for fermentation control **3**, 345, 392
– – Bio-SCAN **3**, 346f
– – real-time – **3**, 346
– for fermentation processes **4**, 628ff

– for use in computer integrated fermentation **4**, 588
– in automation **4**, 589ff
– knowledge acquisition process **4**, 593f
– knowledge base, rule memory **4**, 590
– – working memory **4**, 590
– messages during an actual fermentation **4**, 633
– operation with real-time data **4**, 627
– overall configuration **4**, 628f
– process state identification **4**, 591
– program development **4**, 629f
– real-time **4**, 589ff
– routes in the development **4**, 590
– software base, computer languages **4**, 589f
– – hybrid software **4**, 590f
– supervisory rule sets **3**, 347
– underlying automation systems **4**, 592
– use of **4**, 626
– validation **4**, 627
explosives *see also* TNT
– bioremediation with the heap technology **11B**, 326
– extraction of **11B**, 499
EXPR *see* phosphorus removal
expressed sequence tags (EST), databanks **5B**, 126f
– of crop plants **5B**, 149
– production of **5B**, 178
– use in genome annotation **5B**, 370ff
expression profile **5B**, 125
expression studies **5B**, 437
expression systems, *Bacillus brevis* **9**, 79
– *Bacillus subtilis* **9**, 79
– baculovirus – **9**, 85ff
– *E. coli* **9**, 75ff
– *Hansenula polymorpha* **9**, 85
– *Pichia pastoris* **9**, 85
– *Yarrowia lipolytica* **9**, 85
– yeast **9**, 80ff
expression vector **2**, 411ff, 561
– bicistronic **2**, 710, 722f
– – gene expression **2**, 710, 723
– *E. coli* **2**, 411ff
– – construction **2**, 411
– – promoters **2**, 411
– in yeast, integrative vectors **1**, 499
– – *see also* vectors
– with T7-promoters **2**, 721
ex situ remediation, use of heap technique **11B**, 321
ex situ treatment of soil **11B**, 307f, 313f
– physical properties **11B**, 57f
extended Kalman Filter (EKF) **3**, 335, 380f, **4**, 549ff
– algorithm **4**, 231f
– application, baker's yeast fermentation **4**, 234ff
– – cellulose hydrolysis **4**, 232ff
– assumptions **4**, 549

– definitions **4**, 549
– filter tuning **4**, 232
extensible markup language (XML) **5B**, 396
extracellular domain **5B**, 137
extracellular enzymes *see also* enzymes
– bacterial **1**, 27f
– of bacilli **1**, 384
– of clostridia **1**, 310
– of streptomycetes **1**, 451f
extracellular polymeric substances (EPS) **11A**, 74ff, **11C**, 447f
– binding of metals **10**, 234
– bridging agents **11A**, 75f
– compaction of **11C**, 447
– composition of **11A**, 74
– definition of **11C**, 447
– in activated sludge flocs **11A**, 69
– in biodeterioration **10**, 271
– in biofilms **11C**, 446ff
– – as structural elements **11C**, 448
– – function of **11C**, 448
– metal leaching **10**, 200
– role in biofilms **10**, 290ff
– – biofilm formation **10**, 292
– role in flocs and biofilms **11A**, 74ff
– role in metal precipitation **11A**, 438
extractant diluent **3**, 568ff
– methyl-isobutyl ketone (MIBK), effects on distribution coefficients **3**, 571
– modifiers **3**, 569
– non-polar – **3**, 569
– paraffine/kerosene (ISOPAR H), effects on distribution coefficients **3**, 571
– polar – **3**, 569
extractants *see* solvents
extraction **3**, 594ff
– – *see also* protein partitioning
– bioproduct separation **4**, 610f
– carrier **3**, 566
– – Cyanex 923 **3**, 527f
– – Hostarex A 327 + isodecanol **3**, 527f
– – tri-*n*-octyl-methylammonium chloride (TO-MAC) **3**, 576
– *in situ* – **3**, 584f
– – *Clostridium acetobutylicum* products **3**, 585
– – medium-solvent separation **3**, 584
– – secondary metabolites **3**, 586
– – solvent criteria **3**, 584
– liquid-liquid – *see* liquid-liquid extraction
– of fermentation products **3**, 43
– performance **3**, 596
– systems **3**, 595ff
extractor **3**, 559
– centrifugal – **3**, 559
– – Podbielnak **3**, 564
– column **3**, 559
extreme environments, for microorganisms **10**, 63ff

extremophiles, biocatalysis with –   **10**, 61ff
– cultivation of **10**, 70f
– – dialysis fermenter **10**, 71
– – microfiltration bioreactor **10**, 71
– definition of **10**, 63
– listing of representatives **10**, 65f
– production of, cellulases **10**, 83
– – chitinase **10**, 83
– – enzymes **10**, 79f
– – proteolytic enzymes **10**, 86
– – starch-hydrolyzing enzymes **10**, 79
– – xylanases **10**, 83
– screening for enzymes in **10**, 73f

F

F2 generation **2**, 147ff
F-12 fibroblast medium, composition **3**, 143f, 154
F plasmid **2**, 49ff
– *E. coli* **2**, 52
– homology regions **2**, 52
– map **2**, 52ff
Fab fragments **5**, 224
faciliated diffusion, substrate uptake **9**, 190
facility, changes **3**, 766
– closed systems **3**, 745
– construction documentation **3**, 749
– construction inspections **3**, 749
– environmental monitoring **3**, 763
– multiple-use – **3**, 744
– – concerns for **3**, 744
– – contamination consideration **3**, 744
– – product changes **3**, 765
– – types **3**, 744
– personnel, monitoring **3**, 764
– pre-license inspection **3**, 743
– process changes, regulatory impacts **3**, 765
– regulatory authorities **3**, 746
– system piping **3**, 758
facility design **3**, 739ff
– construction validation **3**, 748
– expansion planning **3**, 748
– master validation plan (MVP) **3**, 741
– production water systems **3**, 755ff
– validation, benefits **3**, 741
– – program **3**, 742
facility validation **3**, 739ff, 760
– – *see also* process validation
– air leakage **3**, 749
– cleaning **3**, 62
– clean steam systems **3**, 751
– compressed air **3**, 751
– computer systems **3**, 764
– decontamination **3**, 763
– deionized (DI) systems **3**, 750
– equipment **3**, 748, 759f

– facility changes, revalidation **3**, 765f
– heating, ventilation and air conditioning (HVAC) systems **3**, 751
– medical gasses **3**, 751
– monitoring frequency **3**, 751
– multiple-use facilities **3**, 745
– nitrogen systems **3**, 751
– personnel **3**, 764
– plant steam **3**, 750
– purified water **3**, 750
– room finishes **3**, 749
– ultilities **3**, 750
– waste treatment **3**, 763
– water for injection (WFI) systems **3**, 750
– welding **3**, 749
factor IX **5A**, 283
σ factors **7**, 61ff
– initiation of transcription **7**, 62
– regulation of – **7**, 64
– structure **7**, 62
factor VIII **5A**, 283
factor VIII protein **5B**, 72
factory operation **5B**, 432
FAD **9**, 33
– – *see* flavin adenine dinucleotide
– – *see* flavin adenine dinucleotide
– reduction of **4**, 87
FAD, microbiological preparation **6**, 602f
falvin adenine dinucleotide (FAD) **8A**, 549ff
– diastereotopic faces of **8A**, 550
– structure **8A**, 536
FAO (Food and Agriculture Organization) **12**, 363, 619ff
Faraday's law **10**, 293
farmers' privilege **12**, 285, 287, 289
farnesamide N-oligopeptides, regio- and enantio-selective epoxidation of **8B**, 507
farnesene, biotransformation of **10**, 407f
– hydroxylation of **8A**, 493
farnesyl pyrophosphate synthetase, from pig liver **8B**, 365
fasciculols, plant growth inhibitors, structures **7**, 515
fasmid **2**, 269
FASTA **5B**, 130
FAST (Forecasting and Assessment in Science and Technology) **12**, 529f, 535f, 543f, 640, 657
Fast protein liquid chromatography (FLPC), on-line **4**, 167
Fast protein liquid chromatography (FPLC) **4**, 208
fat, anaerobic digestion of **11C**, 19ff
fat production, wastewater of **11A**, 203ff
fats, anaerobic degradation of **11A**, 34f
– – thermophilic process **11A**, 34
– fermentation of **11A**, 36f
– – biogas amounts and composition **11A**, 36f

– lipase-catalyzed processes for specialty fats **8A**, 134
fatty acid methyl ester analysis **11C**, 292f
fatty acids **1**, 87f, **7**, 146ff
– as nutrients, for mammalian cell cultures **3**, 146
– biosynthesis **1**, 88
– – in oleaginous molds **1**, 523f
– chlorinated – **8B**, 186
– degradation of **1**, 87
– esterification **7**, 146
– hydroxylation of **8A**, 483
– in bacteria, polyunsaturated fatty acids **7**, 147f
– in chemitaxonomy **1**, 405, 407
– in fungi, polyunsaturated fatty acids **7**, 151
– in lipoglycopeptides **7**, 386f
– in microalgae, arachidonic acid (ARA) **7**, 167f
– – docosahexaenoic acid (DHA) **7**, 169f
– – eicosapentaenoic acid (EPA) **7**, 168f
– – γ-linolenic acid (GLA) **7**, 167
– in molds, arachidonic acid (ARA) **7**, 161f
– – dihomo-γ-linolenic acid (DHGLA) **7**, 160
– – docosahexaenoic acid (DHA) **7**, 162f
– – eicosapentaenoic acid (EPA) **7**, 162
– – eicosatrienoic acid (ETA) **7**, 163
– – γ-linolenic acid (GLA) **7**, 159f
– – polyunsaturated fatty acids **7**, 158
– in mushrooms **9**, 225
– *in vivo* synthesis of **8A**, 373
– in yeasts **9**, 225
– microbial metabolism **9**, 186
– selectivities of lipases, saturated fatty acids **8A**, 142ff
– – unsaturated fatty acids **8A**, 143f
– structures **7**, 139
fatty acids synthetase complex, of yeast, reducing enzymes **8A**, 373f
fatty acid synthases, crosstalk to polyketide synthases **10**, 365f
– – in *Mycobacterium* **10**, 366
– – in *Streptomyces* **10**, 365
fatty acid synthetase cycle, schematic representation **8A**, 373
fault detection, computer control **4**, 571
fault diagnosis, in fermentation processes **4**, 632f
– – use of expert systems **4**, 591, 632
– pH control **4**, 630f
– – logic **4**, 631
– – rules **4**, 631
– – use of expert systems **4**, 630ff
Fault-tolerant programming **4**, 571
FCCSET (Federal Coordinating Council for Science, Engineering and Technology) **12**, 570ff
FDA *see* Food and Drug Administration
FDA (U.S. Food and Drug Administration) **12**, 41, 53, 144, 150, 198, 207, 241, 243, 248f, 349, 461, 568ff

– Statement of Policy, on foods derived from new plant varieties **12**, 176ff
FDH *see* formate dehydrogenase
FDP aldolase *see* fructose-1,6-diphosphate aldolase
feature extraction, in bioprocesses, discriminating power **3**, 391
– – trend analysis **3**, 389ff
Fed-batch bioreactors, adaptive optimization **4**, 557f
– optimization **4**, 530ff
fed-batch culture **1**, 154, **3**, 28, 100, **9**, 204ff
– circulation time distribution **3**, 95
– definition **4**, 530
– dilution rate **3**, 365
– glucose uptake rate **3**, 365
– mass balances **3**, 100
– model simulation **1**, 155
– penicillin production **1**, 156
– production criteria **3**, 101
fed-batch experiment, feed profile **4**, 258
– indentification functional **4**, 258
fed-batch fermentation, baker's yeast, simulation of **4**, 341
– sequential control **4**, 514
Federal Food and Cosmetic Act **5A**, 498
Federal Ministry for Research and Technology *see* BMFT
Federal Pharmaceutical Industry Association *see* BPI
Federal Plant Pest Act (FPPA) **12**, 201ff
Federal Technology Transfer Act **5A**, 522
Federal Virus, Serum and Toxin Act **5A**, 498
feed, ethanol **6**, 198
– stillage **6**, 104ff
– use of, amino acids **6**, 468
Feedback control, definition **4**, 513
– schematic diagram **4**, 513
feedback inhibition, cooperative – **9**, 130ff
– – tryptophan biosynthesis **9**, 130
– in metabolic regulation **9**, 46, 125, 132
feedforward control **4**, 519ff
– schematic diagram **4**, 519
feedforward-feedback control, schematic diagram **4**, 519
feed industry, use of phytase **5A**, 210ff
feed products **9**, 769ff
– antibiotics **9**, 780
– *Aspergillus oryzae* fermentation extract **9**, 781
– bacterial direct-fed microbials **9**, 781
– enzymes **9**, 782
– yeast **9**, 780f
feeds **9**, 769ff
– enhancement of nutritive value **9**, 779f
– fermentation by-products **9**, 771ff
– – of the brewing industry **9**, 771f
– – of the distilling industry **9**, 772ff
– oil extraction by-products **9**, 721

– – protein meals **9**, 725
– silage preservation **9**, 775ff
feedstock, for composting, biodegradability of **11C**, 130
– – listing of **11C**, 128
feed use, of microbial products, safety **9**, 227ff
fenchone, Baeyer-Villiger reaction, *Corynebacterium* mediated – **8A**, 545
Ferguson plot, in polyacrylamide gel electrophoresis **3**, 722
fermentable substance, determination in grain **6**, 110
fermentation **1**, 56, 86, 381, **3**, 8ff, 79, 321ff, 355f
– – *see also* bioprocess, bioreactor
– by recombinant organisms, reactor design **3**, 412ff
– end products **1**, 86
– for therapeutic products **3**, 63
– heat evolution **1**, 142
– processes in clostridia **1**, 297ff
– water quality **3**, 755
fermentation analysis **3**, 324
– data analysis **3**, 330
– data logging **3**, 330
– off-line – **3**, 324
fermentation broth, analysis **3**, 324
– apparent viscosity **3**, 474
– cell removal **3**, 473
– conditioning **3**, 474
– density, effects on separation **3**, 475
– element balancing, composition matrix **3**, 372
– foam control **3**, 422f
– mass transfer coefficients, dimensionless correlation **3**, 203f
– mixing *see* bioreactor mxing
– rheological properties **3**, 411
– solid content **3**, 480
– solid phase, specific gravity **3**, 474
– viscosity, effects on separation **3**, 475
fermentation control **3**, 355ff, **4**, 88f
– – *see also* bioprocess control, bioreactor control
– adaptive – **3**, 340ff
– adaptive linearizing – **3**, 343
– based on off-gas measurements **3**, 329
– computer applications **4**, 564ff
– – criteria **4**, 564
– – disadvantages **4**, 565
– computer-based – **3**, 330
– control loops **3**, 324
– derived variables **3**, 329
– design, model-based – **3**, 327
– energy balancing **3**, 330
– expert systems **3**, 345
– feedback – **3**, 328
– feedforward **3**, 325f
– fixed-step-ahead controller **3**, 341
– General Model Control (GMC) **3**, 343
– glucose analyzer **4**, 88f

– industrial trends **3**, 327
– inferential – **3**, 340f
– L/A controller **3**, 343
– mass balancing **3**, 330
– mathematical optimization **3**, 344f
– multi-step-ahead controller **3**, 341
– non-linear – **3**, 343
– on-line – **3**, 321ff
– open-loop – **3**, 325
– optimizing **3**, 344f
– PI controller **3**, 340
– probe failure **3**, 322
– proportional-integral-derivative – (PID) **3**, 340
– – autotuning strategy **3**, 340
– quality **3**, 328
– sensor reliability **3**, 322
– set-point profiles **3**, 325
– system improvement **3**, 326f
– triggers **3**, 326
fermentation data analysis **3**, 355ff
fermentation estimation *see* process estimation
fermentation gas analysis, instruments **4**, 30ff
– methods **4**, 30ff
fermentation instrumentation **3**, 321f, 327
– – *see also* bioreactor instrumentation
– anaerobic processes **3**, 313
– contamination risk **3**, 321
– for closed loop control **3**, 322
– for status monitoring **3**, 322
– on-line – **3**, 321
fermentation media **3**, 15, 127ff
– – *see also* medium composition, medium sterilization
– antibiotics **3**, 150
– antioxidants **3**, 152
– attachment proteins **3**, 149f
– buffers **3**, 130, 151
– carbon levels **3**, 131
– carbon sources **3**, 128
– cell yield coefficient **3**, 131
– chemically defined – **3**, 11, 154
– complex – **3**, 132ff
– – carbon sources **3**, 132
– – content **3**, 132f
– – ingredient analysis **3**, 132
– – nitrogen sources **3**, 132
– – raw materials **3**, 132, 146
– components **3**, 142
– – costs **3**, 155
– composition **3**, 143ff
– conditioned medium **3**, 143
– contaminants **3**, 143
– Coon's medium **3**, 143f, 154
– design **3**, 153f
– Dulbecco's modified Eagle's medium (DMEM) **3**, 143f, 154
– economics **3**, 131, 137
– environmental requirements **3**, 130

– for mammalian cell cultures **3**, 142ff
– formulations **3**, 130f
– – reduced serum – **3**, 153f
– – serum-free – **3**, 153f
– F-12 fibroblast medium **3**, 143f, 154
– growth factors **3**, 147
– heparin effects **3**, 153
– hormones **3**, 150
– low molecular weight nutrients **3**, 145
– minerals **3**, 129
– nitrogen sources **3**, 129
– nitrogen supply **3**, 131
– nutrient levels **3**, 131
– oxygen supply **3**, 130
– pH vcalues **3**, 130
– polymer effects **3**, 235f
– products **3**, 142
– protease inhibitor **3**, 152
– protective agents **3**, 151
– protein content **3**, 146f
– selection **3**, 153f
– specific nutrients **3**, 130
– supplements **3**, 147ff
– – non-nutritional – **3**, 150ff
– surface tension **3**, 235f
– temperature **3**, 130
– transport proteins **3**, 149f
– viscosity **3**, 235f
– water quality **3**, 143ff
– with serum, constituents **3**, 146
– – effects on cell cultures **3**, 147
– – factors **3**, 147
– with tissue extracts **3**, 146
fermentation media optimization *see* medium optimization
fermentation modeling **3**, 319ff, 330ff
– – *see also* bioreactor modeling, process estimation
– for control **3**, 330ff
– high cell density fermentation **3**, 331
– industrial trends **3**, 327
– models **3**, 331ff
– – black box – **3**, 331
– – input-output – **3**, 331
– – kinetic – **3**, 331
– – linear black box – **3**, 332
– – mathematical – **3**, 330f
– – non-linear black box – **3**, 333
– – physiological – **3**, 331
– – segregated – **3**, 332
– – structured – **3**, 331
– – traditional process – **3**, 331
– – unstructured – **3**, 331
– optimization algorithms **3**, 345
– optimum quality control **3**, 345
– optimum temperature profile **3**, 344
– Pontryagin's Maximum Principle **3**, 344

fermentation monitoring **3**, 16, 121, 319ff, 355ff, 360ff
– – *see also* bioprocess analysis
– anaerobic processes **3**, 313
– computer-based – **3**, 330
– derived variables **3**, 322
– devices **3**, 359, 386
– expert systems **3**, 345
– heat transfer coefficients **3**, 323
– instrumentation **3**, 322
– off-gas measurements **3**, 329
– off-line – **3**, 329
– sampling frequency **3**, 324
– variables, correlations **3**, 328
fermentation parameters, estimation techniques **4**, 546f
– measurement errors **4**, 546
fermentation probes *see* fermentation instrumentation
fermentation process **3**, 94, 297ff
– – *see also* bioprocess
– anaerobic – *see* anaerobic fermentation
– automation **3**, 16
– batch operation **3**, 15
– bioreactor design **3**, 404
– continuous operation **3**, 15
– development **3**, 18
– effects of broth viscosity **3**, 202
– gas balance parameters **3**, 368
– mass transfer coefficient, correlation **3**, 198
– mathematical modeling **3**, 327
– microorganism selection **3**, 19
– on-off control **4**, 514
– product purification **3**, 17
– scale-up **3**, 185ff
– self-regulation **3**, 324
– time constants **3**, 325
– type I, computation of feed rate profile **4**, 536f
– – optimization **4**, 533ff
– type II, computation of feed rate profile **4**, 537f
– – optimization **4**, 535ff
– type III, computation of feed rate profile **4**, 538
– – optimization **4**, 535ff
– validation *see* process validation, facility validation
– with recombinant cells **3**, 283ff
fermentation process parameters **4**, 492ff
fermentation products **3**, 8ff, 41
– – *see also* biotechnological products
– expression systems **3**, 41
– extracellular – **3**, 41
– heterologous proteins **3**, 42
– intracellular – **3**, 41
– sources **3**, 40
fermentation types, definitions **4**, 533ff
fermentation vessel **3**, 410

– – *see also* bioreactor
– codes **3**, 418
– diameter **3**, 418
– dimensions **3**, 418
– geometry **3**, 410
– laboratory –, aeration **3**, 240
– material **3**, 419
– mechanical details **3**, 418
– welds **3**, 420
fermentative bacteria **11C**, 153
fermentative production, of natural aroma chemicals **8B**, 333ff
fermented milk *see* dairy products
fermenter design, cooling **9**, 333f
– for baker's yeast production **9**, 331f
– oxygen dispersion **9**, 332f
fermentor *see* bioreactor
ferribactin, peptide siderophore of *Pseudomonas fluorescens,* non-fluorescent – **7**, 214
ferrichromes, fungal siderophores **7**, 225f
– structure **7**, 225
ferric uptake regulation proteins **7**, 200
ferrimycins **7**, 211
ferrioxamines, hydroxamate siderophores of *Streptomyces pilosus* **7**, 209f
– structures **7**, 209f
ferritin **5A**, 19, 34
ferritin/transferrin receptor, iron response element (IRE) **2**, 679, 686
– synthesis regulation **2**, 679, 686
ferritin/transferrin synthesis, regulation **2**, 686
– – iron response element (IRE) **2**, 686
ferrocenes, lipase-catalyzed reaction of **8A**, 100
ferrocins **7**, 212f
– peptide siderophore of *Pseudomonas fluorescens* **7**, 213
– structure **7**, 212
ferrosamine A, catecholate siderophore of *Erwinia rhapontici* **7**, 208
fertility factor **2**, 50, 55ff
fertilizer, use of stillage **6**, 106ff
fertilizers, use in oil decontamination **11B**, 432
ferulic acid, as substrate for vanillin production **8B**, 338ff
– fed-batch formation **8B**, 338f
– production of **8B**, 336ff
– – as intermediate for vanillin production **8B**, 336ff
– – by *Pseudomonas* sp. **8B**, 337
– – from eugenol **8B**, 337
FET (Foundation on Economic Trends) **12**, 463ff
FIA *see* Flow-injection analysis
fiber-optical technique, bubble sensors **4**, 131f
fiber optics **4**, 85f
fibrin **5A**, 34
fibrinogen **5A**, 34
fibrinogen receptor antagonists **7**, 115
fibroblast growth factors **5A**, 133

– acidic – **5A**, 162
– androgen-induced – **5A**, 162
– basic – **5A**, 162
– glia-activating – **5A**, 162
– in therapy **5A**, 162
– keratinocyte growth factor **5A**, 162
– receptors **5A**, 162
– use in therapy **5A**, 131
fibronectin **5A**, 33
fibrosis **5A**, 132f
Fick's first law of diffusion, pollutant flow per time **11B**, 72
Fick's law **4**, 110f
field composting **11C**, 117f
field release, of genetically engineered microorganisms **11B**, 434
field trials **12**, 149, 250ff, 260ff, 359, 463ff, 555ff, 570ff
– cost **12**, 472
– history **12**, 463
– ice-minus/Frostban bacteria **12**, 463ff
– in Japan **12**, 260ff
– in the USA **12**, 250ff
– notification **12**, 205f, 250, 252
– permits **12**, 52, 202, 205f, 250ff, 466, 472
– – analysis **12**, 487
– – for microorganisms **12**, 206
– – for plants **12**, 206
– petition **12**, 205, 250
– risk assessment **12**, 46ff, 463f
FIFRA (Federal Insecticide, Fungicide, and Rodenticide Act) **12**, 53, 462ff, 571
filamentous fungi **1**, 515ff, **2**, 530ff
– – *see also* fungi
– characteristics **1**, 516
– classification **1**, 516
– cultivation **1**, 520ff
– culture collections **1**, 520
– dikaryotes **1**, 519
– distribution **1**, 518
– fatty acid synthesis **1**, 523f
– fermentation broth, rheological properties **3**, 89f
– genetic diversity **1**, 518
– genetic engineering **1**, 536
– haplo-dikaryotes **1**, 518
– immobilization **1**, 522
– in bioreactors, morphological characteristics **3**, 89f
– industrial application **2**, 530
– isolation **1**, 520f
– lactic acid production **1**, 523
– life cycle **1**, 518ff
– lipid production **1**, 523f
– metabolite production **1**, 522
– morphology **1**, 517
– oleaginous **1**, 523
– procurement **1**, 520f

– production of, fumaric acid **1**, 523
– – lactic acid **1**, 523
– – lipids **1**, 523f
– – metabolites **1**, 522f
– propagation **1**, 517
– solid-state fermentation **1**, 521
– spore **1**, 517
– submerged culture **1**, 521
– submerged pellet **1**, 116
– surface culture **1**, 521
– taxa **1**, 517
– white-rot fungi **1**, 534f
filtering **5B**, 132
filter media **11A**, 338
filters, absolute – **3**, 161f
– depth – **3**, 160
filtrate volume, biomass determination **4**, 191
filtration **3**, 439ff
– cross-flow **4**, 608, 618f
– hollow-fiber, cell separation **4**, 608
– of fermentation products **3**, 43ff
– rapid –, in drinking water treatment **11C**, 389ff
– tangential flow **4**, 609
filtration mechanisms, absolute size exclusion **3**, 162
– bridging **3**, 162
– diffusion **3**, 161, 177
– electrostatic interaction **3**, 161
– gravitational sedimentation **3**, 161, 177
– impaction **3**, 161
– – inertial – **3**, 177
– interception **3**, 161, 177
– particle adsorption **3**, 162
filtration membranes **3**, 161, 488, 495
– boundary layer **3**, 44
– – dynamic formation **3**, 44
– cleaning **3**, 496
– diffusion coefficient, shear-enhanced – **3**, 492
– dynamic membranes **3**, 488
– flow resistance **3**, 490
– flux direction **3**, 440
– for air filtration **3**, 177
– for cell retention **3**, 115
– for ultrafiltration **3**, 44
– fouling **3**, 496
– – minimization **3**, 497
– gel polarization **3**, 490
– macromolecular concentration polarization **3**, 490
– materials **3**, 44, 440
– pore size **3**, 44, 440
– pore structure **3**, 439
– secondary membranes **3**, 488
– selection **3**, 498
– structural changes, control **3**, 497
filtration methods, for biomass determination **4**, 190ff
filtration modules **3**, 498f

– configuration **3**, 498f
– flat sheet – **3**, 500
– flow characteristics **3**, 499
– hollow fiber – **3**, 500
– scale-up **3**, 499
– spiral wound – **3**, 500
– tubular – **3**, 500
filtration probe, computer-controlled **4**, 191
fimbriae **1**, 17, 32
– structure **1**, 18
fimicolon **7**, 502f
– cytotoxic activity **7**, 503
– structure **7**, 502
FIMM **5B**, 407
Finding of No Significant Impact (FONSI) **12**, 204
finish-line technique **5B**, 200, 272
FIRDI (Food Industry Research and Development Institute) **12**, 404
fish embryo transformation **2**, 811f
Fisher information matrix **4**, 255, 261
fishes, antifreeze proteins from **10**, 467
fish sauces, fermented – **9**, 536ff
fixed-bed bioreactor **3**, 118, 432
– anaerobic wastewater purification **4**, 450
– – temperature optimization **4**, 479f
fixed-bed loop reactor, scheme **6**, 213
fixed bed reactors **11A**, 400f
– – *see also* bioreactors
– anaerobic wastewater treatment **11A**, 495ff
– biomass concentration **11A**, 502f
– for solid state processing **11B**, 333f
– – scheme **11B**, 333
– metal ion removal **11A**, 445, 447
– mobile – **11A**, 359f
– operation of **11A**, 507ff
– – results **11A**, 508f
– – start-up procedure **11A**, 507f
– reactor geometry **11A**, 506
– scheme **11A**, 401
– *see also* submerged fixed bed reactors **11A**, 400
– support materials **11A**, 502ff
– technology aspects **11A**, 506
– treatment of agricultural wastewater and slurry **11A**, 226f
fixed bed systems **11A**, 508f
– biofilm characteristics **11A**, 497ff
– – external mass transfer **11A**, 497f
– – internal mass transfer **11A**, 498
– biofilm formation **11A**, 496f
– industrial-scale results **11A**, 508f
– pilot-scale results **11A**, 508f
– support characteristics **11A**, 499
fixed-film bioscrubbers *see* trickling filters
fixed film reactors **11A**, 470
– – *see also* biofilm reactors
– stationary – **11A**, 493ff
– – supports **11A**, 499

– – suspended solids **11A**, 505
– wastewater treatment **11A**, 504ff
fixed film stationary bed reactors **11A**, 493ff
fixed film systems, anaerobic wastewater treatment **11A**, 495ff
– reactor design **11A**, 500ff
– scale up **11A**, 500f
fixed growth reactors *see also* rotating biological contactors, trickling filters
– denitrification **11A**, 335ff
– nitrification **11A**, 335ff
FK506 **7**, 546ff
– immunosuppressive properties **7**, 546f
– structure **7**, 546
FK506 binding proteins **7**, 547f
flacine adenine dinucleotide *see* FAD
flagella, mobility system **5A**, 32
Flame ionization detector, gas analysis **4**, 69
flavidulols, structures **7**, 509
flavin adenine dinucleotide (FAD) **5A**, 20f
– production of **10**, 320ff
flavin coenzyme **1**, 71
flavin dependent monooxygenases (FMO), in mammalian tissue, pig liver **8A**, 575
– – pig lung **8A**, 575
flavine adenine dinucleotide (FAD) **1**, 71
flavin hydroxyperoxide, structure **8A**, 537
flavin mononucleotide (FMN), structure **8A**, 536
– synthesis of **10**, 320
flavin monooxygenases **8A**, 535ff
– identification of **8A**, 541ff
– isolation of **8A**, 541ff
flaviolin, biosynthetic pathway **10**, 352
Flavobacterium, in activated sludge **11A**, 97f
flavoenzymes, catalyzing Baeyer-Villiger reactions, *N*-terminal sequences of **8A**, 549
– use in biotransformations **8A**, 536
flavonoid degradation, in clostridia **1**, 303
flavoproteins **5A**, 20ff
– cofactor **5A**, 20
– structure **5A**, 20
flavor-active alcohols, production of **8B**, 345
flavor-active aldehydes, production of **8B**, 345f
flavor-active lactones, fermentative production of **8B**, 340ff
– sensoric description **8B**, 341
flavor compounds, biogeneration of **10**, 383
– natural –, biocatalytic synthesis of **10**, 374
– terpenoid –, production of **10**, 373ff
flavor enhancers **9**, 702ff
– amino acids **6**, 467f
– glutamate **9**, 702, 704
– meat hydrolysates **9**, 704f
– nucleotides, guanylate (5′) **6**, 563
– – inosinate (5′) **6**, 563
– – xanthylate (5′) **6**, 563
– peptides **9**, 702, 704
– wheat gluten **9**, 704

– yeast extracts **9**, 702f
flavor nucleotides *see* nucleotides (5′)
flocculating agents *see* biosurfactants, surfactants
flocculation, effect on BOM **11C**, 463
– in drinking water treatment **11C**, 383ff
– model of **11A**, 76
– processes during **11C**, 386
flocculation/flotation plant, scheme **11C**, 389
flocculation plant, scheme **11C**, 387
flocs **11A**, 65ff
– – *see also* activated sludge flocs, biofilms
– bacteria involved in the formation of oligonucleotide probes **11A**, 94
– binding forces **11A**, 74
– cohesiveness **4**, 436
– composition of **11A**, 67ff
– definition of **11A**, 66
– development of **11A**, 67f
– dispersion forces **11A**, 74
– enzyme activities in **11A**, 76
– extracellular polymeric substances in **11A**, 74ff
– floc size **4**, 431, 436
– heterogeneous spatial arrangement of microorganisms **11A**, 77
– immobilization of microbial cells **11A**, 76
– in an activated sludge system **4**, 431f
– mass transfer resistance **4**, 431
– settling velocity **4**, 435f
– size of **11A**, 67
– structure of **11A**, 68f
– – analysis methods **11A**, 68f
– – heterogenous distribution of microbial aggregates **11A**, 68
flotation, in drinking water tretment **11C**, 386ff
flour, gelatinized – **9**, 260
– quality criteria **9**, 250ff
– storage **9**, 265f
flow cytometer **1**, 121
– schematic diagram **4**, 197
flow cytometry **10**, 52
– cell-size measurement **4**, 201
– cell-viability determination **4**, 202
– determination of cell parameters **4**, 196ff
– DNA measurement **4**, 204f
– lipid monitoring, metabolic intermediates **4**, 212
– measurable cellular parameters **4**, 198
– pH$_i$ measurement **4**, 203
– protein monitoring **4**, 206f
– RNA measurement **4**, 204f
– single-beam flow cytometer **4**, 201
– split-beam flow cytometer **4**, 201
flow-follower techniques, estimate of circulation time **4**, 308
– in stirred tanks **4**, 307ff
– in two-impeller systems **4**, 315ff
– magneto-flow-follower **4**, 124f, 316f

– magneto-flow-followwert, equipment layout **4**, 308
– radio-flow-follower **4**, 124, 308
flow-injection analysis (FIA) **4**, 159ff, 199f, 207, **10**, 51
– analytes measured off-line **4**, 161f
– analyzer system EVA **4**, 160
– determination of, ammonia **4**, 163
– – ethanol **4**, 163
– – glucose **4**, 163
– – phosphate **4**, 163
– for sucrose **4**, 90
– manifolds **4**, 163
– off-line analysis **4**, 163
– on-line **4**, 159ff
– on-line analysis **4**, 163
– on-line sampling unit **4**, 162
– on-line systems **4**, 164
– – advantages **4**, 165
– – disadvantages **4**, 165
– reliability **4**, 173
– stopped flow approach **4**, 162f
flow injection ELISA **4**, 93f
flowmeter, differential pressure **4**, 12
– floating body **4**, 11f
– magnetic-inductive **4**, 12
flow modelling **4**, 352ff
– air lift reactor flow **4**, 359ff
– bubble column flow models **4**, 354ff
– flow configuration, churn-turbulent regime **4**, 352f
– – homogeneous bubble flow regime **4**, 352f
– – slug flow **4**, 352f
– flow regime transitions **4**, 353f
flow patterns, biofilters **11A**, 370
flow sheeting **4**, 621
flow technologies, miniaturized – **10**, 52
flow visualization, techniques **4**, 127f
fluid flow **3**, 170
– axial dispersion **3**, 171f
– pattern **3**, 170
– Peclet number **3**, 171ff
– Reynolds number **3**, 170
fluidized-bed bioreactor **3**, 28f, 116, 120, 432
– bed diameter **4**, 428
– bed height **4**, 428
– design problems **4**, 427ff
– Verax system **3**, 116
– wastewater treatment **4**, 450
– – modelling **4**, 427
fluidized bed reactors **11A**, 470, 493ff
– – *see also* biofilm reactors
– bed height **11A**, 507
– flow distribution system **11A**, 507
– fluxes **11A**, 501
– metal ion removal **11A**, 447
– modeling of **11A**, 501f
– operation of **11A**, 508ff

– – results **11A**, 511
– – start-up **11A**, 508
– reactor geometry **11A**, 506
– scheme **11A**, 401
– specific limitations **11A**, 512
– support, fluidization of **11A**, 507
– – loss of **11A**, 507
– – material **11A**, 499ff
– support materials **11A**, 502ff
– suspended solids **11A**, 505f
– technological aspects **11A**, 506
– treatment of agricultural wastewater and slurry **11A**, 226f
– treatment of sugar factory wastewater **11A**, 201
– wastewater treatment **11A**, 504ff
– – anaerobic – **11A**, 495ff
fluidized bed systems **11A**, 510f
– anaerobic – **11A**, 521f
– biofilm characteristics **11A**, 497ff
– – external mass transfer **11A**, 497f
– – internal mass transfer **11A**, 498
– biofilm formation **11A**, 496f
– industrial-scale results **11A**, 511
– pilot-scale results **11A**, 510f
– support characteristics **11A**, 499f
– support materials **11A**, 504f
fluid mosaic model **1**, 23, **5A**, 7
fluids, characteristic shear rates **3**, 204f
– Newtonian **3**, 89, 193
– – gas residence time **3**, 193
– – mixing **3**, 206ff
– – oxygen transfer correlations **3**, 197
– – oxygen transfer rates **3**, 197
– non-Newtonian **3**, 89, 193
– – gas residence time **3**, 193
– – mass transfer coefficients **3**, 202
– – mixing **3**, 208f
– – mixing zones **3**, 210
– – xanthan solution **3**, 208
fluoranthene, complete metabolization and mineralization **11B**, 225f
– degradation by bacteria, *Mycobacterium* sp. **11B**, 225ff
– – pathways **11B**, 227
– dihydroxylation of **8A**, 511
– hydroxylation of **8A**, 501
– physical and chemical properties of **11B**, 218
– structure of **11B**, 216
fluorene, complete metabolization and mineraliaztion **11B**, 225
– degradation by bacteria **11B**, 225
– physical and chemical properties **11B**, 218
– structure of **11B**, 216
fluorescence, sensor **4**, 209f
fluorescence-based dideoxy sequencing **5B**, 46
fluorescence detection, charge-coupled detectors **5B**, 199
– in DNA sequencing **5B**, 187ff

– photodiode detectors **5B**, 199
– photomultiplier tubes **5B**, 199
fluorescence detection methods, use in drug screening **10**, 45ff
fluorescence dye chemistry **5B**, 186ff
fluorescence emission **5B**, 204
– energy transfer **5B**, 189
fluorescence instrumentation, design of **5B**, 203
fluorescence intensity detection, in drug screening **10**, 48
fluorescence labeling, in drug screening **10**, 45
– 2D gel electrophoresis **5B**, 248
fluorescence lifetime **5B**, 190ff
fluorescence microscopy **1**, 8
fluorescence monitoring, for biomass concentration **4**, 185ff
– – of bacteria **4**, 187
– – of molds **4**, 187
– – of yeasts **4**, 187
– of viable biomass **4**, 188
fluorescence polarization **10**, 45f
– in drug screening **10**, 45, 48
fluorescence probe, determination of cell concentration, detection with two different wavelengths **4**, 186
fluorescence resonance energy transfer **10**, 45f
– in drug screening **10**, 45
fluorescent dyes, in DNA sequencing **5B**, 187ff
– properties **5B**, 192, 199
– spectral discrimination **5B**, 201
fluorescent *in situ* hybridization **5B**, 83
fluorescent labels **5B**, 250
fluorine **8B**, 184
fluorine metabolism **8B**, 199ff
– enzymes of **8B**, 199ff
fluoroacetate, formation by *Streptomyces catteleya* **8B**, 199f
– plant and microbial metabolite **8B**, 190
fluoroacetone, production by plant homogenates **8B**, 190
fluorobenzene, metabolism, by benzene degrading organisms **11C**, 284
– – hypothetical – **11C**, 285
fluorochromes **4**, 203ff
– DNA-binding **4**, 204f
– pH-sensitive **4**, 203
– protein-binding, FITC **4**, 206
– RNA-binding **4**, 204f
fluorocitrate, biosynthesis of, catalyzed by aconitase **8B**, 110
– citrate synthase **8B**, 199
– formation by **8B**, 199
fluorogenic reagents **5B**, 272
β-fluoroketones, *syn*-elimination of, use catalytic antibodies **8B**, 429
fluorometer **4**, 185
fluorophores **5B**, 187ff
– characteristics **5B**, 187

– commercial – **5B**, 187ff
fluoropyruvate, decarboxylation of, by pyruvate decarboxylase **8B**, 54, 56
fluorothreonine, biosynthesis of **8B**, 199f
fluvibactin **7**, 205f
– catecholate siderophore of *Vibrio fluvialis* **7**, 206
– structure **7**, 205
flux control analysis, metabolic flux **1**, 165
fly ash, bioleaching of **10**, 210ff, 216
– leaching of **10**, 196f
– solubilized metals from **10**, 196f
FlyBase **5B**, 285
FMN *see* flavin mononucleotide
FMO *see* flavin dependent monooxygenases
FMR1 gene **5B**, 73
FNM **9**, 33
foam control **4**, 56, 121
– antifoam agents **4**, 15
– foam breakers **4**, 15
– in bioreactors **4**, 14f
foaming, foam control **3**, 422f
– in bioreactors **3**, 409
foaming bacteria involved in **11A**, 94
fobrosis, medical applications of recombinant proteins **5A**, 132
folate and tetrahydrofolate, interconversion of **8B**, 387
folding, motifs in enzymes **9**, 15
– of cytoplasmic protein **9**, 19ff
folding catalysts, use in inclusion body folding **5A**, 120f
folding enzymes, molecular chaperones **9**, 22f
– proline isomerases **9**, 23
– protein disulfide isomerases **9**, 23
folding types, of proteins **5A**, 64
– – prediction of **5A**, 66
fold recognition methods **5B**, 139
folic acid, chemical production of **10**, 331
folic acid metabolism, inhibition by thimidylate synthase **7**, 649
follicle stimulating hormone **5A**, 164
fomannosin, plant growth inhibitor **7**, 514
– structure **7**, 514
food, annual consumption **11C**, 107
– removal of bacteria **5A**, 285
Food Additive Act **12**, 478
Food Additive Provisions of the Food, Drug and Cosmetic Act **12**, 466
food additives, of marine origin **10**, 462ff
– – pigments **10**, 466
– – PUFA **10**, 462ff
– – vitamins **10**, 466
Food and Drug Administration Export Reform Act **5A**, 528
Food and Drug Administration (FDA) **5A**, 445f, 499ff, 534, 540
– adenovirus vectors, testing of **5A**, 411

– advisory committees **5A**, 501
– approval times **5A**, 535
– clinical studies **5A**, 504f
– drug master file **5A**, 457f, 461
– gene therapy regulation **5A**, 523f
– good manufacturing practice **5A**, 460ff
– – definition **5A**, 460
– guidelines **5A**, 523f
– investigational new drugs **5A**, 504f
– plasmid DNA, quality issues **5A**, 460
– plasmid DNA therapeutics, guidelines **5A**, 445
– plasmid vectors **5A**, 452
– priority review **5A**, 501f
– process for filing **5A**, 500
– regulation of biotechnology products **5A**, 519f
– regulation of gene therapy **5A**, 521
– retrovirus vectors, testing of **5A**, 404
– structure of **5A**, 499f
Food and Drug Administration Modernization Act **5A**, 500, 503ff
food biotechnology, public opinion **12**, 453
– risks **12**, 465ff
food emulsifiers, functional properties **6**, 685
food fermentation, by filamentous fungi **1**, 522
food ingredients, production using lipases, cocoa butter substitutes **8A**, 146
– – flavor esters **8A**, 146
food processing, enzymes **9**, 663ff
food production, with lactic acid bacteria **1**, 327, 362f
food products **12**, 564
– components **12**, 179
– content of **5B**, 15
– decision trees **12**, 175ff
– derived from, GMOs **12**, 182ff
– – new plant varieties **12**, 176ff
– – transgenic animals **12**, 181
– – transgenic plants **12**, 180
– "humanization" of **5B**, 15
– labeling **12**, 465f
– microorganisms **12**, 168f
– quality **12**, 620ff
– risk assessment **12**, 469f
– safety **12**, 616f, 620ff
– with rDNA-encoded materials **12**, 177ff
food proteins *see* proteins
food safety **12**, 465
food use, of microbial products **9**, 226f
– – safety **9**, 227ff
force field methods, energy calculations **5A**, 93f
foreign genes, expression in protists **10**, 146f
foreign protein *see* heterologous protein
formaldehyde dehydrogenase **1**, 275
formaldehyde fixation **1**, 275
– dissimilatory RuMP cycle **1**, 275
– linear oxidation pathway **1**, 275
formate, formation in anaerobic metabolism **11A**, 460

– interspecies transfer **11A**, 28
– syntrophic metabolism, interspecies electron transfer **11A**, 465
– use by methanogenic bacteria **11A**, 459f
formate dehydrogenase (FDH), from *Candida boidinii* **8A**, 403
– from *Pseudomonas* sp. **8A**, 403
– NADH regeneration **3**, 454f
– reaction rate **3**, 455
formate oxidation **1**, 275
formycin production, by *Streptomyces lavendulae*, effect of ppGpp **7**, 77
N-formyl-1-amino-1-deoxy-D-sorbitol, biotransformation of, at industrial scale **8B**, 303f
– – with *Gluconobacter oxydans* cells **8B**, 303f
FORTEC process **11B**, 338
fortimicin family, structure of – **7**, 465
fortimicins **7**, 429ff
– biosynthetic pathway **7**, 429
– enzymes, involved in production of – **7**, 431
– gene products, involved in production of – **7**, 431
forward genetics **5B**, 151
fossil fuels, biodesulfurization of **11B**, 452
founder effects **5B**, 78
four-color discrimination **5B**, 189
four-dye/one-lane approach **5B**, 193
Fourier spectral analysis, automated evaluation **5A**, 58
Fourier transformation, for trend analysis, of bioprocesses **3**, 391
Fourier transform infrared spectrometers **4**, 171f
– analysis of aqueous-phase analytes **4**, 171
Fourier transform mass spectrometer **5B**, 255
FPLC *see* Fast protein liquid chromatography
fragment calculation **5B**, 394
fragrance compounds, biogeneration of **10**, 383
– natural –, biocatalytic synthesis of **10**, 374
– terpenoid –, production of **10**, 373ff
frameshifts **12**, 165f
– minus-one **12**, 166
– plus-one **12**, 165f
France, Biomolecular Engineering Committee **12**, 590
– biotechnology regulatory policy **12**, 587ff
– Committee on Genetic Engineering **12**, 589f
– contaminated site management, financing schemes **11B**, 15
– – legal development **11B**, 9
– – principles **11B**, 11
– – registration **11B**, 17
– – risk assessment **11B**, 19
– – technical development **11B**, 9
– identification of contaminated sites **11B**, 17
– public opinion surveys, on biotechnology **12**, 449ff
– regulatory concerns, action thresholds **11B**, 21
– – conflict management **11B**, 28

– – enforcement **11B**, 26
– – in soil contamination **11B**, 7ff
– – liability rules **11B**, 13
– – organization **11B**, 26
– – priority setting **11B**, 23
– – remediation targets **11B**, 24
– – verification of clean-up **11B**, 24
frankobactin, structure **7**, 219
Frateuria, formation of acetic acid **6**, 384
– phylogenetic status **1**, 404
free access to genomic data **5B**, 439
Freedom of Information Act **5A**, 524
free energy of binding **5A**, 95
freeze-concentration, effects on, biomolecules **3**, 703
– – pH values **3**, 698
– – reaction rates **3**, 699
– – solute concentration **3**, 698
freeze-drying **3**, 696ff
– controlled evaporative – **3**, 707
– economics **3**, 706
– effects of, amino acids **3**, 705
– – polyols **3**, 705
– – sugars **3**, 705
– energy consumption **3**, 707
– excipients **3**, 703ff
– freezing pattern **3**, 700
– heat transfer **3**, 700
– in vials **3**, 701
– mass transfer **3**, 700
– of living cells **3**, 705
– – cryoprotectants **3**, 705
– of multicomponent mixtures **3**, 702
– plant components **3**, 707
– primary – **3**, 700ff
– process control **3**, 707
– protein stability **3**, 703ff
– secondary – **3**, 702
– sublimation rate **3**, 701
– temperature measurement **3**, 702
– time **3**, 701f
– – pressure effects **3**, 702
– undercooling effects **3**, 699
freezing **3**, 697, 699ff
– behavior of solutions **3**, 698
– glass transition temperature **3**, 702
Freundlich adsorption equation **4**, 618
Freundlich equation **11C**, 395f
Freundlich isotherm, application to sorption processes **11B**, 53f
Friedels-Craft acylation **8A**, 539
frogs, as sources of ribosomal peptides **7**, 301
front-end processor **4**, 574, 582
Frostban bacteria **12**, 463ff
Froude number **4**, 369
– of spargers **3**, 261
frozen yogurt, composition of **9**, 410
– definition **9**, 388ff

– manufacture of **9**, 407ff
– – single-stream process **9**, 411
– – two-stream blending process **9**, 412
fructose, applications **9**, 757
– commercial products, crystalline – **9**, 756
– – syrups **9**, 756
– properties **9**, 756
fructose production **3**, 456, **9**, 671ff, 747f, 755ff
fructose syrups, composition **6**, 41
– production of **6**, 40
– uses **6**, 41
fructose-1,6-bisphosphatase, carbon catabolite inactivation **1**, 489
fructose-1,6-diphosphate (FDP) aldolase **8B**, 7ff
– – *see also* rabbit muscle aldolase
– aldol condensation **8B**, 14
– resolution of racemic aldehydes **8B**, 9
fruit acids, microbial production **6**, 368ff
– – of fumaric acid **6**, 369
– – of malic acid **6**, 371f
– – of tartaric acid **6**, 373f
– ways of production, fumaric acid **6**, 369
fruit dehiscence **5B**, 154
fruit juice industry, wastewater, amounts **11A**, 210
– – concentrations **11A**, 210
– – treatment **11A**, 209f
fruit juice production **9**, 712ff
– apple **9**, 713f
– banana **9**, 717f
– carrot **9**, 716
– citrus fruits, enzymatic peeling **9**, 715f
– coconut milk **9**, 717
– gel formation **9**, 718f
– pear **9**, 713f
fruit preparations, in cultured dairy products **9**, 403
fruit processing, typical wastes **11A**, 224
fruit ripening modification **2**, 649f
– ethylene formation inhibition **2**, 649
– in tomato **2**, 649
– polygalacturonase inhbition **2**, 649
fucoidans, structure of **10**, 459
fucosyltransferases **8B**, 259
fucoxanthin, structure **7**, 173
fuculose 1-phosphate aldolase, overexpression of **8B**, 14
fuel alcohol *see* ethanol
fuel cells *see also* bioelectrochemical fuel cells
– efficiency of **10**, 10
– hydrogen-oxygen fuel cells **10**, 9f
– – electricity output **10**, 10
– – overall reaction **10**, 10f
– – principle **10**, 9f
fuel cell type sensor, for biomass determination **4**, 193
full-length cDNA molecules **5B**, 372
fulvic acids **11B**, 102f

– oxidation by, laccasses **11B**, 98
– – peroxidases **11B**, 98
fulvoferruginin, structure **7**, 505
fumarase, in industrial biotransformations **3**, 457
fumarases **8B**, 100ff
– biotechnology of **8B**, 106ff
– class I **8B**, 101ff
– – structure **8B**, 101
– class II **8B**, 103ff
– – of *Thermus thermophilus* **8B**, 109
– enzymology **8B**, 104
– hydration of acetylene dicarboxylate **8B**, 102
– in immobilized *Brevibacterium flavum* cells **8B**, 107ff
– non-natural substrates **8B**, 105f
– of *E. coli* **8B**, 101ff
– role in anaerobic metabolism **8B**, 101
– stereochemistry **8B**, 104f
– structures of **8B**, 104
– substrate specificity **8B**, 102
fumaric acid, biosynthetic pathway **6**, 364
– specification **6**, 369
– uses **6**, 370
fumaric acid production **6**, 369ff
– by filamentous fungi **1**, 523
– by *Rhizopus* sp. **6**, 370
– chemical synthesis **6**, 369
fumaronitrile, conversion into metabolizable products **8A**, 313
functional activity, of a protein **5B**, 124
functional assignment **5B**, 135ff, 386
functional food, from microalgae **10**, 127
functional genomics **5B**, 107, 151f
function from structure **5B**, 437
fungal growth *see* mycelial growth
fungal pathogens **5B**, 154
– in household wastes **11C**, 232
fungal plasmid **2**, 74
fungal protein production, by *Saccharomyces cerevisiae* **1**, 498
fungal siderophores **7**, 225ff
– coprogens **7**, 227
– ferrichromes **7**, 225f
– fusarinines **7**, 228f
– rhizoferrins **7**, 229f
– rhodotorulic acid **7**, 228
fungal strain improvement **2**, 114ff
– edible mushrooms **2**, 120
fungi **1**, 515ff, **2**, 74ff, 111ff
– – *see also* filamentous fungi yeasts
– – *see also* mushrooms
– – *see also* oleaginous molds
– alkaloids from **8A**, 328
– allergies to **12**, 90
– as sources of nonribosomal peptides **7**, 309
– basidiomycetous –, degradation of lignin **11B**, 161ff
– – degradation of organopollutants **11B**, 163f

– – degradation of xenobiotics **11B**, 160ff
– biocontrol of plant pathogenic – **1**, 423
– biodegradation of nitro-substituted compounds **11B**, 288f
– biodeterioration of wood **10**, 306
– biosorption of, metals **10**, 229f
– – radionuclides **10**, 229f
– β-lactam producing –, actinomycetes **7**, 250
– – eubacteria **7**, 250
– chromosome segregation **2**, 111f
– citric acid production **6**, 330
– classes of **11C**, 64
– classification **12**, 76f
– – risk groups **12**, 77
– cometabolic degradation of aromatic compounds **11B**, 158ff
– – principle **11B**, 158
– composition **6**, 173
– composition of **9**, 171
– cyanide hydratase **8A**, 299f
– degradation of xenobiotics **11B**, 156ff
– ecological groups **11C**, 64
– ectomycorrhizal – **10**, 579
– edible – **1**, 534
– effect on cell cultures **10**, 34
– effect on wine quality **9**, 497ff
– essential amino acids in **9**, 174, 224
– fluorescence monitoring for biomass concentration **4**, 187
– for human consumption, clinical tolerance studies **9**, 232
– formation of PUFA **7**, 151, 158
– from biowaste bins, total count in the air **11C**, 239
– genetic fruiting control **2**, 82
– – *Agrocybe aegerita* **2**, 82
– haploidization **2**, 111f
– hematophagous – **10**, 489
– heterokaryon formation **2**, 113
– immobilization of spores **8A**, 488
– in biodeterioration **10**, 272
– in bioleaching **10**, 205f
– in biomechanical pulping **10**, 533
– in composting **11C**, 64f
– – growth of **11C**, 67
– industrial application **2**, 114ff
– insect pathogenic –, in plant protection **10**, 496
– in sewage **11C**, 217
– in sewage sludge **11C**, 217
– in soil **11B**, 95
– interspecies incompatibility **2**, 117ff
– leaching of metals by **10**, 246f
– life cycle **2**, 75ff, 111
– – dikaryotic **2**, 76
– – diploid **2**, 75
– – haplo-dikaryotic **2**, 76
– – haplo-diploid **2**, 76
– – haploid **2**, 75

– – schemes **2**, 75
– lignin degradation **6**, 147
– marine – **10**, 445
– – antibacterial compounds of **10**, 458
– – antifungal compounds of **10**, 458
– – antiinflammatory compounds of **10**, 456
– – cytostatic/antitumor compounds of **10**, 451
– – DHA-synthesizing – **10**, 464
– – EPA-synthesizing – **10**, 464
– metal ion transport, transport systems **10**, 237
– metal removal by **11A**, 440f
– method of resistance to their own fungicide **7**, 501
– minimum water potential **10**, 273
– modification of brown coal **10**, 161
– modification of hard coal **10**, 157ff
– mycorrhizal – **11B**, 374
– – cultivation of **10**, 578f
– nitrilase activity **8A**, 298f
– nitrile hydratase activity **8A**, 295
– occurence of PHA, in *Physarum polycephalum* **6**, 413
– – in *Saccharomyces cerevisiae* **6**, 413
– occurrence of β-carotene **7**, 174f
– of the composting process **11C**, 54ff
– oil product containing γ-linolenic acid **7**, 158
– oxygen transfer to **4**, 331
– parasexual cycle **1**, 519, **2**, 111, 115
– – non-disjunction **2**, 111
– plant pathogenic –, in organic wastes **11C**, 235
– predominant in compost plant aerosols **11C**, 75
– producing, oudemansins **7**, 499
– – strobilurins **7**, 499
– scleroglucan production **6**, 645
– selectable marker **1**, 536
– signaling molecules **7**, 14
– spores of, as respiratory allergens and pathogens **11C**, 76
– spore treated analysis **2**, 80f
– thermophilic –, isolation from compost **11C**, 66
– – isolation from self-heated hay **11C**, 66
– – temperature optimum **11C**, 40
– transformation sytems, for β-lactam production **7**, 253
– translocation of metals **10**, 247
– used for inoculation of biological waste gas reactors **11C**, 278
– used in plant disease control **10**, 488
fungicides, monitoring with biosensors **11A**, 131f
fungi imperfecti **2**, 74
fungus breeding *see* concerted fungus breeding
fungus genetic engineering, defined mutation **2**, 535
– heterokaryon disruption **2**, 536
– – selection **2**, 536
– repeat-induced point maturation (RIP) **2**, 536
funnel-and-gate **11B**, 473
– in soil remediation **11B**, 363f

– process scheme **11B**, 364
furan, bacterial degradation of **11B**, 205
furans, chlorinated – **8B**, 187
furanyl acrolein, yeast biotransformation of **8B**, 387
fursarinines **7**, 228f
– structure **7**, 228
Fusarium, in ogi fermentation **9**, 545
fusel oil *see* higher alcohols
fusel oils, as by-products in ethanol fermentation **6**, 172f
fusion protein *see* heterologous protein
fusion proteins, medical applications **5A**, 169f
fusogen **2**, 97f
– lipids **2**, 97
– NaNO₃ **2**, 96
futile cycle **1**, 180ff
– proton leakage **1**, 181
fuzzy matching **5B**, 332
fuzzy methods **4**, 595f
– incorporation of heuristic knowledgement into automation systems **4**, 595

G

GAC *see* granular activated carbon
GAD *see* glutamate decarboxylase
Gaeumannomyces graminis, biocontrol **1**, 423
D-galactonate dehydratase **8B**, 119
galactopyranosidase **8B**, 434
galactose-inducible gene **1**, 493
– regulation **1**, 493
galactose metabolization, in lactic acid bacteria **1**, 354
– in yeasts **1**, 493f
α-galactosidase, from pseudomonads **1**, 417, 421
β-galactosidase, crossflow microfiltration **3**, 488
– purification **3**, 601
galactosyltransferases **8B**, 259f
– from bovine colostrum **8B**, 265f
gallidermin, medical uses **7**, 359
– primary structure **7**, 333
– solution structure, stereorepresentation of – **7**, 339
gametophytic incompatibility, in plants **2**, 149, 154
gamma radiation, determination of gas holdup **4**, 137
gancyclovir **5A**, 482
gangliosides **8B**, 262f
GAO (General Accounting Office) **12**, 24
gap junction **1**, 40
gap penalties **5B**, 132, 303f
garbage *see also* waste
– material recycling from **11C**, 10f

– uncontrolled long-term bioconversion by deposition in sanitary landfills   **11C**, 11ff
garbage waste fractions, semi-dry –, anaerobic fermentation of   **11C**, 151ff
– wet –, anaerobic fermentation of   **11C**, 151ff
gari, production of   **9**, 543f
gas analysis, accuracy   **4**, 45
– calculation of errors   **4**, 48
– CO_2 production rate (*CPR*)   **4**, 34
– elemental balancing methods   **4**, 41
– gas flow measurement   **4**, 58f
– infrared analyzer   **4**, 56f
– listing of mass spectrometers   **4**, 67
– mass balancing   **4**, 31ff
– – basic equations   **4**, 31f
– mass spectrometry   **4**, 57
– measurement requirements   **4**, 45
– oxygen transfer rate (*OTR*)   **4**, 35
– oxygen uptake rate (*OUR*)   **4**, 33
– paramagnetic analyzer   **4**, 56f
– reliability   **4**, 45
– sample valves   **4**, 57f
gas balance   **4**, 31ff
– inert gas   **4**, 33
– steady-state   **4**, 33ff
gas balance equations   **4**, 31f
– heat production rate   **4**, 32
– oxygen uptake rate   **4**, 32
– partial pressure   **4**, 31f
– simplifications   **4**, 46
– temperature effects   **4**, 31f
gas balancing, error analysis   **4**, 44ff
– – ethanol in yeast cultures   **4**, 70
– – measurement errors   **4**, 49ff
– – schematic of measurements   **4**, 47
– – simplification of balncing   **4**, 46ff
– – steady-state assumption   **4**, 47
gas bubble *see* bubble
gas chromatography, column switching   **4**, 68f
– conductivity detector (TCD)   **4**, 66
– equipment   **4**, 66
– flame ionization detector (FID)   **4**, 66, 68f
– in fermentation gas analysis   **4**, 68
– in wastewater purification   **4**, 451
– thermal conductivity detector (TCD)   **4**, 68
gas composition analysis, instruments   **4**, 60ff
gas exchange, in drinking water treatment   **11C**, 399f
gas flow control   **4**, 514
gas flow measurement, instruments   **4**, 58f
– – positive displacement devices   **4**, 58f
– – rotameter   **4**, 59
– – thermal mass flow monitors   **4**, 59
gas flow patterns   **4**, 38ff
gas holdrup, in bioreactors   **3**, 409, 412
gas holdup, determination of   **4**, 119f
– – by pressure differences   **4**, 120
– – volume expansion method   **4**, 119f

– measurement, by single-point bubble probes   **4**, 129f
gas linear viscosity *see* linear gas velocity
gas-liquid interface, effects on cell death   **3**, 231
– oxygen profile   **3**, 190
gas/liquid interfacial area, determination, by light scattering methods   **4**, 133f
– – by ultrasound pulse reflection methods   **4**, 137
– – by ultrasound transmission methods   **4**, 136
– – chemical method   **4**, 120f
– mass transfer   **4**, 110ff
gas/liquid mass transfer   **3**, 84
gas mass flow, external oxygen supply   **3**, 254
gasohol *see* ethanol
gasoline vapors, treatment of   **11C**, 286, 365
gas residence time, distribution, in highly viscous fluids   **3**, 193
– – in low viscous fluids   **3**, 193
gas residence times   **4**, 121f
– in bioreactors   **4**, 121
gastification, of coal   **10**, 169f
gas transfer coefficient   **3**, 189
– – *see also* oxygen transfer coefficient, mass transfer coefficient
gas volume flow, effects on linear gas velocity   **3**, 409
gateway sensors, definition   **4**, 543
GATT (General Agreement on Tariffs and Trade)   **12**, 144, 350, 545, 587f, 619
– agreement on patent protection   **12**, 289f
Gaucher's disease, use of recombinant enzymes   **5A**, 168
GAWTS *see* genomic amplification with transcript sequencing
G + C content, of clostridia   **1**, 286ff
– of euactinomycetic genera   **1**, 435
GD.pm   **5B**, 381
gearbox promoters, expression at low growth rates   **7**, 64
gel electrophoresis   **3**, 720ff
– two-dimensional –   **3**, 725
gel filtration chromatography (GFC), in protein purification   **3**, 48
gel formation, in fruit juices   **9**, 718
gellan, consumption   **6**, 627
– food uses   **6**, 643
– price   **6**, 627
– production by *Sphingomonas paucimobilis*   **6**, 629, 640
– structure   **6**, 640f
gellan gum, gel formation   **6**, 642
– – recovery   **6**, 626
Gelsolin   **5A**, 167
gemini virus   **2**, 564, 567f
– bipartite   **2**, 569
– genetics   **2**, 568
– genome organization   **2**, 568f

– monopartite **2**, 569
– of dicotyledonous plants **2**, 568
– of monocotyledonous plants **2**, 568
– vector design **2**, 569
Gen-Bank **5B**, 283, **12**, 326
gene, constitutive **2**, 250
– definition **2**, 666
– investigation of function **2**, 517
– selection **2**, 702
– size in eukaryotes **2**, 155
– transcriptional activity **2**, 780
gene activation **2**, 250ff
– inducer molecule **2**, 251
gene activity **2**, 595ff
– – *see also* transcription regulation
– dominant control region (DCR) **2**, 595
– minilocus **2**, 595
– position effect **2**, 595
– – avoidance **2**, 595
gene addition therapy **5A**, 384, 392
gene amplification **2**, 702, 711ff
– amplified DNA structure **2**, 712
– application in animal host cells **2**, 711
– coamplification **2**, 712
– double minute (DM) **2**, 712
– extended chromosome region (ECR) **2**, 713
– influencing parameters **2**, 714
– in manufacturing of antibodies **5A**, 251f
– – cell lines **5A**, 251ff
– – DHFR **5A**, 251
– – glutamine synthetase **5A**, 251f
– – markers **5A**, 251ff
– – methotrexate **5A**, 251f
– – selection procedures **5A**, 251
– marker gene **2**, 712f
– mechanisms **2**, 712f
– natural **2**, 711
– onion skin model **2**, 713
– sister chromatid exchange **2**, 713
gene banks **12**, 144
gene-based therapies **5B**, 77, 119
– cystic fibrosis (CF) **5B**, 84
gene characterization **2**, 444ff
– expression analysis **2**, 449
– gene orientation **2**, 449
– molecular characterization after cloning **2**, 401ff
– transcriptional unit indentification **2**, 449
gene chips **5B**, 89, 110, 116
– – *see also* DNA microarrays
– applications of **5B**, 228
gene cluster **2**, 146
gene clusters, biosynthetic – **7**, 17ff
– – analysis of – **7**, 20ff
– – listing **7**, 18f
– – organization of – **7**, 21
– in streptomycin production **7**, 420
gene construction **2**, 342ff

– oligonucleotde ligation, chemical **2**, 343ff
– oligonucleotide ligation **2**, 342ff
– strategies **2**, 342ff
gene deletion **2**, 516, 535
gene delivery **5A**, 385ff, 438
– approaches **5A**, 385f
– barriers to **5A**, 429
– vehicles **5A**, 386f
– – non-viral – **5A**, 386, 429
– – physical – **5A**, 386
– – viral – **5A**, 386
gene density **5B**, 35
gene design **2**, 346
gene-directed enzyme prodrug therapy **5A**, 387
gene discovery **5B**, 92
gene disruption **2**, 468, 515ff, 535
– application **2**, 516
– one-step **2**, 515f
– two-step **2**, 515f
gene disruption mutant **2**, 469
gene dosage **2**, 595
gene expression **1**, 97ff, 496ff, **2**, 233ff, 346ff, **12**, 164
– – *see also* heterologous gene expression, synthetic gene expression
– – *see also* heterologous gene expression, transcription regulation, translation regulation
– bacteriophage genes **1**, 554f
– barriers to **5A**, 429
– efficiency **2**, 780f
– expression cassette **2**, 519
– – vectors **2**, 521
– in animal cells, endogenous genes **2**, 665
– – heterologous genes **2**, 665
– in bacteria **1**, 97ff
– in *E. coli*, maxicell system **2**, 405
– – minicell system **2**, 405
– – Zubay system **2**, 406
– in *Escherichia coli*, human insulin **5A**, 112
– – somatostatin **5A**, 112
– in eukaryotes **1**, 107f
– in fungi **2**, 537
– – fungal enzymes **2**, 539
– inhibition **2**, 636
– – anti-sense RNA **2**, 636f
– – by oligonucleotides **2**, 352
– – co-suppression **2**, 636
– in *Lactococcus* **1**, 361f
– in yeast **1**, 491ff, **2**, 519
– – translation initiation **2**, 520
– non-viral – **5A**, 429
– of foreign genes, in plants **2**, 632ff, 640
– overexpression **2**, 783
– parameters **2**, 666
– post-transcriptional gene regulation **1**, 494
– quantitative control **2**, 593
– regulation, anti-sense RNA **2**, 784
– regulation by DNA binding proteins **1**, 492

– regulatory sites **1**, 104
– stability of, cell line dependence **5A**, 254
– – influencing factors **5A**, 254
– synthetic regulatory unit **2**, 348
– tissue specificity **2**, 719, 779
– – investigation **2**, 779
– transient nature of **5A**, 437
gene expression amplification **3**, 31
– DHFR expression vector **3**, 31
– metallothionein promoter **3**, 32
– protein secretion **3**, 32
gene expression factor **2**, 593ff
– monocot-dicot barrier **2**, 594
– phylogenetic distance **2**, 593
– phylogenetic promoter specificity **2**, 593
gene expression profile **5B**, 92, 228
gene expression rate, in recombinant fermentation **4**, 499ff
gene expression system **2**, 405
– *in vitro* translation system **2**, 406
– maxicell **2**, 405
– minicell **2**, 405
– prokaryotic **2**, 721f
– synthetic **2**, 721
– *trans*-acting factor **2**, 721
– two-cistron **2**, 414
– with phage RNA polymerase **2**, 406
– Zubai **2**, 406
gene expression systems **5A**, 250f
– in manufacturing of antibodies **5A**, 250
– protists **10**, 146f
gene expression vector **2**, 717f
– – *see also* expression vector, vector
– elements **2**, 717ff
– in animals **2**, 717ff
– promoter/enhancer for mammals **2**, 719f
– termination signal **2**, 719
gene families **5B**, 376
gene farming **2**, 800ff
gene finding programs **5B**, 128
gene finding tools **5B**, 363
gene flow, by pollen **12**, 164
– horizontal **12**, 161ff, 191
– – conjugation **12**, 163
– – mating **12**, 163f
– – pollen exchange **12**, 164
– – transduction **12**, 162f
– – transformation **12**, 162
– vertical **12**, 161ff, 191
gene fusion **2**, 410, 415, **5B**, 370
gene gun **5A**, 386, 388, 392, 466
gene homology **2**, 170ff
gene identification, based on *de novo* protein sequencing **5B**, 263f
gene inactivation **2**, 535, 589
– – *see also* heterologous gene suppression
– action mechanism **2**, 590
– anti-sense RNA **2**, 590

– functional **2**, 589
– homologous co-suppression **2**, 597ff
– – models **2**, 598
– mutational **2**, 537
– physical disruption **2**, 589
– regulation **2**, 590
– sense RNA **2**, 590f
– sense transgenic construct **2**, 597f
gene inactivation system **5B**, 155
gene indentification **2**, 394ff, 444ff, 535
– gene localization **2**, 447f
– mutagenesis **2**, 446
– mutant complementation **2**, 446
– mutant isolation **2**, 446
– mutated gene cloning **2**, 446
– transconjugant selection **2**, 446
gene isolation **2**, 534
– transforming plasmid **2**, 535
gene library **2**, 387ff
– cDNA library **2**, 390f, 399
– – construction scheme **2**, 391
– – fragment orientation **2**, 391f
– cloning strategies **2**, 389f
– construction **2**, 387, 389
– genomic DNA library **2**, 388f
– vectors for expression libraries **2**, 401
gene library screening **2**, 394ff
– amino acid sequence based **2**, 398
– clone selection in *E. coli* **2**, 394f
– expression library **2**, 399
– for cloned DNA homology **2**, 397
– hybrid arrested translation **2**, 399
– hybrid selected translation **2**, 399
– phenotype selection **2**, 394
– selection after clone transfer **2**, 395
– strategies **2**, 397f
– with cDNA probes **2**, 397ff
gene linkage **2**, 147ff
– disruption frequency **2**, 177
– evolution **2**, 177
– homology **2**, 172
– – cattle/human **2**, 173
– – mouse/human **2**, 173ff
– – Poaceae **2**, 174
– – potato/tomato **2**, 173ff
– – X chromosome **2**, 174f
– in organelles **2**, 167ff
gene linkage estimation **2**, 147
– marker gene locus **2**, 147
– quantitative trait locus **2**, 154
– undefined loci **2**, 154
gene localization **2**, 403, 446ff
– by Southern hybridization **2**, 403
– chromosomal **2**, 64
– induced genes **2**, 398ff
– lipopolysaccharide biosynthesis **2**, 446ff
– methods **2**, 64ff
– plasmidal **2**, 64

gene loci **2**, 143ff
– linkage disequilibrium **2**, 153
gene manipulation technique **2**, 508ff
gene map **2**, 149
– distance calculation **2**, 149
– genetic **2**, 160ff
– mitochondrial **2**, 165f
– nuclear, in animals **2**, 164ff
– – in plants **2**, 164ff
– physical **2**, 155ff
– plastidal **2**, 165
– – *Nicotiana tabacum* **2**, 171
gene mapping, chromosome aberration **2**, 159
– comparative **2**, 172ff
– DNA techniques **2**, 161
– dosage mapping **2**, 159
– dot-blot **2**, 161
– in animals **2**, 141ff
– in plants **2**, 141ff
– overlapping restriction analysis **2**, 167
– physical **2**, 155
– somatic cell methodology **2**, 158
gene model **5B**, 363
gene mutation **2**, 7ff
– base pair substitution **2**, 9
– – transition **2**, 9
– – transversion **2**, 9
– frameshift mutation **2**, 9
– frequency **2**, 9
– hot spot **2**, 11
– macrolesion **2**, 8
– mechanism **2**, 9
– microlesion **2**, 9
– processing **2**, 145
– silent mutation **2**, 10
GeneNet **5B**, 408
Gene Ontology (GO) **5B**, 291
gene overexpression, foreign genes, in bacilli **2**, 462
– genetic stabilization **2**, 418
– heterologous genes, in *E. coli* **2**, 410ff
– in animals **2**, 711
gene prediction **5B**, 128, 365
gene product **2**, 176
– homology **2**, 176
– – mouse/human **2**, 176
gene products, commercialization of **4**, 487
– formation kinectis **4**, 490f
– in streptomycin production **7**, 418f
gene product screening, immunodetection **2**, 399f
Generally Recognized as Safe (GRAS) **5B**, 15
gene refinement **5B**, 369f
gene regulation **2**, 250ff
– dominant control region (DCR) **2**, 595
– heterologous gene **2**, 718
– in *Streptomyces* **2**, 473
– – *afs* genes **2**, 473
– – antibioic production **2**, 473f

– – transcription regulation **2**, 474
– – translation **2**, 474
– – tRNA Leu-like gene (*bldA*) **2**, 473
– leucine operon regulation **2**, 209
– locus control region (LCR) **2**, 779f
gene replacement **2**, 436, 470, 515ff, 535
– recombinase driven – **5A**, 253
– selection method **2**, 437f, 447
– truncation integration **2**, 518
gene replacement therapy **5A**, 384, 392
gene repression **2**, 250ff
generic drugs, market share **5A**, 535
Generic Model Control (GMC) **3**, 343
gene(s), amylase – **1**, 200f
– antibiotic – **1**, 493ff
– biosynthetic –, idenfication of – **7**, 20
– for glycolytic enzymes **1**, 480ff
– heterologous expression, in yeasts **1**, 491f
– – *see* heterologous gene expression
– in bacilli, sporulation – **1**, 390ff
– probing **1**, 411f
– *RAS* – **1**, 487f
– species-specificity **12**, 78ff
– transfer **1**, 195f, 545
– – horizontal **1**, 559f
– varibale penetrance **5B**, 111
GeneSmith **5B**, 17
gene structure **2**, 176
– analysis **2**, 401ff
– homology **2**, 176
– in eukaryotes **2**, 778
– – *cis*-acting control elements **2**, 779
– – coding region **2**, 778
– prediction of **5B**, 128, 371
– protein coding gene **2**, 668
gene synthesis **2**, 342ff, 392
– – *see also* gene construction
– direct cloning **2**, 344
– fill-in method **2**, 343
– insertional synthesis **2**, 345
– PCR-mediated **2**, 345
– shotgun method **2**, 345
– solid-phase synthesis **2**, 345
– strategies **2**, 392
GeNet **5B**, 407
gene targeting **2**, 587
– animal cells **2**, 714ff
– applications **2**, 715f
– conditional positive selection **2**, 715ff
– parameters influencing efficiency **2**, 715
– positive/negative selection (PNS) system **2**, 714ff
– strategies **2**, 588, 716
– targets **2**, 718
gene targeting in plants **5B**, 155f
gene targeting technology **5B**, 152
gene targets **5B**, 92
gene therapy **5A**, 383ff, **12**, 568

– access to target cells **5A**, 387f
– anionic liposomes **5A**, 434
– antibody-targeted – **5A**, 234
– approved clinical protocols **5A**, 385f
– Archer Report **5A**, 526
– bioethical conflicts **12**, 139ff
– calcium phosphate precipitation **5A**, 430f
– candidate diseases **5A**, 385f
– cationic liposomes **5A**, 432ff
– – adult respiratory distress syndrome **5A**, 433
– – cystic fibrosis **5A**, 433
– – limitations **5A**, 433
– – stability **5A**, 433
– – toxicity **5A**, 433
– clinical status **5A**, 386f
– cost issues **5A**, 391f
– definitions **5A**, 384
– delivered DNA components **5A**, 437
– development **5A**, 387ff
– direct gene transfer **5A**, 429ff
– – expression **5A**, 430
– electroporation **5A**, 431
– ethical limits **12**, 150
– *ex vivo* **5A**, 384ff, 522
– for HIV infection **5A**, 471ff
– – future concepts **5A**, 485ff
– – optimization of – **5A**, 484ff
– – phase I clinical trials **5A**, 482ff
– future of **5A**, 392f
– gene delivery **5A**, 385ff
– – approaches **5A**, 385f
– – costs **5A**, 392
– – vehicles **5A**, 386f
– gene expression **5A**, 437
– guidance for industry **5A**, 524f
– guidelines **5A**, 461, 523f
– history **5A**, 386
– immune responses **5A**, 391
– in animals **5A**, 170f
– inflammatory reactions **5A**, 387, 391
– *in vivo* **5A**, 384ff, 521
– liposome-HVJ complexes **5A**, 434
– locus control regions **5A**, 389f
– long-term transgene persistence **5A**, 390
– microinjection of DNA **5A**, 430
– neutral liposomes **5A**, 434
– non-targeted cationic polymers **5A**, 431
– non-viral vectors **5A**, 427ff
– of AIDS **5A**, 405
– of cystic fibrosis **5A**, 413, 421
– of germ cells **5A**, 521
– of inoperable gliomas **5A**, 405
– of somatic cells **5A**, 521
– of tumors **5A**, 405f, 413
– overview **5A**, 383ff
– particle bombardment, DNA immunization
  5A, 432
– patentability **12**, 288

– plasmid vector design **5A**, 452
– promoters used **5A**, 437
– public opinion **12**, 133, 453
– receptor-mediated endocytosis **5A**, 434f
– – DNA-binding **5A**, 435
– – DNA-compaction **5A**, 435
– – efficiencies **5A**, 437
– – endosome release **5A**, 436
– – targeting ligands **5A**, 436
– – targeting moiety **5A**, 435
– regulating acts **5A**, 522f
– regulations **5A**, 460f, 517ff
– – history **5A**, 520f
– – in the European Union **5A**, 526ff
– regulatory committees **5A**, 520f
– replicating vectors **5A**, 389
– safety issues **5A**, 391f
– target cells **5A**, 385, 390
– technology transfer approaches **5A**, 522
– trafficking of vectors **5A**, 388
– transfection efficiency **5A**, 388
– transfection systems **5A**, 429ff
– transgene expression **5A**, 389f
– – level **5A**, 437
– transgene size limitation **5A**, 389
– vector systems **5A**, 386
– – for *in vivo* delivery **5A**, 438
– – non-viral – **5A**, 427ff
– – viral – **5A**, 395ff
– White Paper **5A**, 526
Gene Therapy Advisory Committee, guidelines
  5A, 527
gene therapy products, registration **5A**, 498ff
– – autorities **5A**, 498ff
– – in Europe **5A**, 498
– – in the USA **5A**, 498ff
– – laws **5A**, 498ff
gene therapy research, germ line experiments
  5A, 524
– somatic cells **5A**, 524
genetically engineered microorganisms (GEM)
  11B, 441ff
– – *see also* degradative genes
– and bioremediation **11B**, 441ff
– degradation of, branched aromatic hydrocarbons **11B**, 445
– – chlorinated compunds **11B**, 445ff
– – chlorobenzoates **11B**, 445f
– – chlorobiphenyls **11B**, 446
– – nitroaromatic compounds **11B**, 452
– – polychlorinated biphenyls **11B**, 446f
– – sulfur compounds **11B**, 452
– – trichloroethylene **11B**, 449ff
– – 2,4-dichlorophenoxyacetic acid **11B**, 451
genetically engineered organisms *see* recombinant
  organisms
genetically modified organisms see GMOs
genetic code **2**, 240f, **5A**, 49f

– codon usage **2**, 240ff
– degeneration **2**, 240
– evolution of **5A**, 3, 50
– supression **2**, 243
genetic colonization **2**, 557
genetic counselling **5B**, 117
genetic disease **2**, 786
genetic disease model, autoimmune disease **2**, 791
– dominant negative mutation **2**, 782
– genetic ablation (amputation) **2**, 784
– giant mice **2**, 783, 794, 798
– in mice **2**, 781ff
– insertion mutation **2**, 781
genetic diseases **5B**, 61ff
– – *see also* diseases
– age of onset **5B**, 70f
– chromosome disorders **5B**, 74ff
– genomics of **5B**, 105ff
– impact of the Human Genome Project **5B**, 88ff
– individualized therapy **5B**, 92ff
– management of **5B**, 83ff
– Mendelian – **5B**, 63ff
– mitochondrial disorders **5B**, 62
– monogenic – **5B**, 62ff
– – finding disease genes for **5B**, 77ff
– – molecular diagnosis of **5B**, 112ff
– multifactorial – **5B**, 65ff, 111
– – finding disease genes for **5B**, 79ff
– oligogenic – **5B**, 62
– penetrance of **5B**, 70f
– – in complete – **5B**, 70
– phenotypic variation **5B**, 69ff
– – sex-specific – **5B**, 72ff
– polygenic – **5B**, 62, 65, 111
– – molecular diagnosis of **5B**, 117f
– recurrence risk **5B**, 66f
– treatment of **5B**, 118ff
genetic drift **5B**, 78
genetic engineering **2**, 375, 550ff, 836
– – *see also* genetically engineered microorganisms (GEM)
– – *see also* rDNA technology
– – *see also* strain improvement
– cassette insertion **2**, 470
– creation of pathogens **12**, 44f
– for biodegradation of contaminants **11B**, 444
– – process efficacy **11B**, 444
– galactose operon transfer **1**, 195f
– hazards **2**, 837
– history **12**, 41ff, 509ff
– hydroxylase enzymes **8A**, 482f
– limitations in fungi **1**, 88
– novel organisms **12**, 44
– of bacilli **1**, 394ff
– of filamentous fungi **1**, 536
– of lactic acid bacteria **1**, 360ff
– of microalgae **10**, 133f

– of plant cells **1**, 601f, **2**, 550
– of vertebrates **2**, 746ff
– of yeast **8A**, 383
– of yeasts **1**, 476f
– press coverage, in Germany **12**, 495ff
– public opinion **12**, 125, 482
– – in the U.S. **12**, 659
– representation **12**, 496ff
– – journalists' attitudes **12**, 501ff
– risks **12**, 101ff, 474
– techniques **2**, 747
genetic epidemiology, databanks **5B**, 140
genetic exchange **2**, 47ff
– frequency in nature **2**, 48
– in bacteria **2**, 48
– – mechanisms **2**, 105
– lower eukaryotes **2**, 74ff
– lysogenic conversion **2**, 61
– potential **2**, 67
– prokaryotes **2**, 48ff
genetic exchanges, between microbial species **7**, 15
genetic immunization **2**, 807
genetic linkage **5B**, 42
genetic manipulation, public opinion **12**, 128ff, 451ff, 473
– – benefits **12**, 130f, 133
– – consumption of products **12**, 135f
– – reasons for unacceptability **12**, 128f
– – risks **12**, 130, 132f
– risk assessment **12**, 46
– U.K. regulations **12**, 517, 576ff
genetic map **2**, 48, 63ff, 106
– bacterial species **2**, 63
– mapping techniques **2**, 63ff
– – methylotrophs **2**, 63ff
– *Pseudomonas putida* **2**, 65f
– TOL plasmid **2**, 65
genetic mapping **5B**, 42, 91, 108ff
genetic mosaic animal **2**, 775
genetic recombination *see* recombination
– in fungal parasexual cycle **1**, 519
genetic resources, exploitation **12**, 295
genetic revolution **12**, 510ff
genetic screening **5B**, 84ff
– colon cancer **5B**, 87
– cystic fibrosis (CF) **5B**, 84
genetic testing **5B**, 112ff
genetic variability, of crop plants **5B**, 146
genetic variants, low-penetrance **5B**, 117
genetic variation **5B**, 178
gene transfection efficiency, non-viral vectors **5A**, 388
– proliferating cells **5A**, 388
gene transfer **2**, 577, 702ff, **5B**, 119, **12**, 93f
– – *see also* DNA transfer, transformation
– animal host cell lines **2**, 703f
– – COS cells **2**, 724

– – murine C127   **2**, 725
– by large DNA viruses   **2**, 729f
– direct –   **5A**, 429f
– efficiency in chicken   **2**, 810
– efficiency in fish   **2**, 812
– gene expression level   **2**, 709ff
– horizontal   **12**, 81f, 94
– mechanisms   **12**, 80ff
– sperm-mediated   **2**, 771f, 811
– – stable integration   **2**, 768
– – stem cell transformation   **2**, 762
– – transcription efficiency   **2**, 750
– – transkaryotic implantation technique   **2**, 774
– – *Xenopus*   **2**, 747f
– stable transfection   **2**, 702
– techniques in fish   **2**, 811
– to animal cells   **2**, 747ff
– – embryo cells   **2**, 748ff
– – history   **2**, 747f
– – liposome-mediated   **2**, 771
– – membrane receptor-mediated   **2**, 774
– – methods   **2**, 771
– – mice   **2**, 747ff
– – somatic cells   **2**, 772f, 774f
– to COS cells   **2**, 724
– to plants   **2**, 629ff
– – criteria for stable integration   **2**, 577
– – transient expression   **2**, 576
– – vectorless   **2**, 575ff
– transient expression   **2**, 702, 724
– vectorless   **2**, 705
gene transfer agent   **2**, 61
gene transfer technique   **2**, 575ff
– clacium phosphate coprecipitation   **2**, 578
– laser microbeam   **2**, 580
– lipofection   **2**, 582
– liposome-cell fusion   **2**, 581
– liposome injection   **2**, 582
– membrane-destabilizing chemicals   **2**, 577
– silicon carbide fiber-mediated   **2**, 580
– sonication   **2**, 583
– target/DNA-incubation   **2**, 583
– – imbibition   **2**, 583
– – macro-injection   **2**, 583
Genewise   **5B**, 375
Genewise suite   **5B**, 136
genome, amplification stability   **2**, 714
– isochore   **2**, 177
– size   **2**, 63
– – in eukaryotes   **2**, 155
genome annotation, automated –   **5B**, 361ff
– process   **5B**, 363ff
– re-annotating *Drosophila melanogaster*   **5B**, 375f
– re-annotating *E. coli*   **5B**, 373ff
genome cross-referencing   **5B**, 175f
Genome Database   **5B**, 285
genome evolution   **2**, 167

– speed   **2**, 167
genome mapping, methods   **5B**, 169
– molecular markers   **5B**, 167f
genome mutation   **2**, 7f, 436ff
genome projects, of model organisms   **5B**, 5ff
– – *Arabidopsis thaliana*   **5B**, 24ff
– – *Archaeoglobus fulgidus*   **5B**, 16ff
– – *Bacillus subtilis*   **5B**, 13ff
– – *Caenorhabditis elegans*   **5B**, 28ff
– – *Drosophila melanogaster*   **5B**, 33ff
– – *Escherichia coli*   **5B**, 8ff
– – *Saccharomyces cerevisiae*   **5B**, 20ff
genome research, commercialization of   **5B**, 425f
genomes, annotated sequences   **5A**, 49
– *Escherichia coli*   **5A**, 3
– human   **5A**, 52
– large-scale sequencing projects   **5A**, 51
– of archaebacteria   **5A**, 51
– of *Caenorhabditis elegans*   **5A**, 51
– of eubacteria   **5A**, 51
– of eukaroytes   **5A**, 51
– of *Haemophilus influenzae*   **5A**, 51
– phylogenetic tree   **5A**, 51
– sizes   **5A**, 51
– yeast   **5A**, 3
genome sequencing, in Japan   **12**, 385
genome-wide modeling   **5B**, 352
genomic amplifications with transcript sequencing (GAWTS)   **2**, 287
genomic data base   **2**, 163f
– development   **2**, 163
– organization   **2**, 163f
genomic databases   **5B**, 285f
genomic data representation   **5B**, 379ff
– hierarchical –   **5B**, 380ff
– visual –   **5B**, 380ff
genomic features   **5B**, 381
genomic inserts, cloning of   **5B**, 172ff
genomic mapping, of plants   **5B**, 165ff
genomics, ethical aspects   **5B**, 417ff, 439
– future developments   **5B**, 431ff
– government programs   **5B**, 438
– level of acceptance   **5B**, 439
– public perception   **5B**, 439
– risks   **5B**, 438
genomic sequence comparison, maize   **5B**, 176
– rice   **5B**, 176
– wheat   **5B**, 176
genomic sequence data, pre-processing   **5B**, 363
genomic sequencing   **2**, 263, 283ff
– dot blot hybridization of ASO probes   **2**, 341
– ligation mediated PCR   **2**, 284
genomics laboratories, future toolset   **5B**, 436f
– organization of   **5B**, 438
genomics toolkits   **5B**, 434f
genotype classification   **2**, 148
genotype multiplication   **2**, 154
– in animals   **2**, 154

– in plants **2**, 154
GENSCAN **5B**, 375
gentamicin family, structure of – **7**, 468
gentamicins, biosynthetic pathway **7**, 433
geochemical models, interaction between contaminants and subsurface matrix **11B**, 55f
geochemical processes, effect on transport of polutants **11B**, 49ff
geological situation, effects on soil remediation **11B**, 43ff
Geotrichum candidum, glycerol dehydrogenase of **8A**, 443
– in gari fermentation **9**, 543
– in kanji fermentation **9**, 541
– in poi production **9**, 548
– in pozol fermentation **9**, 549
geraniol, biotransformation of **10**, 388ff
– hydroxylation of **8A**, 491
– transesterification of **10**, 401
geraniol acetate, biotechnological production of **10**, 397
geranylacetone, dihydroxylation of **8A**, 513
GERI (Genetic Engineering Research Institute) **12**, 390ff
German Epidemics Law **12**, 71
German Federal Supreme Court **12**, 282
– Red Dove decision **12**, 284
German Gene Technology Act **12**, 73f, 102f, 274, 584ff
German Genetic Engineering Law **12**, 486f, 645
German Human Embryo Protection Act **12**, 288
German Patent Act **12**, 288
Germany, biotechnology regulatory policy **12**, 580ff
– contaminated site management, financing schemes **11B**, 15
– – legal development **11B**, 9
– – principles **11B**, 11
– – registration **11B**, 17
– – risk assessment **11B**, 19
– – technical development **11B**, 9
– identification of contaminated sites **11B**, 17
– public opinion survey, on gene technology **12**, 660
– regulatory concerns, action thresholds **11B**, 21
– – conflict management **11B**, 28
– – enforcement **11B**, 26
– – in soil contamination **11B**, 7ff
– – liability rules **11B**, 14
– – organization **11B**, 26
– – priority setting **11B**, 23
– – remediation targets **11B**, 24
– – verification of clean-up **11B**, 24
germ cell therapy **5A**, 521
germline intervention **5B**, 420
germline mutations **5B**, 87f
– novel – **5B**, 88
– somatic – **5B**, 88

germs, emissions from biological waste air filtration plants **11C**, 294f
GFP *see* fluorescent protein
Giardia, in drinking water **11C**, 374
– in river water **11C**, 381
Giardia lamblia, elimination from drinking water **11C**, 420ff
– transmission via drinking water **11C**, 414ff
– – listing of outbreaks **11C**, 418f
gibberellins, hydroxylation of **8A**, 494
Gibbs free energy **5A**, 89f
ginseng, from plant cell cultures **7**, 611
ginsenosides, production, by cell cultures **1**, 589
GLA *see* γ-linolenic acid
glass, collection system **11C**, 9ff
glass transition temperature **3**, 706
– effects of additives **3**, 705
– in freezing **3**, 698, 702
Glaucocystophyta, characteristics of **10**, 111
glial cell gowth factor 2 **5A**, 160
glial-derived neurotrophic factor **5A**, 159
glial maturation factor **5A**, 160
GLIMMER 2.0 **5B**, 373
gliomas, gene therapy with retroviruses **5A**, 405
globalization **5B**, 419
β-globin protein **5B**, 64
GLP *see* good laboratory practice
glucan phosphorylase, purification **3**, 602
β-glucans, in wine **9**, 727
glucoamylase **6**, 143
– – *see* amyloglucosidase
– gene (*GAM1*), *Schwanniomyces occidentalis* **1**, 201
– in clostridia **1**, 310
– of *Aspergillus niger* **6**, 71
– of *Rhizopus* sp. **6**, 71
glucoamylases **10**, 74ff
– action of **10**, 75
– heat and acid stable – **10**, 77
– heat-stable – **10**, 76f
glucocerebrosidase, production **3**, 473
glucokinase, gene (*GLK*) **1**, 481
glucomannan, action of enzymes on **10**, 532
– composition of **10**, 527
– content in wood **10**, 527f
gluconate production **1**, 178
gluconeogenesis **1**, 85, 484
gluconeogensis, long-term regulation **4**, 395
gluconic acid, industrial production **6**, 352ff
– production figures **6**, 357
– uses **6**, 357
gluconic acid fermantation **6**, 347ff
– history **6**, 348
gluconic acid fermentation, mass transfer coefficient, correlation **3**, 199
gluconic acid formation, by *Aspergillus niger* **6**, 349ff
– – biochemical pathway **6**, 349

– – localization of enzymes **6**, 349
gluconic acid production, by *Aspergillus niger* **6**, 352ff
– by *Aureobasidium pullulans* **6**, 355
– by bacteria **6**, 355f
– – *Acetobacter methanolicus* **6**, 352, 356
– – *Gluconobacter oxydans* **6**, 351, 355
– – *Zymomonas mobilis* **6**, 352, 356
– calcium gluconate process **6**, 352f
– – production medium **6**, 353
– product recovery **6**, 356
– sodium gluconate process **6**, 353f
– – production medium **6**, 353
– – sporulation medium **6**, 353
– – with *Aspergillus niger* **6**, 353
– use of enzymes **6**, 354f
– use of immobilized cells **6**, 354
Gluconobacter, Entner-Doudoroff pathway **6**, 387
– formation of acetic acid **6**, 384
Gluconobacter oxydans **8B**, 305
– biotransformation of *N*-formyl-1-amino-1 deoxy-D-sorbitol, at industrial scale **8B**, 303f
– cells, fermentative production at industrial scale **8B**, 304ff
– continuous fermentation of **8B**, 306
– D-sorbitol dehydrogenases **8B**, 306
– gluconic acid production **6**, 351, 355
– in the synthesis of vitamin C **8B**, 297
– oxidation of 1-amino-1-deoxy-D-sorbitol **8B**, 300ff
– – protection groups **8B**, 302
– – pyridine formation from 6-amino-6-deoxy-L-sorbose **8B**, 302
– regioselective oxidation of aminosorbitol **8B**, 295ff
Gluconobacter sp., tartaric acid formation **6**, 374
δ-gluconolactone, precipitation **6**, 356
– uses **6**, 357
γ-gluconolactone, precipitation **6**, 356
β-D-glucopyranoside, as affinity cosurfactant, for reversed micellar extractions **3**, 609
glucose, aerobic metabolism, energy balance **11C**, 37
– aerobic respiration **11A**, 21ff
– anaerobic degradation **11A**, 23ff
– as nutrient, for mammalian cell cultures **3**, 145
– crystalline **6**, 41f
– – composition **6**, 42
– degradation of, heterofermentative **1**, 170f
– – homofermentative **1**, 170f
– determination, with hexokinase **9**, 141
– – with hexokinase/glucose-6-phosphate dehydrogenase **9**, 144ff
– energy yield **1**, 141
– exogonic oxidation of **8B**, 222
– fermentation **1**, 182
– – in clostridia **1**, 299f
– – nutrient limitation effect **1**, 173f

– – pathways **1**, 170ff
– – pH dependence **1**, 182
– – product quantity **1**, 182
– mass and energy balance **11A**, 21ff
– selective acylation by PPL **8A**, 119
– uptake, by *Aspergillus oryzae* **3**, 82
– – in bioprocesses **3**, 365
– – in yeast **1**, 487
– – specific rate **3**, 385
glucose consumption rate, cephalosporin production **4**, 150f
glucose dehydrogenase **1**, 173ff, 178f, 182
– of Acetobacteraceae **6**, 351
glucose effect, during yeast growth **4**, 388
glucose fermentation, mass transfer coefficient, correlation **3**, 199
glucose hydrolyzates *see* glucose syrups, maltodextrins
glucose isomerase **2**, 34
– in industrial biotransformations **3**, 456
– producing strain **2**, 34
glucose isomerases, thermostable – **10**, 91
glucose isomerization **6**, 40, **9**, 671ff, 747
glucose metabolism, regulation of **4**, 150
– yeast **4**, 150
glucose monitoring **4**, 86ff
– enzyme-based **4**, 86ff
– – glucose oxidase **4**, 88f
– – hexokinase **4**, 88
– – PQQ-enzymes **4**, 88
– fermentation control **4**, 88f
glucoseoxidase **5A**, 209f, **10**, 491
– exposure to hydrogen peroxide **4**, 87ff
– immobilization **4**, 87
– induction **6**, 350
– location in *Aspergillus niger* **6**, 349
– monitoring of activity **4**, 87ff
– overexpression in *Aspergillus niger* **6**, 350
– recombinant – **6**, 350
– use in biosensors **4**, 87
glucose oxidase/catalase, use in baking **9**, 675
glucose syrup, as brewing adjunct **9**, 431
glucose syrups, composition **6**, 37f
– production of **6**, 37
– uses **6**, 37f
glucose-6-monocorynomycolate, from *Corynebacterium diphteriae* **6**, 663
glucose-6-phosphate, enzymatic synthesis of **8B**, 228f
glucose-6-phosphate dehydrogenase, gene (*ZWF*) **1**, 480
– solubilization **3**, 509
glucosidase, and glycosyltransferase, consecutive glycosylation reactions **8B**, 268
α-glucosidase **10**, 74f
– action of **10**, 75
– from thermophiles **10**, 77
– in clostridia **1**, 310

– production **1**, 239
α-glucosidase inhibitors **8B**, 297f
– deoxynojirimycin **8B**, 297ff
– Miglitol **8B**, 298ff
– nojirimycin **8B**, 297ff
β-glucosidase, production **1**, 239
β-glucosidase (cellobiase) **5A**, 203
β-glucosidases, thermostable – **10**, 81f
glucurate dehydratase **8B**, 119
glues, biodeterioration of **10**, 311
glutamate **9**, 702
glutamate decarboxylase (GAD) **8B**, 76ff
– mechanism of decarboylation **8B**, 77
glutamate dehydrogenase **1**, 93
glutamate dehydrogenases, applications of **8A**, 445
– characteristics of **8A**, 444f
– reductive amination of 2-oxoglutarate to L-glutamate **8A**, 444f
– regeneration of NAD(P) **8A**, 406f
glutamate production, *in vivo* NMR studies **4**, 213
glutamate racemase **6**, 536f
glutamate synthase **1**, 93
glutamic acid, lipase-catalyzed deprotection of carboxylate groups **8A**, 133
L-glutamic acid, biosynthesis **6**, 469f
– – enzymes **6**, 471
– – in *Corynebacterium glutamicum* **6**, 471
glutamic acid production **1**, 206
– in *Corynebacterium glutamicum* **1**, 456
L-glutamic acid production **6**, 469ff
– by coryneform bacteria **6**, 506f
– downstream processing **6**, 472
– effect of biotin **6**, 506
– fermentation technology **6**, 472f
– strain development **6**, 471f
glutamine, as nutrient, for mammalian cell cultures **3**, 145
L-glutamine production, by microbial fermentation **6**, 489
glutamine synhetase **5A**, 250ff
– expression systems **5A**, 250
– vector system **5A**, 249
glutamine synthetase **6**, 542f
– in streptomycetes **1**, 443
– operon **1**, 387
– regulation **1**, 387
glutarate diesters, hydrolysis of, use of pig liver esterase **8B**, 353f
glutarates, prochiral –, asymmetrization by pig liver esterase **8A**, 195f
glutarimide antibiotics, redifferentiation of cancer cells **7**, 683
glutathione *S*-transferase (GSH), transformation of TCE in plant tissues **11B**, 375
glutathione S transferase inhibitors, structures **7**, 673

gluten **9**, 250f, 265ff, 678, 704
– enzymatic hydrolysis, whipping properties **9**, 698f
– enzymatic modifications **9**, 679
N-glycans, biosynthesis of **8B**, 263f
O-glycans, biosynthesis of **8B**, 263
glyceraldehyde, specific phosphorylation of **8B**, 230
glyceraldehyde-3-phosphate dehydrogenase, gene (*TDH*) **1**, 481
glycerides, 1,3 regioselective reactions **8A**, 134
glycerine esters, as nonionic surfanctants **6**, 673
– use in cosmetics **6**, 673
glycerol, as protein folding aid **3**, 542
– esterification with fatty acids **8A**, 142
– formation in wine making **9**, 475f
– formation of DAGs **8A**, 142
– formation of MAGs **8A**, 142
– production volumes 1910–1929 **8A**, 9
– specific phosphorylation of **8B**, 230
– uses **6**, 219
glycerol dehydrogenases, applications of **8A**, 443f
– characteristics of **8A**, 443f
– *Geotrichum candidum* **8A**, 443
– substrates accepted by, oxidative direction **8A**, 443
glycerol formation, fructose bisphosphate pathway **6**, 207f
– – scheme **6**, 208
– microorganisms **6**, 207
– under alkaline conditions **6**, 208f
– – scheme **6**, 209
glycerol kinase, phosphorylation by **8B**, 224
glycerol metabolism, in yeasts **9**, 182
glycerol metabolization, in clostridia **1**, 298, 301
glycerol production **6**, 205ff
– alkaline fermentation route **6**, 214ff
– continuous fermentation **6**, 215
– CO_2 process **6**, 212
– economics **6**, 223
– history **6**, 206f
– semicontinuous fed-batch fermentation **6**, 214
– sulfite process **6**, 210
– synthetic pathway **6**, 206
– vacuum process **6**, 212
– with algae **6**, 210f, 217
– – process flow sheet **6**, 218
– with bacteria **6**, 217
– with yeasts, free *Saccharomyces cerevisiae* cells **6**, 210ff
– – immobilized *Saccharomyces cerevisiae* cells **6**, 213f
– – osmotolerant yeasts **6**, 216f
glycerol recovery, methods **6**, 219
glycerol trinitrate *see* nitroglycerine
glycerolysis, of triacylglycerides **8A**, 141
glycidol butyrate, resolution of, using porcine pancreatic lipase **8B**, 283

(*R*)-glycidyl butyrate **3**, 453
– production **3**, 458
glycine, conversion into hydrogen cyanide **8A**, 279
– food uses **6**, 468
glycocalyx **1**, 38
glycoconjugates **5A**, 9
– synthesis of **8B**, 251ff
– – by glycosyltransferases **8B**, 261ff
glycogen polymerization **1**, 95
glycogen synthesis **1**, 174, 485
glycolipids **6**, 662f
– anaerobic degradation of **11A**, 34f
– degradation efficiency in soils **6**, 695
– overproducing microbial strains **6**, 696
– synthesis of **8B**, 262f
– – solid-phase methods **8B**, 264
glycolysis **1**, 102, 479ff
– allosteric control **1**, 102
– Embden-Meyerhof-Parnas pathway **6**, 125
– enzyme coding genes **1**, 480
– *in vivo* NMR studies **4**, 213
glycolysis pathway **9**, 182f
glycolytic enzymes *see* enzymes
glycopeptide antibiotics *see* dalbaheptides
glycopeptides, enzymatic synthesis, using transglycosylation **8B**, 252
– glycosylation sites, *N*-linkage **8B**, 263
– – *O*-linkage **8B**, 263
– synthesis of **8B**, 263f
glycoproteins **3**, 734, **5A**, 9ff
– carbohydrate composition determination **3**, 734
– function of **5A**, 9
– glycosylphosphatidyl inositol membrane-anchored proteins **5A**, 13
– microheterogeneity **3**, 32, 734
– N-linked – **5A**, 12
– – primary structure **5A**, 13
– O-glycosidic attachment **5A**, 10
– production **3**, 32
– therapeutic – **5A**, 14
glycosidase, in cocoa fermentation **9**, 570
glycosidase antibody **8B**, 434
glycosidases **8B**, 245ff
– availability of **8B**, 246
– immobilization of **8B**, 268f
– specificity for glycosyl donor **8B**, 246
– synthesis of small oligosaccharides **8B**, 249
– use in consecutive glycosylation reactions **8B**, 268
glycosides, cleavage of, use of catalytic antibodies **8B**, 432ff
– cyanogenic – **8A**, 280ff
– synthesis of, multiglycosylated substrates **8B**, 255
glycosidic linkages, formation of **8B**, 243ff
– – use of organic solvents **8B**, 254

glycosphingolipids, biosynthetic pathways **8B**, 262f
glycosylation **12**, 166f
– – *see also* transglycosylation
– by reverse hydrolysis **8B**, 253ff
– – aqueous systems **8B**, 253f
– – use of organic solvents **8B**, 254
– diastereoselective – **8B**, 248f
– disaccharide formation, regioselectivity of **8B**, 250
– glycosides as acceptors **8B**, 249
– of cell surfaces **8B**, 264
– of dalbaheptides **7**, 389
– of non-natural disaccharides **8B**, 266
– whole-cell systems **8B**, 266f
glycosylation reactions, isolation of products **8B**, 269f
glycosylation site binding protein (GSBP) **3**, 534
glycosyl donors, activated –, commonly used – **8B**, 247
– anomeric configuration **8B**, 247
– availability to glycosyltransferases **8B**, 256f
glycosylic bond, cleavage of, use of catalytic antibodies **8B**, 433f
glycosyl phosphatidylinositols, resemblance to aminoglycosides **7**, 402
– structure **7**, 402
glycosyltransferase activity, assay of **8B**, 269
glycosyltransferases **8B**, 256ff
– acceptor-donor analogs **8B**, 264ff
– availability of, enzymes **8B**, 256
– – glycosyl donors **8B**, 256f
– expression in insect cells **8B**, 268
– expression in yeast **8B**, 268
– heterologous expression of **8B**, 266f
– immobilization of **8B**, 258, 268f
– in glycoconjugate synthesis **8B**, 261ff
– in oligosaccharide synthesis **8B**, 258ff
– use in consecutive glycosylation reactons **8B**, 268
– using non-nucleotide donor substrates **8B**, 267f
glycoxylate cycle **2**, 799
glyoxisome **1**, 33
– enzyme set **1**, 33
glyoxylate cycle **1**, 85, 484
– genetics in *Saccharomyces cerevisiae* **1**, 484
– key enzymes, inhibitors **7**, 513
– microbial metabolism **9**, 184
glyphosate degradation, by pseudomonads **1**, 423
GMAC (Genetic Manipulation Advisory Committee) **12**, 429, 516f, 524, 573ff
– risk assessment **12**, 517
GMOs (genetically modified organisms), contained use **12**, 555ff, 578
– definition **12**, 555, 570
– field tests **12**, 260ff
– – permits **12**, 250ff
– – regulations **12**, 149

– – risk assessment **12**, 46ff
– food products **12**, 182ff
– labelling **12**, 622
– large-scale use **12**, 614
– NIH guidelines **12**, 241
– public opinion **12**, 134f, 148
– regulatory policy **12**, 49ff
– release **12**, 269f, 555ff, 570ff
– – approval **12**, 183ff
– – GDP **12**, 615f
– – German regulations **12**, 584
– – in Europe **12**, 550ff
– – large scale **12**, 616f
– – risk assessment **12**, 184ff
– – risks **12**, 527f
– – transgenic plants **12**, 198ff, 251f
– safety categories **12**, 49ff
GMP *see* good manufacturing practice
– – *see* guanylate (5')
GNE (Groups of National Experts on Safety in
  Biotechnology) **12**, 613ff
Gnu Database Manager **5B**, 381
GOGAT *see* glutamate synthase
gold, bioleaching of **10**, 208, 213
golden algae **10**, 113
Golgi apparatus **1**, 39
(+)-goniofufurone, preparation of using poly-leucine catalyzed epoxidation **8B**, 501f
(+)-goniopypyrone, preparation of, using poly-
  leucine catalyzed preparation **8B**, 501f
(+)-goniotriol, preparation of, using poly-leucine
  catalyzed preparation **8B**, 501f
Good Clinical Practice (GCP) **12**, 227
Good Developmental Practice (GDP) **12**, 572
Good Developmental Principles (GDP) **12**, 208f,
  615f
good industrial large scale practice (GILSP) **2**,
  837, 847, **3**, 775, **12**, 73, 260, 264, 614f
– engineered organism **3**, 776
– host organisms **3**, 775
– vectors **3**, 776
good laboratory practice (GLP) **5A**, 307f, **12**, 69,
  73, 227
Good Large-Scale Practice (GLSP) **12**, 246f
Good Manufacturing Practice (GMP) **3**, 743, **5A**,
  510f, **12**, 227
– cell banking **5A**, 257f
– in manufacture of therapeutic antibodies **5A**,
  268f
– – quality assurance **5A**, 268
– – quality control **5A**, 268
– in manufacturing of antibodies **5A**, 261
– manufacturing facilities **5A**, 269
– – inspection of required interactions **5A**, 269
– plasmid DNA **5A**, 445, 452
– regulatory authorities **5A**, 269
Good Manufacturing Practice Regulations
  (GMPRs) **3**, 743

Good Microbiological Practice (GMP) **12**, 517
good microbiological technique (GMT) **2**, 844ff
good occupational safety and hygiene **2**, 846
Gordal, olive variety **9**, 599
Gordon Conference on Nucleic Acids **12**, 511f
Gouda cheese **9**, 376f
G-Protein Coupled Receptors **5B**, 135
gp120-CD4 binding inhibitors, structures **7**, 125ff
grade efficiency curve, of sedimenting centrifuges
  3, 478
grain, determination of fermentable substance **6**,
  110
– in industrial ethanol production, flow diagram
  6, 159
– processing in ethanol production **6**, 90ff
– – barley **6**, 93
– – corn **6**, 92f
– – rye **6**, 92, 96
– – triticale **6**, 92, 95
– – wheat **6**, 91f, 95
– starch content **6**, 109
grain processing, typical wastes **11A**, 224
gram-negative bacteria **1**, 19ff, **2**, 427ff
– capsule **1**, 17
– envelope **1**, 19
– facultative methylotrophs **1**, 270
– methanogens **1**, 257
– molecular envelope composition **1**, 19
– nitrilotriacetic acid metabolization **1**, 417f
– obligate methylotrophs **1**, 269
– outer membrane **1**, 19
– periplasmic space **1**, 19
– phage receptors **1**, 552f
gram-positive bacteria **1**, 20, **2**, 457ff
– capsule **1**, 17
– cell wall assembly **1**, 21
– envelope **1**, 20
– methanogens **1**, 256
– methylotrophs **1**, 270
– peptidoglycan layer **1**, 20
– phage receptors **1**, 522f
granular activated carbon (GAC) **11C**, 397f
– as carrier materials of biofilters **11C**, 466
– filter types **11C**, 398f
– operation of GAC filter **11C**, 397
granulation, in anaerobic sludge bed reactors
  11A, 522f
granulocyte colony stimulating factor **5A**, 156
– clinical trials **5A**, 157
– production by *in vitro* folding **5A**, 113
granulocyte macrophage colony stimulating factor,
  in therapy **5A**, 157
grape must, acid content **9**, 489f
– clarification **9**, 484
– from *Botrytis*-infected berries, compunds **9**, 497
– yeasts in – **9**, 466ff
– – spontaneous fermentation **9**, 468f
graph abstraction **5B**, 401

graphical display system **5B**, 380ff
graphical user interfaces **5B**, 339
GRAS (generally regarded as face) organisms
  5A, 237
grasses, genome cross-referencing **5B**, 175f
– use in phytoremediation **11B**, 379
grass family, comparison of orthologous alleles
  5B, 176
gravel, as carrier materials of biofilters **11C**, 466
gravity *see also* microgravity
– as abiotic factor **10**, 508ff
– perception, in higher pants **10**, 513ff
– – signal transduction chain **10**, 513ff
gravity thickeners, in sludge treatment **11A**, 306f
green algae **10**, 111
green fluorescent protein (GFP) **10**, 470
grinding, of raw materials **6**, 77ff
– – use of dispersing machines **6**, 79f
– – use of mills **6**, 78
griseofulvin **8B**, 178f
griseoviridin, enzymatic synthesis of **8B**, 361
Große-Lohmann-Spradau process **6**, 85f
ground filtration, effect on BOM **11C**, 463
groundnut oil, genetic engineering of – **7**, 138
groundwater *see also* subsurface water
– consistency of **11C**, 378
– definition of **11B**, 45
– divisions of **11B**, 45
– nitrate in **11C**, 379
– pollutant loads **11C**, 377ff
– pollutant transport **11B**, 44
– saturated flow **11B**, 47
– treatment of, geogeneous conditions **11C**, 377f
– – processes **11C**, 378f
– treatment of contaminated –, use of sulfate-reducing bacteria **10**, 243
– unsaturated flow **11B**, 46
– virus survival **3**, 783
groundwater movement **11B**, 47ff
– effect of geochemical processes **11B**, 56f
– effect of permeability **11B**, 48
– effect of porosity **11B**, 48
groundwater recharge **11C**, 379ff, 425ff
– efficiency, assessment of **11C**, 441
– filter layers, algae in **11C**, 432
– – bacteria in **11C**, 433f
– – biology of **11C**, 431
– – metazoa in **11C**, 434
– – protozoa **11C**, 434
– for drinking water production **11C**, 427ff
– infiltration systems, classification of **11C**, 426
– – scheme **11C**, 427
– infiltration technqiues **11C**, 427ff
– process scheme **11C**, 381
– purification efficiency, effects on **11C**, 437
– purification mechanisms **11C**, 435f
– – assessment of **11C**, 441
– redox succesion during **11C**, 434ff

– removal capacity **11C**, 436ff
– removal of, microorganisms **11C**, 438ff
– – organic matter **11C**, 438
– – pathogens **11C**, 438ff
– – trace elements **11C**, 438
– – viruses **11C**, 440
– – xenobiotics **11C**, 438
– scheme **11C**, 380
groundwater velocity **11B**, 49
groundwater wells, in soil remediation **11B**, 358f
group transfer potential **1**, 54
growth *see* microbial growth
– balanced – **3**, 86
– doubling time **1**, 118, 139, 585
– of microorganisms **1**, 111ff
– – *see also* microbial growth, mycelial growth
– of plant cell cultures **1**, 585f
growth and differentiation factor-1 **5A**, 160
growth energetics, of methylotrophs **1**, 278
growth factors, interaction with cell surface receptors **5A**, 39
– therapeutic use **5A**, 155ff
growth hormone cascade **2**, 793
– regulation **2**, 793
growth hormone releasing factor **5A**, 165
growth hormones **5A**, 163f, 283
growth hormone (STH) **2**, 793ff
– amino acid sequence **2**, 795
– production in transgenic mammary glands **2**,
  803ff
– – yield **2**, 805f
growth kinetics **3**, 85ff, 299
– anaerobic – **3**, 302
– for the specific growth rate **4**, 274
– in anaerobic wastewater processes **4**, 457ff
– in cell models **4**, 274f
– parameter estimation **3**, 302
– product inhibition **4**, 275
– substrate-independent **4**, 275
– substrate inhibiton **4**, 275
growth media *see* media, microbial growth media
– Coon's medium **3**, 143f, 154
– Dulbecco's modified Eagle's medium (DMEM)
  3, 143f, 154
– for animal cell cultures **3**, 31
– for fibroblasts, F-12 – **3**, 143f, 154
– for mammalian cell cultures **3**, 142ff
– nitrogen supply **3**, 131
growth rate, maximum specific – **3**, 302
– medium osmolality effect **1**, 130
– specific – **1**, 118, 585, **3**, 85, 385
– – of microorganisms **3**, 299
– – temperature sensitivity **3**, 163
– temperature dependence **1**, 128
growth regulation **2**, 793ff
– gene regulation elements **2**, 793, 797
– growth hormone releasing hormone (GHRH)
  2, 794

– hypothalamus **2**, 793f
– somatomedin (IGF) *see* insulin-like growth factor
– somatostatin (SRIF) **2**, 794
growth regulation alteration, in transgenic mice **2**, 794, 797
– in transgenic pigs **2**, 794
– in transgenic sheep **2**, 796f
GSH *see* glutathione *S*-transferase
GTZ (Deutsche Gesellschaft für Technische Zusammenarbeit) **12**, 440
guaiacol, formation from vanillic acid **8B**, 346
guanine **2**, 194, 320
guanosine, production by *Bacillus subtilis* **6**, 586
– – recombinant strains **6**, 589
guanosine 3′,5′ monophosphate *see* cyclic guanosine 3′,5′ monophosphate (cGMP)
guanylate (5′) (GMP), fermentative production **6**, 577
– flavor compounds **6**, 563
Guidance for Industry, somatic cell therapy **5A**, 524f
gum production **3**, 11
gums *see* polysaccharides
gushing **9**, 428
gylcerol monostearate, use in cosmetics **6**, 676
gypsum, as by-product in citric acid production **6**, 333
gyrase **2**, 207f
– binding site **2**, 207
– gene expression regulation **2**, 209

H

habitats, of bacilli **1**, 374
– of lactic acid bacteria **1**, 328ff, 341, 344
– of methanogens **1**, 255f
– of methylotrophs **1**, 268
– of pseudomonads **1**, 411
– of streptomycetes **1**, 436
– of yeasts **1**, 471f
– soil **1**, 244
Haemophilus influenzae genome **5A**, 51
β-hairpins **5A**, 62
hairy root cultures, fermentation techniques **7**, 620
– production of tropane alkaloids **7**, 619
– scopolamine secretion **7**, 620
– secondary product formation **7**, 619f
– transformation with *Agrobacterium rhizogenes* **7**, 719
hairy root disease **2**, 549
– auxin overproduction **2**, 556f
– physiological causes **2**, 556f
Haldane equation **8A**, 431

halichondrin, antitumor compound of marine origin **10**, 453
halide methyltransferase **8B**, 198f
– biosynthesis of chloromethane **8B**, 198f
halides, physicochemical properties **8B**, 189
halidohydrolases, classification of **8B**, 205
haloalcohol dehalogenases, formation of epoxides **8B**, 205
haloaliphatics, priority pollutants **8B**, 203
haloalkaliphiles **10**, 69
haloalkanoic acid, dehalogenation by hydrolytic dehalogenases **11B**, 256f
haloaromatic compounds, degradation pathways **11B**, 257
halobacteria, cell wall **1**, 22
Halobacterium **10**, 66
halobenzoic acids, dehalogenation mechanisms **11B**, 261f
halocompounds **8B**, 173ff
– anthropogenic sources **8B**, 175f
– biogenic origins **8B**, 184ff
– biosynthesis **8B**, 173ff
– biotransformation **8B**, 173ff
– distribution **8B**, 175
– function **8B**, 175ff
– of basidiomycetes **8B**, 176
– of marine algae **8B**, 178
halocompound transforming enzymes, synthetic importance **8B**, 179ff
Haloferax sp., biopolymer formation **6**, 617
halogenated compounds *see also* chlorinated compounds
– abiotic degradation of **11B**, 250
– biodegradation of aromatic – **11B**, 257ff
– – technical applications **11B**, 263ff
– degradation of organics **11B**, 185
– natural occurrence **11B**, 245
– physicochemical data **11B**, 244
– properties of **11B**, 243f
– reductive dehalogenation **11B**, 433
halogenated hydrocarbons *see also* chlorinated hydrocarbons
– degradation, by pseudomonads **1**, 415
– dehalogenation by intramolecular substitution **11B**, 249
– dehydrohalogenation **11B**, 249f
– detoxification **1**, 313f
– hydrodehalogenation **11B**, 249f
– hydrolytic dehalogenation **11B**, 248f
– oxygenolytic dehalogenation **11B**, 248
– reductive dehalogenation, by anaerobic microorganisms **11B**, 247
– – enzymes **11B**, 247f
– thiolytic dehalogenation **11B**, 249
halogenated organic compounds, monitoring with biosensors **11A**, 133f
halogenation, enhancement of bioactive efficacy **8B**, 179

halomon, antitumor compound of marine origin
  10, 450ff
– structure **10**, 454
haloorganics, recalcitrance of **8B**, 177
haloperoxidases **8B**, 181ff
– assay methods **8B**, 180
– bacterial –, catalytic mechanism **8B**, 182
– basic mechanism **8B**, 192
– classification of **8B**, 193
– enantioselectivity of **8B**, 180f
– heme, mechanism of **8B**, 193, 195
– immobilization of **8B**, 202
– isolated enzymes **8B**, 194
– natural sources of **8B**, 192f
– non-heme non-metal –, mechanism **8B**, 195
– of *Caldariomyces fumago* **8B**, 197
– pH optima **8B**, 180
– producing cells **8B**, 194
– reactions of **8B**, 196ff
– regiospecific biotransformations **8B**, 198
– stereoselective biotransformations **8B**, 198
– temperature optima **8B**, 180
– vanadium **8B**, 193ff
halophiles, biotopes of **10**, 69
– characteristics of **10**, 69f
– classification of **10**, 69
– enzymes of **10**, 70
halorespiration, chlorinated hydrocarbons **11B**,
  246f
– chlorinated hydrocarbons, reaction scheme
  11B, 247
HAMA *see* human anti-mouse antibody
hammer mills **6**, 77ff
– disintegration of organic wastes **11C**, 114f
Hammond postulate **8B**, 406
Hanseniaspora uvarum **9**, 466f
– in grape must **9**, 467
Hansenula **1**, 269
Hansenula anomala, in kanji fermentation **9**, 541
– in olive fermentation **9**, 607, 612
Hansenula ciferri, formation of ceramides **7**, 184
Hansenula polymorpha **1**, 272, 280, 470ff, **2**, 523ff
– expression of heterologous proteins **9**, 85
– life cycle **1**, 474
– peroxisomes **1**, 34
Hansenula subpelliculosa, in olive fermentation **9**,
  607
HAO *see* hydroxylamine oxidoreductase
haplodization **2**, 85
haplotype **5B**, 108
hapten, phosphonamide **8A**, 269f
– transition state mimic **8A**, 269
hapten affinity binding sites, of monoclonal antibodies **8A**, 269
haptens, boronic acid, generation of catalytic antibodies **8B**, 438f
– charged, generation of catalytic antibodies **8B**,
  412ff

– design –, strategies **8B**, 410ff
– dialkylphosphinic acid, generation of catalytic
  antibodies **8B**, 438
– for generating catalytic antibodies, for phosphodiester cleavage **8B**, 436f
– generation of catalytic antibodies, bait and
  switch strategy **8B**, 412ff
– – carrier proteins for **8B**, 409
– – design of **8B**, 410ff
– – Diels-Alder cycloaddition **8B**, 417f
– – entropic trapping **8B**, 415ff
– – tetrahedryl intermediates **8B**, 410f
– – transition state analogs **8B**, 410ff
– heterologous immunization strategy **8B**, 414f
– reactive immunization **8B**, 439ff
– zwitterionic –, generation of catalytic antibodies **8B**, 414f
hard coal *see also* coal
– modification of, by bacteria **10**, 157
– – by fungi **10**, 157ff
hardwood, pulp composition **10**, 526
harpins **10**, 495
Harvard Onco Mouse **12**, 290, 293f
harvesting residues, cereal straw **11A**, 223
– maize straw **11A**, 223
– potato haulm **11A**, 223
– wastes from **11A**, 222f
hash tables **5B**, 391
hatomarubigins, inhibitors of multi-drug resistant
  tumor cells **7**, 672
Hatta number **4**, 379
hay, isolation of thermophilic actinomycetes **11C**,
  61
– isolation of thermophilic fungi **11C**, 66
– self-heating of **11C**, 66
hazard, definition **12**, 101f, 160
hazardous substances, in wastewater **11A**, 197
hazardous waste **3**, 771
HCH *see* hexachlorocyclohexane
HDC *see* histidine decarboxylase
HDL **5A**, 15
Health and Safety at Work Act **12**, 517, 522
health care, in biotechnology **2**, 838
health care biotechnology, biomedicinal product
  development **12**, 213ff, 352f
– in developing countries **12**, 353ff
– – entry barriers **12**, 344f, 356
– public opinion **12**, 484
– technology assessment **12**, 352f
– technology transfer **12**, 354ff
– technology trends **12**, 354
health surveillance, for workers **2**, 840
heap technique, covers, treatment systems **11B**,
  323
– pretreatment of soil **11B**, 321f
– principles of **11B**, 321ff
– – Terraferm system **11B**, 321
– soil monitoring, treatment systems **11B**, 323

– use in biomediation **11B**, 319ff
– – degradation rates **11B**, 325
– – economy of **11B**, 325ff
– – efficiency of **11B**, 325ff
– – oxygen supply **11B**, 325
– use of additives **11B**, 322
heaptitis A, as waterborne disease, outbreaks in the 19th and 20th century **11C**, 415f
heat evolution, microbial **1**, 141
heat exchangers, plate-type – **6**, 81
– spiral-plate – **6**, 81
– spiral-tubular – **6**, 81
– tubular – **6**, 81
heating, code adherence **3**, 752
– contamination **3**, 753
– control considerations **3**, 754
– design **3**, 752
– economics **3**, 753
– environmental control **3**, 753
– maintenance training **3**, 755
– manufacturing environment **3**, 752
– of biotechnological facilities **3**, 752ff
– recirculating systems **3**, 754
– single-pass systems **3**, 754
– system flexibility **3**, 752
– validation **3**, 752
– ventilation and air conditioning (HVAC) systems, airflow **3**, 753
heat production, by microorganisms, with glucose as substrate **11C**, 37ff
heat-pulse-TOF technique, measurement of liquid flow **4**, 139f
heat-shock proteins **10**, 72
heat shock proteins (Hsp) **3**, 537f
– in *Escherichia coli*, functions in protein folding **3**, 539
heat stability, of hyperthermophiles **10**, 72f
heat transfer, from bioreactors **3**, 411f, 417
– in bioreactors **4**, 119
– in freeze-drying **3**, 701f
– in spray-drying **3**, 709
– log mean temperature driving force **3**, 411
– overall coefficient **3**, 411, 414
heat transfer coefficient **4**, 119
heavy metal removal, by bioaccumulation **11A**, 441
– by biosorption **11A**, 441
heavy metals, accumulation in plants **11A**, 243f
– accumulation of, by non-viable algae **10**, 133
– content in plant residues **11C**, 105
– in composts **11C**, 145
– in sewage sludge, accumulation factors **11A**, 184
– – prescribed limits **11A**, 183
– in soil, extraction/elution procedures **11B**, 498
– limits for land application **11A**, 229
– monitoring of **11A**, 127f
– sorption of, by non-viable algae **10**, 133

heavy metal sulfides, solubility products of **10**, 242
Helianthus protoplasts, electrofusion of **10**, 515f
– stimulation by microgravity **10**, 514f
Helicobacter pylori varrine **5A**, 139
α-helix **5A**, 59ff
helix types, in protein structure **5A**, 60ff
helophytes **11A**, 244ff
– inhibition of *Escherichia coli* **11A**, 244
– in wetland systems **11A**, 242ff
– root exudation of **11A**, 244f
– thermoosmotic gas transport **11A**, 245
helper proteins **12**, 168
hematopoietic growth factors, in tumor therapy **5A**, 137
heme **9**, 37
heme haloperoxidases **8B**, 193
– mechanism **8B**, 195
hemicellulases **9**, 709, **10**, 527f
– in silage **9**, 778
hemicellulolytic enzymes **10**, 529
hemicellulose, microbial metabolism **9**, 185
– structure **6**, 144
hemicelluloses, organization of **10**, 531
hemimycin **7**, 502f
– cytotoxic activity **7**, 503
– structure **7**, 502
hemlock alkaloid **8A**, 328
hemoglobin **5A**, 20
– purification **3**, 602
hemophilia A **5B**, 72
hemoproteins **5A**, 20ff
hemostasis protein, posttranslational modification **2**, 699
Henry coefficient **4**, 18f, **11C**, 350
Henry constant **4**, 111, **11A**, 266
Henry constant of pollutants **11C**, 266
Henry-Dalton law **4**, 465
Henry isoterm, application to sorption processes **11B**, 53f
Henry's equation **4**, 618
Henry's law **3**, 191, **4**, 31, 38, 393
Henze cooker, dimensions **6**, 82
– processing of distillers' mashes **6**, 82ff
– scheme **6**, 83
heparin **5A**, 371, 374, **6**, 643f
– in bleeding events **5A**, 369, 376
– removal from blood **3**, 454
hepatitis A virus, monoclonal antibody therapy **5A**, 317
hepatitis B surface proteins, genetic engineering of **5A**, 540
hepatitis B vaccination **5A**, 137f
hepatitis B virus, monoclonal antibody therapy **5A**, 317
hepatitis B virus (HBV) **2**, 787
– expression in transgenic mice **2**, 787
hepatitis virus vaccines **5A**, 139

hepatocyte growth factor **5A**, 163
heptapeptide structure, of dalbaheptides **7**, 372
herbicidal compounds, from basidiomycetes **7**, 513ff
– – structures **7**, 514f
herbicide, action mechanism **2**, 639f
– detoxifying enzymes **2**, 640
– modification of the molecular target **2**, 639f
herbicide resistance, introduction into plants **2**, 638ff
– – glyphosate **2**, 640
– – imidazolinones **2**, 639
– – introduction of detoxifying enzymes **2**, 640
– – phosphinotricine **2**, 639f
– – sulfonylureas **2**, 639
– – triazines **2**, 639
– of crop plants **5B**, 154
herbicides, application of, formation of bound residues **11B**, 109ff
– degradation **1**, 201
– detoxification, by pseudomonads **1**, 423
– monitoring with biosensors **11A**, 131f
– resistance to **10**, 493f
herbimycin, immunostimulation **7**, 675
– structure **7**, 686
herbimycin A, angiogenesis inhibitor **7**, 674
– contact inhibition of tumor cells **7**, 680
– phosphoinositol kinase inhibition **7**, 681
hereditary nonpolyposis colorectal cancer (HNPCC) **5B**, 54, 86, 110, 127
Herhof boxes **11C**, 134
– loading **11C**, 143
heritable gene silencing **5B**, 155
heroin, microbial transformations of **8A**, 351
– potent analgesic **8A**, 340
heroin esterase, heroin metabolites formed by **8A**, 351
– overexpression in *Escherichia coli* **8A**, 351
herpes simplex vaccines **5A**, 138f
herpes simplex virus (HSV), monoclonal antibody therapy **5A**, 317
Herpetosiphon auranticus **1**, 114
Herstein's matching law **4**, 291
heterocycles, chlorinated – **8B**, 187
heterocyclic biotransformation products **8B**, 384ff
heterocyst **10**, 472f
heteroduplex analysis **5B**, 114f
heterogenic incompatibility *see* vegetative incompatibility
heterokaryon formation **2**, 84
heterologous gene **2**, 584, 719
– – *see also* transgene
– construction **2**, 749
– design **2**, 750
– establishment **2**, 778
– hybridization probes **2**, 758
– integration pattern analysis **2**, 757
– intron **2**, 749

– localization in target DNA **2**, 757
– multiple copy integration **2**, 757, 759
– splicing **2**, 722
– structure of MThGH **2**, 759
– transcriptional fusion **2**, 410
– translational fusion **2**, 410, 415
heterologous gene expression **2**, 462, 473ff, 519ff, 524, 539, 585, 665
– – *see also* enzyme production, gene expression, protein production
– – *see also* gene expression
– additive product effects **2**, 783
– analysis strategies **2**, 585
– cDNA constructs **2**, 780
– characteristics **2**, 726
– disease resistance gene **2**, 806ff
– expression vector system **2**, 411
– for optimization of enzyme activity **8A**, 26
– in bacilli **1**, 396
– in baculovirus vectors **2**, 731
– in Corynebacteria **2**, 484f
– inducible **2**, 720ff
– in fish **2**, 812
– in plants **2**, 633ff
– in streptomycetes **1**, 450f
– in transgenic animals **2**, 778
– intron effect **2**, 781
– in yeast **1**, 496ff
– mammalian protein **2**, 539
– – bovine chymosin **2**, 539
– – therapeutics **2**, 540
– mammary glands **2**, 800ff
– milk protein gene **2**, 799ff
– overexpression **2**, 521
– – genes in plants **2**, 562
– promoters in Corynebacteria **2**, 485
– promoters in *Streptomyces* **2**, 475
– protein inclusion bodies **1**, 26f
– regulatable promoters **2**, 462ff
– – *E. coli* systems in bacilli **2**, 462
– – endogenous bacillar **2**, 463
– – phage lambda in bacilli **2**, 462
– regulation **2**, 562, 804
– – in animals **2**, 718ff
– regulatory elements **2**, 779
– requirements **2**, 410
– – for translation **2**, 142f
– secreted products of bacilli **2**, 463
– – animal **2**, 463
– – bacterial **2**, 463
– – human **2**, 463
– systems, in animals **2**, 726
– termination in Corynebacteria **2**, 485
– transcriptional terminator **2**, 521
– transient expression assay **2**, 586
heterologous gene suppresion **2**, 596f
– chromatin structure **2**, 597
– methylation **2**, 596

– regulatory region **2**, 597
heterologous gene transmission, in animals **2**, 775ff
heterologous nuclear RNA **2**, 675f
heterologous protein, compartmentalization **2**, 589
– inclusion body formation **2**, 416
– inhibition of proteolysis **2**, 416
– overexpression **2**, 416
– solubility **2**, 416
– stabilization **2**, 346f
heterologous protein expression, in goat milk **2**, 805
– in mouse milk **2**, 803
– in pig milk **2**, 805
– in rabbit milk **2**, 805
– in sheep milk **2**, 804
– monoclonal antibodies **2**, 807f
heterologous protein production **2**, 410, 800f
– animal host cell lines **2**, 703f
– – improvement **2**, 704
– expression requirements **2**, 410
– human β-globin in mice **2**, 769
– in mammary glands **2**, 800
– – efficiency **2**, 801f
– – foreign protein types **2**, 801f
– – promotor construction **2**, 801f
– – protein processing **2**, 800
– – purification **2**, 800
– – tissue expression **2**, 801
heterologous proteins, production, inclusion body formation **3**, 535
– – in eukaryotes **3**, 531
heterologous protein stability **2**, 636f
heterologous RNA stability **2**, 636
heteropolysaccharides *see* polysaccharides
heterotrophs **3**, 128
heterotrophy **1**, 50
– definition of **10**, 275
HETP **4**, 616ff
heuristic clustering **5B**, 327
HEW (Department of Health, Education and Welfare) **12**, 514
hexachlorocyclohexane (HCH), aerobic biodegradation **11B**, 256f
– persistence **11B**, 256
– simulation of biodegradation **11B**, 77
hexacyanoferrate, use in citric acid production **6**, 320
hexadecyl trimethylammonium bromide (CTAB), for lysate clarification **9**, 93
n-hexane, scrubbing liquid quantities for the absorption of **11C**, 312
hexane squalene **8B**, 506
hexokinase **4**, 88
– activity **1**, 236
– gene (*HXK1*) **1**, 480
– glucose determination **9**, 141f

– phosphorylation by **8B**, 226
hexokinase/glucose-6-phosphate dehydrogenase, glucose determination **9**, 144
hexosamine pathway **7**, 423
hexosamines, as moieties of antibiotics **7**, 457
β-hexosaminidase, production **1**, 239
hexose fermentation, in bacilli **1**, 385
– in clostridia **1**, 300
– in lactic acid bacteria **1**, 353
hexose metabolism **9**, 182ff
– in yeast **9**, 182
hexose monophosphate pathway **9**, 183
hexose phosphate metabolization, in clostridia **1**, 298
hexoses, amination of – **7**, 415
– as nutrient, for mammalian cell cultures **3**, 145
HFV *see* human foamy retrovirus
Hidden Markov Models **5B**, 129, 134f
– alignments **5B**, 307
Highbie's penetration theory **11A**, 555
high-cell-density cultivation **10**, 574f
higher alcohols, formation, in wine making **9**, 477
– in brandy **9**, 482
high gradient magnetic separation **10**, 244
Highly Advanced National Project (HAN) **12**, 388ff
high-performance bioreactor *see* bioreactor
high-performance computing environments **5B**, 435
high performance filters **11C**, 328
High performance liquid chromatography (HPLC) **4**, 165ff, **5B**, 257f
– analytes monitored on-line **4**, 166
– equipment **4**, 165
– high-pressure systems **5B**, 258
– low-pressure systems **5B**, 258
– on-line **4**, 165
– – advantages **4**, 167
– – disadvantages **4**, 167
high pressure cooking process **6**, 82ff
– scheme **6**, 83
high pressure liquid chromatography **9**, 65
– in enzyme purification **9**, 56
high-speed agitator bead mill *see* bead mill
high temperature reverse transcription, use of thermostable DNA polymerases **10**, 88f
high-throughput DNA synthesis **5B**, 230ff
high-throughput screening (HTS), for drug discovery, automation **10**, 44
– – miniaturization **10**, 44
high voltage gradients **5B**, 202
hispidospermidins **7**, 680
histidase **8B**, 132ff
– catalytic activity **8B**, 134
– deamination of labeled L-hididine, mechanism **8B**, 137
– deamination of L-histidine, mechanism **8B**, 138
– inhbition of **8B**, 135

– mutants **8B**, 135, 139
histidine decarboxylase (HDC) **8B**, 81f
L-histidine, decarboxylation of, by histidine decarboxylase **8B**, 81
L-histidine production **6**, 513
– by microbial fermentation **6**, 489
histone **2**, 673f
– histone-like protein H1 effect **2**, 211
HIV *see* human immunodeficiency virus
HIV gene therapy **5A**, 471ff
– antiviral constructs, antisense RNA **5A**, 475
– – design of **5A**, 473ff
– – protein inhibitors **5A**, 476ff
– – RNA decoys **5A**, 476
– – RNA inhibitors **5A**, 474ff
– future concepts **5A**, 485ff
– – next generation of clinical trials **5A**, 486
– immunotherapies **5A**, 481f
– – expression of immunogenic HIV proteins **5A**, 482
– – selectable marker genes **5A**, 482
– optimization of – **5A**, 484ff
– – tissue-specific gene expression **5A**, 484
– – transduction protocols **5A**, 485
– – vectors **5A**, 484
– phase I clinical trials **5A**, 482ff
– – biosafety issues **5A**, 483
– – ethical issues **5A**, 483f
– vector design **5A**, 479ff
HIV (human immunodeficiency virus) *see also* AIDS
– infection **5A**, 471ff
– life cycle **5A**, 473f
– protein **5A**, 473f
HIV Molecular Immunology Database **5B**, 408
HIV protease **5A**, 96ff
HIV protease-inhibitor complex **5A**, 97f
HIV Sequence Database **5B**, 284
HLADH *see* horse liver alcohol dehydrogenase
HLE *see* horse liver esterase
HNPCC genes **5B**, 83
hnRNA **2**, 675f
hog kidney acylase (HKA) **8B**, 356f
– stereocontrolled synthesis of a penicillin **8B**, 356f
hollow-fiber bioreactor **3**, 28f, 116f, 121
– cell viability **3**, 117
hollow-fiber modules, for liquid-liquid extraction **3**, 605, 611
homogeneous time-resolved-fluorescence detection, in drug screening **10**, 46f
homogenic incompatibility **2**, 78
– gene **2**, 78
– – idiomorph **2**, 78
homogenization *see* cell disruption
homogenizers, cell disruption valve **3**, 508ff
– – materials **3**, 510
– for small sample volumes **3**, 510

– high-pressure – **3**, 508ff
– – automation **3**, 522
– – construction **3**, 508
– – containment **3**, 520
– – cylinder block assembly **3**, 521
– – electric pressure control (EPC) **3**, 522
– – feeding control (FC) **3**, 522
– – homogenizing pressure **3**, 509
– – hydraulic valve actuator (HVA) **3**, 522
– – operating pressure **3**, 522
– – safety control **3**, 521
– – sealing **3**, 521
– operating parameters **3**, 508
– pilot-scale – **3**, 511
homogenotization **2**, 436
Homohiochii medium **1**, 348
homologous recombination **2**, 587, 785
– – *see also* gene targeting
– extrachromosomal recombination **2**, 588
– frequency **2**, 588
– in embryonic stem cells **2**, 763
– selection method **2**, 588
– structural requirements **2**, 588
homology-based searches **5B**, 363
homology-modeled protein structures **5B**, 354
homology modeling **5B**, 347
– of proteins **5B**, 351f
homology searches, in protein databases **5A**, 57ff
homopolysaccharides *see* polysaccharides
homoserine lactones **7**, 62f
L-homoserine, aspartate derived amino acids, in corynebacteria **8B**, 323
hop products **9**, 425f
hops **9**, 423ff
– bitter compounds **9**, 423ff
– – *iso*-α-acids **9**, 426
– botany **9**, 423f
– chemical composition **9**, 424
– harvesting **9**, 424
horizontal filters **11C**, 328f
horizontal flow reactors, with agitated solids bed, use in composting **11C**, 123
– with static solids bed, use in composting **11C**, 123
horizontal gene transfer, in biofilms **11C**, 450
– in flocs and biofilms **11A**, 78
horizontal transfer hypothesis, β-lactam biosynthetic pathway **7**, 262f
hormone-dependent tumors, treatment of – **7**, 685
hormones **5A**, 163ff
– interaction with cell surface receptors **5A**, 39
– therapeutic recombinant, under development **5A**, 164
horse liver alcohol dehydrogenase, applications of **8A**, 439
– Baeyer-Villiger biotransformations **8A**, 565

– reduction of four- to nine-membered cyclic ke-
    tones **8B**, 379f
horse liver alcohol dehydrogenase (HLADH)
    8A, 437ff, **8B**, 362f
– characteristics of **8A**, 437
– oxidation reactions **8B**, 362f, 372f
horse liver esterase (HLE) **8A**, 209ff
– asymmetrization of prochiral organosilyl-substi-
    tuted esters **8A**, 211f
– kinetic resolution of, racemic lactones **8A**, 210
– – racemic methyl 2-alkyl-2-aryl esters **8A**, 211
horseradish peroxidase, commercial availability
    8A, 482
– hydroxylation by **8A**, 504f
host cells, genetic engineering **3**, 31
– selection **3**, 284
host organism, GILSP recommendations **3**, 776
household wastes *see* waste
HOX Pro **5B**, 408
HPLC, separation processes **11B**, 503
HPLC chromatography, of polyamines **1**, 409
HPLX *see* High performance liquid chromatogra-
    phy
HP medium **1**, 348
HSC (Health and Safety Commission) **12**, 575ff
HSDH *see* hydroxysteroid dehydrogenases
HSE (Health and Safety Executive) **12**, 517,
    575ff
HSV *see* herpes simplex virus
HTML files **5B**, 381, 436
html linking **5B**, 292
HTS *see* high-throughput screening
HTU-NTU model **11C**, 310
Huber process **11B**, 338
human anti-mouse antibody (HAMA) **5A**, 290
– development of **5A**, 248
human cloning **5B**, 420
human dignity **5B**, 419
human embryonic kidney cell line **5A**, 249
human factor, in laboratory automation **5B**, 435
human foamy retrovirus (HFV), use as vector
    5A, 481
human genes, cDNA sequences **12**, 142f, 291, 305
human genome, annotation of **5B**, 49ff
– – cataloguing the genes **5B**, 51ff
– common heritage of **5B**, 425
– disease gene identification **5B**, 53ff
– DNA sequencing **5B**, 45ff
– – cDNA sequencing **5B**, 47
– – systematic mapping **5B**, 47
– number of genes **5B**, 53
– patenting of **5B**, 425
– physical map **5B**, 139
– sequence variation **5B**, 55
– working draft DNA sequence **5B**, 48f
Human Genome Organisation, Statement on
    DNA Sampling **5B**, 421
Human Genome Organization **5A**, 528

Human Genome Project **5A**, 520ff, **5B**, 41ff, 127f,
    12, 143, 305, 385
– and genetic diseases **5B**, 53, 88ff
– annotating the DNA sequence **5B**, 42
– chromosome maps **5B**, 42
– – genetic maps **5B**, 43f
– – physical maps **5B**, 44f
– large-scale DNA sequencing **5B**, 42
– web resources **5B**, 52
– whole genome shotgun strategy **5B**, 48
human growth hormone, genetic engineering of
    5A, 540
– production by *in vitro* folding **5A**, 113
human growth hormone (hGH), storage, after
    freeze-drying **3**, 704
human immunodeficiency virus *see* HIV
human immunodeficiency virus (HIV), anti-HIV
    activity assay **7**, 124
– blocking of HIV entry **7**, 124
– HIV disease, use of cyclosporins **7**, 540ff, 550f
– inhibition of recplication **7**, 125
human immunodefiency virus (HIV) **2**, 787
human insulin like growth factor I **5A**, 134
human leukocyte antigen (HLA) **5B**, 117
human nutrition, role of cyanobacteria, *Spirulina*
    10, 126f
– role of microalgae, *Chlorella* **10**, 126f
– – *Dunaliella* **10**, 126f
human papilloma virus vaccines **5A**, 139
human protein production, by *Saccharomyces cer-
    evisiae* **1**, 496
Human Research Ethics Committee, requirement
    of consent **5B**, 423
human rights **12**, 118
human T-lymphotropic virus (HTLV) **2**, 787
humic acids **11B**, 102f
– as terminal electron acceptors **11B**, 97
– models of **11B**, 105
– oxidation by, laccases **11B**, 98
– – peroxidases **11B**, 98
– reductive processes **11B**, 98
Humicola lanuginosa, lipase, expression in *Asper-
    gillus oryzae* **9**, 728
humic particles **11B**, 104f
humic substances **11B**, 93, 102ff
– and the "humification" process **11B**, 102ff
– binding to clay minerals **11B**, 94
– bipolar charge transport **11B**, 105
– carbon distribution **11B**, 103
– cell free –, enzymes in **11B**, 98
– classification of **11B**, 102
– composition of **11B**, 102f
– consistence of **11B**, 105
– distribution of carbon atoms **11B**, 103
– formation of **11B**, 89ff, 99f
– – aggregative models **11B**, 106
– – composting experiment **11B**, 107
– – degradative models **11B**, 106f

– – model reactions **11B**, 108
– in coal, depolymerization of **10**, 167ff
– incorporation of nitrogen **11B**, 108
– interaction with xenobiotic compounds **11B**, 104
– intermolecular interactions of **11B**, 108
– models **11B**, 103ff
– – based on amphiphilic properties **11B**, 108
– – degradative – **11B**, 106f
– – of aggregation **11B**, 104
– nitrogen content **11B**, 103
– turnover of **11B**, 114ff
humification **11B**, 89ff, 99ff, 474
– definition of **11B**, 100, 106, 467
– formation of bound residues **11B**, 109ff
– – covalent bonds **11B**, 112f
– – reaction mechanisms **11B**, 112ff
– formation of humic substances **11B**, 106ff
– hypothesis on **11C**, 70
– in biological waste gas purification **11C**, 283f
– of pollutants **11B**, 68
– rate, of plant material **11B**, 101
humins **11B**, 102f, 106
humulene, biotransformation of **10**, 408
humulone **9**, 423
humus, formation of **11C**, 69f
Huntington disease **5B**, 109
– trinucleotide repeat expansion **5B**, 71
HVAC systems *see* heating, ventilation and air conditioning (HVAC) systems
hyaluronic acid **6**, 643f
– formation by, *Streptococcus equi* **6**, 644
– – *Streptococcus zooepidemicus* **6**, 644
– medical uses **6**, 644
– structure **6**, 644
hybridization **2**, 74
– dot blot – **11A**, 91, 97f
– – method **11A**, 97ff
– in microbial community analysis **11A**, 91ff, 97ff
– *in situ* **11A**, 98ff
– – analytical tools **11A**, 100f
– – detection of *Zoogloea ramigera* **11A**, 101
– – flow chart **11A**, 99
– – with a fluorescently labeled oligonucleotide probe **11A**, 100
– of brewer's yeast **9**, 436
– parasexual **2**, 74
– sexual **2**, 74
hybrid mass spectrometers **5B**, 255
hybridoma cell **2**, 130
– parental species **2**, 130
hybridomas **5A**, 128
– in manufacturing of antibodies, serum-free media **5A**, 257
hybrid RNA/DNA molecules **5B**, 156
hybrid somatic cell **2**, 156f
– application in mapping **2**, 158
– chromosome segregation **2**, 156

– formation **2**, 156
– from mutant cell lines **2**, 156
– karyotypic analysis **2**, 158
– multiple chromosome mapping **2**, 157
– single chromosome mapping **2**, 157
hycoscyamine **8A**, 332
hydantoinase **3**, 449
– distribution in microorganisms **6**, 516f
– enzymatic hydrolysis **6**, 516
– production of non-natural amino acids **8B**, 285f
hydantoinases **8A**, 267
hydantoin racemase **6**, 517f
– of *Arthobacter* sp. **6**, 517
hydantoins **3**, 449
– as precursors of amino acids **6**, 516ff
– racemic –, conversion to L-amino acids **8A**, 267
hydantoin substrates, hydrolysis by, allantoinase **8A**, 267
– – carboxymethylhydantoinase **8A**, 167
hydraulic circuits, in soil remediation **11B**, 357f
hydrocarbon, degradation, by pseudomonads **1**, 413
– estimation methods **1**, 126
hydrocarbon metabolism **1**, 89
– aliphatic **1**, 89
– aromatic **1**, 89
hydrocarbons, aerobic degradation of **11B**, 146
– aliphatic –, anaerobic degradation of **11B**, 171f
– aromatic – **11B**, 183ff
– – benzylic hydroxylation **8A**, 479f
– – biochemically oxidizable –, in waste gas **11C**, 283
– biodegradation of, aliphatic – **6**, 687
– – alkenes **6**, 687
– – aromatics **6**, 686f
– – chlorinated aromatics **6**, 687ff
– biodeterioration of **10**, 311
– extraction of **11B**, 499
– in biosurfactant production **6**, 667f
– in municipal wastewater **11A**, 175ff
– in oily sludge, degradation rate **11B**, 419f
– in petroleum, bioremediation **11B**, 422
– microbial oxidation **9**, 176ff
– unsaturated –, anaerobic degradation of **11B**, 172
hydrocarbon uptake, by bacteria, function of biosurfactants **11B**, 152
hydrocodone **8A**, 341, 347
– biological production of **8A**, 350f
hydrocyanation **8B**, 493ff
– asymmetric –, of aldehydes **8B**, 496
– stereoselective –, of benzaldehyde **8B**, 493ff
– – of benzaldehyde derivatives **8B**, 493ff
– – use of cyclic dipeptides, aminol mechanism **8B**, 498f
– – autoinduction **8B**, 497
– – imidazole/cyanide salt mechanism **8B**, 498
– – mechanism **8B**, 497ff

hydrodehalogenation **11B**, 249f
hydrodynamic injection **5B**, 271
hydrogen, use by methanogenic bacteria **11A**, 459f
hydrogenation, stereoselective – **8B**, 507
hydrogen cyanide **8A**, 279f, 299f
– bacterial metabolism **8A**, 300f
– synthesis **8A**, 279f
– – by fungi **8A**, 279
– – by photosynthetic microbes **8A**, 279f
hydrogen-oxygen fuel cells *see* fuel cells
hydrogen peroxide, as source of oxygen **11B**, 430
hydrogen production, by *Anabaena azollae* **1**, 237
– by marine microorganisms **10**, 471ff
– – optimization of **10**, 472f
hydrogen sulfide, as source of biogenic sulfuric acid **10**, 300
– corrosion by **10**, 283
– dissociation of, as a function of pH **11A**, 538f
– formation in wine making **9**, 480
– in biogas reactors **11A**, 535
– inhibition of methane formation **11A**, 541
– oxidation of, in biofilters **11C**, 287
– – pathways **11C**, 287
hydrogeological situation, effects on soil remediation **11B**, 43ff
hydrol **6**, 17
hydrolase **1**, 66
hydrolase-catalyzed reactions, examples **8A**, 27ff
α/β-hydrolase fold, of lipases **8A**, 45
hydrolases **5A**, 25f, **9**, 8
– in industrial biotransformations **3**, 448ff
– in lignite solubilization **10**, 164f
– in yeast **8A**, 379
hydrologic cycle **11B**, 44
hydrolyases **8B**, 46, 98f
hydrolysis, of polymers, by immobilized enzymes **3**, 454
– of triacylglycerides **8A**, 141
hydrometallurgy, definition of **10**, 192
hydromorphone **8A**, 341, 347
– biological production of **8A**, 350f
hydromycin, antitumor agent, structure **7**, 662
hydroperoxide formation, by lipoxygenases **8A**, 502ff
hydrophobic compounds, as nutrients, for mammalian cell cultures **3**, 146
hydrophobic interaction chromatography, in enzyme purification **9**, 58f, 103f
hydrophobic interaction chromatography (HIC) *see* chromatography
hydroquinone, anaerobic degradation of **11B**, 179f
– monitoring with biosensors **11A**, 130
hydroxamate siderophores **7**, 209ff
– albomycins **7**, 211f
– alcaligin **7**, 210f
– bisucaberin **7**, 210f
– ferrimycins **7**, 211
– ferrioxamines **7**, 209f
hydroxyacid esters, chiral –, synthesis of **8A**, 411ff
– enantioselective reduction **8A**, 412
– racemic –, enzymatic hydrolysis of **8A**, 413f
– – selective acylation **8A**, 418f
– – transesterification reactions **8A**, 416ff
hydroxyacids, enantioselective reduction **8A**, 412
hydroxyacyl coenyzme A thioester **6**, 414
α-hydroxy acid production **3**, 456
hydroxyalkanoic acid constituents, of PAH **6**, 407ff
– – branched – **6**, 408
– – unsaturated **6**, 407
– – with ester groups **6**, 409
– – with halogen atoms **6**, 409
hydroxyapatite chromatography, in enzyme purification **9**, 55, 109
p-hydroxybenzoate, bacterial degradation of **11B**, 153
hydroxybenzoates, anaerobic degradation of **11B**, 177f
hydroxycinnamoyl putrescines **7**, 600f
hydroxy compounds *see also* hydroxyacid esters
– chiral –, preparation by *PADH* **8A**, 453
– – resolution with alcohol dehydrogenases **8A**, 434ff
– – synthesis of **8A**, 411ff
– enzymatic hydrolysis of racemic acyl derivatives **8A**, 415f
– – racemic ether derivatives **8A**, 414f
hydroxy esters, chiral – **8A**, 373
7-hydroxyguanine, antitumor activity **7**, 647
hydroxyhydroquinone, as intermediate in anaerobic degradation of aromatic compounds **11B**, 181ff
hydroxyisocarproate dehydrogenases, applications of **8A**, 451
– characteristics of **8A**, 448ff
hydroxylamine oxidoreductase (HAO) **11A**, 56
hydroxylase enzymes, mammalian –, expression in microbial systems **8A**, 482
hydroxylation **8A**, 475ff
– – *see also* aromatic hydroxylation, benzylic hydroxylation
– by *Beauveria sulfurescens*, alicyclic – **8A**, 486
– – stereochemical model **8A**, 479
– by chloroperoxidase **8A**, 504
– by cyclooxygenase **8A**, 504
– by genetically engineered organisms **8A**, 482f
– by horseradish peroxidase **8A**, 504f
– by lipoxygenases **8A**, 502f
– by peroxidases **8A**, 504f
– enzymes **8A**, 477f
– enzymic –, in drug synthesis **8B**, 390, 392
– formation of phenols **8A**, 499ff
– history **8A**, 477

– microbial –, preparation of chiral alcohols **8A**, 486
– occuring with rearrangement **8A**, 492
– of α-damascone **8A**, 493
– of α-ionone **8A**, 493
– of alkaloids **8A**, 495f
– of amides **8A**, 479
– of antibiotics **8A**, 496
– of aromatic rings **8A**, 499ff
– of avermectin derivatives **8A**, 484
– of bioactive compounds **8A**, 484f
– of biphenyl **8A**, 500
– of citral **8A**, 491
– of citronellol **8A**, 491
– of codeine **8A**, 496
– of cyclic terpenes **8A**, 492
– of diterpenes **8A**, 494f
– of farnsene **8A**, 493
– of fatty acids **8A**, 483
– of fluoranthene **8A**, 501
– of geraniol **8A**, 491
– of giberellins **8A**, 494
– of isobutyric acid **8A**, 483
– of labdane diterpenes **8A**, 495
– of milbemycins **8A**, 496
– of monensin **8A**, 496f
– of monoterpenes **8A**, 491f
– of morphine **8A**, 496
– of nicotinic acid **8A**, 501f
– of phenanthrene **8A**, 501
– of polycyclic aromatic hydrocarbons **8A**, 501
– of progesterone, by *Rhizopus arrhizus* **8A**, 477
– of pyrazines **8A**, 502
– of pyrene **8A**, 501
– of pyridines **8A**, 501f
– of quinoline **8A**, 502
– of sesquiterpenes **8A**, 492ff
– of steroids **8A**, 479
– – by plant cell cultures **8A**, 489
– – estrogens **8A**, 489f
– – progesterone **8A**, 488ff, 505
– – testosterone **8A**, 489
– of substituted adamantanes **8A**, 487
– of terpenes **8A**, 479, 491ff
– of terphenyl **8A**, 500
– of vitamin D_3 **8A**, 485
– regioselectivity of **8A**, 478ff
– use of isolated enzymes **8A**, 482
– use of whole-cell catalysts **8A**, 481f
hydroxylation and dihydroxylation, parallel pethways for **8A**, 514f
hydroxylation reactions, biotransformation with isolated enzymes **8A**, 502ff
– by microbial cultures **8A**, 482
– whole-cell biotransformations **8A**, 483ff
hydroxyl groups, in non-sugars **8A**, 129ff
– primary –, lipase-catalyzed reactions **8A**, 118ff

– secondary –, lipase-catalyzed deacylations **8A**, 124ff
– – lipase-catalyzed regioselective acylation **8A**, 126ff
– – model of acylation **8A**, 127
– selective protection of **8A**, 118
hydroxyls, lipase-catalyzed deacylation, of aliphatic – **8A**, 130f
– – of phenolic – **8A**, 129
hydroxymandelonitrile lyase **8B**, 90
hydroxynitrilase (NHL), synthesis of pyrethroid precursors **8B**, 282
(*R*)-2-(4-hydroxyphenoxy)propionic acid, synthesis of, use of *Beauveria bassiana* **8B**, 282
p-hydroxyphenylglycine, biosynthesis **7**, 385
hydroxysteroid dehydrogenases, applications of **8A**, 448
– characteristics of **8A**, 447f
hydroxysteroid dehydrogenases (HSDH), regiospecific reduction of ketones **8B**, 378f
2-hydroxy-1,4-naphtoquinone (HNQ), as redox mediator in bioelectrochemical fuel cells **10**, 13ff
3-hydroxy-γ-decalactone, formation of **8B**, 341ff
hygromycin, structure **7**, 464
hygromycin A, streptomycin-related aminoglycosides **7**, 428
hygromycin family, structure of – **7**, 469
hyoscyamine, biosynthesis of **8A**, 332
– from plant cell cultures **7**, 613
– 6-β-hydroxylase **8A**, 332ff
Hyoscyamus albus, root culture **1**, 601
hyperaccumulation, of metals **11B**, 389ff
hyperfiltration *see* reverse osmosis (RO)
hypersensitive reaction, in plants **10**, 494f
hyperthermophiles **10**, 64ff, 274
– biotopes of **10**, 64f
– cellulases, xylanases and chitinase from **10**, 83
– characteristics of **10**, 66ff
– enzymes from **10**, 72f
– heat resistance, biochemical basis of **10**, 72f
– proteolytic enzymes from **10**, 86
– reverse gyrase **10**, 72
– – starch hydrolyzing enzymes from **10**, 79f
– thermosome **10**, 72
hypha **1**, 116, 517
– branching **1**, 116
– chain elongation **1**, 116
hyphodontal **7**, 511f
– structure **7**, 511
Hyphomicrobium **1**, 267, 270
Hyphomicrobium sp., in wastewater treatment **11A**, 102
hypnophilin, cytotoxic activity **7**, 507
– structure **7**, 505
hypoparathyroidism, allotransplantation of parathyroid tissue **10**, 557
– cell implantation **10**, 548

– encapsulated-cell therapy **10**, 548
– xenotransplantation of parathyroid tissue **10**, 556f
hypoxanthine **6**, 587

I

IARCs (International Agricultural Research Centers) **12**, 364f, 626
IAWPRC model, for wastewater treatment **8A**, 410ff, 431, 433
IBA (Industrial Biotechnology Association) **12**, 535
IBC (Institutional Biosafety Committee) **12**, 243ff
IBD *see* inflammatory bowel disease
ibotenic acid, insecticidal compound **7**, 515f
– structure **7**, 516
IBPGR (International Board for Plant Genetic Resources) **12**, 365
IC *see* Ion chromatography
ICBG (International Cooperative Biodiversity Group), program **12**, 295f
ICCBD (Intergovernmental Committee on the Convention on Biological Diversity) **12**, 631
ice-cream plant, use of submerged fixed bed reactors in **11A**, 359ff
ice crystal, formation **3**, 697f
ICETT (International Center for Environmental Technology Transfer) **12**, 386
ICGEB (International Centre for Gentic Engineering and Biotechnology) **12**, 330, 625f
ICPS (International Programme on Chemical Safety) **12**, 620ff
ICSU (International Council of Scientific Unions) **12**, 46, 527
idarubicin, antitumor agent, structure **7**, 658
idli **9**, 538ff
– nutritional characteristics **9**, 539
– production of **9**, 538
IDPC *see* indolepyruvate decarboxylase
IDRC (International Development Research Center of Canada) **12**, 440
IgG immunoglobulin, structure of **8B**, 404f
IgGκ **5A**, 225ff
– domain **5A**, 225
ILCA (International Livestock Center for Africa) **12**, 440
illudins **7**, 491ff
– antitumor activity **7**, 503
– biosynthesis **7**, 492
– toxic principle **7**, 493
ILO (International Labour Organization) **12**, 623f
image distortions **5B**, 211
– band spacing **5B**, 213

– band spreading **5B**, 213
– band tilt **5B**, 213
– deconvolution techniques **5B**, 213f
image maps **5B**, 385
image wraping **5B**, 328
IMCB (Institute of Molecular and Cell Biology of Singapore) **12**, 419
IMGT **5B**, 284
Imhoff tank **11A**, 8
imidazoleglycerol phosphate dehydratase **8B**, 114
imidodiacetic acid (IDA), in affinity partitioning **3**, 602
immobilization **3**, 432
– – *see* cell immobilization
– coupling methods for **8A**, 407ff
– in yeast biotransformation **8A**, 369
– of *Acinetobacter* **8A**, 298
– of coenzymes, coupling to cross-linked enzymes **8A**, 407f
– of enzymes **8A**, 18
– – in peptide synthesis **8A**, 255
– of fungal spores **8A**, 488
– of glycosidases **8B**, 268f
– of glycosyltransferases **8B**, 258, 268f
– of haloperoxidases **8B**, 202
– of lipases **8A**, 52
– of microalgae **10**, 119
– of microbial biomass, biosorption of metals and radionuclides **10**, 231ff
– of microorganisms, by adsorption **11B**, 417
– – by entrapment **11B**, 417f
– – on different carriers **11B**, 421
– of nicotinamide coenzymes, by entrapment methods **8A**, 409
– – coupling on a carrier **8A**, 408f
– of penicillin G acylase, formation of loracarbef **8A**, 259f
– of pollutants **11B**, 68
– of yeast **8A**, 369
– on membranes **3**, 441
immobilization of antibodies **5A**, 278ff
immobilization of biomass **11A**, 439f
Immobilization of biomolecules **4**, 81ff
– coupling methods **4**, 81
– crosslinking **4**, 82
– entrapment **4**, 81
– enzymes **4**, 81
– glucose oxidase **4**, 87
– proteins **4**, 81
– reversible **4**, 82
immobilization reactors **3**, 431f
– enzyme reactors **3**, 434
– optimal conditions **3**, 438
immobilized cells, biosorption of, metals **10**, 231ff
– – radionuclides **10**, 231
immobilized cell systems, in acetic acid production, acetic acid bacteria **6**, 395
– in acetone/butanol fermentation **6**, 252ff

– – *Clostridium acetobutylicum* **6**, 252f
– in ethanol formation, effect of dilution rate **6**, 134
– – effect of pH **6**, 134
– – effect to temperature **6**, 134f
– in ethanol production **6**, 133ff
– in gluconic acid production, *Aspergillus niger* **6**, 354
– in glycerol production, *Saccharomyces cerevisiae* **6**, 213f
– in lactic acid production **6**, 302
– in L-asparate production, *E. coli* **6**, 473
– in 2,3-butanediol fermentation **6**, 286
immobilized enzymes **3**, 432
– carrier properties **3**, 435
– classification **3**, 434
– immobilization effects **3**, 438
– immobilization mode **3**, 434
– in bioreactors *see* enzyme reactors
– reaction rate, controlling steps **3**, 435
immobilized microorganisms **1**, 223ff
– – *see also* cell immobilization
– alterations of, macromolecular composition **1**, 236
– – morphology **1**, 233
– – physiology **1**, 234ff
– – product formation **1**, 239ff
– colonies, formation **1**, 230
– – growth **1**, 230
– – morphology **1**, 231ff
– *Cryptococcus elinovii* **1**, 242f
– development **1**, 226
– enzyme activity **1**, 236
– for process improvement **1**, 226
– fungi **1**, 522
– growth **1**, 235
– in industrial biotransformations **3**, 457f
– in natural habitats **1**, 244
– in soil remediation **11B**, 415ff
– – chlorophenol degradation **11B**, 418f
– – dichloroacetic acid degradation **11B**, 419f
– – phenol degradation **11B**, 418f
– – results **11B**, 418ff
– interface effect **1**, 235
– membrane composition **1**, 243
– microfilms **1**, 231
– oxygen availability **1**, 228
– oxygen limitation effect **1**, 234
– oxygen supply **1**, 226ff
– phenol effect **1**, 242
– poison resistance **1**, 242
– – protection mechanism **1**, 243
– productivity prolongation **1**, 242
– *Pseudomonas putida* **1**, 242f
– viability **1**, 239ff
– water activity effect **1**, 230
immune deviation, medical applications of recombinant proteins **5A**, 142

immune responses **5A**, 35f
– cellular – **5A**, 35
– humoral – **5A**, 35
– in gene therapy **5A**, 391
immune system **2**, 790ff
– action machanism **2**, 790f
– B cells **2**, 790f
– MHC class I molecule **2**, 791
– T cell receptor (TCR) **2**, 791
– T cells **2**, 790
immune system analysis **2**, 789ff
– granulocyte-macrophage colony-stimulation factor (GM-CSF) **2**, 792
– heterologous immunoglobulin **2**, 791
– in transgenic animals **2**, 790ff
– lymphokine action **2**, 790
immunization, heterologous – **8B**, 414ff
immunoaffinity chromatography **5A**, 276, 280f
immunoassay **3**, 730ff
– antibody labeling **3**, 731
– – conjugation methods **3**, 732
– enzyme-linked immunosorbent assay (ELISA) **3**, 730
– – direct antigen – **3**, 730f
– – indirect antigen – **3**, 731
– – sandwich – **3**, 731
– host cell impurity **3**, 732
– radioimmunoassay (RIA) **3**, 730
immunoassays **4**, 80
– enzymatic **9**, 151ff
– for preclinical testing of antibodies **5A**, 291
– heterogeneous – **9**, 152
– homogeneous – **9**, 152f
– in drug screening **10**, 47f
– on-line monitoring of proteins **4**, 172
– reversible **4**, 94ff
– – reflectometry **4**, 96
– – streaming potential **4**, 95f
immunoconjugates *see also* calicheamicin, CMA-676, CMB-401
– in therapy **5A**, 356ff
immunoelectrode **4**, 94
immuno flow injection analysis **11A**, 133
immunofluorescence, detection of plant pathogens **10**, 499
immunofluorescence microscopy **1**, 7
immunogenicity, of antibodies, preclinical testing **5A**, 289ff
– of CDP571 **5A**, 348ff
immunoglobulin G, structure **5A**, 12
immunoglobulin genes **2**, 146
immunoglobulin (Ig), structure **2**, 790
– – H chain **2**, 790
– – L chain **2**, 790
immunoglobulins **5A**, 36
– classes **5A**, 37
immunoisolated transplantation **10**, 548
immunomodulators, antitumor effects of – **7**, 673f

– structures **7**, 674
immunomycin, demethylation of **8A**, 497
immunophilins **7**, 547f, 559ff
– 3D-structure **7**, 561f
immunopurification, choice of antibodies **5A**, 276f
– clinical applications **5A**, 285
– detection, aflatoxin contamination **5A**, 284
– – cannabis metabolites **5A**, 284
– for substance measurements **5A**, 284
– immobilization of antibodies **5A**, 277f
– – chemical coupling techniques **5A**, 278
– – properties of desirable matrices **5A**, 278
– immunoaffinity chromatography procedures **5A**, 280ff
– monoclonal antibodies **5A**, 276
– of growth hormones **5A**, 283
– of interferons **5A**, 283
– of tumor necrosis factor **5A**, 283
– of whole cells **5A**, 284
– polyclonal reagents **5A**, 276
– removal of bacteria **5A**, 285
– ricin removal **5A**, 284
– site specific immobilization of antibodies **5A**, 279f
– specific removal of substances **5A**, 284
– use of antibodies **5A**, 275ff, 283
– – applications **5A**, 282f
– – clinical applications **5A**, 285
– – single domain antibodies **5A**, 283f
– use of antibody fragments **5A**, 283f
immunoregulants, FK506 **3**, 10
immunosensor **4**, 93ff
– dip-stick type **4**, 93f
– reversible type **4**, 94ff
immunosuppressant macrolides, model of the mode of action **7**, 548
immunosuppressants, of mibrocial origin, dual domain concept **7**, 546
immunosuppression **5A**, 317f
– by cyclosporins **7**, 539ff
– – mode of action **7**, 542ff
– – side effects **7**, 540
– – use in transplantation surgery **7**, 539f
– by FK506 **7**, 546ff
– by rapamycin **7**, 546ff
– drug-immunophilin complexes **7**, 549f
– monoclonal antibody therapy **5A**, 317
– polyclonal antibody therapy **5A**, 317
immunosuppressive agents, from basidiomycetes, structures **7**, 509
immunosuppressors, from secondary metabolites **7**, 7
immunotherapies for HIV, gene therapy-based – **5A**, 481f
IMP *see* inosinate (5′)
impeller **3**, 200
– aeration number **3**, 406
– axial flow – **3**, 83
– axial flow hydrofoil – **3**, 13
– bubble dispersion **3**, 265
– bubble entrainment **3**, 244
– "cavern", formation **3**, 210
– cavity configurations **3**, 200
– design **3**, 13, 267, 406f
– – baffles **3**, 410
– – for bulk mixing **3**, 410
– – for surface aeration **3**, 245
– diameter **3**, 407
– disc turbine – **3**, 243
– effects of, baffles **3**, 92
– – impeller number **3**, 407
– – spacing **3**, 407
– – speed **3**, 407
– effects on, mass transfer coefficients **3**, 201, 241ff
– flooding **3**, 409
– for surface aeration **3**, 240
– gas dispersion **3**, 201
– mass transfer coefficient, of aeration membranes **3**, 251f
– mechanical stress, on cells **3**, 407
– parameter values **3**, 91
– power input **3**, 405ff, 407
– – Prochem – **3**, 93
– power number **3**, 91, 406
– Reynolds number **3**, 92, 406
– Rushton turbine – **3**, 406
– – power input **3**, 92f
– – radial flow – **3**, 13
– Sherwood number **3**, 240
– sizing **3**, 415
– spacing **3**, 417
– subsurface –, for sparged aeration **3**, 243
– surface – **3**, 243
– surface agitation **3**, 243
– turbine – **3**, 83
– vibration potential **3**, 407
impeller configuration, two-impeller system **4**, 317
– – effect of liquid height **4**, 318
Imperial Epidemics Law **12**, 71
imprinting **5B**, 69, 72ff
InBase **5B**, 408
INBio (Instituto Nacional de Biodiversidad) **12**, 295
incineration, of biological waste **3**, 780
– – implications on public health **3**, 784
inclusion bodies **1**, 26
– formation **3**, 535
– – environmental effects **3**, 535
– heterologous protein **1**, 27
– of *Bacillus thuringiensis* **1**, 26
– protein characteristics **3**, 535
– protein inclusion **1**, 26
– protein recovery **3**, 530

– separation **3**, 42
inclusion body proteins **5A**, 111ff
– formation, role of overexpression **5A**, 113
– – suppression of **5A**, 113
– in cytosol **5A**, 113
– *in vitro* aggregation **5A**, 115
– *in vitro* disulfide bond formation **5A**, 115f
– *in vitro* folding **5A**, 111ff, 115ff
– – effect of denaturant concentrations **5A**, 118
– – effect of L-arginine **5A**, 117
– – effect of Na_2SO_4 **5A**, 118
– – effect of Tris buffer **5A**, 117
– – immobilized catalysts **5A**, 120
– – improvement of **5A**, 116f
– – low-molecular weight additives **5A**, 116
– – pH dependence **5A**, 116
– – transfer optimization **5A**, 117f
– – transfer procedures **5A**, 118
– – use of folding catalysts **5A**, 120f
– – use of molecular chaperones **5A**, 120
– – yield **5A**, 118f
– isolation **5A**, 113f
– – by cell lysis **5A**, 113
– solubilization of **5A**, 114f
– – by chaotrophs **5A**, 114
inclusion body renaturation **5A**, 121
incompatibility locus, in plants **2**, 149
indentification functional, batch process **4**, 257f
– fed-batch process **4**, 258
– model indentification **4**, 255
India, biotechnology regulations **12**, 605
Indian fermented foods **9**, 538ff
– dhokla **9**, 540
– idli **9**, 538ff
– jalebies **9**, 541
– kanji **9**, 541
– kenima **9**, 540f
– khaman **9**, 540
– kurdi **9**, 541
– papadam **9**, 540
– waries **9**, 540
indigenous fermented foods **9**, 505ff, 542ff
– nutritional aspects **9**, 551
– of Africa **9**, 542ff
– of India **9**, 538ff
– of the orient **9**, 507ff
– overview **9**, 508ff
– public health aspects **9**, 551f
indoleacetic acid, synthesis by agrobacteria **8A**, 294
indole alkaloid, production by *Catharanthus roseus* **1**, 596
indolepyruvate decarboxylase (IDPC) **8B**, 87ff
indoles, chlorinated – **8B**, 186
– halogenation of **8B**, 196
indolocarbazole compounds, antitumor activity **7**, 679
– bactericidal activity **7**, 679

– fungicidal activity **7**, 679
Indonesia, biotechnology **12**, 409ff
– – companies **12**, 411
– – industry **12**, 410f
– – institutions **12**, 409f
– – research institutes **12**, 421f
– – universities **12**, 412, 415
– structure of S&T **12**, 414
induced fit hypothesis **5A**, 22
industrial bioreactor *see* bioreactor
industrial enzymes **5A**, 189ff
– – *see* enzymes
– α-amylases **5A**, 192ff
– cellulases **5A**, 202ff
– chymosin **5A**, 209
– lipases **5A**, 199ff
– market **5A**, 191, 214
– oxidoreductases **5A**, 209ff
– phytase **5A**, 210f
– proteases **5A**, 205ff
industrial fermentations, bioreactor design **3**, 401ff
industrial property rights **12**, 294
industrial revolution, first wave **12**, 27
– fourth wave **12**, 28
– second wave **12**, 27f
– third wave **12**, 28
industrial wastewater, aerobic treatment, history **11A**, 376
– composition of **11A**, 434
– emitting industries, with mainly inorganic pollutants **11A**, 194
– – with mainly organic pollutants **11A**, 193f
– flow fractions, cooling water **11A**, 195
– – production wastewater **11A**, 196
– – rain water **11A**, 194f
– – wastewater from in-plant water preparation **11A**, 195
– – wastewater from sanitary and social facilities **11A**, 195
– residence times for degradation **11A**, 567
– sources **11A**, 191ff, 433f
– specific amount **11A**, 434
– treatment concepts **11A**, 198ff
– – production-integrated – **11A**, 199
– treatment strategies **11A**, 191ff
industrial wastewater treatment, aerobic biological posttreatment **11A**, 490f
– completely stirred tank reactors (CSTR) **11A**, 479ff
– contact process **11A**, 479ff
– mass balance concept **11A**, 485f
– one-stage vs. two-stage anaerobic system **11A**, 486f
– process configurations **11A**, 480ff
– reactor design **11A**, 483ff
– – anaerobic contact process **11A**, 484f
– – CSTR process **11A**, 484

– use of covered anaerobic lagoons **11A**, 483
infectious disease, mAb products **5A**, 222
– monoclonal antibody therapy **5A**, 316
infectious diseases, and wastewater **11C**, 218f
infiltration, redox succession during **11C**, 434ff
infiltration techniques, bank filtration **11C**, 428, 431
– classification of **11C**, 426f
– comparsion of **11C**, 428
– for drinking water production **11C**, 427ff
– subsruface infiltration, use of seepage treches **11C**, 31
– subsurface infiltration, **11C**, 428, 430f
– – use of horizontal infiltration galleries **11C**, 430
– – use of infiltration wells **11C**, 430
– surface infiltration **11C**, 428ff
– – recharge through overgrown soil **11C**, 428f
– – use of infiltration basins **11C**, 429
– – use of through-trench systems **11C**, 429f
infiltration wells, scheme **11C**, 430
inflammation, use of recombinant enzymes **5A**, 167
inflammatory bowel disease (IBD) **5A**, 144f
– medical applications of recombinant proteins **5A**, 144
– role of tumor necrosis factor **5A**, 344, 346
influenza hemagglutinin proteins, use as vaccines **5A**, 138
informed consent **5B**, 419
infusion processes, cold mash process, scheme **6**, 85
– Große-Lohmann-Spradau process **6**, 85
– – scheme **6**, 86
– milling and mashing at higher temperatures **6**, 86
in-gel-digestion **5B**, 251
inheritance of foreign genes **2**, 775
– inheritance pattern in mice **2**, 776
inhibition, of enzymes, irreversible **9**, 41ff
– – reversible **9**, 43ff
inhibition kinetics **4**, 275f
inhibitors, as protein folding aids **3**, 541
injera, production of **9**, 546
ink particles, detachment by enzyme treatment **10**, 539f
inoculation *see* bioaugmentation
inosinate (5′) (IMP), flavor compounds **6**, 563
– isomers **6**, 568
– production, by *Bacillus subtilis* **6**, 570
– – by *Brevibacterium ammoniagenes* **6**, 571ff
inosine **6**, 587
– production, by *Bacillus subtilis* **6**, 582f, 589
– – by *Brevibacterium ammoniagenes* **6**, 583ff
inosinemonophosphate (5′-IMP) **2**, 34f
– biosynthesis **2**, 34
– production yield **2**, 35
inositol, enantiomerically pure derivatives **8A**, 81

Inoviridae **1**, 550
insect antennae, receptor-based sensors **4**, 96
insect cell culture **3**, 107
– shear sensitivity, effects of impeller type **3**, 231
– – effects of viral infection **3**, 231
insect cells, baculovirus expression **9**, 85ff
– expression of glycosyltransferases **8B**, 268
insecticidal compounds, from basidiomycetes **7**, 515f
– – structures **7**, 516
insecticides, monitoring with biosensors **11A**, 131f
– products of marine biotechnology **10**, 468
insect pests, control by pheromones **10**, 498f
insect resistance, introduction into plants **2**, 644ff
– – ecological aspects **2**, 644
– – insecticidal proteins **2**, 644
– – protease inhibitors **2**, 646
– – resistance degree **2**, 645
insects, as sources of ribosomal peptides **7**, 300
insect toxin, from *Bacillus thuringiensis* **1**, 377
insertional mutagenesis **5B**, 151
insertion sequence (IS) **2**, 29, 51, 440
– IS21 **2**, 51ff
in silico biology **5B**, 436
in situ extraction *see* extraction
in situ hybridization **5B**, 44
in situ remediation **11B**, 426ff
– activated zones, process scheme **11B**, 361
– adhesion-deficient strains **11B**, 410f
– advanced – **11B**, 399ff
– bacterial activity enhancement, methods **11B**, 428ff
– bioaugmentation **11B**, 356, 408ff, 426ff, 433ff
– – trichloroethylene (TCE) as model target **11B**, 409
– biochemical processes **11B**, 350
– biogeochemical conditions **11B**, 403
– bioscreens **11B**, 361ff
– – as denitrification zones **11B**, 363
– – composition of **11B**, 362
– – process scheme **11B**, 362
– bioslurping **11B**, 359f
– – treatment wells **11B**, 360
– biosparging **11B**, 359f
– – process scheme **11B**, 360
– biostimulation **11B**, 406ff, 426ff
– – aerobic conditions **11B**, 406
– – anaerobic conditions **11B**, 406
– – cost **11B**, 408
– – nutrient addition **11B**, 406
– – use of oxygen **11B**, 406
– bioventing **11B**, 356f
– – process scheme **11B**, 357
– condition parameters **11B**, 353
– degradability of contaminants **11B**, 351
– evolving technologies **11B**, 364ff
– funnel and gate **11B**, 363f

– – process scheme **11B**, 364
– hydraulic circuits **11B**, 357f
– – process scheme **11B**, 358
– hydrogen release compounds **11B**, 362
– indigenous microflora **11B**, 353
– influence parameters **11B**, 353
– *in situ* measurements **11B**, 367
– *in situ* respiration test **11B**, 367
– investigations **11B**, 353ff
– laboratory test methods **11B**, 353f
– measurement of microbial populations **11B**, 403
– metal bioprecipitation **11B**, 365
– metal solubilization, bioleaching **11B**, 365
– – polysulfide mechanism **11B**, 365
– – thiosulfate mechanism **11B**, 365
– monitoring **11B**, 366ff
– natural attenuation **11B**, 363f
– – contaminant concentrations **11B**, 402f
– – documentation of **11B**, 405f
– – modeling of **11B**, 405
– – plume size **11B**, 402f
– – receptor analysis **11B**, 405
– of soil **11B**, 56f, 308f, 312ff, 330ff, 349ff
– oxygen demand **11B**, 352
– oxygen release compounds **11B**, 362
– oxygen supply **11B**, 351f
– parameters to assess natural attenuation **11B**, 404
– passive technologies **11B**, 361ff
– process at different scales **11B**, 351f
– site model **11B**, 355
– target parameters **11B**, 353
– technologies **11B**, 355ff
– treatment of metals **11B**, 364f
– up-scaling process **11B**, 355
– use of groundwater wells **11B**, 358f
installation qualification (IQ) **3**, 741
instrument drift, source of systematic errors **4**, 51
instruments, for process analysis **4**, 6ff
– for process control **4**, 6ff, 20, 21
– – accuracy **4**, 21
– – precision **4**, 21
– – reliability **4**, 20f
– – resolution **4**, 21
insulin **5A**, 163
– expression in *E. coli* **5A**, 112
– genetic engineering of **5A**, 540
– production **3**, 106
– – of recombinant human – **3**, 473
– production by *in vitro* folding **5A**, 113
insulin-dependent diabetes mellitus, medical applications of recombinant proteins **5A**, 144
insulin-like growth factor I (IGF-I) **2**, 794
– amino acids sequence **2**, 796
insulin like growth factors **5A**, 133, 163f
insulin resistance **5A**, 134f

– medical applications of recombinant proteins **5A**, 134
intasome **2**, 214
integrated databank systems **5B**, 126
integrin antagonists **5A**, 134
integrin receptor **5A**, 130
integron, in pseudomonads **1**, 410
intellectual property **12**, 281ff, 350f
– protection **12**, 142f, 301, 350f, 531
– rights **12**, 282f, 350, 538, 639
– trade-related **12**, 350
intellectual property issues, and genomes **5B**, 96
Interact **5B**, 408
interaction data **5B**, 404
– – *see* molecular interaction data
interaction databases **5B**, 399ff
– Aminoacryl-tRNA synthetase Database **5B**, 405
– ASN.1 **5B**, 403f
– BIND **5B**, 403f
– Biocarta **5B**, 405
– Biocatalysis/Biodegradation Database **5B**, 405
– BRENDA **5B**, 405
– BRITE **5B**, 405
– COMPEL **5B**, 406
– computer-aided design (CAD) **5B**, 403
– CSNDB **5B**, 406
– Curagen Pathcalling **5B**, 406
– DIP **5B**, 406
– DPInteract **5B**, 406
– DRC **5B**, 407
– EMP **5B**, 407
– ENZYME **5B**, 407
– Eocyte **5B**, 407
– FIMM **5B**, 407
– GeneNet **5B**, 408
– GeNet **5B**, 407
– graph abstraction **5B**, 401
– HIV Molecular Immunology Database **5B**, 408
– Hox Pro **5B**, 408
– InBase **5B**, 408
– Interact **5B**, 408
– KEGG **5B**, 408
– MDB **5B**, 409
– Metacyte **5B**, 407
– MHCPEP **5B**, 409
– MIPS Yeast Genome Database **5B**, 409
– MMDB **5B**, 409
– ooTFD **5B**, 409
– ORDB **5B**, 409
– PFBP **5B**, 410
– PhosphoBase **5B**, 410
– ProNet **5B**, 410
– REBASE **5B**, 410
– RegulonDB **5B**, 410
– Selex_DB **5B**, 410
– SoyBase **5B**, 411
– SPAD **5B**, 410

– STKE **5B**, 410
– SYFPEITHI **5B**, 411
– TRANSFAC **5B**, 411
– TRANSPATH **5B**, 411
– TRRD **5B**, 411
– WIT **5B**, 411
– YPD **5B**, 411
interactive commitment building, characteristics **12**, 22ff
intercalating drugs **7**, 650ff
– actinomycins **7**, 650ff
– ellipticin **7**, 650
– structures **7**, 651
intercellular adhesion molecule-1 **5A**, 131
interconnected suspension bioreactor (ISB) cascade **11B**, 339ff
– scheme **11B**, 336
interesterification of margarine **8A**, 137
interfacial activation, of lipases **8A**, 40, 45f
interferon α, genetic engineering of **5A**, 540
α-interferon, production **3**, 25
interferon β, in therapy of multiple sclerosis **5A**, 142
interferon γ, in tumor therapy **5A**, 136
interferons **5A**, 283
– in therapy **5A**, 147ff
– type I **5A**, 148f
– – indications for – **5A**, 147
– type II **5A**, 148
interleucin production **1**, 396
interleukin **2**, 792
– overexpression effects **2**, 792f
interleukins **5A**, 135f, 149ff
– antagonists **5A**, 150, 152
– antibodies to **5A**, 135
– genetic engineering of **5A**, 540
– in therapy **5A**, 135, 144, 148ff
– in tumor therapy **5A**, 136
– receptors **5A**, 149
– recombinant –, indications for – **5A**, 149
– – under development **5A**, 149
– therapeutic potential **5A**, 149
– use in therapy **5A**, 130
– – of osteoporosis **5A**, 133
internal labeling **5B**, 197
International Agreement on Sanitation **12**, 71
International Bioethics Survey **12**, 121f, 124ff, 137, 143, 146
International Committee on Bioethics **12**, 510
international harmonization of research ethics **5B**, 418
International Patent Bureau **12**, 314
International Patent Cooperation Treaty (PCT) **12**, 314
International Potato Center *see* CIP
International Tropical Bioresources Institute **12**, 386, 443ff
– concept **12**, 444

International Union for the Protection of New Varieties of Plants *see* UPOV
InterPro **5B**, 289f
InterProScan **5B**, 297
interspecific communication **7**, 13, 15
intestinal flora **1**, 330f
– *Enterococcus* **1**, 346
intracellular enzymes *see* enzymes
intramolecular processes, use of catalytic antibodies **8B**, 468ff
introgression of genes **5B**, 152
intron **1**, 97, **2**, 143, 676
– intron/exon junction **2**, 677
– SV40 small t-intron **2**, 722
intron–exon boundaries **5B**, 365, 368
introns **5B**, 365
inulin, production of fructose **9**, 674
inverse folding problem **5B**, 352
invertase, **9**, 674
– in cocoa fermentation **9**, 570
– lytic enzyme-dependent release **3**, 518
inverted repeat sequence, effect **2**, 215
investigational new drugs **5A**, 504f
– clinical studies **5A**, 504f
– regulatory aspects **5A**, 504f
investigation strategy, for physico-chemical soil assessment **11B**, 481
in vitro fertilization, public opinion **12**, 124ff
iodine **8B**, 190
iodomethane, production by marine algae **8B**, 190
iodoperoxidases, isolated enzymes **8B**, 194
– producing cells **8B**, 194
ion chromatography (IC) **4**, 167ff
– ion-exchange (HPJC) **4**, 168f
– ion exclusion (HPJCE) **4**, 168
– ion-line **4**, 167
– ion pair formation (MPJC) **4**, 168f
– use of exchange columns **4**, 168
– use of potentiometric detectors **4**, 168
ion exchange, expanded bed – **3**, 47
– in protein fractionation **3**, 47
ion exchange chromatography, automation **9**, 102
– in enzyme purification **9**, 53ff, 100ff
ion exchange chromatography (IEC) *see* chromatography
ion exchange columns **5B**, 233
ion exchanger **3**, 728
– anion exchanger **3**, 729
– cation exchanger **3**, 729
– functional groups **3**, 729
– strong – **3**, 729
– supports, conventional – **3**, 729
– – high performance – **3**, 729
– weak – **3**, 729
ionization techniques **5B**, 252
ionizing radiation **5B**, 151
α-ionone, hydroxylation of **8A**, 493

ionones, biotransformation of **10**, 381f, 411
– odor threshold **10**, 411
ionophores, use in feeds **9**, 780
ion-selective membranes, in electrodes, for process control **4**, 22
ion trap mass spectrometers **5B**, 255
IPGRI (International Plant Genetic Resources Institute) **12**, 440
IPNS *see* isopenicillin N synthase
iron, as electron acceptor for anaerobic respiration **11A**, 43
– corrosion of **10**, 279f
– removal by biosorption **11A**, 42f
iron cycle **10**, 279f
iron(III) ions **10**, 280
iron(II) ions, microbial oxidation of **10**, 280
iron limitation, metabolic effect **1**, 179
iron nutrition, role of siderophores **7**, 235
iron transport, siderophore-mediated systems **7**, 233ff
– – in enterobacteria **7**, 234
iron uptake **7**, 200ff
irreversible thermodynamics **1**, 52ff
IRRI (International Rice Research Institute) **12**, 626
ISAH process, in enhanced biological phosphorus removal **11A**, 298
ischemia, use of ReoPro **5A**, 377f
IS element **2**, 29
– in *Lactococcus* **1**, 361
isepamicin, semisynthetic aminoglycosides **7**, 461
ISNAR (International Service for National Agricultural Research) **12**, 611, 626
isoamylase **9**, 669, 746
– action of **10**, 75f
isobutyric acid, hydroxylation of **8A**, 483
isochalciporon, antimicrobial activity **7**, 495
isochromophilones **7**, 124ff
– inhibition of HIV replication **7**, 125
– structures **7**, 125ff
isocitrate lyase **8B**, 90
isocodeine, structure **8A**, 348
isoelectric focusing **3**, 724, **5B**, 272
– Immobiline **3**, 725
– synthetic carrier ampholyte **3**, 724
isoelectric point, prediction of **5B**, 333f
isoeugenol, as substrate for vanilline production **8B**, 335
ISO (International Organization of Standardization), guidelines **12**, 227
isolation media, for lactic acid bacteria **1**, 347ff
L-isoleucine, aspartate derived amino acids, in corynebacteria **8B**, 323
– biosynthetic pathway **6**, 509
L-isoleucine production **6**, 510
– by microbial fermentation **6**, 492
isomerase **1**, 66
isomerases **5A**, 25ff, **9**, 8

– carbon-carbon bond formation **8B**, 6
– in industrial biotransformations **3**, 456
isomerizations, use of catalytic antibodies **8B**, 468f
isometric blocks **5B**, 387
isooctane, as reversed micellar medium **3**, 606
isopenicillin N synthase (IPNS) **7**, 256f
isoprene, bacterial metabolism of **11B**, 204f
– degradation by *Rhodococcus* **11B**, 204f
– emission from plants **11B**, 195
– environmental relevance **11B**, 195
isoprenoid lipids, structures **7**, 182
isopropanol, formation by clostridia **6**, 236
– – pathway **6**, 238
– physical properties **6**, 230
– production **6**, 229ff
– production by *Clostridium beijerincki* **8B**, 62
isoquercitin, selective acylation by lipase **8A**, 124
isovelleral, antimicrobial activity **7**, 495
isovellerol, antimicrobial activity **7**, 495
istamycin family, structure of – **7**, 465
istamycins **7**, 429ff
– biosynthetic pathway **7**, 429
itaconic acid, biosynthetic pathway **6**, 364
– recovery **6**, 367
– uses **6**, 368
itaconic acid production **6**, 366ff
– by *Aspergillus itaconicus* **6**, 366f
– by *Aspergillus terreus* **6**, 366f
ITRI (Industrial Technology Research Institute of Taiwan) **12**, 404

J

jackets, for bioreactors **3**, 420
Jacob-Monod kinetics *see* Monod kinetics
jalebies **9**, 541
Japan, bio-products **12**, 377
– biotechnology, associations **12**, 381f
– – companies **12**, 376
– – ERATO projects **12**, 383f
– – Frontier Research Program **12**, 384
– – history **12**, 375
– – industry **12**, 376f
– – research associations **12**, 381f
– – research companies **12**, 381ff
– – universities **12**, 380f
– biotechnology regulatory policy **12**, 55, 241ff, 253ff, 349, 378, 385, 592ff
– – institutions **12**, 253f
– contaminated site management, financing schemes **11B**, 15
– – legal development **11B**, 9
– – principles **11B**, 12
– – registration **11B**, 17
– – risk assessment **11B**, 19

– – technical development **11B**, 9
– Environment Agency **12**, 596f
– identification of contaminated sites **11B**, 17
– public opinion surveys, on biotechnology **12**, 124ff, 146ff
– – on genetic manipulation **12**, 128ff
– regulatory concerns, action thresholds **11B**, 21
– – conflict management **11B**, 28
– – enforcement **11B**, 26
– – in soil contamination **11B**, 7ff
– – liability rules **11B**, 14
– – organization **11B**, 26
– – priority setting **11B**, 23
– – remediation targets **11B**, 24
– – verification of clean-up **11B**, 24
– structure of S&T **12**, 379
Java-based display tools **5B**, 396, **5B**, 396
JBA (Japan Bioindustry Association) **12**, 255, 376, 434
Jerusalem artichoke, in industrial ethanol production **6**, 141, 159f
– – economic analysis **6**, 178ff
– – flow diagram **6**, 160
Johannesburg (JHB) process **11A**, 303
J. R. Simplot Anaerobic Bioremediation **11B**, 292f
judgement-based sampling **11B**, 483f
Jülich wastewater treatment process (JARV) **11A**, 276f
justice, and ethics **12**, 119f
juvenile polyposis **5B**, 86

K

kaanga-kopuwai, production of **9**, 548
KAERI (Korea Atomic Energy Research Institute) **12**, 393
kaffir beer, production of **9**, 546
kainic acid, structure **10**, 459
KAIST (Korea Advanced Institute of Science and Technology) **12**, 393
Kalman filter **4**, 230ff
– – *see also* Extended Kalman filter
– discrete form, assumptions **4**, 547f
– – definitions **4**, 548
– estimates, of cell mass **4**, 550
– – of model parameters **4**, 551
– – of substrate concentration **4**, 550
– extended – (EKF) **3**, 335
– for bioprocess state estimation **3**, 380f
– on-line data processing algorithm **4**, 230
– principle **4**, 469f
– use in wastewater process **4**, 469ff
kanamycin family, structure of – **7**, 467
kanamycin resistance, of tansgenic plants **2**, 560
kanamycins, biosynthetic pathway **7**, 433

kanji **9**, 541
Kaplan-Meier event rates, in myocardial infarction **5A**, 378
karyogamy **2**, 75
kasugamycin, structure **7**, 464
kasugamycins, streptomycin-related aminoglycosides **7**, 427
katsuobushi, production of **9**, 537
KDPG pathway **1**, 332
– – *see* Entner-Doudoroff pathway kefir
– microflora **1**, 332
Kearns–Sayre disease **5B**, 74
kefir, definition **9**, 394
Kemp decarboxylation, of 6-nitro-3-carboxybenzisoxazole **8B**, 419
kenima, production of **9**, 540
α-keratin **5A**, 33
kerosene, degradation of, revolving tubular reactors **11B**, 344
ketal, hydrolysis of, use of catalytic antibodies **8B**, 465
ketalization, use of catalytic antibodies **8B**, 465
α-keto acids, reductive amination, in bioreactors **3**, 454
2-ketogluconate production **1**, 178
keto hydroxy bidentates **7**, 225
2-keto-L-gulonic acid biosynthesis **1**, 211
ketones, conversion of **11B**, 200
– degradation of **11B**, 174f
ketoprofen, enantioselective amide hydrolysis **8A**, 310
ketoreductase domain, of polyketide synthases **10**, 361
ketoses, synthesis from amino acid synthons **8B**, 10
ketosynthase domain, of polyketide synthases **10**, 359f
Key components, concentration control **4**, 151f
khaman, nutritional characteristics **9**, 539
– production of **9**, 540
kimchi **9**, 537f, 645ff
– consumption **9**, 645
– ingredient composition **9**, 650
– processing **9**, 645f
– sensory properties **9**, 649f
– spoilage **9**, 650f
– – over-acidification **9**, 650
– – softening **9**, 651
– storage of **9**, 646
kimchi fermentation **9**, 645ff
– effect of temperature **9**, 648
– *Lactobacillus plantarum* **9**, 647f
– *Leuconostoc mesenteroides* **9**, 647f
– raw materials **9**, 645f
kindreds **5B**, 110
kinetics, of antibodies **5A**, 289ff, 297f
– – preclinical testing **5A**, 298
– of microbial growth **3**, 85

– of product formation **3**, 85
King-Altman method **9**, 38
kishk, production of **9**, 547
KIST (Korean Institute of Science and Technology) **12**, 392
kitchen waste, composition of **11C**, 107
Kjiedahl conversion factors, for food proteins **9**, 688
$K_L a$ value **4**, 34f, 39f
– CO_2 **4**, 9
– determination of, dynamic method **4**, 35
– indirect measurement **4**, 543f
– in multi-compartment models **4**, 321f
– oxygen **4**, 8, 35, 39
Klebsiella, in olive fermentation **9**, 605
Klebsiella aerogenes, glycerol-limited culture, growth yield **4**, 281
Klebsiella oxytoca, ethanol production **1**, 206
– 2,3-butanediol formation **6**, 270f, 273f
– – effect of initial sugar concentration **6**, 280
– – isomeric forms **6**, 273
– 2,3-butanediol productivity **6**, 276, 284
– 2,3-butanedoil yield **6**, 276
Klebsiella ozaenae, nitrilase activity **8A**, 298
Klebsiella planticola, ethanol production **1**, 204
– pullulanase **9**, 746
– strain improvement **1**, 204
Klebsiella pneumoniae **1**, 168ff
– cell immobilization **6**, 286
– *dha* regulon **1**, 206
– glucose conversion **1**, 172
– glucose dehydrogenase **1**, 178
– malonate decarboxylation mechanism **8B**, 71
– malonyl-CoA decarboxylase **8B**, 69ff
– nutrient limitation effect **1**, 172
– protein separation **3**, 602
– specific growth rate **4**, 282
– 1,3-propanediol formation **10**, 582
– 2,3-butanediol productivity, effect of dilution **6**, 285
Klebsiella terrigena, diauxic grwoth on glucose and maltose, cybernetic model **4**, 288
– – formal kinetic model **4**, 286
– – two-compartment model **4**, 287
– – unstructured model **4**, 279
Klenow enzyme *see* DNA polymerase I
Klenow fragment **5B**, 196
Klinefelter syndrome **5B**, 75
Kluyver effect, in yeast **1**, 486
Kluyveromyces fragilis, fermentation of sugars **6**, 123
Kluyveromyces lactis **1**, 470ff
– chymosin formation **9**, 680
– galactose metobilism, life cycle **1**, 473
– – regulation **1**, 494
kneading *see* dough mixing, mixers
Kneer system **11C**, 143
knockout studies **5B**, 437

Knoll process, for the manufacture of L-phenyl carbinol **8B**, 53f
knowledge-based systems (KBS) *see* expert systems
KOCH's postulates, application to biodeterioration **10**, 268f
KOGERA (Korea Genetic Research Association) **12**, 388ff
kojic acid, biosynthetic pathway **6**, 364
kojic acid production, by *Aspergillus flavus* **6**, 372
– by *Aspergillus oryzae* **6**, 372
Koji process **6**, 319f
– in miso production **9**, 518ff
– in soy sauce production **9**, 515f
Kolmogoroff's theory **4**, 115
Kolmogorov microscale **3**, 229
Korea, bio-products **12**, 388
– biotechnology **12**, 386ff
– – associations **12**, 394f
– – industry **12**, 387f
– – patents **12**, 395
– – universities **12**, 394
– biotechnology regulatory policy **12**, 388ff
– education **12**, 386f
– structure of S&T **12**, 391
Korea Academy of Industrial Technology **12**, 393
Korea Ginseng and Tobacco Research Institue **12**, 393
Korea Institute of Energy Research **12**, 393
Korea Research Institute of Chemical Technology **12**, 393
Korup Project, Cameroon **12**, 439, 443
KOSEF (Korea Science and Engineering Foundation) **12**, 392
koumiss, definition **9**, 394
Kozak sequence **5A**, 402
kraft pine pulp, bleaching of, laccase-mediator treatment **10**, 537
Krebs cycle *see* tricarboxylic acid (TCA) cycle
– microbial metabolism **9**, 184
krestin, antitumor polysaccharide **7**, 509
Kuehneromyces mutabilis **1**, 534f
kuehneromycins **7**, 511f
– structure **7**, 511
kurdi **9**, 541
kushik, production of **9**, 547
Kyoto Encyclopedia of Genes and Genomes (KEGG) **5B**, 290, 408
kyotorphin, large-scale production **8A**, 252
kyotorphin production **3**, 453

L

labdane diterpenes *see* diterpenes
labeling chemistry **5B**, 272
laboratory composting, heat loss control **11C**, 53ff

– material composted – **11C**, 52ff
– performance of **11C**, 52ff
– strategies of **11C**, 50f
– studies on **11C**, 49ff
– temperature regime **11C**, 52ff
– thermal balance **11C**, 85
– thermogenesis in, use of an adiabatic apparatus **11C**, 49ff
laccase, apple juice stabilization **9**, 713
– bleaching of paper pulp **10**, 536
– mediated oxidation of lignin, scheme **10**, 531
laccase-mediator system **10**, 537f
laccases, depolymerization of coal and coal humic substances **10**, 168f
– depolymerization of lignin **11B**, 161f, 230
– modification of lignin **10**, 530f
Lacea® **10**, 433
lacquers, biodeterioration of **10**, 311
β-lactam antibiotics **7**, 247ff
– biosynthesis of carbapenem **7**, 65
– biosynthetic clusters **7**, 21
– biosynthetic enzymes **7**, 253ff
– – compartmentalization of – **7**, 261f
– biosynthetic genes **7**, 251
– – cloning of – **7**, 259
– – clustering of – **7**, 259ff
– – DNA sequences **7**, 260
– – orientation of transcription **7**, 261
– – regulation of gene expression **7**, 265ff
– biosynthetic pathway, cephalosporin C **7**, 252
– – evolution of – **7**, 262f
– – penicillin G **7**, 252
– genetic transformation systems, for fungi **7**, 253
– – for prokaryotes **7**, 253
– history **7**, 249f
– penicillin biosynthesis, enzymes **7**, 253ff
– producing organisms **7**, 250, 253
– use of recombinant DNA technology **7**, 264f
β-lactamase, crystal structure **5A**, 78
β-lactam, enzymatic synthesis of **8B**, 360f
β-lactam antibiotics, commercial synthesis, use of lipases **8A**, 147f
– manufacture **8A**, 15ff
lactam hydrolysis, by microbial acylases **8A**, 257
β-lactams, amide hydrolysis of **8A**, 19
– ester hydrolysis of **8A**, 19
lactase **9**, 680f
lactate, export, in *Lactococcus* **1**, 350
– formation by *Clostridium acetobutylicum* **6**, 239
– metabolization, in *Lactobacillus* **1**, 352
– production volumes 1910–1929 **8A**, 9
lactate dehydrogenase, purification **3**, 600
lactate dehydrogenase (LDH), of yeast **8A**, 372
D-lactate dehydrogenases **8A**, 442f
L-lactate dehydrogenases **8A**, 442
lactate monitoring **4**, 89
lactic acid, break-even sugar price **6**, 52
– conversion cost **6**, 50

– extraction **3**, 561f
– – equilibria **3**, 570
– – re-extraction degrees **3**, 572, 575
– – with Cyanex 923 **3**, 575
– – with Hostarex A327 **3**, 574
– fermentative production of **10**, 425
– formation, in sour doughs **9**, 293, 297
– – in wine making **9**, 477
– industrial production **6**, 297ff
– – scheme **6**, 299
– market prices **6**, 51
– partition coefficient, in liquid-liquid extraction **3**, 561
– production figures **6**, 303
– production of technical polymers **10**, 425
– product recovery **6**, 299f
– purification **6**, 299f
– qualities, food grade – **6**, 303
– – pharmacopoeia grade – **6**, 303
– – plastic grade – **6**, 303
– – technical grade – **6**, 303
lactic acid bacteria **1**, 325ff
– arginine diphydrolase pathway **1**, 358
– as starter cultures, for vegetable fermentations **9**, 651ff
– bacteriocins **1**, 358ff
– bacteriophages **1**, 566
– bioenergetics **1**, 350f
– biotopes **1**, 328ff
– coccoid **6**, 295
– cultivation **1**, 347ff
– evolution **1**, 328
– for cultured dairy products **9**, 394ff
– for cultures dairy products, growth and inhibition **9**, 395ff
– formation of, bacteriocins **9**, 652
– – nisin **9**, 652
– for vitamin determination **1**, 364
– functional classification **1**, 328, **6**, 295
– galactose metabolization **1**, 353f
– genetically marked cultures **9**, 653
– genetic engineering **1**, 360ff, **9**, 399
– genetic exchange **2**, 56
– genetics **1**, 360ff
– growth kinetics **9**, 654f
– growth requirements **1**, 330
– heterofermentative **6**, 294f
– hexose fermentation **1**, 353
– homofermentative **6**, 294f
– – *Lactobacillus delbrueckii* **6**, 296
– in animals **1**, 330
– in biological research **1**, 328
– in cabbage fermentation **9**, 642
– in cheese making **9**, 360f, 369ff
– – culture propagation **9**, 371ff
– – functions **9**, 370
– industrial application **3**, 307
– inhibitory compounds in olives **9**, 617

– in historical food preparations **1**, 327f
– in man **1**, 326
– in modern food **1**, 327, 362f
– in olive fermentations, heterofermentative **9**, 618f
– – homofermentative **9**, 618f
– in peanut milk fermentation **9**, 550
– in plants **1**, 330
– in silage **9**, 776ff
– in soybean milk fermentation **9**, 550
– in table olive processing **9**, 617ff
– isolation **1**, 347ff
– lactose metabolization **1**, 354
– malolactic fermentation **9**, 490ff
– mesophilic **6**, 294
– metabolism **1**, 350ff
– – amino acid – **1**, 354
– – carbohydrate – **1**, 350ff
– – oxygen – **1**, 352
– occurrence **1**, 328ff
– phage resistance **1**, 566
– phage-resistant strains **6**, 296
– rod-shaped **6**, 295
– specific isolation media **1**, 347ff
– taxonomy **1**, 328ff
– thermophilic **6**, 294
– vegetable fermentations **9**, 635
– wine spoilage by – **9**, 493
lactic acid fermentation **3**, 306f, **6**, 293ff
– ammonium lactate process **6**, 300ff
– by *Bacillus coagulans* **6**, 303
– by *Lactobacillus delbrueckii* **6**, 296ff
– by *Rhizopus oryzae* **6**, 302
– by *Sporolactobacillus inulinus* **6**, 303
– classical calcium lactate processes **6**, 298ff
– continuous fermentation processes **6**, 302
– fermentation parameters **6**, 299
– history **6**, 294
– mineral nutrients **6**, 298
– nitrogenous nutrients **6**, 298
– of soybeans **3**, 307
– phage infections **6**, 296
– production strains **6**, 295f
– raw materials **6**, 296ff
– – glucose **6**, 297
– – maltose **6**, 297
– – sucrose **6**, 297
– – whey **6**, 297
– reactors **6**, 298
– use of immobilized cells **6**, 302
lactic acid production, by filamentous fungi **1**, 523
lactic acid tone, of wine **9**, 494
lactic dehydrogenase, *in vitro* folding **5A**, 118
lactic fermentation **1**, 327ff
– history of exploration **1**, 328
– starter culture **1**, 328
lacticin 481, primary structure **7**, 332

– structure **7**, 335
lactobacilli, in pre-ferments **9**, 277
– in sour doughs **9**, 293ff
– in sour dough starters **9**, 255ff
Lactobacillus **1**, 329ff
– acid tolerant – **1**, 335
– facultative heterofermentative – **1**, 335ff
– food preserve spoilage **1**, 335
– genetic engineering **2**, 489
– growth conditions **1**, 336ff
– habitats **1**, 330ff
– in beer **1**, 334
– in diet **1**, 331
– in meat **1**, 331
– in milk products **1**, 332
– in sausage fermentation **3**, 311
– in the human vagina **1**, 331
– in the intestinal tract **1**, 330
– in the oral cavity **1**, 330
– in the stomach **1**, 331
– isolation media **1**, 347ff
– lactic acid production strains **6**, 296
– obligately heterofermentative – **1**, 335ff
– obligately homofermentative – **1**, 335ff
– pathogenicity **1**, 331
– species characteristics **1**, 336ff
– substrates **1**, 336ff
– wine production **1**, 334
Lactobacillus acidophilus, as direct feed additives **9**, 781
– in cultured dairy products, characteristics **9**, 397
– in sour doughs **9**, 293f
Lactobacillus amylophilus **6**, 296f
– in lactic acid production **6**, 296
Lactobacillus amylovorus **6**, 296f
– in lactic acid production **6**, 296
Lactobacillus brevis **1**, 333, 335
– in kishk production **9**, 548
– in sour doughs **9**, 256f, 293f
– in wine spoilage **9**, 496
– malolactic fermentation **9**, 491
Lactobacillus buchneri, in sour doughs **9**, 294
Lactobacillus casei, in cultured dairy products, characteristics **9**, 397
– in kishk production **9**, 548
– in sour doughs **9**, 293f
– in tempeh fermentation **9**, 530
– malolactic fermentation **9**, 491
Lactobacillus delbrueckii **6**, 296ff
– in cheese making **9**, 370
– in cultured dairy products, characteristics **9**, 397
– in idli fermentation **9**, 539
– in lactic acid production **6**, 296
– in olive fermentation **9**, 607
– in poi production **9**, 548
– in sour doughs **9**, 293f

Lactobacillus delbrueckii ssp. *bulgaricus* **1**, 332, 335
Lactobacillus fermentum **1**, 333
– in idli fermentation **9**, 539
– in sour doughs **9**, 293f
Lactobacillus fructivorans, in sour doughs **9**, 293f
Lactobacillus helveticus **1**, 333, 335
– in cheese making **9**, 369f
Lactobacillus kefir alcohol dehydrogenase (*LKADH*), applications of **8A**, 454
– characteristics of **8A**, 454
– synthesis of chiral hydroxy compounds **8A**, 454
Lactobacillus lactis, in cheese making **9**, 369f
– in cultured dairy products, chracteristics **9**, 396
– in idli fermentation **9**, 539
– in poi production **9**, 548
– in sour doughs **9**, 293f
– pediocin formation **9**, 652
Lactobacillus pastorianus, in poi production **9**, 548
Lactobacillus pentacetius, in poi production **9**, 548
Lactobacillus plantarum **1**, 333
– in cabbage fermentation **9**, 642
– in cucumber fermentation **9**, 636ff
– in gari fermentation **9**, 543
– in kimchi fermentation **9**, 647f
– in kishk production **9**, 548
– in ogi fermentation **9**, 545
– in olive fermentation **9**, 606f, 615
– in sour doughs **9**, 293f
– silage inoculants **9**, 776f
Lactobacillus rhamnosus, in sour doughs **9**, 293f
lactocin S, prepeptide structure **7**, 345
– primary structure **7**, 332, 335
– structural gene **7**, 344
lactococcin, prepeptide structure **7**, 345
– structural gene **7**, 344
lactococcin A, nucleotide sequence **1**, 359f
Lactococcus **1**, 329, 343
– DNA transfer **1**, 361
– energy consumption **1**, 351
– energy generation **1**, 351
– gene expression **1**, 361f
– genetic engineering **2**, 489
– habitats **1**, 343
– isolation media **1**, 348
– lactate gradient **1**, 351
– membrane potential **1**, 351
– peptidase **1**, 355f
– pH gradient **1**, 351
– plasmid coded functions **1**, 361
– plasmid profiles **1**, 361
– proton motive force **1**, 351
– species chracteristics **1**, 343
– substrates **1**, 343
Lactococcus lactis **1**, 328, 339, 343
– cell wall proteinase, genes **1**, 355

– – properties **1**, 355
– – specificity **1**, 354f
– molecular genetics **1**, 360
– nisin production **7**, 330
– prolyl dipeptidyl-aminopeptidase **8A**, 256
lactonase, optical resolution of pantolactone **10**, 326
lactone precursors, lipase-catalyzed resolution **8A**, 114
lactones, flavor active –, production of **8B**, 340ff
– lipase-catalyzed enantioselective reactions **8A**, 112f
lactones (reacemic-), kinetic resolution by horse liver esterase **8A**, 209f
lactonohydrolase, stereospecific hydrolysis of lactones **10**, 325f
lactose, determination with biosensors n4, 78
– metabolization, by improved cultures **1**, 195
– – inlactic acid bacteria **1**, 354
– microbial metabolism **9**, 185
lactose operon **2**, 250ff
– dual control **2**, 250f
– transfer from *Escherichia coli* **1**, 195f
lag phase, requirement for structured models **4**, 285
lakes, treatment of water from **11C**, 381ff
– – degree of trophication **11C**, 382
– – process schemes **11C**, 382
Lambert-Beer's law **4**, 85, 183
lamellar settler **3**, 476
land disposal, of solid wastes **11C**, 195ff
landfill, base sealing system **11C**, 180
– engineering features, profile of **11C**, 196ff
– waste pretreatment **11C**, 175ff
– – landfill simulation experiments **11C**, 175ff
– – scheme **11C**, 175
– – simulation experiments **11C**, 175ff
landfill control, biostabilization **11C**, 178
– water input **11C**, 178
landfilling, of contaminated soil **11B**, 315
landfill reactor, control of **11C**, 178
landfills *see also* sanitary landfills
– age of influence on leachate composition **11C**, 173
– – deposition **11C**, 186
– biological waste disposal **3**, 782
– – public health implications **3**, 784
– classes of **11C**, 174
– cleaning of gases, use of biofilters **11C**, 330
– common configurations **11C**, 198
– environmental monitoring **11C**, 181
– leachate control **11C**, 174
– leachate discharge control **11C**, 180
– leachates, long-term problems with **11C**, 182ff
– – nitrogen concentration **11C**, 187
– – problems of **11C**, 173f
– leachate treatment **11C**, 187ff
– – biological – **11C**, 188

– – chemical-physical – **11C**, 189
– long-term emissions **11C**, 182
– LSR leachate concentrations **11C**, 183
– production of **11C**, 181f
– simulation reactor **11C**, 176ff
landfill simulation reactors (LSR) **11C**, 176ff
– leachate concentrations **11C**, 184
– lysimeter tests **11C**, 182ff
– test for leachate parameters **11C**, 176ff
landfill system **11C**, 167ff
landmarking **5B**, 328
lane detection **5B**, 211
lane tracking **5B**, 211f
Langerhans islets *see* pancreatic islets
Langmuir isotherm, application to sorption processes **11B**, 54
lanolin **6**, 672
lanosterol, structure **7**, 171
– structure of **8B**, 28
lanosterol cyclase **8B**, 382
lanosterol cyclization, mediated by yeast sterol cyclase **8A**, 378
lanthionine, biosynthesis **7**, 328f
– biosynthetic pathway **7**, 329
– non-protein amino acid **7**, 347
– structure **7**, 327
lantibiotic **1**, 359
lantibiotic biosynthesis, epidermin **7**, 342
– genetic regulation **7**, 341ff
– leader peptide, role of – **7**, 346f
– modification events, sequence of – **7**, 353
– nisin, gene clusters **7**, 343
– regulation, by proteins **7**, 350f
– structural genes **7**, 341ff
lantibiotic prepeptides, leader peptide **7**, 345ff
– primary structure **7**, 345
– structural genes, transcription of – **7**, 341ff
lantibiotics **7**, 323ff
– – *see also* antibiotics
– actagardine **7**, 337f, 357
– ancovenin **7**, 336f
– applications of – **7**, 358f
– – as food/beverage preservatives **7**, 358
– – medical/paramedical – **7**, 359
– – veterinary – **7**, 359
– biological activities **7**, 353ff
– carnocin UI49 **7**, 336
– chemical characteristics **7**, 326
– chemistry of – **7**, 325ff
– cinnamycin **7**, 336f, 340, 357
– compilation of – **7**, 326f
– cytolysin/bacteriocin **7**, 335f
– definition **7**, 325
– disruption of energy transduction **7**, 355
– duramycins **7**, 336f, 340, 357
– effect on cytoplasmic membrane of susceptible bacteria **7**, 354
– epidermin **7**, 332ff

– epilancin K7 **7**, 332ff
– gallidermin **7**, 333, 339
– lacticin 481 **7**, 332, 335
– lactocin S **7**, 332, 335
– leader peptidases **7**, 348ff
– mersacidin **7**, 337f, 357
– modified amino acids in – **7**, 327ff, 347f
– – D-alanine **7**, 330
– – didehydroalanine **7**, 328
– – didehydrobutyrine **7**, 328
– – lanthionine **7**, 328
– – methyllantionine **7**, 328
– – structures **7**, 327
– mutacin **7**, 336
– nisin **7**, 330ff, 339
– non-disulfide cyclic peptides **7**, 281
– Pep 5 **7**, 332ff, 340
– physical characteristics **7**, 326
– presursor peptide **7**, 328
– producing organisms **7**, 326
– ribosomal biosynthesis **7**, 325
– salivaricin A **7**, 332, 335
– streptococcin A-FF22 **7**, 334
– subtilin **7**, 331ff, 339
– transport proteins, genes **7**, 348
– type A, induced pore formation **7**, 356
– – induction of autolysis **7**, 356
– – modes of action **7**, 353ff
– – primary structures **7**, 330ff
– – solution structures **7**, 339f
– type B, mode of action **7**, 357f
– – primary structures **7**, 336ff
– – role in human immune and blood pressure regulatory systems **7**, 336
– – solution structures **7**, 340f
lantibiotic synthesizing cells, self-protection mechanisms **7**, 351ff
lao-chao, production of **9**, 525
large-scale epidemiology **5B**, 422
large-scale factories **5B**, 438
Laser Doppler technique **4**, 138
laser excitation **5B**, 199, 204
laser scanning microscopy **1**, 8
laser technology **5B**, 432
laundry detergents, use of lipases **5A**, 200f
N-laurylbiotinamide, as affinity cosurfactant, for reversed micellar extractions **3**, 609
lauryl maltoside, as protein folding aid **3**, 542
LC *see* Liquid chromatography
LCDR/MerMade **5B**, 231
LDH *see* lactate dehydrogenase
LDL **5A**, 15
leachate control, in landfills **11C**, 174
leachates, from biowaste windrow composting, composition of **11C**, 132
– from landfills **11C**, 198ff
– – biological treatment of **11C**, 188
– – characteristics of **11C**, 199f

– – chemical-physical treatment of **11C**, 189
– – composition of **11C**, 169ff, 199f
– – constituents **11C**, 172
– – contaminants **11C**, 200
– – discharge control **11C**, 180
– – emission potentials **11C**, 185
– – long-term problems with **11C**, 167ff, 182ff
– – nitrogen conentration **11C**, 187
– – production of **11C**, 181
– – recirculation of **11C**, 189, 200
– – treatment of **11C**, 180f, 187ff
– in a landfill cell, developments in **11C**, 170
leaching, of contaminants in soil **11B**, 490f
– of metals **10**, 312, **11A**, 435f
– – autotrophic – **10**, 245f
– – heterotrophic – **10**, 246f
lead, as electron acceptor for anaerobic respiration **11A**, 43
– biorecipitation **11B**, 365
– removal by rhizofiltration **11B**, 392f
leader peptidases **7**, 348ff
leader peptide, of secreted proteins **3**, 531
leader peptides, in lantibiotic biosynthesis **7**, 345ff
– – conserved motifs **7**, 346f
– – roles **7**, 346f
lead structures **10**, 43
– optimization of **10**, 55
leaianafulvene, cytotoxic activity **7**, 508
– structure **7**, 506
leather, biodegradation of **10**, 307
leavening, of doughs **9**, 253ff, 272ff
leavening agents, biological –, sour dough starters **9**, 254ff
– – yeast **9**, 253f
– chemical –, baking powder **9**, 258
– – CO_2 **9**, 258
– – pottash **9**, 258
– physical – **9**, 259
lecithin, use in baked goods **9**, 261
lectin **1**, 35
Legionella, in composting material **11C**, 71
Leishmania, expression of foreign genes **10**, 147
leishmaniasis **10**, 144
Leloir glycosyltransferase **8B**, 267
lentinan **7**, 508
– as anti-tumor agent **6**, 646
lentinellic acid, antibacterial activity **7**, 501
– structure **7**, 502
Lentinus edodes **1**, 534f
leonardite, solubilization of, chelating agents **10**, 162ff
less-developed countries (LDCs) **12**, 342ff
leucine dehydrogenase **6**, 537ff
– co-immobilization **6**, 539
– in industrial biotransformations **3**, 455f
– purification **6**, 538
– reaction rate **3**, 455f

leucine dehydrogenases, application of **8A**, 445f
– characteristics of **8A**, 445
– reductive amination by **8A**, 445f
L-leucine aminopeptidase **8A**, 246f
L-leucine production **6**, 510
– by precursor fermentation **6**, 492
L-*tert*-leucine production **3**, 454f
leucine production, with leucine dehydrogenase **8A**, 446
leucomelone, iron-binding pigments **7**, 231
– structure **7**, 231
Leuconostoc **1**, 329, 340ff
– genus differentiation **1**, 342
– growth conditions **1**, 342
– habitats **1**, 341
– in olive fermentation **9**, 606
– isolation media **1**, 348
– pathogenicity **1**, 341
– psychrophilic **1**, 341
– species characteristics **1**, 342
– substrates **1**, 342
Leuconostoc mesenteroides **1**, 339, 341
– dextran production, fed-batch fermentation **6**, 638
– growth kinetics **9**, 655
– in cabbage fermentation **9**, 642
– in cheese making **9**, 370
– in cultured dairy products, characteristics **9**, 396
– in idli fermentation **9**, 539
– in kimchi fermentation **9**, 647f
– in puto fermentation **9**, 528
Leuconostoc oenos **1**, 341
– malolactic fermentation **9**, 491
Leuferon **5A**, 151
leukemia, AML **5A**, 358ff
– monoclonal antibody therapy **5A**, 320
leukemia inhibitory factor **5A**, 158
Leukine **5A**, 157
Leukoctropin **5A**, 157
leukotriene analogs **8B**, 380f
leukotriene B_4 receptor antagonists **7**, 114
leukotriene biosynthesis, inhibitors of –, from basidiomycetes **7**, 522f
levan production, beet molasses as raw material **6**, 25
Leviviridae **1**, 550
levodopa **6**, 468, 522
lexical ordering **5B**, 391
liability, in contaminated site management **11B**, 13f, 31f, 36f
licensing, of patentable inventions **12**, 320
lichens, in biodeterioration **10**, 272
life cycle, of bacteriophages **1**, 558
– of *Claviceps purpurea* **1**, 526
– of filamentous fungi **1**, 518f, 523, 526, 533
– of streptomycetes **1**, 436
– of yeasts **1**, 472ff

Li–Fraumeni syndrome **5B**, 82
LIGAND **5B**, 291, 353
ligand binding assays, in drug screening **10**, 47f
ligand-binding sites **5B**, 351
ligand receptors, antagonists **7**, 117ff
– – producers **7**, 117
ligands, radio-labeled, radioimmunoassay **7**, 110
– – screening of receptor-active compounds **7**, 110
– to enzymes **9**, 48f
ligase **1**, 66
ligase chain reaction **10**, 89ff
ligases **5A**, 25ff, **9**, 8
– acid-base catalysis **5A**, 25
– carbon-carbon bond formation **8B**, 6
– thermostable –, applications of **10**, 90
light emission, by the autoinducer of *Vibrio fischeri* **7**, 66
light microscopy **1**, 7ff
– confocal *see* laser scanning microscopy
– conventional **1**, 7
– Nomarski interference contrast **1**, 7
– phase contrast **1**, 7
– sample preparation **1**, 7
– sample visualization **1**, 7ff
lignin **11A**, 25f
– action of enzymes on **10**, 532
– biosynthesis of **10**, 524
– degradation by microorganisms **6**, 146f
– degradation of **11C**, 69f
– depolymerization of **11B**, 230
– – by ligninolytic enzymes **11B**, 161ff
– formation of **11B**, 100
– hydrolysis of **11A**, 25f
– in wood **10**, 306
– microbial degradation of **11B**, 100
– microbial metabolism **9**, 185
– modifying enzymes **10**, 529ff
– – laccase **10**, 530f
– – peroxidases **10**, 530
– – phenol oxidase **10**, 530
– non-specific oxidation by radicals **11B**, 230
– organization of **10**, 531
– structure **6**, 144, **11B**, 100, 230
lignin degradation, by filamentous fungi **1**, 534f
lignin peroxidase **10**, 530
– biodegradation of nitroaromatics **11B**, 289f
– degradation of lignin **11B**, 161f, 230
– depolymerization of lignite **10**, 165f
– generation of radicals **11B**, 232
– gene regulation **1**, 534f
– oxidation of benzo(a)pyrene **11B**, 231
– – mechanism **11B**, 232
– reductive dehalogenation **11B**, 246
lignite *see also* brown coal, coal
– biotechnologically liquefied –, for production of bioplastics **10**, 170ff

– depolymerization, by manganese peroxidase **10**, 167
– – by peroxidases **10**, 165ff
– depolymerization of **10**, 160f
– modification by soybean hull peroxidase **10**, 168
– solubilization of **10**, 160ff
– – by micro-fungi **10**, 161
– – involvement of hydrolases **10**, 164f
– – non-enzymatic effectors **10**, 162ff
– utilization of **10**, 160f
lignite depolymerization products, bacteria growing on –, isolation of **10**, 172
– – screening of **10**, 171
– polyhydroxyalkanoic acid synthesis from, by *Pseudomonas oleovorans* **10**, 171f
– – by *Rhodococcus ruber* **10**, 171f
lignocellulose, anaerobic degradation of **11A**, 30f
– degradation by actinomycetes **11C**, 69
– in industrial ethanol production **6**, 143ff, 166ff
– milling **6**, 145
– solvent pretreatment **6**, 145f
– steam explosion **6**, 145
– – effect on enzymatic hydrolysis **6**, 146
– structure **6**, 144
– swelling **6**, 146
lignolytic enzymes **11B**, 161ff, 230ff
– oxidation of polycyclic aromatic hydrocarbons **11B**, 230ff
lignozyme process **10**, 537f
limonene **10**, 378f, 390ff
– biotransformation of **10**, 378, 379
– microbial conversion of **10**, 391ff
– – main products **10**, 391ff
– – pathways **10**, 391f
– – to α-terpineol **10**, 391ff
– – to carvone **10**, 393ff
– – to perillic acid **10**, 393ff
– – to perillyl alcohol **10**, 393f
linalool, bioconversion by *Aspergillus niger* **10**, 390
– biotransformation of **10**, 378
lincomycin A, biosynthetic gene, "sugar" subcluster **7**, 453
– hypothetical biosynthetic pathway **7**, 452
lincosamides **7**, 451ff
– structure **7**, 473
– sugar containing moieties **7**, 451
lindane, dehydrochlorination of **8B**, 206
– dehydrohalogenation of **8B**, 205f
lindane dechlorination **1**, 313
linear dipeptides, use in stereoselective synthesis **8B**, 508ff
linear gas velocity **3**, 408f
– gas volume flow **3**, 409
– in bioreactors **3**, 412
Lineweaver-Burk method **4**, 458
Lineweaver-Burk kinetics **1**, 69

Lineweaver-Burk plot **9**, 35
linkage analysis **5B**, 77, 108
linkage disequilibrium **5B**, 78f, 111f
linkage groups **5B**, 43
linkage mapping **5B**, 90, 108, 168ff
linkage maps **5B**, 175
linkage testing **5B**, 113
linkers, in multimodular proteins, polyketide synthases **10**, 362ff
γ-linoleic acid, production of **10**, 335
γ-linolenic acid (GLA), in microalgae **7**, 167
– medicinal uses **7**, 159
– production by phycomyceters **7**, 159
linseed oil, genetic engineering of – **7**, 138
lipase **10**, 400
– bleaching of paper pulp **10**, 536
– from pseudomonads **1**, 421
– in detergent formulations **1**, 421
– stereoselectivity **1**, 421
lipase-catalyzed acylations, of hydroxyl groups **8A**, 126ff
– of modified sugars **8A**, 121ff
– of unmodified sugars **8A**, 119ff
– use of acid anhydrides **8A**, 58
lipase-catalyzed amidation **8A**, 265
lipase-catalyzed aminolysis **8A**, 263ff
– formation of acyl-enzyme intermediate **8A**, 266
– formation of amides **8A**, 263ff
– of diethyl fumarate **8A**, 267
– of succinic acid diesters **8A**, 267
lipase-catalyzed asymmetric syntheses, by PCL **8A**, 67f
– by PPL **8A**, 67f
– enantiomeric purity **8A**, 66ff
lipase-catalyzed deacylations, of aliphatic hydroxyls **8A**, 130f
– of amines **8A**, 131f
– of hydroxyl groups **8A**, 124ff
– of phenolic hydroxyls **8A**, 129
– of polyhydroxylated benzenes **8A**, 129
lipase-catalyzed enantioselective reactions, involving the lactone ring **8A**, 112f
– *myo*-inositol derivatives **8A**, 87
– of acyclic alcohols **8A**, 79ff
– of alcohols, with remote stereocenters **8A**, 97
– of axially-disymmetric alcohols **8A**
– of carboxylic acids **8A**, 103ff
– – with a stereocenter at the β-position **8A**, 108
– – with quanternary stereocenters **8A**, 109
– – with remote stereocenters **8A**, 110
– – with sulfur stereocenters **8A**, 110
– of cyclic secondary alcohols **8A**, 82ff
– of primary alcohols **8A**, 89ff
– – with quaternary setereocenters **8A**, 95
– of secondary alcohols, RML **8A**, 85
– of spiro alcohols **8A**, 96
– of tertiary alcohols **8A**, 94
– of thiols **8A**, 102

lipase-catalyzed reactions, asymmetric synthesis **8A**, 66ff
– β-blockers **8A**, 84
– catalytic mechanism **8A**, 45, 47
– enantiomeric purity **8A**, 63ff
– formation of macrolactones **8A**, 115
– in enzyme reactors **3**, 453
– *in situ* recycling **8A**, 64
– of ferrocenes **8A**, 100
– of triacylglycerides **8A**, 135
– polymerizations, condensation reactions **8A**, 144
– – ring-opening polymerizations **8A**, 145
– – transesterification reactions **8A**, 144f
– reaction media, aqueous solutions **8A**, 47ff
– – organic solvents **8A**, 50ff
– – reverse micelles **8A**, 48f
– – supercritical fluids **8A**, 60ff
– ring opening of 2-phenyloxazolin-5-ones **8A**, 116
– sequential kinetic resolution **8A**, 64
– synthesis of, diacylglycerols **8A**, 138
– – MLM lipids **8A**, 136f
– – monoacylglycerides **8A**, 140f
– use of acyl donors **8A**, 56ff
– water content of reaction mixture, thermodynamic water activity **8A**, 59
– – total water content **8A**, 58f
lipase CLECs **8A**, 52
lipase enantiopreference, empirical rules **8A**, 69, 88
– size-based rules **8A**, 68
lipase hydrolysis, in aqueous media **8A**, 47f
– in organic solvents **8A**, 48ff
– in reverse micelles **8A**, 48f
lipases **5A**, 199ff, **8A**, 28f, 37ff, 250, **8B**, 368ff, 385f, **9**, 187
– acyl chain binding site **8A**, 47f
– alcohol binding site **8A**, 47f
– availability **8A**, 40ff
– bacterial – **8A**, 42f
– biology **5A**, 199
– catalytic amino acid triad **8A**, 70
– characteristics **5A**, 200
– chemoselective reactions **8A**, 118ff
– classification, by protein sequence **8A**, 43
– commercial –, protein sequence alignments **8A**, 43
– – source and names **8A**, 42
– – world market **8A**, 41
– commercial applications, enantiomerically pure chemical intermediates **8A**, 146
– – enantiomerically pure pharmaceutical intermediates **8A**, 146f
– – food ingredients **8A**, 146
– commercial synthesis of diltiazem **8B**, 279
– crude preparations **8A**, 43f
– developments **5A**, 202

– directed evolution, random mutagenesis **8A**, 149f
– – screening **8A**, 149f
– enantiomeric ratio **8A**, 63
– enantioselective reactions involving the lactone ring **8A**, 112f
– enantioselectivity toward triglycerides **8A**, 91ff
– entrapment of **8A**, 51f
– fatty acid selectivity, saturated fatty acids **8A**, 142ff
– – unsaturated fatty acids **8A**, 143f
– function **5A**, 199f
– fungal – **8A**, 42f
– immobilization **3**, 453
– in cheese making **8A**, 40f
– in commercial synthesis of diltiazem **8A**, 147
– increase of enantioselectivity in organic solvents **8A**, 54ff
– indirect resolutions of lactones **8A**, 114
– industrial applications **5A**, 199ff
– in industrial biotransformations **3**, 453, 458
– in laundry detergents **5A**, 200f, **8A**, 40f
– in organic solvents, catalytic activity **8A**, 51ff
– – CLECs **8A**, 52
– – covalent immobilization **8A**, 52
– – covalently modified – **8A**, 53
– – enantioselectivity **8A**, 54ff
– – lipid-coated – **8A**, 53f
– – recovery from **8A**, 51
– – surfactant-coated – **8A**, 53f
– in peptide synthesis **8A**, 262
– in supercritical fluids **8A**, 60ff
– interfacial activation of **8A**, 40, 45f
– in tricyclic β-lactam antibiotic synthesis **8A**, 147f
– large-scale application, optimization of **8A**, 148f
– mammalian – **8A**, 42f
– nomenclature **8A**, 41ff
– nonselective – **8A**, 134ff
– of *Mucor* sp. **8B**, 385f
– of *Pseudomonas* sp. **8B**, 385
– primary alcohols as substrates for **8A**, 87ff
– purification **3**, 611
– reaction catalyzed by – **5A**, 200
– recombinant – **5A**, 200
– regioselective reactions **8A**, 118ff
– regioselective ring opening of anhydrides **8A**, 111
– resolution of amines **8A**, 101
– resolution of metal carbonyl complexes **8A**, 99
– secondary alcohols as substrates for **8A**, 68ff
– solvent-free ester synthesis **10**, 400ff
– sources **5A**, 199f
– stability **5A**, 202
– stability in anhydrous solvents **8A**, 262
– structure, α/β-hydrolase fold **8A**, 45
– supercritical CO_2 reactor **8A**, 62

– synthesis of, cocoa butter equivalents **8B**, 289
– – pharmaceutical precursors, at kilogram-scale routes **8B**, 284
– – structured triglycerides **8B**, 289
– synthesis of monoterpenyl esters **10**, 397ff
– use in baking **9**, 675
– use of household detergents, effect of wash cycles **5A**, 201
– X-ray crystal structures **8A**, 46
– 1,3-selective – **8A**, 134ff
lipase selectivity, modeling of **8A**, 149
– mutating of **8A**, 149
lipid accumulation, biochemistry **7**, 143ff
– culture conditions **7**, 140ff
– effect of carbon-to-nitrogen ratio **7**, 141
– efficiency **7**, 142ff
– – in molds **7**, 142f
– – in yeasts **7**, 142f
– in continuous culture **7**, 141
– patterns **7**, 139ff
– – in molds **7**, 140
– – in yeasts **7**, 140ff
– yields, on carbohydrates **7**, 142
– – on ethanol **7**, 142
lipid-coated lipases **8A**, 53f
lipid emulsions, for drug delivery **6**, 680
lipid-linked proteins **5A**, 16ff
lipid modification **8B**, 289
lipid modifications, lipase-catalyzed – **8A**, 134ff
– 1,3-regioselective reactions of glycerides **8A**, 134f
– – acyl migration in mono- and diacylglycerides **8A**, 139
– – cocoa butter substitutes **8A**, 135f
– – diacylglycerides **8A**, 137ff
– – esterification of glycerol with fatty acids or fatty acid esters yielding 1(3)-MAGs **8A**, 142
– – glycerolysis of triglycerides to 1(3)-MAGs **8A**, 141
– – hydrolysis or alcoholysis of triglycerides to 2-MAGs **8A**, 141
– – modified triglycerides **8A**, 135
– – monoacylglycerides (MAGs) **8A**, 139
– – synthesis of MLMs **8A**, 136f
– – triacylglycerides containing polyunsaturated fatty acids (PUFAs) **8A**, 137
lipid monitoring, by staining methods **4**, 212
lipids **7**, 133ff
– anaerobic degradation of **11A**, 34f
– biosurfactants **7**, 180f
– chlorinated – **8B**, 186
– classification **7**, 138f
– enzymatic modification **9**, 728
– ether lipids **7**, 181ff
– fatty acid profile, in molds **7**, 157f
– – in yeasts **7**, 148f
– microbial metabolism **9**, 186ff
– nomenclature **7**, 138f

– phospholipids **7**, 183f
– prodcutin of structures – **8B**, 289
– prostanoid-type – **7**, 184f
– sphingolipids **7**, 183
– substrates for biomass production **9**, 213
lipid synthesis, by filamentous fungi **1**, 523f
– in yeast **1**, 490f
LIPI (R&D Center for Biotechnology of Indonesia) **12**, 412
lipoamide **9**, 34
lipocalins **5A**, 29
lipoglycopeptides, fatty acids, origin **7**, 386
– naturally occurring – **7**, 375
– producing organisms **7**, 380
lipoic acid, chemical production of **10**, 331
lipopolysaccharide-binding protein **5A**, 135
lipoproteins **5A**, 14ff
– chylomicrons **5A**, 15
– function **5A**, 14
– high density – (HDL) **5A**, 15
– low density – (VDL) **5A**, 15
– non-covalent aggregates **5A**, 14
– of human plasma, major classes of **5A**, 15
– very low density – (VLDL) **5A**, 15
liposomes, as transfection systems **5A**, 432ff
Lipothrixviridae **1**, 550
lipoxygenases **8B**, 363ff
– catalysis of hydroxylations **8A**, 502f
– definition of **8A**, 502ff
– mushroom – **8A**, 503f
– porcine – **8A**, 503f
– potato tuber – **8A**, 503
– soybean – **8A**, 502f
– use in baking **9**, 675
liquefied starch **9**, 668ff, 744f
– for sweetener production **9**, 748ff
– saccharification **9**, 669ff, 745ff, 750
liquid chromatography (LC) **4**, 611
– for bioproduct separation **4**, 614f
liquid circulation, flow visualization **4**, 127
– time **4**, 123f
liquid flow control **4**, 514
liquid flow properties, in bioreactors **4**, 138ff
– – determination by heat pulse technique **4**, 139
– – determination by hot film anemometers **4**, 138
– – determination by laser Doppler technique **4**, 138
– – mechanical fluctuation probes **4**, 139
liquid handling systems, miniaturized – **10**, 52
liquid level, instruments for determination of **4**, 10f
liquid-liquid extraction **3**, 558ff, 593ff
– aqueous phase **3**, 558
– continuous countercurrent –, mass balance **3**, 559
– countercurrent –, decanter scheme **3**, 565
– equilibrium solute concentration **3**, 559

– equipment **3**, 559
– extraction factor **3**, 558
– fat-soluble product **3**, 565
– hollow fiber modules **3**, 605, 611
– organic phase **3**, 558
– solvent selection **3**, 561
– – for ethanol **3**, 560
– techniques **3**, 558
liquid-liquid extraction extractants *see* solvents
liquid manure *see* manure
liquid membrane extration (LM), emulsion LM (ELM) **3**, 581
– solid supported LM (SLM) **3**, 581
– technique **3**, 581
liquid pre-ferment process **9**, 274ff
– flow sheet **9**, 276
liquid residence time **4**, 123f
– in bioreactors **4**, 123
liquids *see* fluids
liquid-solid separation, fixed film reactor **11A**, 120
liquid throughput, instruments for determination of **4**, 11
– measurement **4**, 11
liquid viscosity, flow equation **4**, 13f
– instruments for determination of **4**, 13f
– measurement **4**, 13f
– shear diagram **4**, 13f
liquid volume, measurement **4**, 10
Lithospermum erythrorhizon, accumulation of shikonins **7**, 601f
– commercial shikonin production **1**, 589, 592f, 604
lithotrophs **3**, 128
lithotrophy **1**, 50
– definition of **10**, 275
livestock facility, odor control in **11C**, 363
livestock farming, wastes **11A**, 219ff
– – characteristics **11A**, 220f
– – manure **11A**, 220
– – quantities **11A**, 220f
LKADH see Lactobacillus kefir alcohol dehydrogenase
local area networks, in automation systems **4**, 577f
lock and key hypothesis **5A**, 22
locus heterogeneity **5B**, 65, 68, 110
LOD scores **5B**, 108
logical groups **5B**, 380
London energy **5A**, 89
look back time **5B**, 132
loop reactor, determination of *OUR* **4**, 36f
loracarbef, formation by immobilized penicillin G acylase **8A**, 259f
lovastatin **3**, 9
– biosynthesis of **10**, 345ff
– – hypothetical pathway **10**, 347
– chemical structure **10**, 342

lovastatin diketide synthase **10**, 346
lovastatin nonaketide synthase **10**, 346
lowfat cheeses **9**, 379
Lowry assay **8A**, 43
low-temperature enzymology **9**, 20
LSR *see* landfill simulation reactors
luciferase **5B**, 210
– bacterial – **11B**, 453
– – oxidation of sulfide to sulfoxide **8A**, 574
– bioluminescence reaction **10**, 469f
Luedeking-Piret equation **4**, 283, 491
Luenberger observer, principle **4**, 469
lupinic acid amide, asymmetric hydrolysis **8A**, 246
L-lupinic acid **8A**, 245f
lupus erythematodes, antibody therapy **5A**, 314, 319
Lutefisk **5B**, 263
luteinizing hormone **5A**, 164
lyase **1**, 66, **2**, 16
– AP lyase **2**, 16
lyases **5A**, 25f, 26, **8B**, 41ff, **9**, 8
– acting of poly-D-galacturonates **8B**, 123
– acting on polysaccharides **8B**, 122ff
– bacterial production of **8B**, 321ff
– – pathway engineering **8B**, 321ff
– classes **8B**, 44ff
– decarboxylation of, by lysine decarboxylase **8B**, 80
– definition **8B**, 43
– in industrial biotransformations **3**, 457ff
– in the citric acid cycle **8B**, 48
– nomenclature **8B**, 43
– systematic names **8B**, 6
Lyme disease, vaccine **5A**, 138, 140
lymphomas, monoclonal antibody therapy **5A**, 321
lyophilization *see* freeze-drying
lyoprotectants **3**, 703ff
lysate clarification, solid filtering aids **9**, 92f
– with CTAB **9**, 93f
– with PEI **9**, 93
– with streptomycin sulfate **9**, 94
lysergic acid, biosynthesis **1**, 530
lysine, biosynthetic pathway **1**, 207
– secretion **1**, 209
lysine biosynthesis **2**, 485ff
– feedback inhibition prevention **2**, 487
– gene cloning methods **2**, 487
– overproduction-affecting genes **2**, 488
– pathway **2**, 486ff
lysine decarboxylase **8B**, 79f
L-lysine, aspartate derived amino acids, in coryne-bacteria **8B**, 323
– biosynthesis **6**, 475ff
– – genes **6**, 481f
– – in *Corynebacterium glutamicum* **6**, 476
– – in *Escherichia coli* **6**, 485

– biosynthetic pathway **6**, 509
– feed uses **6**, 468
L-lysine production **3**, 449, **6**, 475ff, 507f
– beet molasses as raw material **6**, 25
– downstream processing **6**, 477
– fermentation technology **6**, 477
– from racemic ACL **6**, 514
– mechanisms **6**, 507
– strain development **6**, 476f
L-lysine production, pathway control **1**, 206
lysogen *see* prophage
lysozyme **2**, 103f, **3**, 517, **10**, 491

M

mAb *see* monoclonal antibodies
mabinlin, proteinaceous sweetener **10**, 577f
machine learning **5B**, 327
macroalgae *see also* algae
– antiinflammatory compounds of **10**, 456
– antiviral compounds of **10**, 460
– cell culture **10**, 448
– cytostatic/antitumor compounds of **10**, 451f
– production **10**, 448
macrocapsules, encapsulated-cell therapy **10**, 548
macro elements, for microbial growth **9**, 199f
– – nitrogen **9**, 199
– – phosphorus **9**, 199
macrolactones, lipase-catalyzed formation **8A**, 115
macrolides, immunosuppressant –, model of the mode of action **7**, 548
– structures **7**, 34
macrophage colony stimulating factor, clinical trials **5A**, 157
– in therapy **5A**, 157
macrophytes, in wetland systems **11A**, 242ff
mad cow disease **5A**, 513
maduraferrin **7**, 219f
– oligopeptide siderophore of *Actinomadura maduurae* **7**, 219f
– structure **7**, 200
MAFF (Ministry of Agriculture, Forestry and Fisheries of Japan) **12**, 254ff, 378ff, 595
MAF (Ministry of Agriculture and Fishery of Korea) **12**, 392
magnesium limitation, effect of calcium **1**, 179
– metabolic effect **1**, 179
magnesium transport system **1**, 180
magnetic susceptibility, oxygen **4**, 60ff
magnetosomes, bacterial – **10**, 469
– biotechnological applications of **10**, 469, 576f
– of marine bacteria **10**, 468f
Magnetospirillum gryphiswaldense, magnetosomes **10**, 469, 576
magnetotactic bacteria **10**, 469, 576f

magnolin, demethylation of **8A**, 498

Magnus test, screening of receptor-active compounds **7**, 109f

MAGPIE **5B**, 229, 236, 373ff, 379ff

– display of **5B**, 381

– dynamic mode **5B**, 381

– hierarchical display system **5B**, 382ff

– static graphical mode **5B**, 381

– types of information **5B**, 381

MAGs *see* monoacylglycerides

Maillard reaction, during break baking **9**, 307

– formation of fumic acid-like residues **11A**, 26

maintenance energy **1**, 141

maintenance energy requirement **1**, 168

maize *see* corn

– male sterility **2**, 648

MaizeDB **5B**, 285

maize streak virus (MSV) **2**, 574

major histocompatibility complex (MHC) **5A**, 141

major non-immune-based diseases **5A**, 129f

– therapeutic uses of recombinant proteins **5A**, 129

malaria vaccine **5A**, 139

malate, routes to **8B**, 105

Malaysia, biotechnology **12**, 417

Malaysian Industrial Development Authority **12**, 418

Malaysian Technology Development Corporation **12**, 418

MALDI *see* matrix-assisted laser desorption ionization

maleate hydratase, of *Pseudomonas pseudoaligenes* **8B**, 114

male sterility, introduction into plants **2**, 648

– – by *Bacillus amyloliquefaciens* RNAse **2**, 648

malic acid, biosynthetic pathway **6**, 364

malic acid production **6**, 371f

– by *Aspergillus flavus* **6**, 371

– by enzymatic conversion of fumaric acid **6**, 372

L-malic acid production **3**, 457

malolactic fermentation **9**, 489ff

– biochemistry **9**, 492

– in cucumbers **9**, 636, 652

– inhibition of **9**, 491

Malomonas rubra, by *Malonomonas rubra*, reaction mechanisms **8B**, 71

– malonate decarboxylation mechanism **8B**, 71

– malonyl-CoA decarboxylase **8B**, 69ff

– synthesis of chiral – **8B**, 106

malonate decarboxylation, by *Klebsiella pneumoniae*, reaction mechanism **8B**, 71

malonates, prochiral –, asymmetrization by pig liver esterase **8A**, 195

malonic acid, decarboxylation of, abiotic mechanism **8B**, 46

malonyl-CoA, decarboxylation of, by malonyl-CoA decarboxylase **8B**, 69

malonyl-CoA decarboxylase, of *Klebsiella pneumoniae* **8B**, 69ff

– of *Malomonas rubra* **8B**, 69ff

malonyl-CoA decarboxylase (MCD) **8B**, 68ff

malt **6**, 68ff, **9**, 426ff

– amylolytical components **6**, 68f

– barley **9**, 426f

– green malt **6**, 69

– kiln-dried malt **6**, 69f

– sprouts, as feed **9**, 771

– use in baked goods **9**, 263f

malting grain, barley **6**, 70

– triticale **6**, 70

– wheat **6**, 70

malting process **9**, 427

– scheme **9**, 428

maltodextrins **6**, 35ff, **9**, 669

– applications **9**, 761

– composition **6**, 36

– production of **6**, 35, **9**, 760f

– properties **9**, 761

– types **6**, 36

– uses, production of antibiotics **6**, 36

maltose permease, of brewer's yeast **9**, 441

maltose syrups, composition **6**, 39

– in antibiotic fermentations **6**, 39

– production of, use of β-amylase **6**, 38

maltotetraose, in maltose syrups **6**, 39f

maltotriose, in maltose syrups **6**, 39f

mammalian cell, nutrient elements, physiological function **3**, 145

mammalian cell culture **3**, 8, 26, 106ff

– antibody manufacture **5A**, 249

– bioreactors **3**, 110ff

– control **3**, 121

– media **3**, 142ff

– operation **3**, 121

– post-translational product modification **3**, 155

– – medium effects **3**, 155

mammalian drug metabolism, aromatic hydroxylation **8A**, 500

mammalian protein production, by *Saccharomyces cerevisiae* **1**, 497

mammalian tissues, flavin dependent monooxydases **8A**, 575f

mammals, as sources of ribosomal peptides **7**, 302f

mammary gland, heterologous protein production **2**, 800f

mandelonitrile lyase **8B**, 90

manganese, as electron acceptor for anaerobic respiration **11A**, 43

manganese-oxidizing bacteria, metal corrosion by **10**, 298

manganese peroxidase **10**, 166ff

– biodegradation of nitroaromatics **11B**, 289f

– bleaching of paper pulp **10**, 536

– catalytic cycle of **10**, 166

– depolymerization of coal humic substances **10**, 167
– gene regulation **1**, 535
– modification of lignin **10**, 530
manganese peroxydase, catalytic cycle **11B**, 161ff
– degradation of lignin **11B**, 161ff, 230
Manhattan project **12**, 28
manioc *see* cassava (*Manihot esculenta*)
– analysis of **6**, 67
– as raw material for ethanol production **6**, 66
– processing in ethanol production **6**, 94
– stillage, nutrient content **6**, 108
mannanase, bleaching of paper pulp **10**, 536
– mode of action of **10**, 528f
mannitol, formation by yeasts **6**, 222
mannitol spoilage, of wine **9**, 494f
mannose syrups **6**, 41
β-mannosidase **10**, 529
mannosyltransferase **8B**, 260f
manoalid, antiinflammatory compound of marine origin **10**, 455ff
– structure **10**, 457
manufacturing *see* biotechnological manufacturing
manufacturing belts **5B**, 434
manure **11A**, 219ff, 227ff, 232f
– aerobic treatment **11A**, 232f
– – thermophilic processes **11A**, 232f
– anaerobic treatment **11A**, 226ff
– nutrient contents **11A**, 220f
– quantities **11A**, 220
– spreading of, sanitary aspects **11C**, 225
– treatment plant **11A**, 227ff
– – scheme **11A**, 227f
manure piles, self-ignition of **11C**, 41
Manzanilla, olive variety **9**, 599
map distance **5B**, 169
mapping method, Haldane's **5B**, 169
– Kosambi's **5B**, 169
– Morgan's **5B**, 169
marasmic acid **7**, 503f
– antitumor activity **7**, 503
– structure **7**, 504
margarine, interesterification of **8A**, 137
margarine production, wastewater concentrations **11A**, 205
marine animals, adhesive proteins from **10**, 467f
– antibacterial compounds of **10**, 459
– antifreeze proteins from **10**, 467
– antifungal compounds of **10**, 458f
– antiinflammatory compounds of **10**, 456
– antiparasitic compounds of **10**, 458
– antiviral compounds of **10**, 460
– aquaculture of **10**, 448f
– cnidarians **10**, 445f
– crustaceans **10**, 445f
– cytostatic/antitumor compounds **10**, 453
– cytostatic/antitumor compounds of **10**, 452
– molluscs **10**, 445f

– sponges **10**, 445
– tunicates **10**, 445
marine biotechnology **10**, 441ff
– fields of **10**, 443
– genetic engineering in **10**, 450
– product discovery **10**, 446f
– – steps **10**, 446f
– production processes **10**, 446ff
– products of **10**, 450ff
– – antitumor compounds **10**, 450ff
– – cytostatic compounds **10**, 450ff
– – pharmaceuticals **10**, 450ff
marine environment, diversity of **10**, 443f
– – analytic techniques **10**, 444
– types of organisms **10**, 443f
marine microorganisms **10**, 444f
– antibacterial compounds of **10**, 458
– antifungal compounds of **10**, 458
– antiinflammatory compounds of **10**, 456
– antiviral compounds of **10**, 460
– bacteria **10**, 444
– biopolymers from **10**, 466
– bioremediation with **10**, 470f
– biosurfactant-producing – **10**, 470f
– cultivation **10**, 447ff
– cytostatic/antitumor compounds of **10**, 451f
– DHA-synthesizing – **10**, 464
– EPA-synthesizing – **10**, 464
– fermentation **10**, 447ff
– fungi **10**, 444
– hydrogen-producing – **10**, 471
– petroleum-degrading – **10**, 470
– protists **10**, 444
– PUFA production by – **10**, 142, 463
– – in aquaculture **10**, 465
– – molecular biology of microbial PUFA synthesis **10**, 465
– – optimization of **10**, 464f
– saline environment **10**, 447
marine organisms, biopolymers from **10**, 466f
marine plants, algae **10**, 445
– antibacterial compounds of **10**, 458
– antifungal compounds of **10**, 458
– antiinflammatory compounds of **10**, 456
– antiviral compounds of **10**, 460
– biopolymers from **10**, 466
– cytostatic/antitumor compounds of **10**, 451f
– seaweeds **10**, 445
marker-assisted breeding **5B**, 152
marker density **5B**, 112
marker gene **2**, 144ff, 439, 558
– allelic variation **2**, 147
– amplification marker **2**, 711f
– antibiotic resistance gene **2**, 558f
– application, in breeding **2**, 178f
– – in population genetics **2**, 178
– classes **2**, 152
– cotransfer **2**, 710

– criteria **2**, 144
– defective mutant complementation **2**, 707
– detectable marker **2**, 630, 710
– differences **2**, 152
– DNA maker **2**, 145f
– – detection **2**, 146
– – fingerprint **2**, 146
– – restriction fragment length polymorphism **2**, 146
– – variable number of tandem repeats **2**, 146
– drug resistance gene **2**, 707f
– enzyme maker **2**, 145
– for gene amplification **2**, 708
– formation of foci **2**, 709
– for plants **2**, 630
– hygromycin phosphotransferase (*hpt*) gene **2**, 558
– immunological marker **2**, 145
– in Corynebacteria **2**, 481
– in plants **2**, 558ff
– in *Saccharomyces* **2**, 509
– – counterselectable **2**, 509
– – dominant **2**, 509
– – URA3 **2**, 509
– locus **2**, 147
– mapping application **2**, 178
– – gene tapping **2**, 179
– – marker-based selection **2**, 179
– – outcrossing **2**, 178
– marker of chromosome structure **2**, 144
– morphological marker **2**, 144
– – mutation localization **2**, 144
– neomycin posphotransferase (*npt II*) gene **2**, 558
– phosphinotricine acetyltransferase gene **2**, 558
– physical linkage **2**, 710
– polymorphism information content (PIC) **2**, 144
– protein marker **2**, 145
– regulatory DNA locus **2**, 145
– selectable **2**, 382, 439, 558f, 630, 707ff
– – *Bacillus subtilis sacB* **2**, 439f
– – dominant **2**, 707ff
– – in fungi **2**, 531ff
– – in yeast **2**, 522, 524
marker locus **5B**, 108
marker segregation population **2**, 153
– in animals **2**, 153
marker segregation ratio **2**, 149f
– tests **2**, 149f
market approval *see* registration
market assessment **12**, 216
market prices, citric acid **6**, 51
– ethanol **6**, 51
– lactic acid **6**, 51
– single-cell protein **6**, 51
market research **12**, 216ff
– background information **12**, 217f

– cost-benefit analysis **12**, 218ff
– product profile **12**, 218ff
markets, for biotechnological products **6**, 48f
Mascot **5B**, 262
mash distillation column, schemes **6**, 100f
mash hydrosizing **6**, 111ff
– apparatus **6**, 112
– flow rates of water **6**, 112
mashing processes **6**, 77ff
– equipment **6**, 77
– – dispersing machines **6**, 79f
– – mills **6**, 78
– pressure boiling processes **6**, 82ff
– – high pressure cooking process **6**, 82ff
– – Verlinden process **6**, 84
– pressureless breakdown of starch **6**, 85ff
– – infusion processes **6**, 85ff
– recycling processes, dispersing mash process **6**, 88f
– – stillage recycling process **6**, 87f
– using autoamylolytical activities **6**, 94ff
– – in corn **6**, 96
– – in potatoes **6**, 96
– – in rye **6**, 96
– – in triticale **6**, 95
– – in wheat **6**, 95
mash tub **6**, 80f
– scheme **6**, 81
mass balance, in bioprocesses, general balance equation **3**, 364
– in fermentation control **3**, 330
mass balance concept, in industrial wastewater treatment **11A**, 485f
mass balance equations, assumptions for modelling **4**, 415f
mass distribution, in stirred tanks **4**, 304ff, 321f, 333ff
– – five-compartment model **4**, 321
mass spectrometer **4**, 63ff
– capillary gas inlet **4**, 65
– comparison with other gas analysis instruments **4**, 66
– fermentation gas analysis **4**, 63
– ion detectors, Faraday cup **4**, 64
– – secondary-electron-multiplier (SEM) **4**, 64
– ionization **4**, 63
– ion separation **4**, 63
– listing for gas analysis **4**, 67
– magnetic sector analyzer **4**, 64
– practical problems **4**, 66
– quadrupole analyzer **4**, 64
– rotary valve system **4**, 58
– sample stream valves **4**, 57f
– solenoid valves **4**, 57f
– spectral analysis **4**, 64
mass spectrometers, types of **5B**, 255
mass spectrometry **3**, 323, **5B**, 89, 274
– for proteomics **5B**, 241ff, 250ff

mass-to-charge ratios **5B**, 330
mass transfer, across gas/liquid interfacial areas
  4, 110ff
– characteristic time **3**, 212
– effectiveness factor **3**, 94
– effects on reaction parameters **3**, 438
– film diffusion **3**, 435
– gas-liquid – **3**, 89ff, 366
– – viscosity effects **3**, 91
– in bioreactors **3**, 89ff, 190ff
– – immobilization effects **3**, 435ff
– – molecular diffusion **3**, 89ff
– in soil, kinetic data **11B**, 73ff
– in spray-dyring **3**, 709
– in tower reactors, structured models **4**, 370f
– in two-phase systems **3**, 605
– pore diffusion **3**, 435
– resistance **4**, 111
– – in the liquid phase **3**, 190
mass transfer coefficient **3**, 190f, 324, 405ff, **4**,
  111, 120, 368ff, 393f
– – *see also* oxygen transfer coefficient
– diffusion membranes **3**, 251
– – impeller system effects **3**, 251f
– in aeration membranes **3**, 245, 247
– in immobilization beads **3**, 435
– in Newtonian fluids **3**, 196ff
– – dimensionless correlation **3**, 203
– – in stirred tank reactors **3**, 206
– in non-Newtonian fluids **3**, 202ff
– – dimensionless correlation **3**, 203
– – in stirred tank reactors **3**, 206
– in surface aeration **3**, 239
– specific service area **3**, 191f
– steady-state experiments **4**, 368
– unsteady-state experiments **4**, 368
– volumetric – **3**, 90, 192ff
– – chemical determination methods **3**, 196
– – correlation **3**, 198
– – determination **3**, 192ff
– – dynamic determination methods **3**, 192ff
– – in stirred tank reactors **3**, 196
– – parameter correlations **3**, 196
mass transfer modelling, in tower reactors **4**,
  368ff
– of baker's yeast production **4**, 393f
mass transfer rate **4**, 111, 369ff, 393
– fluid dynamic characteristics **4**, 369
– prediction of **4**, 369
master cell bank **5A**, 452f
– quality control **5A**, 453
Master Cell Bank (MCB) **10**, 30
master validation plan (MVP) **3**, 741ff
mat composting **11C**, 139
mathematical models *see* fermentation modeling
– for biotechnological processes, elements **4**, 272
– – structure **4**, 272
– metabolic redulator approach **4**, 292

– of biotechnological processes **4**, 269ff
math-function modules, in process control systems **4**, 583
mating **12**, 163f
mating disruption technique **10**, 498f
mating type **2**, 77
– Basidiomycetes **2**, 78
– gene **2**, 77
– – minus (α)-allele **2**, 77
– – plus (α)-allele **2**, 77
– switching **2**, 77
matrices, BLOSUM62 **5B**, 132
– PAM **5B**, 132
– PAM250 **5B**, 132
matrix-assisted laser desorption ionization (MAL-
  DI) **5B**, 329
matrix-assisted laser desorption ionization time of
  flight (MALDI-TOF) **5B**, 233, 251
– bioinformatics for **5B**, 253
matrix-assisted-laser-desorption/ionization-time-of-
  flight (MALDI-TOF) mass spectrometry **5B**,
  210, 252f
matrix water potential, in composting **11C**, 46
Matthiola incana **2**, 179
Maxam and Gilbert sequencing **5B**, 185
maximum likelihood approach **5B**, 169
Maximum Principle **4**, 525ff
maytansine, inhibition of the microtubular system
  7, 668
– structure **7**, 668
MDB (Metalloprotein Database) **5B**, 409
mead acid *see* eicosatrienoic acid
measurement, information content **4**, 255f
– noise **4**, 254ff
– oprimization, criteria **4**, 256
– optimization **4**, 254ff
– relative sensitivity **4**, 256
– semi-relative sensitivity **4**, 256
measurement errors **4**, 49ff
– gas flow **4**, 49
– oxygen gas analysis **4**, 50f
meat, hydrolysates **9**, 704f
– microflora **1**, 331
– spoilage **1**, 341
– tenderization **9**, 698
– use of blood cell hydrolysate **9**, 700
meat processing wastes **11A**, 219, 221
mechanical-biological treatment, of municipal solid waste **11C**, 204ff
mechanical stress, effects of, impellers **3**, 274
– – reactor geometry **3**, 273f
– – reactor size **3**, 273
– – reactor volume **3**, 274
– maximum energy density **3**, 273
– on cell cultures **3**, 272ff
medermycin **1**, 215
– antitumor agent, structure **7**, 662
mederrhodin **1**, 215

media **1**, 143ff, 170ff
– – *see also* microbial growth media
– – *see* fermentation media, growth media
– for callus cultures **1**, 582
– for clostridia 293ff
– for *Escherichia coli* **1**, 143, 173
– for lactic acid bacteria isolation **1**, 347ff
– for plant cell cultures **1**, 590, 594
– M17 **1**, 348
– optimization **1**, 145, 590
– pH value **1**, 180
medial waste, microbiological activity **3**, 783
medium composition, for mammalian cell cultures **3**, 143ff
– – amino acids **3**, 144
– – attachment compounds **3**, 149f
– – buffers **3**, 143
– – bulk ions **3**, 143
– – carbohydrates **3**, 144
– – coenzymes **3**, 144
– – cytokines **3**, 148
– – growth factors **3**, 147, 149
– – interleukins **3**, 148
– – nucleotides **3**, 144
– – protease inhibitors **3**, 152
– – substratal proteins **3**, 149
– – trace elements **3**, 143
– – transport proteins **3**, 149f
– – vitamins **3**, 144
medium optimization **3**, 133ff
– biochemical methods **3**, 136
– statistical-mathematical approaches **3**, 134ff
medium sterilization **3**, 157ff
– – *see also* sterile filtration
– adsorption **3**, 160
– batch mode **3**, 164
– by inactivation **3**, 162ff
– by removal **3**, 159ff
– continuous – **3**, 170ff
– – steam injection **3**, 175
– filtration **3**, 159f
– heat exchangers **3**, 175
– heating up **3**, 175
– high temperature short time (HTST) treatment **3**, 170
– modes **3**, 159
– sedimentation **3**, 160
– thermal – **3**, 159, 162ff
– validation **3**, 183
MEDLINE **5B**, 283
Meisenheimer complex **11B**, 279, 281ff
meitauza, production of **9**, 525
melanoma, familial – **5B**, 82
melanoma vaccines **5A**, 138
melatonin **5A**, 165
membrane **1**, 23
– – *see* biological membrane
– bacterial plasmalemma, fluidity **1**, 23

– – membrane protein **1**, 23
– – molecular composition **1**, 23
– composition, in immobilized *Escherichia coli* **1**, 243
– – lipid-protein ratio **1**, 243
– – of methanogens **1**, 259
– – phenol effect **1**, 243
– ethanol effect **1**, 136
– intracellular **1**, 24
– – dinitrogen fixer **1**, 25
– – dynamic changes **1**, 24
– – non-phototrophic bacteria **1**, 25
– – phototrophic bacteria **1**, 24
membrane aeration **3**, 33, 237, 245ff
– diffusion membranes **3**, 250
– *ex situ* – **3**, 237, 254f
– *in situ* – **3**, 237, 248, 252f
– mass transfer **3**, 249
– – coefficients **3**, 270
– – reactor type effects **3**, 249
– oxygen input **3**, 269
– silicone tubing – **3**, 237
membrane bioreactor **3**, 432, 434, 439ff
– aminoacylase process **3**, 448
– configurations **3**, 441
– driving force **3**, 439
– enzyme reactors, process design **3**, 452
– membrane function **3**, 439
– multiphase reactors **3**, 458
– new designs **3**, 458f
– separation **3**, 439
– stirred tank – **3**, 432
membrane bioreactors, in waste gas purification **11C**, 261
– schematic representation **11C**, 262
– semi-technical –, cleaning performance **11C**, 351
membrane channel **1**, 73
membrane filtration **3**, 440
– bioproduct separation **4**, 609ff
– in drinking water treatment **11C**, 405ff
– membrane structures **11C**, 406f
– – scheme **11C**, 406
– pore size **4**, 610
– principle of **11C**, 407
– process –, parameters **11C**, 407
membrane flow **1**, 31
membrane fouling **3**, 44
membrane fusion **2**, 95ff, 101, 131
– fusion agents **2**, 102
– membrane dehydration **2**, 102
– membrane perturbation **2**, 102f
– molecular mechanism models **2**, 101f
– repulsive forces **2**, 101
– spontaneous **2**, 96
– viral fusion protein **2**, 101
membrane gradient, in *Lactococcus* **1**, 351

membrane materials, diffusion coefficients of some compounds **11C**, 267
membrane permeabilization, in plant cells **1**, 598
– physical methods **1**, 600
membrane processes **11C**, 345ff
– and alternative techniques **11C**, 345ff
– in surface water treatment **11C**, 405ff
– – microfiltration **11C**, 405ff
– – nanofiltration **11C**, 405ff
– – ultrafiltration **11C**, 405ff
– use of capillary modules **11C**, 408
membrane protein anchoring **5A**, 7f
membrane reactor **11A**, 568ff
– concentration profile **11A**, 568f
– scheme **11A**, 568
membrane reactors, in biological waste gas purification, biological air filter (BAF) **11C**, 349f
– – dense membranes **11C**, 349ff
– – microporous membranes **11C**, 349
membranes *see* aeration membranes, cell membranes, filtration membranes
– use in biological waste gas purification **11C**, 348
membrane separation, flux rate decline **4**, 618f
membrane separation process *see* crossflow filtration (CFF)
membrane skleleton **1**, 38
membrane structure, fluid mosaic model **5A**, 7
membrane transport **1**, 72ff
– energetics **1**, 73
– group translocation **1**, 75
– in lactic acid bacteria **1**, 350
– kinetics **1**, 76
– macromolecules **1**, 79
– mechanism **1**, 73
– models **1**, 73ff
– primary – **1**, 73f
– proton leakage **1**, 181
– rate **1**, 78
– secondary – **1**, 73ff
– studies by NMR **1**, 193
– substrate-carrier interaction **1**, 73
membrane transport system **1**, 74ff
– binding protein dependent **1**, 75
– classification **1**, 74
membrane transport systems, PEP-dependent sugar transport **1**, 75
ME (Ministry of Education of Korea) **12**, 393
memprane proteins **5A**, 6ff
memraben technology, in wastewater treatment **11A**, 565f
menaquinone **1**, 404
– production of **10**, 335f
– structure **10**, 336
MENDEL **5B**, 285
Mendelian traits, dominant – **5B**, 63
– independent assortment **5B**, 63
– recessive – **5B**, 63

– segregation **5B**, 63
menoxymycin, antitumor agent, structure **7**, 662
mental retardation **5B**, 73
Mentha piperita **10**, 396
menthol, biotransformation of **10**, 396ff
– enantiomers **8A**, 69
– enantioselective – **10**, 402ff
menthone, biotransformation of **10**, 396ff
mercuric reductase **10**, 248
mercury, as electron acceptor for anaerobic respiration **11A**, 43
– biosensors for depletion of **11B**, 455
– monitoring with biosensors **11A**, 127f
merissa beer, production of **9**, 547
meristeme **1**, 34, 579
MEROPS database **5B**, 288
mersacidin, action on cell wall of susceptible bacteria **7**, 357
– prepeptide structure **7**, 345
– primary structure **7**, 337f
– structural gene **7**, 344
merulidial **7**, 504f
– structure **7**, 504
mesophilic microorganisms, specific growth rate **1**, 128
mesophyll cell **1**, 580
mesosome **1**, 25
messenger RNA *see* mRNA
messenger RNA (mRNA) **1**, 97
– half-life **1**, 97
– splicing **1**, 97
messenger tm **10**, 495
metabolic branchpoint **1**, 167
metabolic circles, carbon cycle **10**, 279
– iron cycle **10**, 279f
– nitrogen cycle **10**, 278f
– sulfur cycle **10**, 276ff
metabolic control theory **9**, 124ff
metabolic design **1**, 189ff
metabolic energy turnover, in *Lactococcus* **1**, 350
metabolic engineering **1**, 164, 190ff
– nitrile transformations **8A**, 314
– of secondary pathways, nicotine biosynthesis **7**, 624f
– – scopolamine production **7**, 625f
– – serotonin biosynthesis **7**, 624
– – use of plant cell cultures **7**, 623ff
metabolic flux control **1**, 104
– modification **1**, 216
metabolic flux control analysis **1**, 165f
– summation theorem **1**, 165f
metabolic fluxes **9**, 132ff
– manipulation of **9**, 133ff
metabolic flux manipulation **1**, 183
metabolic pathway **1**, 165
– rate-limiting step **1**, 166
metabolic pathways, of microorganisms **9**, 174ff
metabolic rate, calculations **3**, 364ff

– in bioprocess analysis **3**, 362
metabolic regulation **9**, 124ff
– cooperativity **9**, 47, 125f, 132
– feedback inhibition **9**, 46, 125, 132
metabolic regulator model, acetone-butanol production **4**, 292
– butanol-diol production **4**, 292
– schematic diagram **4**, 294
– yeast growth **4**, 292ff, 394f
metabolism **1**, 15, 47ff, 442ff
– – *see also* carbohydrate metabolism
– central **1**, 81
– compartmentation **1**, 478
– concept **1**, 50
– coordination **1**, 100
– flux control **1**, 104
– of lactic acid bacteria **1**, 350ff
– of methanogens **1**, 255ff
– of methylotrophs **1**, 266ff
– of plant cells **1**, 580
– of pseudomonads **1**, 414f
– of *Saccharomyces cerevisiae* **1**, 478
– of streptomycetes **1**, 442ff
– pathway redesign **1**, 214
– regulation **1**, 100
– secondary *see* secondary metabolism
metabolite, overproduction **1**, 163ff, 450
– production improvement **1**, 204
– quantification **1**, 190ff
– screening **1**, 447f
– – target **1**, 449
metabolite production, primary metabolites **3**, 9
– regulations **3**, 20
– secondary metabolites **3**, 9
Metacyc **5B**, 407
meta-data **5B**, 292
metal accumulation, biosorption, by immobilized biomass **10**, 231ff
– – by microorganisms **10**, 228ff
– by microorganisms **10**, 225ff
– metal desorption **10**, 233f
metal-affinity chromatography **3**, 50
metal-binding proteins **10**, 234ff
metal carbonyl complexes, resolved by lipases **8A**, 99
metal chelates, use in phytoremediation **11B**, 387f
metal chelators **11B**, 390
metal-contaminated sites, bioremediation of **10**, 215ff
metal-contaminated waters, treatment of, use of bacterial sulfate reduction **10**, 242f
metal desorption, from activated sludge **10**, 234
– from microbial biomass **10**, 233f
metal detoxification, by marine *Bacillus* spores **10**, 471
metal hyperaccumulators **11B**, 389f
metal immobilization **10**, 227ff

metal ion accumulation, from an artificial groundwater **11B**, 393
metal ion removal, from an artificial groundwater **11B**, 393
metal ions, as electron acceptors for anaerobic respiration **11A**, 43
– bioaccumulation of **11A**, 422, 435, 441
– biosorption of **11A**, 42, 441ff
– precipitation of **11A**, 42ff
– – sulfate reducing bacteria **11A**, 44
metal ion transport **10**, 236ff
– in fungi **10**, 237
– toxic metals **10**, 237
metallic raw materials, microbial leaching of **10**, 192
metalloid oxyanions, microbial reduction of, applications for bioremediation **10**, 241
metalloids, microbial dealkylation **10**, 248f
– microbial methylation **10**, 248f
metalloid transformations and bioremediation **10**, 249
metalloproteases **5A**, 207f
– characteristics **8A**, 248
metalloproteins **5A**, 19ff
metallothioneins **10**, 235, 238
– of *Saccharomyces cerevisiae* **10**, 239
metal mobilization **10**, 245ff
– reductive – **10**, 247f
metal phosphates, formation of **11A**, 309
– solubility of **11A**, 309
metal precipitation, as phosphates **10**, 244
– bioremediation applications **10**, 239f
– by metal-reducing bacteria **10**, 239f
– by sulfate-reducing bacteria **10**, 241f
metal-reducing bacteria, metal precipitation by **10**, 239f
metal removal **11A**, 42ff, 431ff
– by biomass **11A**, 431ff
– – adsorption experiments **11A**, 442f
– – bacteria **11A**, 437ff
– – biosorption **11A**, 435
– – desorption **11A**, 445
– – fungi **11A**, 440f
– – kinetics **11A**, 444f
– – leaching **11A**, 435
– – mechanism **11A**, 441ff
– – microalgae **11A**, 439f
– by high gradient magnetic separation **10**, 244
– in a fixed bed column **11A**, 445f
– physico-chemical elimination methods **11A**, 431ff
– – electrolysis **11A**, 448
– – ion exchange absorption membrane plants **11A**, 448
– – precipitation **11A**, 447f
– – spray evaporation **11A**, 448
metals, autotrophic leaching of **10**, 245f
– binding of, by biopolymers **10**, 234ff

– – by metallothioneins **10**, 235
– – by polysaccharides **10**, 235
– – by siderophores **10**, 235
– biocorrosion of **10**, 293ff
– bioleaching of **10**, 191ff
– bioreporter systems for **11B**, 455f
– biosorption of **10**, 228ff
– – by algae **10**, 230f
– – by fungi **10**, 229f
– chelate-assisted phytoextraction, field applications **11B**, 388
– continuous phytoextraction **11B**, 388ff
– – field application **11B**, 390ff
– heterotrophic leaching of **10**, 246f
– induced phytoextraction, protocol **11B**, 387f
– in municipal wastewater **11A**, 179f
– *in situ* remediation **11B**, 364ff
– – bioleaching **11B**, 365
– – bioprecipitation **11B**, 365
– – biosorption **11B**, 365
– – phytoremediation **11B**, 365, 385ff
– monitoring with biosensors **11A**, 127f
– production by bioleaching **10**, 312
– recycling of **11C**, 8ff
– removal by phytoextraction **11B**, 389
– removal by rhizofiltration **11B**, 392ff
– world production, by bioleaching **10**, 214
metal sulfide precipitation **10**, 242ff
metal sulfides, microbial attack on **10**, 174
metal transformation, by microorganisms **10**,
225ff
metal uptake, microbial –, phases **10**, 227f
metastasis, prevention of – **7**, 673f
metazoa, in artificial groundwater recharge **11C**,
434
– in biofilms **11A**, 66
– in biofilters **11C**, 468
methane *see also* biogas
– anaerobic degradation of **11B**, 172
– anerobic formation **4**, 448ff
– as product of coal bioconversion **10**, 169
– concentration in biogas **4**, 454
– energy-rich fuel **11A**, 456
– formation, anaerobic conversion of biopolymers **11A**, 30ff
– – rate-limiting steps **11A**, 30
– heat energy **11A**, 24
– infrared analyzer **4**, 62
– pathways of oxidation **11B**, 197
– scheme **4**, 448
– utilization by *Methylosinus trichosporium* **11B**,
203
methane bacteria, identification of **4**, 450
– Methanosarcinae **4**, 449
methane combustion, energy gains **4**, 457
methane flow, in anaerobic wastewater processes
4, 461
methane formation reactions **1**, 260f

– autotrophic CO_2 fixation **1**, 261
– cofactors **1**, 260f
methane monooxygenase **11B**, 196f
– membrane-bound – **11B**, 203
– structure **11B**, 196
methane monooxygenase (MMO) **1**, 271, 280
methane oxidation, in methanotrophic organisms
1, 271
methane production **3**, 304
– inhibition by, hydrogen sulfide **11A**, 541
– – propionic acid **11A**, 541
– rate, of a biogas reactor **11A**, 538ff
methane reactors, competition of sulfate reducers
with methanogens **11A**, 35f
– design **11A**, 484ff
Methanobacteriales **1**, 253
Methanobacterium **11A**, 460
– anaerobic wastewater process **4**, 447
Methanobacterium thermoautotrophicum **1**, 261
– genome **1**, 262
– methane formation **1**, 261
Methanobacterium uliginosum, electron micrograph **1**, 256
Methanobrevibacter smithii, electron micrograph
1, 256
methanochondroitin **1**, 258
Methanococcales **1**, 253
Methanococcus janaschii, electron micrograph **1**,
257
Methanocorpusclaceae **1**, 255
methanogenesis **11A**, 456ff
– acetate inhibition **11A**, 469
– effect of sulfate **11A**, 461
– hydrogenotrophic –, effect of H_2 partial pressure **11A**, 464f
– propionate inhibition **11A**, 469
methanogenic bacteria **1**, 251ff, **11C**, 153f
– bacteriophages **1**, 565
– carbon metabolism **1**, 261
– cell envelope **1**, 21, 257
– coenzymes **1**, 72
– culture **1**, 259ff
– – handling **1**, 260
– – maintenance techniques **1**, 260
– DNA-assiciated proteins **1**, 262
– ecology **1**, 255
– energy metabolism **1**, 260
– environmental distribution **1**, 255
– enzymes cofactors **1**, 259
– extreme habitats **1**, 256
– Gram-negative **1**, 257
– Gram-positive **1**, 256
– in animals **1**, 255
– medium optimization **1**, 260
– membrane **1**, 259
– metabolic pathways **1**, 255
– molecular biology **1**, 262
– morphology **1**, 256

– nutrient requirements **1**, 260
– restriction enzymes **1**, 263
– ribosome structure **1**, 262
– substrates **1**, 255
– taxonomy **1**, 252
– temperature requirements **1**, 260
methanogenic granular sludge, cell densites **11A**, 467
methanogenic substrates **11A**, 457
Methanogenium frigidum **11A**, 468
methanogens **11A**, 28f, 35f, 457ff
– affinity for H₂ use **11A**, 458f
– competition with sulfate reducers in methane reactors **11A**, 35f
– effect of temperature **11A**, 467f
– groups of **11A**, 457
– in anaerobic processes **11A**, 98
– oligonucleotide probes **11A**, 94
– pH optima **11A**, 468
– physiology of **11A**, 458ff
– reactions catalyzed by **11A**, 28
– synthrophic metabolism **11A**, 459
– use of, acetate **11A**, 459f
– – formate **11A**, 459f
– – hydrogen **11A**, 459f
methanol, gas analysis in bioreactors **4**, 60
– partition coefficients, lin liquid-liquid extraction **3**, 560
– substrates for biomass production **9**, 213
methanol dehydrogenase (MDH), in methylotrophic bacteria **1**, 273
methanol fermentation, in clostridia **1**, 300
methanol metabolism, in yeasts **9**, 180f
Methanolobus vulcani, electron micrograph **1**, 257
methanol oxidation, in microorganisms **1**, 271ff
– in yeasts **1**, 271, 471
Methanomicrobiaceae **1**, 254
Methanomicrobium mobile, electron micrograph **1**, 257
Methanoplanaceae **1**, 254
Methanoplanus limicola, electron micrograph **1**, 257
Methanopyrus **1**, 255
Methanopyrus kandleri, temperature optimum **11A**, 467
Methanosaeta sp. **11A**, 460
– temperature optimum **11A**, 467
– use of acetate, kinetic data **11A**, 459
Methanosarcina, anaerobic wastewater process **4**, 447
Methanosarcina barkeri, degradation of acetic acid **4**, 457
– – experimental values **4**, 458
Methanosarcinaceae **1**, 254
Methanosarcina sp. **11A**, 460
– temperature optimum **11A**, 467
– use of acetate, kinetic data **11A**, 459

Methanoshaera stadtmanae, electron micrograph **1**, 256
Methanospirillum hungatei **11A**, 460
– electron micrograph **1**, 257
Methanothermus fervidus, electron micrograph **1**, 256
Methanothrix, anaerobic wastewater process **4**, 447
methanotrophic bacteria, classes **11B**, 203
– cometabolic degradation of TCE **11B**, 152f
methanotrophic microorganisms *see also* methylotrophic microorganisms
– characteristics **1**, 269
– methane oxidation **1**, 271
methanotrophs, oligonucleotide probes **11A**, 93
methionine aminopeptidase (map) **2**, 249
D,L-methionine, chemical synthesis **6**, 475
– feed uses **6**, 468
L-methionine, biosynthetic pathway **6**, 509
L-methionine γ-lyase **6**, 528f
– reactions catalyzed by – **6**, 529
L-methionine production, by racemic resolution **6**, 492
L-methionione, aspartate derived amino acids, in corynebacteria **8B**, 323
methionine-tRNA **2**, 245
methotrexate **5A**, 250
methoxycarbonyl phosphate, phosphorylation potential **8B**, 231
methoxy-propanol, degradation rate in bioscrubbers **11C**, 309
8-methoxypsoralen (MOP) **2**, 21
methylacetate, biotransformation of **10**, 379
methyl alkylphenylphosphinoylacetates (racemic –), kinetic resolution by pig liver esterase **8A**, 206
methylamine dehydrogenase **1**, 274
methylamine oxidation **1**, 274
methyl aromatics, degradation of **1**, 202f
β-methylaspartase **8B**, 129ff
– addition of amines to substituted fumarates **8B**, 132
– addition of ammonia to substituted fumarates **8B**, 131
– bearing the Ser-Gly-Asp sequence **8B**, 130
– inhibitors of **8B**, 130
– mechanism of action **8B**, 133
– of *Clostridium tetanomorphum* **8B**, 129
β-methyldigitoxin (Mdt) **1**, 597
– biotransformation **1**, 596f, 599
β-methyldigoxin (Mdg) **1**, 597
methylenomycin biosynthesis, by *Streptomyces coelicolor* **7**, 74f
methylenomycin production, in *Streptomyces coelicolor*, negative regulation of – **7**, 91f
methyl isobutyl ketone, scrubbing liquid quantities for the absorption of **11C**, 312

methyl ketones, formation by *Penicillium roque-forti* **8B**, 347
methyllantionine **7**, 328
N-methyl-L-glucosamine, biosynthetic pathway **7**, 424
methyl methanesulfonate (MMS) **2**, 24
N-methyl-N'-nitro-N-nitrosoguanidine (MNNG) **2**, 24ff
– comutation **2**, 26
– lesion repair **2**, 26
– mutagenic effect **2**, 25
– sequential mutagenesis **2**, 26
N-methyl-N-nitrosourea **2**, 24
Methylobacillus **1**, 267, 269
Methylobacter **1**, 267, 269
Methylobacterium **1**, 267, 270
Methylobacterium extorquens, methylamine dehydrogenase **1**, 274
Methylococcus **1**, 267, 269
Methylococcus capsulatus, non-heme iron protein methane monooxygenase **8A**, 482
Methylocystis **1**, 267, 269
Methylomonas **1**, 267, 269
Methylophaga **1**, 267, 269
Methylophilus **1**, 267, 269
α-methalornithine, decarboxylation of, by ornithine decarboxylase **8B**, 79
Methylosinus **1**, 267, 269
Methylosinus trichosporium, methane utilization by **11B**, 203
methylotrophic bacteria **1**, 269
– bacteriophages **1**, 566
– oxidation of landfill gas **11C**, 14
– production of poly(3HB) **6**, 441
methylotrophic microorganisms **1**, 265ff
– amino acid synthesis **1**, 279
– definitions **1**, 268
– ecology **1**, 268
– growth energetics **1**, 278
– industrial applications **1**, 278
– isolation **1**, 268
– metabolism **1**, 266, 271
– methylamine oxidation **1**, 274
– non-methane-utilizing **1**, 269
– physiology **1**, 271
– products **1**, 278
– rRNA sequence data **1**, 271
– single cell protein (SCP) production **1**, 278
– species characteristics **1**, 266f
– substrate pathways **1**, 266
– substrates **1**, 266f
– symbioses **1**, 268
– taxonomy **1**, 268ff
methylotrophic yeast **2**, 523ff
– alcohol oxidase **2**, 523
methylotrophic yeasts **1**, 269
Methylovorus **1**, 267, 269

methyl propanoate, laboratory-scale production of **10**, 403
6-methylsalicylic acid synthase **10**, 344f
methyl-*t*-butyl ether (MTBE), degradation in biotrickling filters **11C**, 286
– metabolism of **11C**, 281
methyl *tert*-butyl ether (MTBE), bacterial degradation of **11B**, 205
– pathway **11B**, 205f
Metschnikowia pulcherrima, in grape must **9**, 467
mevinolin, synthesis by yeast alcohol dehydrogenase **8B**, 360
MHC *see* major histocompatibility complex
MHCPEP **5B**, 409
MHSW (Ministry of Health and Social Welfare of Korea) **12**, 393f
MHW (Ministry of Health and Welfare of Japan) **12**, 149, 253ff, 378ff, 595f
MI *see* myocardial infarction
Michael addition, by bakers' yeast **8B**, 31f
Michaelis constant **1**, 69
– mass transfer effects **3**, 438
Michaelis-Menten equation **1**, 68, **5A**, 23f, **9**, 33ff
Michaelis-Menten kinetics **1**, 132
– plot **1**, 68
microalgae *see also* algae, protists
– – *see also* oleaginous microalgae
– adsorption equilibrium of immobilized – **11A**, 443f
– antibacterial compounds of **10**, 458
– antifungal compounds of **10**, 458
– aquaculture of **10**, 128
– biology of **10**, 108ff
– biotechnological use of **10**, 105ff
– blooms, therapeutical value of toxins **10**, 131
– carotenoid contents **7**, 172ff
– cultivation of **10**, 117, **11A**, 439
– definition of **10**, 108f
– ecological applications **10**, 132f
– exploitation of **10**, 126ff
– genetic engineering of **10**, 133f
– heavy metal sorption by **10**, 133
– immobilization of **10**, 119
– in human diet **10**, 126f
– kinetics of heavy metal loading **11A**, 444
– mass cultivation **10**, 119ff
– – inclined surface plant **10**, 121f
– – in closed and semi-closed photobioreactors **10**, 122f
– – in tubular photobioreactors **10**, 123ff
– – plate type reactors **10**, 125f
– – ponds **10**, 119ff
– metal removal by **11A**, 439f
– production of biomass **10**, 115ff
– – animal feed **10**, 127f
– – aquaculture products **10**, 128
– – biofertilizer **10**, 128f
– – human diet **10**, 126f

– – market estimations **10**, 109
– production of polyunsaturated fatty acids **10**, 129f
– products from, market estimations **10**, 109
– regeneration of **11A**, 445
– stable isotopes in **10**, 131f
– use as biofertilizer **10**, 128f
microarraying robots **5B**, 234
Microbacterium lacticum, uridine production **6**, 592f
microbes, pollutant degrading –, measurement in subsoil **11B**, 403
microbial activity, in soil **11B**, 96ff
microbial aggregates, characteristics of life in **11A**, 77
– heterogeneous spatial arrangement of microorganism **11A**, 68, 77
– life in **11A**, 76ff
– role of extracellular polymeric substances **11A**, 74f
– survival benefits **11A**, 76
microbial biomass **9**, 223
– accetability, as feed **9**, 228
– – as food **9**, 228
– allergenicity of **9**, 232f
– fatty acid composition **9**, 225
– for animal consumption **9**, 227ff
– for human consumption, clinical tolerance studies **9**, 229ff
– nucleic acids, limitations for human food **9**, 233f
– nutritional value **9**, 173f, 221ff
– oxygen demand **9**, 195f
– oxygen transfer **9**, 198
– protein determination **9**, 173f
– safety of **9**, 221ff, 227ff
– substrates **9**, 222
– toxicological status **9**, 173f
– vitamins **9**, 225f
– yield **9**, 192ff
– – calculation of **9**, 194
– – energy requirements **9**, 192ff
microbial biomass production **9**, 167ff
– batch culture **9**, 201f
– carbohydrate substrates, molasses **9**, 207ff
– – starch **9**, 211f
– – sulfite waste liquor **9**, 211
– – whey **9**, 209ff
– carbon sources **9**, 175ff
– continuous culture **9**, 202ff
– – chemostat **9**, 202f
– fed-batch culture **9**, 204ff
– – culture volume **9**, 205
– – input rate **9**, 204f
– – substrate concentration **9**, 205
– improvement of medium composition **9**, 200f
– mixed cultures **9**, 206f

– non-carbohydrate substrates, carbon dioxide **9**, 214f
– – lipids **9**, 213
– – methanol **9**, 213f
– – *n*-alkanes **9**, 214
– – photosynthesis **9**, 175ff
– strain selection **9**, 172f
microbial cell rupture *see* cell rupture
microbial cells, lysis **3**, 517
– recovery, from bioreactors **3**, 473
– suspension viscosity, homogenization effects **3**, 485
microbial communities, diversity **11A**, 88ff
– – in wastewater treatment plants **11A**, 91
– dynamics of **11A**, 88ff
– fingerprints of **11A**, 91
– *Hyphomicrobium* **11A**, 102
– in biofilms **11C**, 277ff
– in biofilters **11C**, 450ff
– in biological waste gas purification reactors **11C**, 276ff
– in biosuspensions **11C**, 277
– in wastewater treatment **11A**, 85ff
– – analysis of **11A**, 85ff
– *Nitrospira* **11A**, 102
– *Paracoccus* **11A**, 102
– proteobacteria **11A**, 102
– structure of **11A**, 91ff
– *Zoogloea ramigera* **11A**, 102
microbial community analysis, by amplified ribosomal DNA restriction analysis **11A**, 87
– by denaturing gradient gel electrophoresis **11A**, 88ff
– – applications **11A**, 88
– – flow chart **11A**, 89
– – limitations **11A**, 88ff
– – method **11A**, 88f
– by hybridization **11A**, 91ff
– – oligonucleotide probes **11A**, 91ff
– – probe selection and designs **11A**, 91ff
– by restriction fragment length polymorphisms **11A**, 87
– by terminal restriction fragment length polymorphisms **11A**, 90f
– community structure **11A**, 91ff
– diversity and population dynamics **11A**, 88
– dot blot hybridization, advantages and limitations **11A**, 98
– – applications **11A**, 97f
– – method **11A**, 97
– full cycle RNA approach **11A**, 96
– *in situ* hybridization, analytical tools **11A**, 100f
– – applications **11A**, 101f
– – combined with other methods **11A**, 103
– – detection of *Zoogloea ramigera* **11A**, 101
– – limitations **11A**, 103f
– – method **11A**, 98ff
– – population analyses **11A**, 101

– – with a fluorescently labeled oligonucleotide probe **11A**, 100
– molecular characterization of isolates **11A**, 87f
microbial conversion *see* fermentation
microbial corrosion *see also* biocorrosion
– by biofilm/biofouling **10**, 284
– by CO **10**, 283f
– by complexation **10**, 282
– by emulsifying agents **10**, 285
– by exoenzymes **10**, 285
– by H₂S **10**, 283f
– by inorganic acids **10**, 281f
– by NO **10**, 283f
– by NO₂ **10**, 283f
– by organic acids **10**, 282
– by organic solvents **10**, 283
– by salt stress **10**, 283
– inhibition of **10**, 265ff
– mechanisms **10**, 280ff
microbial culture, oxygen demand **3**, 190
– scale-up **3**, 189
microbial death **3**, 163
microbial deterioration *see* biodeterioration
microbial drugs **7**, 2ff
– catabolite regulation **7**, 11
– phase-dependency of biosynthesis **7**, 11
microbial enhanced oil recovery **10**, 313
microbial enzymes, oxidation reactions **8B**, 373f
microbial esterases **8B**, 357f
microbial extracellular polysaccharide (EPS) *see* exopolysaccharides
microbial growth **1**, 111ff, **3**, 163, 331f
– – *see also* mycelial growth, bacteria
– acceleration phase **1**, 152
– adaptive non-linear observers **3**, 337
– anaerobic –, kinetics **3**, 299ff
– bacterial – **1**, 113
– balanced – **1**, 139
– carbon- and energy source **10**, 275
– characterization **1**, 118ff
– clostridia **1**, 293ff
– conditions, for lactic acid bacteria **1**, 336ff
– – for streptomycetes **1**, 438
– continuous culture, steady state level **1**, 156f
– countermeasures **10**, 285ff
– – biological – **10**, 290
– – chemical treatment **10**, 289f
– – constructive – **10**, 290
– – physical treatment **10**, 287ff
– death phase **1**, 153
– diauxic growth **1**, 154
– doubling time **1**, 118, 139, 585
– effect of, pH **9**, 198
– – temperature **9**, 198f
– effects of, alkanol **1**, 135
– – ammonia **1**, 140
– – ammonium **1**, 177
– – carbon dioxide **1**, 135

– – carbon source **1**, 132
– – energy supply **1**, 140
– – environment **1**, 127ff
– – mechanical stress **1**, 129
– – medium pH **1**, 131
– – nitrogen source **1**, 133
– – nutrients **1**, 137, 146
– – organic acids **1**, 136
– – oxygen **1**, 134
– – sodium ion concentration **1**, 131
– – temperature **1**, 127f
– – water activity **1**, 130
– effects on macromolecular composition **1**, 138
– energetics **1**, 168
– – cell integrity maintenance **1**, 169
– – cell material synthesis **1**, 169
– – energy drain manipulation **1**, 169
– – energy-spilling reaction **1**, 169
– energy balance **1**, 142
– energy demand **1**, 141
– energy requirements **1**, 141
– estimators **3**, 336
– exponential – **3**, 97
– – in batch cultures **3**, 380
– exponential growth **1**, 152
– factors influencing – **10**, 272ff
– – minerals **10**, 273
– – oxygen usage **10**, 274
– – pH **10**, 274f
– – redox potential **10**, 274
– – temperature **10**, 273f
– – water activity **10**, 272f
– – water availability **10**, 272f
– immobilized microorganisms **1**, 235
– in biofilms **10**, 275f
– in paper production **10**, 307
– kinetics **1**, 152f
– lag phase **1**, 152
– limiting steps **1**, 139
– maintenance coefficient **3**, 300
– measurement **1**, 118ff
– – *see also* cell number estimation, cell mass estimation
– minimum water potential **10**, 273
– modes **1**, 113ff
– nutritional requirements **9**, 199ff
– – growth factors **9**, 200
– – macro elements **9**, 199f
– – trace elements **9**, 200
– on insoluble substrates **3**, 309
– on various substrates, actinomycetes **9**, 196
– – algae **9**, 196
– – bacteria **9**, 196
– – higher fungi **9**, 197
– – molds **9**, 197
– – saturation constants **9**, 192
– – yeasts **9**, 197
– parameters **9**, 189ff

– physical constraints **1**, 138
– Pirt's model **3**, 301
– product inhibition **3**, 98
– rates in batch culture **1**, 153
– relaxation time **1**, 138
– retardation phase **1**, 152
– segregated model **3**, 332
– specific growth rate **9**, 189
– stationary phase **1**, 152
– substrate inhibition **3**, 98
– substrate uptake **1**, 139, **9**, 190f
– – active transport **9**, 190f
– – facilitated diffusion **9**, 190
– – passive diffusion **9**, 190
– true yield **3**, 300
– variability **1**, 138
– yeasts **1**, 115f
microbial growth media **1**, 143ff
– bagasse **1**, 150
– carbon compounds **1**, 143
– carbon limitation **1**, 170f
– cassava **1**, 149
– cheese whey **1**, 149
– chemostat pulse method **1**, 146
– complex natural media **1**, 148
– computer-assisted optimizing **1**, 147f
– corn steep liquor **1**, 149
– energy limitation **1**, 170f
– factorial design **1**, 146
– for clostridia **1**, 294f
– for methanogens **1**, 260
– for streptomycetes **1**, 436f
– growth factors **1**, 145
– macro-elements **1**, 145
– material balances **1**, 145
– medium optimization **1**, 145
– molasses **1**, 148
– nitrogen compounds **1**, 144
– nutrient limitation **1**, 172ff
– optimization, by neural network **1**, 147
– Placket-Burman two-factorial design **1**, 147
– trace elements **1**, 145
microbial lipids *see* lipids
microbially influenced corrosion *see* microbial corrosion
microbial medium, Newtonian behavior **3**, 89
– non-Newtonian behavior **3**, 89
– power law model **3**, 89f
– rheological properties **3**, 89
microbial metabolism, and biological waste gas purification **11C**, 280ff
– of C1 compounds **9**, 178ff
– of ethanol **9**, 181f
– of glycerol **9**, 182
– of hydrocarbons **9**, 176ff
– of lipids **9**, 186ff
– – energy balance **9**, 188
– – fatty acid degradation **9**, 186f

– of polysaccharides **9**, 185f
– of saccharides **9**, 182ff
– – Entner-Doudoroff pathway **9**, 183
– – glycolysis pathway **9**, 182f
– – glyoxylate cycle **9**, 184
– – hexose monophosphate pathway **9**, 183
– – Krebs cycle **9**, 184
– – phosphoketolase pathway **9**, 184
– regulation of **11C**, 287f, 288
– respiratory chain **9**, 188f
microbial metal leaching *see also* bioleaching
– mechanisms **10**, 191ff
– – direct – **10**, 194ff
– – indirect – **10**, 194ff
– – models **10**, 194ff
microbial pathways, for amino acid production **8B**, 313ff
microbial physiology **1**, 164
microbial polysaccharides *see* polysaccharides
microbial population, detection limit of **11A**, 98
microbial populations, measurement in subsoil **11B**, 403
microbial product, modification **1**, 204ff
– – spectrum **1**, 204ff
– – yield **1**, 204ff
microbial product formation **1**, 170ff
– effects of, carbon limitation **1**, 170f
– – carbon source **1**, 133
– – medium composition **1**, 170ff
– – microbial growth **1**, 169
– – nutrient limitation **1**, 170ff
– – potassium limitation **1**, 178
– energetics **1**, 167f, 169
– energy limitation **1**, 170f
– primary metabolites **1**, 170
microbial production, of acetic acid **6**, 381ff
– of acetone **6**, 229ff
– of butanol **6**, 229ff
– of citric acid **6**, 307ff
– of gluconic acid **6**, 347ff
– of glycerol **6**, 205ff
– of isopropanol **6**, 229ff
– of lactic acid **6**, 293ff
– of organic acids **6**, 363ff
– of PHA **6**, 403ff
– of polyhydroxy alcohols **6**, 220
– of polysaccharides **6**, 613ff
– of poly(3HB) **6**, 403ff
– of 2,3-butanediol **6**, 269ff
microbial products, low-molecular weight – **3**, 560
microbial siderophores *see* siderophores
microbial systems, applications of stirred tank models **4**, 330ff
microbial transformation *see* fermentation
microbodies **1**, 33, 36
microcapsule, for long-term transplantation, design of **10**, 559

microcapsules *see also* alignate beads
– diffusion within **10**, 558ff
– encapsulated-cell therapy **10**, 548ff
– fibrotic overgrowth **10**, 562ff
– for long-term transplantation **10**, 558ff
– permeability **10**, 561f
– size **10**, 558ff
– stability **10**, 560f
– swelling **10**, 560f
– topography **10**, 562ff
– use of highly viscous alginate **10**, 561f
– use of low-viscosity alginate **10**, 562
microcarriers, culture, agitation effects **3**, 228ff
– – bubble-induced cell damage **3**, 231
– in stirred tank bioreactors, hydrodynamic cell damage **3**, 112
– macroporous – **3**, 108f
– solid – **3**, 107
microcin C7, synthesis in *E. coli* **7**, 61ff
microcins **7**, 61
microcystin, inhibition of protein kinase C **7**, 677
microelectrospray technique **5B**, 257
microencapsulated cells, monitoring of the functional state **10**, 564f
– size **10**, 558ff
– transplantation site **10**, 564f
microencapsulation **3**, 441
– in cell therapy **10**, 547ff
– of permanent cell lines **10**, 566
– of tissue **10**, 555ff
– – pancreatic islets **10**, 555f
– – parathyroid tissue **10**, 556f
microencapsulation techniques **10**, 549f
microexamination, of mashes **6**, 114
microfilaments **1**, 30
microfilm, of microorganisms **1**, 231
– – *see also* biofilm
microfiltration, use in water treatment **11C**, 405ff
microfiltration (MF) **3**, 440, 487ff
– effects of solids concentration **3**, 495
microfluidizers, for cell disruption **3**, 518f
– – efficiency **3**, 519
– – residence times **3**, 519
microfungi, modification of brown coal **10**, 161
microgravity **10**, 508ff
– – *see also* gravity
– clinostating **10**, 508
– effects on, cell cultures **10**, 509ff
– – chromosomal anomalies **10**, 509f
– – DNA stability **10**, 517
– – electrofusion of cells **10**, 515f
– – eurkaryotes **10**, 509ff
– – gene expression **10**, 517
– – genetic stability of bacteria **10**, 509
– – *Nicotiana* protoplasts **10**, 511
– – plant cell wall synthesis **10**, 510f
– – plant protoplasts **10**, 510ff
– – prokaryotes **10**, 509

– – sunflower hypocotyl protoplasts **10**, 514f
– – the cytoskeleton system **10**, 512
– – *Vicia faba* protoplasts **10**, 511f
– space flights **10**, 508
– stimulated by **10**, 514f
micro-grooved channel gel electrophoresis **5B**, 209
micro-injection of DNA **2**, 580f, 747ff, 748, **5A**, 430
– concateneric structures **2**, 758f
– DNA size **2**, 749
– DNA solution preparation **2**, 752
– experimental **2**, 754
– frequency of DNA integration **2**, 749
– into early embryos **2**, 748ff
– into pronuclei of fertilized oocytes **2**, 748, 755
– molecular DNA structure **2**, 749f
– target immobilization **2**, 580
– target range **2**, 581
– technical setup **2**, 754, 757
– to fish oocytes **2**, 811f
– to intact plant cells **2**, 581
– to plant embryoids **2**, 581
– to plant protoplasts **2**, 581
micro-injection of mRNA **2**, 747
micromanipulations **5B**, 438
micromixing, in stirred tanks **4**, 313ff, 334, 338f
Micromonospora, chemotaxonomic marker **1**, 435
– G + C content **1**, 435
– morphology **1**, 435
– production of antibiotics **1**, 445
microorganism handling **2**, 844ff
– hygienic measures **2**, 844
– risk group dependent **2**, 850
microorganism oxidation **8B**, 380f
– preparation of leukotriene analogs **8B**, 380f
microorganism pathogenicity **2**, 840ff
– clinial material **2**, 841
– environmental organisms **2**, 841
– plant pathogens **2**, 843
– recombinant-DNA organisms **2**, 843f
– risk groups **2**, 848
– wild-type strains **2**, 841ff
microorganism reduction, of polycyclic compounds **8B**, 378ff
microorganism risk groups **2**, 840
– criteria **2**, 841, 848
– environmental risk **2**, 842
– – classes **2**, 843
– harmless **2**, 842, 848
– high risk **2**, 842
– low risk **2**, 842
– medium risk **2**, 842
– national lists **2**, 842
microorganisms **10**, 461
– – *see also* bioreactor aeration
– – *see also* pathogenic microorganisms
– acidophiles **10**, 68

– active in composting **11C**, 54ff
– – main groups of **11C**, 56ff
– aerobic degradation of organic pollutants **11B**, 145ff
– aerogenic distribution of **11C**, 238
– alkaliphiles **10**, 68f
– autoaggregation of **11A**, 65ff
– batch culture **1**, 150
– biodegradation of nitro compounds **11B**, 273ff
– cell envelope **3**, 506
– cell mass estimation **1**, 121ff
– cell number estimation **1**, 119ff
– cellulolytic – **11A**, 25f
– chemolithoautotrophic – **11C**, 280
– chemoorganoheterotrophic – **11C**, 280
– classification **12**, 244f, 267, 554ff
– classification of, according to their temperature range **11C**, 57
– continuous culture **1**, 155
– cryo-conservation **3**, 705
– culture techniques **1**, 150ff
– degradative –, constitutive variants **11B**, 409f
– degradative reactions **11C**, 280f
– depolymerizing brown coal **10**, 163
– distribution in air **3**, 176
– domestication **12**, 41f
– effects of, ethanol **1**, 136
– – mechanical stress **1**, 129
– – medium pH **1**, 131
– – temperature **1**, 128
– – water activity **1**, 130
– energy demand **1**, 141
– energy metabolism **3**, 129
– environmental requirements, in biotechnological manufacturing **3**, 130
– fed-batch culture **1**, 154
– freeze-drying **3**, 705
– genetically engineered –, in fermentation processes **3**, 283ff
– – stable strain isolation **3**, 289
– GILSP **12**, 614f
– growth *see* microbial growth
– growth kinetics **1**, 133
– halophiles **10**, 69f
– heat evolution **1**, 141
– heat production by **11C**, 37ff
– heat resistance **3**, 164
– high-cell-density cultivation of **10**, 574f
– – overview **10**, 575
– homogenization **3**, 505ff
– – rate constants **3**, 514
– hyperthermophiles **10**, 64ff
– immobilization of **11B**, 417f
– – on different carriers **11B**, 421
– inactivation methods **3**, 162ff
– in biofilms **11C**, 448
– in biofilters **11C**, 323f
– in compost, as biocontrol agents **11C**, 78

– insect pathogenic – **10**, 496ff
– in soil **11B**, 95f
– in solid matrices **10**, 287
– in the atmosphere **10**, 286f
– involved in biodeterioration, algae **10**, 271
– – bacteria **10**, 269ff
– – cyanobacteria **10**, 269ff
– – fungi **10**, 272
– – lichens **10**, 272
– in water **10**, 287
– mechanical strength **3**, 506
– metabolic types **10**, 275
– metal accumulation by – **10**, 225ff
– metal transformation by – **10**, 225ff
– micro-aerophilic – **3**, 306
– mixed culture **3**, 298
– mixed-pure culture **3**, 298
– nutrient requirements **1**, 133
– nutritional groups **3**, 128
– nutritional requirements **3**, 128ff
– of the composting process **11C**, 54ff
– PCP-degrading – **11B**, 260
– phage host groups **1**, 549
– physical properties **3**, 160
– psychrophiles **10**, 63ff
– PUFA-producing – **10**, 463f
– pure culture **3**, 297
– – back inoculation **3**, 298
– – isolation **3**, 297
– – lyophilization **3**, 297
– – stock – **3**, 297
– quantiation methods **3**, 325, 328
– reduction of inorganic substances, thermodynamic sequences **10**, 274
– removal of, during artificial groundwater recharge **11C**, 438
– risk groups **12**, 554
– safety levels **12**, 582
– secretion of biosurfactants **11C**, 290
– sensitivity towards heat **11C**, 57f
– size **3**, 160
– solubilizing brown coal **10**, 163f
– sources of **8A**, 31f
– specific growth rate **1**, 118
– specific substrate uptake rate **1**, 132
– state assessment **3**, 324f
– strain development **3**, 297f
– substrate uptake **1**, 139
– suspension cultures, oxygen supply **3**, 190
– thermal death kinetics **3**, 162
– – Damköhler number **3**, 173f
– utilizing brown coal as growth substrate **10**, 163
microorganisms, thermal treatment *see* sterilization
microphysiometer **10**, 52
microsatellites **5B**, 43, 55
– instability **5B**, 82

– repeats **5B**, 132
microscopy **1**, 7ff
microsequencing using mass spectrometry **5B**, 150
microtubules **1**, 30
Microviridae **1**, 550
microvolume techniques, in drug screening **10**, 47
Miglitol **8B**, 296ff
– α-glucosidase inhibitor **8B**, 298
– structure of **8B**, 296
migraine **5B**, 117
milbemycins, hydroxylation of **8A**, 496
milk, antibiotics in – **9**, 398
– chemicals in – **9**, 398
– cultured –, definition **9**, 393f
– culturing with lactic acid bacteria **9**, 360f
– fermented, consumption of **9**, 387
– for cheese production **9**, 364ff
– – analysis **9**, 358
– – pretreatments **9**, 358ff
– – quality control **9**, 358
– heterologous protein expression **2**, 801ff
– of various species, composition **9**, 404f
milk clotting, effects of caseins **9**, 366f
milk-clotting enzymes **9**, 680
– in cheese making **9**, 373ff
– – functions **9**, 373f
– – types **9**, 374f
– microbial –, *Endothia parasitica* **9**, 374f
– – *Mucor miehei* **9**, 374f
– – *Mucor pusillus* **9**, 374f
– nomenclature **9**, 375
milk components, carbohydrates **9**, 368
– enzymes **9**, 368f
– lipids **9**, 367f
– minerals **9**, 368
– proteins **9**, 364ff
milk fermentation, history **1**, 327f
milk processing **11A**, 208
milk products, cheese production **1**, 332
– *Enterococcus* **1**, 346
– *Lactobacillus* **1**, 332
– lactococcal starter culture **1**, 349
– *Lactococcus* **1**, 343
– legume-based – **9**, 549ff
– *Leuconostoc* **1**, 341
– yoghurt **1**, 332
milk protein **2**, 799ff
– β-casein **2**, 801
– expression in transgenic animals **2**, 801
– – yield **2**, 802
– lactoglobulin **2**, 801
– milk content modification **2**, 800
– whey acidic proteins (WAP) **2**, 801ff
– – developmental regulation **2**, 804
– – expression construct **2**, 803
– – transgene expression level **2**, 804

milling and mashing process, at higher temperatures **6**, 86f
– at saccharification temperature (cold mash process) **6**, 85
mineral content, of bread **9**, 247
mineral materials, biodeterioration of **10**, 299ff
mineral oil, degradation of, slurry decontamination process **11B**, 340ff
– humification process **11B**, 101f
– microbial degradation **11B**, 101f
mineral oils, biodegradation of **11A**, 386ff
– composition of **11A**, 387
– properties **11A**, 386
– use **11A**, 386
mineral rock, bonding capacity for pollutants **11B**, 50
minerals, estimation methods **1**, 126
– in fermentation media **3**, 129
– influence in microbial growth **10**, 273
mineral surfaces, bacterial attachment to **10**, 200
mineral water industry, wastewater, amounts **11A**, 210
– – concentrations **11A**, 210
– – treatment **11A**, 209f
miniaturization, of detection systems in drug screening **10**, 51f
miniaturization of biosensors **11A**, 136
minisatellite DNA **5B**, 78
minosaminomycin, streptomycin-related aminoglycosides **7**, 428
– structure **7**, 464
MIPS Yeast Genome Database **5B**, 285, 409
misfolding, of proteins **9**, 21
mismatch cleavage **5B**, 115
miso **9**, 518ff
– aging **9**, 520f
– composition **9**, 518
– low-salt – **9**, 521
– production of, koji process **9**, 518ff
– – modified procedures **9**, 521f
– – preparation of soybeans **9**, 520
– types **9**, 518
MITI (Ministry of International Trade and Industry of Japan) **12**, 253ff, 376, 378ff, 596
mitochondrial enyzmes, genetics **1**, 485
mitochondrial genome **2**, 165
mitogenic signal **7**, 644
mitomycin C, antitumor agent **7**, 655
mitosis, cellular signal transduction pathway **7**, 677
mitosis spindle **1**, 30
mitoxantron, antitumor agent, structure **7**, 651
mixed *see* bioreactor mixing
mixed liquor, membrane separation of **11A**, 258
mixed-lymphocyte test, mitogenic impurities in alginates **10**, 553
mixed polyposis syndrome (HMPS) **5B**, 83
mixers, for bread doughs **9**, 267ff

– – continuous – **9**, 267ff
– – optimal temperature **9**, 271
– – speeds **9**, 269f
mixing, in bioreactors **4**, 116ff
– of wastes **11C**, 116f
mixing behavior, in bioreactors, balancing random motions **4**, 118
– – deterministic circulatory flow modes **4**, 117
– – deterministic motions **4**, 118
– – random dispersive motions **4**, 118
mixing length theory **4**, 366f
mixing parameters, estimation of the gas phase **4**, 326
– estimation of the liquid phase **4**, 326
mixing time, influence of aeration **4**, 311ff
– in stirred tanks **4**, 306
MJADH see Mucor javanicus dehydrogenase
MLM lipids **8A**, 135ff
– synthesis of **8A**, 136f
MMDB – Molecular Modeling Database **5B**, 409
mniopetals **7**, 510ff
– structures **7**, 511
MNNG *see* N-methyl-N′-nitro-N-nitrosoguanidine
model, definition **4**, 409
model building, for biotechnological processes, principles **4**, 270ff
– parameter identification **4**, 272f
model equations, description of, batch fermentations **4**, 272
– – continuous fermentations **4**, 272
– – fed-batch fermentations **4**, 272
– – semicontinuous fermentation **4**, 272
– for cell models **4**, 271ff
model identification, definition **4**, 254
– identification functional **4**, 255
– measurement noise **4**, 254ff
– optimization of measurements **4**, 254ff
modeling *see* fermentation modeling
– – *see* protein modeling
– of biofilters for BOM removal, empirical steady state **11C**, 471f
– – mechanistic approaches **11C**, 472ff
– of biogas reactors **11A**, 527ff
– of biological trickling **11C**, 339ff
– of natural attenuation **11B**, 405
modelling, of aerobic growth **4**, 259ff
– – tower loop reactor **4**, 259
– of aerobic wastewater processes **4**, 409ff
– of anaerobic wastewater process **4**, 445ff
– of baker's yeast process **4**, 386ff
– of bioreactor systems **4**, 269ff
– of recombinant fermentation **4**, 487ff
– of stirred tanks **4**, 301ff
– of tower reactors **4**, 351ff
model of critical air and water volumes **11A**, 571
model of critical loads **11A**, 572
model organisms, bridging for positional cloning **5B**, 176

– genome projects of **5B**, 5ff
– – *Arabidopsis thaliana* **5B**, 24ff, 147
– – *Archaeoglobus fulgidus* **5B**, 16ff
– – *Bacillus subtilis* **5B**, 13ff
– – *Caenorhabditis elegans* **5B**, 29ff
– – *Drosophila melanogaster* **5B**, 32ff
– – *Escherichia coli* **5B**, 8ff
– – *Saccharomyces cerevisiae* **5B**, 20
model parameters, estimation by EKF **4**, 231ff
Model Reference Adaptive Control (MRAC), in fermentation modeling **3**, 342
models, of biofilms, dense confluent – **11A**, 70f
– – heterogenous mosaic **11A**, 70f
– – mushroom-like – **11A**, 70
Modernization Act **5A**, 528
modification sites, prediction of **5B**, 335
modified T7 DNA polymerase **5B**, 196
modularity, in automation systems, hardware **4**, 568ff
– – software **4**, 569ff, 596f
molasses **1**, 148, **6**, 17ff, **9**, 326ff
– – *see also* beet molasses, cane molasses
– citrus – **6**, 18
– clarification **6**, 22f
– composition **9**, 207ff, 327
– constituents **1**, 148
– dilution **9**, 328f
– fermentation inhibitors **6**, 22
– growth substances **6**, 22f
– high-test –, composition **6**, 18
– mean specific heat **6**, 24
– pasteurization **9**, 328f
– terminology **6**, 17f
– types **6**, 17
– use in citric acid production **6**, 318f
– – batch fermentation **6**, 323
– – fed-batch fermentation **6**, 323
– viscosity **6**, 23
– yield in beet processing **6**, 13
molds *see* fungi
– – *see* oleaginous molds
– immobilized in polymer beads **11B**, 418
– use of aromatic compounds as growth substrates **11B**, 159
molecular biology databases **5B**, 281ff
– bibliographic – **5B**, 283
– genomic – **5B**, 285f
– heterogeneity of the data **5B**, 291f
– InterProf **5B**, 289
– nucleotide sequence – **5B**, 283f
– protein sequence – **5B**, 286ff
– protein signature – **5B**, 289
– taxonomic – **5B**, 283
Molecular Biology Institute of Taiwan **12**, 403
molecular chaperones **3**, 534, 537ff, **9**, 22f
– as folding mediators **3**, 537
– catalytic – **3**, 540
– non-catalytic – **3**, 537ff

– SecB, from *Escherichia coli* **3**, 534
– use in inclusion body folding **5A**, 120f
molecular diffusion **4**, 115f
– – *see* diffusion
– across eddy boundaries **4**, 115
– across interphase boundaries **4**, 110f
– in biolfilms **4**, 418f
molecular evolution **2**, 349ff
– *in vitro* selection cycle **2**, 351
– peptide library **2**, 351
– principle **2**, 350
– random cassette expression **2**, 351
– random oligonucleotide-mediated **2**, 350
molecular genetics, of clostridia **1**, 293
molecular interaction data **5B**, 400, 402
– integration of **5B**, 401
– pairwise interactions **5B**, 401
molecular interaction databases **5B**, 404
molecular machines **5B**, 354f
molecular markers, AFLP **5B**, 168
– RAPDs **5B**, 167f
– RFLPs **5B**, 167
– SSCP **5B**, 168
– SSRs **5B**, 168
molluscs **10**, 445f
– as sources of nonribosomal peptides **7**, 314
– as sources of ribosomal peptides **7**, 301
– cytostatic/antitumor compounds of **10**, 452
– substances of **10**, 446
Moloney murine leukemia virus (Mo-MLV),
    genome structure **5A**, 399
Moloney murine sarcoma virus (Mo-MSV) **5A**,
    397
molten globule, CD spectra **9**, 20
Monascus purpureus, in ang-kak fermentation **9**,
    526
– pigments **9**, 526f
Monbusho (Ministry of Education and Science of
    Japan) **12**, 378ff
monellin, gene expression **10**, 579
– proteinaceous sweetener **10**, 577ff
monensin, hydroxylation of **8A**, 496
– synthesis of **8B**, 358f
Monilia sitophila, bread spoilage **9**, 312
– in pozol fermentation **9**, 549
monitoring, in soil remediation with heap tech-
    nique **11B**, 323
– of biovailability of pollutants **11B**, 82f
– of contaminated soil areas **11B**, 39f
– of environmental processes, cyanide **11A**, 128
– – halogenated organics **11A**, 133f
– – heavy metals **11A**, 127f
– – naphtalene **11A**, 134
– – nitrate **11A**, 128
– – nitrite **11A**, 128
– – pesticides **11A**, 131ff
– – phenolic compounds **11A**, 130f
– – sulfide **11A**, 128f

– – sulfite **11A**, 128f
– – sum parameters **11A**, 134ff
– – use of biosensors **11A**, 125ff
– of *in situ* soil remediation **11B**, 366ff
– of wastewater treatment plants **11A**, 109ff
– – AOX **11A**, 121
– – carbon removal **11A**, 115ff
– – enhanced biological P removal capacity **11A**,
    119
– – heavy metals **11A**, 121
– – influence of treatment systems **11A**, 119
– – liquid-solid separation **11A**, 120
– – nitrogen flows **11A**, 118f
– – PAH **11A**, 121
– – quality control **11A**, 122
monkey kidney cell lines (COS) **5A**, 248f
monoacylglycerides (MAGs), formation from glyc-
    erol **8A**, 142
– formation from triglycerides **8A**, 141
– general features **8A**, 139
– lipase-catalyzed syntheses **8A**, 140
monoclonal antibodies **2**, 807
– hapten affinity binding sites **8A**, 269
– yield in transgenic mice **2**, 808
monoclonal antibodies (MAb) *see also* antibodies,
    therapeutic antibodies
– against tumor antigens **5A**, 294
– anti-endotoxin – **5A**, 331ff
– anti-TNF **5A**, 134, 332f
– CDP571 **5A**, 343ff
– conjugation with calicheamicin **5A**, 358
– cross-reactivity **5A**, 295
– engineering of **5A**, 343ff, 540
– half-life, animal models **5A**, 298
– – in humans **5A**, 298
– – stability **5A**, 298
– human – **5A**, 228ff
– – *in vitro* production **5A**, 229
– – production in transgenic animals **5A**, 228
– human anti-mouse antibody response **5A**, 299
– human/mouse chimeric fragment **5A**, 365ff
– immunogenicity **5A**, 299f
– in therapy **5A**, 21, 290ff, 299
– – animal models **5A**, 296ff
– in xenotransplantation **5A**, 146
– medical applications of recombinant proteins
    5A, 134
– pharmaco-toxicological testing **5A**, 307ff
– preclinical testing, use of transgenic models
    5A, 297
– – use of xenografts **5A**, 297
– reduction of pyrexia **5A**, 294
– registration of **5A**, 498ff
– – authorities **5A**, 498ff
– – in Europe **5A**, 498
– – in the USA **5A**, 498ff
– – laws **5A**, 498ff
– serotype-specific – **5A**, 331

– therapeutic applications **5A**, 311ff
– – targets for intervention **5A**, 314
– therapeutic grade –, purity **5A**, 264
– therapy, clinical trials **5A**, 335
– treatment of chronic diseases **5A**, 290
– use in HIV gene therapy **5A**, 477f
– use in sepsis therapy **5A**, 330ff
– use in tumor therapy **5A**, 135
monoclonal antibody anti-intercellular adhesion molecule-1 **5A**, 131
monoclonal antibody-based products **5A**, 222f
Monod equation **4**, 419, **11C**, 268
Monod kinetics **1**, 132, **4**, 274, 278, 286, 289, 295
Monod model **3**, 86, **4**, 524
monoecism **2**, 77
mono-isotopic standards **5B**, 330
mononitriles, biotransformations of **8A**, 302, 307f
monooxygenase, synthesis of (*R*)-2(4-hydroxyphe-noxy)proprionic acid **8B**, 282
monooxygenases, asymmetric sulfoxidations in bacteria **8A**, 570ff
– bacterial degradation of xenobiotics **11B**, 149
– biodegradation of nitrophenols **11B**, 281
– cytochrome P-450-dependent – **8A**, 477f
– degradation of organic compounds **11B**, 171
– dihydroxylation reactions **8A**, 511ff
– in alkane biodegradation **11B**, 198
– oxygenolytic dehalogenation **11B**, 248
– transformation of TCE in plant tissues **11B**, 375
monosaccharide carboxylic acid dehydratases **8B**, 115ff
– listing of **8B**, 116f
– of *Pseudomonas putida* **8B**, 115
monosodium L-glutamate **6**, 563
– food uses **6**, 467f
monoterpene indole alkaloids, biosynthetic pathway **7**, 609
– from plant cell cultures **7**, 606ff
monoterpene metabolism, of pseudomonads **1**, 415
monoterpenes, alkenic –, dihydroxylation of **8A**, 512f
– biotransformation of, by plant cell cultures **8A**, 492
– hydroxylation of **8A**, 491f
monoterpenoid indole alkaloids **8A**, 352ff
monoterpenoids, bicyclic – **10**, 404ff
– biotransformation of, acyclic – **10**, 384ff
– monocyclic – **10**, 390ff
monoterpenyl esters, stereoselective esterification **10**, 402ff
– synthesis of, by lipases **10**, 397ff
– – by transesterification **10**, 400
monothiophosphate, stereospecific phosphoryla-tion of **8B**, 229f
Monterey cheese **9**, 376
Morchella **1**, 525

morel **1**, 532
Morinda citrifolia, accumulation of anthraqui-nones **7**, 603
– shear stress effect **1**, 587
morphinan alkaloids, from plant cell cultures **7**, 613
morphine, biosynthesis in *Papaver somniferum* **8A**, 342f
– dimerization of **8A**, 345
– hydroxylation of **8A**, 496
– potential analgesic **8A**, 340
– redox reactions of **8A**, 350
– transformation by *Pseudomonas putida* **8A**, 347
morphine alkaloids **8A**, 339ff
– as potent analgesics, codeine **8A**, 340f
– – heroin **8A**, 340
– – morphine **8A**, 340f
– – naloxone **8A**, 345
– – naltrexone **8A**, 345
– biosynthesis in *Papaver somniferum* **8A**, 342f
– microbial metabolism of **8A**, 344ff
– microbial transformations of heroin **8A**, 351
– production of, hydromorphone **8A**, 350
– – hydroxodone **8A**, 350
– transformation by bacteria **8A**, 345
– – *Pseudomonas putida* **8A**, 346ff
morphine dehydrogenase **8A**, 346ff
– of *Pseudomonas putida*, characteristics of **8A**, 348f
– structural gene for **8A**, 349
morphine reductase **8A**, 346f
– constitutive expression of **8A**, 349
morphinone reductase, structural gene for **8A**, 349
morphological differentiation, genes affecting also antibiotic production **7**, 84f
morphology, of bacilli **1**, 368
– of bacteria **1**, 13ff, 114
– of euactinomycetic genera **1**, 435
– of eukaryotic cells **1**, 36f, 578
– of filamentous fungi **1**, 517f
– of immobilized microorganisms **1**, 231ff
– of methanogens **1**, 256f
– of phages **1**, 547
– of plant cells **1**, 36f, 578
– of streptomycetes **1**, 435, 437
mortality **12**, 123
Mortierella alpina, formation of arachidonic acid **7**, 161
– formation of eicosapentaenoic acid **7**, 162
Mortierella isabellina **8A**, 486f
– benzylic hydroxylase model **8A**, 480
– benzylic hydroxylation **8A**, 487f
mosaicism **5B**, 74f
MOST (Ministry of Science and Technology of Korea) **12**, 390ff
motif databanks **5B**, 137

motif search tools **5B**, 366
motilides, receptor agonists **7**, 122f
mouse genetic engineering **2**, 749ff
– application **2**, 871ff
– by retroviral vectors **2**, 768ff
– embryonic stem cells **2**, 760ff
– targeted mutation **2**, 764f
Mouse Genome Database **5B**, 285
mousiness, of wine **9**, 495
mouth flora **1**, 330
Mozzarella cheese **9**, 378
MPKSN (Malaysian National Council for Scien-
    tific Research and Development) **12**, 417
mRNA **2**, 239
– – *see* messenger RNA
– analysis with PCR **2**, 341
– anti-sense RNA **2**, 809
– cytoplasmic turnover **2**, 678f
– degradation **2**, 239
– half-life **2**, 678
– *in situ* detection **11A**, 104
– of ferritin/transferrin receptor **2**, 679
– open reading frame (ORF) **2**, 688
– polycistronic, in animals **2**, 688
– processing **2**, 239
– recording **12**, 165f
– ribosomal binding site (rbs) **2**, 246f
– secondary structure effects **2**, 685
– sequences affecting translation efficiency **2**, 684
– splicing **2**, 143
– translation pattern of polycistronic RNA **2**, 688
– untranslated region (UTR) **2**, 678ff
mRNA processing of heterologous genes *see* tran-
    script processing
MRS agar **1**, 347
MS/MS spectra, definition of **5B**, 255
MTBE *see* methyl-*t*-butyl ether
– – *see* methyl *tert*-butyl ether
MTIE (Ministry of Trade, Industry and Energy of
    Korea) **12**, 393
mucidin, biosynthesis **7**, 500
– structure **7**, 501
mucin **5A**, 12
muconic acid production, by *Pseudomonas putita*
    2, 64ff
Mucor **1**, 522
Mucor circinelloides, fatty acid profiles **7**, 158
Mucor corticolus, in sufu fermentation **9**, 524
Mucor hiemalis, in sufu fermentation **9**, 524
Mucor javanicus alcohol dehydrogenase
    (*MJADH*), applications of **8A**, 452f
– characteristics of **8A**, 452f
Mucor miehei **9**, 682
– milk-clotting enzyme **9**, 375
Mucor praini, in sufu fermentation **9**, 524
Mucor pusillus, milk-clotting enzyme **9**, 375
Mucor racemosus, infection of wine grapes **9**, 500
– in pozol fermentation **9**, 549

– in sufu fermentation **9**, 524
Mucor rouxianus, in pozol fermentation **9**, 549
Mucor silvaticus, in sufu fermentation **9**, 524
Mucor sp. lipases, synthesis of heterocyclic com-
    pounds **8B**, 385f
Mucor subtilissimus, in sufu fermentation **9**, 524
Mülheim process **11C**, 458
mulching machine **11C**, 116
multi-domain proteins **5B**, 133
multidrug resistance **5B**, 93
– of tumor cells **7**, 540, 551, 672
multi-enzyme systems, Baeyer-Villiger biotransfor-
    mations **8A**, 567
– classical regulatory mechanisms **9**, 124ff
– – cooperativity **9**, 125f
– – feedback inhibition **9**, 125
– cooperative feedback inhibition **9**, 130ff
– in biotransformations **8A**, 33
– kinetics **9**, 121ff
– metabolic control analysis **9**, 126ff
– – concentration control coefficient **9**, 127
– – elasticity **9**, 127
– – flux control coefficient **9**, 126
– – response coefficient **9**, 127
– metabolic control theory **9**, 124
– metabolic fluxes, manipulation of **9**, 133ff
– summation relationships **9**, 128f
multifactorial diseases **5B**, 65ff
– criteria for **5B**, 68
– identification of genes **5B**, 79ff
– population incidence **5B**, 66f
multigene families **5B**, 149, 153
multiloop control **4**, 519f
– cascade control, schematic diagram **4**, 520
– ratio control, schematic diagram **4**, 520
multi-media filters, design of **11C**, 391
multiphase bioreactors **3**, 442
multiple domains **5B**, 337
multiple panes **5B**, 386
multiple sclerosis **5A**, 142f
– medical applications of recombinant proteins
    5A, 140, 142
– monoclonal antibody therapy **5A**, 319
– role of tumor necrosis factor **5A**, 344
– vaccines **5A**, 138
multiple sequences alignment **5B**, 130, 133, 338
multiplex DNA sequencing **5B**, 202
Multiplexing **4**, 56ff
Multi-substrate kinetics **4**, 279
multi-substrate reactions, kinetics **9**, 36ff
municipal sewage **11A**, 161ff
– concentrations of detergents **11A**, 176
– effect of sewer **11A**, 167
– imported water **11A**, 165
– major components **11A**, 166
– metal concentrations **11A**, 180
– natural organic substances **11A**, 181
– origin of, halogens **11A**, 179

– – nitrogen **11A**, 179
– – phosphorus **11A**, 179
– – sulfur **11A**, 179
– sources of **11A**, 163ff
– specific daily amount **11A**, 166
– synthetic organic substances **11A**, 181
– TOC **11A**, 173ff
municipal solid waste **11C**, 167ff
– – *see also* solid waste, waste
– anaerobic digestion processes **11C**, 156ff
– collection of **11C**, 9
– combined mechanical and biological treatment
  11C, 203
– composition **6**, 180
– composition of **11C**, 194
– – in Europe **11C**, 204
– composting of **11C**, 15ff
– definition of **11C**, 192
– density of, influence of compaction **11C**, 193
– emission potentials **11C**, 185
– generation rate **11C**, 192f
– in industrial ethanol production **6**, 166
– – economic analysis **6**, 176
– landfill gas, composition of **11C**, 199
– landfills, regulations for **11C**, 197
– landfill systems **11C**, 167
– non-recyclable –, combined mechanical and biological processing of **11C**, 24f
– – direct anaerobic fermentation of **11C**, 24
– pretreatment before landfilling **11C**, 175
– recycling of **11C**, 8ff
– treatment of, biotechnology **11C**, 5ff
municipal waste *see also* waste
– exploitation of **11C**, 7ff
– generation of **11C**, 7f
– – in Germany **11C**, 7
municipal waste incineration, leaching of fly ash
  from **10**, 211
municipal wastewater **11A**, 161ff, 420
– analysis of **11A**, 169ff
– biochemical characteristics **11A**, 169
– chemical characteristics **11A**, 168f
– chlorination of **11A**, 186
– common analytical methods **11A**, 171
– composition of **11A**, 167ff
– – AOX **11A**, 177
– – chlorides **11A**, 178f
– – classification of compounds **11A**, 167ff
– – detergents **11A**, 175
– – hydrocarbons **11A**, 175ff
– – metal compounds **11A**, 179f
– – natural organic substances **11A**, 181
– – nitrogen-containing compounds **11A**, 178
– – non-metal compounds **11A**, 177ff
– – organic nitrogen compounds **11A**, 177f
– – persistent, accumulateable and toxic substances **11A**, 180f
– – phenols **11A**, 177

– – phosphorus compounds **11A**, 178f
– – semimetal compounds **11A**, 179f
– – single organic substances **11A**, 179ff
– – sulfur compounds **11A**, 178f
– – synthetic organic substances **11A**, 181
– decentralized biological treatment **11A**, 351,
  358f
– definition of dangerous substances in origin of
  11A, 163ff
– fractionation of **11A**, 422
– group parameters **11A**, 170, 174ff
– – adsorbable organic chlorinated compounds
  (AOX) **11A**, 177
– – for organic compounds **11A**, 174ff
– – hydrocarbons **11A**, 175ff
– – organic nitrogen compounds **11A**, 177
– – phenols **11A**, 177
– inhabitant-specific influent **11A**, 165f
– inhabitant-specific parameters **11A**, 172
– origins of pollution **11A**, 163f
– parameter categories **11A**, 170
– physical characteristics **11A**, 168f
– physical-chemical parameters **11A**, 171
– pollutants, concentration limits **11A**, 164
– self-monitoring sewage treatment plants **11A**,
  172
– sewage charge **11A**, 185
– sources **11A**, 433f
– specific pollution per person **11A**, 176
– standard operating methods **11A**, 172f
– standards of analysis **11A**, 170f
– sum parameters **11A**, 170
– – values **11A**, 172ff
– treating methods **11A**, 184ff
– – biological processes **11A**, 186
– – chemical processes **11A**, 186
– – combinations of methods **11A**, 186f
– – physical processes **11A**, 186
– typifying physical-chemical methods **11A**, 173
– volume of **11A**, 165f
municipal wastewater treatment, continuous flow
  processes **11A**, 321ff
– on line measurements **11A**, 172ff
– parameters for organic compounds **11A**, 174ff
– sampling **11A**, 172, 175
– sequential processes **11A**, 321ff
Munster cheese **9**, 377f
murein **1**, 19f
murine leukemia virus (MLV) **2**, 767ff
murine mammary tumor virus (MMTV) **2**, 788
MURS (Universal Movement for Scientific Responsibility) **12**, 151
muscarine **7**, 112, 123
muscarinic acetylcholine receptor, agonists **7**, 112
– antagonists **7**, 112ff
muscimol, insecticidal compound **7**, 515f
– structure **7**, 516
muscle contraction, filaments **5A**, 28

– schematic diagram of **5A**, 28
muscle proteins **5A**, 31
mushroom lipoxygenase, hydroperoxide formation **8A**, 503
mushrooms *see* basidiomycetes
– *Agaricus bisporus* genes **10**, 577
– biological properties **10**, 580
– biotechnological capacities of **10**, 577ff
– edible **1**, 534
– fatty acid composition **9**, 225
– medical properties **10**, 578, 580
mussels, adhesive proteins **10**, 467
must *see* grape must
mustard gas **2**, 24
mutacin, structure **7**, 336
mutagen **2**, 20ff
– alkylating agent **2**, 23ff
– base analog **2**, 26
– DNA replication affectant **2**, 26
– dose-response curve **2**, 31
– far ultraviolet radiation **2**, 20
– – base substitution **2**, 20
– – repair rate **2**, 21
– frameshift mutagen **2**, 26
– freezing and thawing **2**, 20
– heat **2**, 20
– hydroxylamine **2**, 23
– ionizing radiation **2**, 21
– near ultraviolet radiation **2**, 21
– nitrous acid **2**, 22f
– – dose-response curve **2**, 32
– PUVA treatment **2**, 21
mutagenesis **2**, 5ff, 438
– – *see also* transposon mutagenesis
– by PCR **2**, 409
– deletions **2**, 406f
– development of phenyl alanine overproducing organisms **8B**, 315ff
– directed in embryonic stem cells **2**, 785
– dominant negative mutation **2**, 782
– gapped duplex method **2**, 407ff
– genetic ablation **2**, 784
– history of investigation **2**, 7
– induction **2**, 20ff
– industrial application **2**, 30ff
– in *Saccharomyces* **2**, 516ff
– – with oligonucleotides **2**, 518
– insertional **2**, 29, 406f
– "Kunkel" method **2**, 407f
– mechanisms **2**, 7
– methods **2**, 439f
– mutant identification **2**, 446
– non-selectable mutation **2**, 438ff
– of Streptomycetes **2**, 468ff
– of synthetic genes **2**, 348f
– – cassette mutagenesis **2**, 349f
– – PCR-mediated **2**, 348f

– – site-specific **2**, 348f
– oligonucleotide-directed **2**, 345, 407
– oligonucleotide-mediated **2**, 345, 408
– optimization **2**, 31
– random **12**, 167
– random insertion **2**, 781
– region-specifc **2**, 409
– selective cell type suppression **2**, 784
– site-directed **12**, 167
– site-directed by PCR **2**, 341
– site-specific **2**, 30, 345, 406ff
– using nucleotide thioanalogs **2**, 407f
– with base specific mutagens **2**, 409
– with sodium bisulfite **2**, 409
mutagenicity testing, of coal derived reference substances **10**, 178ff
mutagenization, of crops plants **5B**, 151f
mutant selection **2**, 32
mutant viability test **2**, 517
– functional redundancy **2**, 517
mutation **2**, 7ff
– generative **2**, 10
– phenotypic expression **2**, 31
– – in Actinomycetes **2**, 31
– rate **2**, 10ff
– somatic **2**, 10
– spontaneity **2**, 9
– type **2**, 7
mutational analysis **5B**, 228
mutational cloning **2**, 469
mutational spectra **2**, 11
– base sequence **2**, 11
mutation repair mechanism **2**, 12
– adaptive response **2**, 19f
– alkylation repair **2**, 19f
– base excision repair **2**, 16
– inhibitor **2**, 29
– nucleotide excision repair **2**, 15
– – incision **2**, 16
– photo-reactivation **2**, 12
– post-replication repair **2**, 17f
– – mismatch repair **2**, 18
– – recombination repair **2**, 17
– radioadaptive response **2**, 22
– recAB pathway **2**, 17f
– recF pathway **2**, 17f
– SOS repair **2**, 18
– SOS response regulation **2**, 18
– Weigle reactivation **2**, 18
mutations, missense – **5B**, 114
– neutral – **5B**, 114
– nonsense – **5B**, 114
mutation scanning **5B**, 114, 116
mutation screening **5B**, 116
– ethical issues **5B**, 117
mutator gene **2**, 11
– action mechanism **2**, 12

myasthenia gravis, monoclonal antibody therapy **5A**, 319
mycel, recovery from bioreactors **3**, 473
mycelial fermentation, impeller design **3**, 13
mycelial growth **1**, 116
– – *see also* microbial growth
– model based simulation **1**, 117
mycelianamide, iron-binding compounds **7**, 233
– structure **7**, 233
Myceliophtora thermophyla, laccase **9**, 713
mycelium **1**, 116, 517
Mycobacterium, homologous genes for polyketide synthases **10**, 366
– in foams **11A**, 98
Mycobacterium sp., degradation of, fluoranthene **11B**, 227ff
– – pyrene **11B**, 228ff
– mycobactins **7**, 217
mycobactins **7**, 217ff
– function as iron shuttle **7**, 217
– structure **7**, 218
mycoherbicides **10**, 490
mycoparasitism **10**, 488
mycoplasm, cell wall **1**, 20
Mycoplasma jannaschii, genome project **5B**, 17ff
mycoplasmas, effect on cell cultures **10**, 34
mycoprotein **3**, 9
mycorrhizae **11B**, 374
– control of plant pathogens **10**, 488f
– cultivation of **10**, 578f
myeloma cells, respiratory quotient **3**, 226
myeloma cells (NSO) **5A**, 248f
– in manufacturing of antibodies **5A**, 252f
– – serum-free media **5A**, 257
myocardial infarction (MI), Kaplan-Meier event rates **5A**, 378
– use of ReoPro **5A**, 367ff, 374, 377f
myoglobin **5A**, 20
myo-inositol derivatives, lipase-catalyzed enantio-selective reactions of **8A**, 87
myo-inositol-D-3-phosphate synthase, substrates **7**, 405
myo-inositol pathway **7**, 403ff
myomycin, streptomycin-related aminoglycosides **7**, 428
myomycins, structure **7**, 464
myosin filaments **5A**, 31
Myoviridae **1**, 550
myrcene **10**, 384ff
– biotransformation by basidiomycetes **10**, 385
– catabolism of **10**, 386
– conversion products of **10**, 385
myrcenol **10**, 384f
Myrothecium verrucaria, nitrile hydratase activity **8A**, 295
myths of biotechnology **12**, 41ff
myxobacterium, sporangiophore **1**, 115
myxochelin A, structure **7**, 203

Myxococcus fulvus, antibiotic production, three-compartment model **4**, 290
myxothiazol production, mathematical model **4**, 288f

N

NAA (The North Atlantic Assembly) **12**, 524f
N-acylamino acids, use in cosmetics **6**, 676
N-acyl-L-homoserine lactones (AHL) **11C**, 450
NAD *see* nicotinamide adenine dinucleotide
NADH
– cofactor regeneration, in enzyme reactors **3**, 454f, 458
– polyethyleneglycol binding **3**, 454
– regeneration of, characteristics of selected enzymes **8A**, 445
NADH assay **4**, 208ff
– NADH-dependent fluorescence **4**, 209ff
– ^{31}P-NMR **4**, 209
NAD(P)$^+$ **9**, 31
NAD(P), regeneration *in situ* **8A**, 405ff
– – use of glutamate dehydrogenase **8A**, 406f
NAD(P)-dependent oxidoreductases, classification of **8A**, 394f
NAD(P)H **9**, 31
– – *see* nicotinamide adenine dinucleotide phosphate
– cofactor regeneration, in enzyme reactors **3**, 459
– regeneration *in situ* **8A**, 403ff
NADPH assay **4**, 208ff
– NADPH-dependent fluorescence **4**, 209ff
– ^{31}P-NMR **4**, 209
NADPH regeneration system, in yeast using carbohydrates as energy source **8A**, 367
NAD$^+$ regeneration cycle, in yeast reduction **8A**, 368
Na$^+$-K$^+$ pump **5A**, 29
n-alkanes, in biosurfactant production **6**, 666f
naloxone, synthesis of narcotic analgesics **8A**, 345
naltrexone, synthesis of narcotic analgesics **8A**, 345
– treatment of severe alcohol addiction **8A**, 342
nannochelins **7**, 223
nanoelectrospray ionization **5B**, 256
nanoelectrospray-tandem mass spectrometry **5B**, 251
nanofiltration, in wastewater treatment **11A**, 565f
– membrane structure **11C**, 406f
– use in water treatment **11C**, 405ff
naphthalene, bioremediation of **11B**, 428
– bioreporter systems **11B**, 454
– complete metabolization and mineralization **11B**, 221f
– degradation by bacteria **11B**, 221f

– – pathway **11B**, 221
– mineralization of, biokinetic limited **11B**, 76
– – mass transfer limited **11B**, 76
– monitoring with biosensors **11A**, 134
– physical and chemical properties of **11B**, 218
– structure of **11B**, 216
naphthalenes, dihydroxylation of **8A**, 509
naphthalene sulfonic acids, biodegradation of **11A**, 390ff
– – stoichiometric and kinetic coefficients **11A**, 391
– – wastewater treatment **11A**, 392
naphthoquinones, from plant cell cultures **7**, 601f
– – yields **7**, 601
(S)-naproxen, synthesis of **8B**, 281f
– with carboxylesterase NP **8B**, 281
naringenin, biosynthetic pathway **10**, 352
naringin, selective acylation by lipase **8A**, 124
NAS (U.S. National Academy of Sciences) **12**, 16, 47, 468, 511f
National Audobon Society **12**, 465
National Biotechnology Policy Board **12**, 469
National Commission for the Preservation of Plant and Animal Genetic Resources of Indonesia **12**, 412
National Competitiveness Technology Transfer Act **5A**, 522
National Institutes of Biotechnology and Applied Microbiology *see* BIOTECH
National Institutes of Health (NIH) **2**, 837
– bio-waste regulations **3**, 772ff
– Division of Biologics Control **5A**, 498
– gene therapy regulations **5A**, 520, 523f
– Guidelines **5A**, 519f, 523f
– regulation of biotechnology products **5A**, 519f
– research guideline **2**, 837
– – recombinant DNA **2**, 837
National Wildlife Federation **12**, 465, 476ff
natto **9**, 522ff
– production of **9**, 522f
Natuphos **5A**, 212
natural aroma chemicals *see* aroma chemicals
natural attenuation **11B**, 363f, 401ff, 473
– contaminant concentrations **11B**, 402f
– documentation of **11B**, 405f
– modeling of **11B**, 405
– monitored –, definition of **11B**, 401, 474f
– – sampling protocol **11B**, 404
– parameters **11B**, 404
– plume size **11B**, 402f
– preliminary site evaluation **11B**, 402
– receptor analysis **11B**, 405
– stages **11B**, 401
natural flavors *see also* aroma chemicals
– definition of **8B**, 334
natural gene transfer, markers **1**, 414
natural products, preparation of, using poly-leucine catalyzed epoxidation **8B**, 501f

– production of, using cyclic dipeptide catalyzed hydrocyanation **8B**, 496
Nature Conservation Foundation **12**, 439
Natureworks™ PLA **10**, 433
NCBI data model **5B**, 404
NCBI toolkit **5B**, 404
NCDB (National Committee on the Development of Biotechnology of Indonesia) **12**, 411f
NCGEB (National Center for Genetic Engineering and Biotechnology of Thailand) **12**, 420f
neamine **7**, 434
near infrared dyes **5B**, 187
Needleman and Wunsch method **5B**, 303
negamycin synthesis **8B**, 353f
– use of pig liver esterase **8B**, 353f
Nelder-Mead method **4**, 320, 327
nematicidal compounds, from basidiomycetes **7**, 515f, 521
– – structures **7**, 516
nematodes, biocontrol of plant pathogenic – **10**, 489
neomycin, biogenesis of components **7**, 435
neomycin family, structure of – **7**, 466
neomycins, biosynthetic pathway **7**, 433
neopinone, transformations by *Trametes sanguinea* **8A**, 344
neoplanocin, antitumor activity **7**, 647
neoplasm inhibitors *see* anticancer drugs
neoplastic cell lines, use in screening for anticancer drugs **7**, 645ff
nephelometer **4**, 182, 201
nephelometric methods, cell-size distribution measurement **4**, 201
– for biomass determination **4**, 182ff
Nernst equation **4**, 9, 15f
Nernst potential **4**, 16
nerol, biotransformation of **10**, 388ff
nerolidol, dihydroxylation of **8A**, 513
nerve gas sensor **4**, 96, 98
nerve growth factor **5A**, 158ff
netropsin, antitumor agents **7**, 650
– structure **7**, 649
networking **5B**, 435
NeuAc aldolase *see* sialic acid aldolase
Neuberg's glycerol fermentation **6**, 208f
Neupogen **5A**, 158
neural network, use in media design **1**, 147
neural network pattern recognition **5B**, 338
neural nework *see* artificial neural network
neurofibromatosis type 1 (NF1) **5B**, 69
neurokinin A, receptor antagonists **7**, 120
neurological diseases, monoclonal antibody therapy **5A**, 319
neuropeptides, receptor antagonists **7**, 120
Neurospira crassa **2**, 530ff, 536
Neurospora intermedia, in oncom fermentation **9**, 535
neurotrophic factor, brain-derived – **5A**, 159

– ciliary – **5A**, 159
– glial-derived – **5A**, 159
neurotrophin-3 **5A**, 160
new drugs applications, statistics **5A**, 503
newspaper, in industrial ethanol production, enzymatic hydrolysis **6**, 165
New Zealand, public opinion surveys, on biotechnology **12**, 124ff, 146f
– – on genetic manipulation **12**, 128ff
NHL *see* hydroxynitrilase
NIABA (Netherlands Industrial and Agricultural Biotechnology Association) **12**, 662
nickel, bioleaching of **10**, 208, 213
– biorecipitation **11B**, 365
– hyperaccumulator **11B**, 389ff
– removal by phytoextraction **11B**, 389ff
– removal by rhizofiltration **11B**, 392f
Nicotiana protoplasts, exposition to microgravity **10**, 511
Nicotiana tabacum, production of caffeoyl putrescine **7**, 599ff
nicotinamide **8A**, 285
– hydrolysis of **8A**, 286
– microbial production of **10**, 323ff
– production of **8B**, 287f
nicotinamide adenine dinucleotide coenzymes, structure **8A**, 395
nicotinamide adenine dinucleotide (NAD) **1**, 71
– *in situ* NADH regeneration **10**, 325
– origin of coenzymes **8A**, 399
– production of **10**, 325
nicotinamide adenine dinucleotide phosphate (NADP) **1**, 71
– *in situ* NADPH regeneration **10**, 325
– production of **10**, 325
nicotinamide adenine dinucleotides *see* NADH and NADPH
nicotinamide coenzymes *see also* NAD(P), NAD(P)H
– half-lives **8A**, 400, 410
– immobilization methods **8A**, 408
– immobilization of, by coupling methods **8A**, 407ff
– – by entrapment methods **8A**, 409f
– prices of **8A**, 398
– regeneration **8A**, 397ff
– – by coupling on a carrier **8A**, 408f
– – by coupling to cross-linked enzymes **8A**, 407f
– – by entrapment methods **8A**, 409f
– – methods **8A**, 401ff
– – of enzyme-coupled – **8A**, 402
– – of immobilized coenzymes **8A**, 407ff
– – of substrate-coupled – **8A**, 401ff
– – principles for nonenzymatic – **8A**, 400ff
– stability **8A**, 400
nicotine, biosynthesis **7**, 624f
nicotinic acid, hydroxylation of **8A**, 501

– microbial production of **10**, 323ff
NICs *see* Asian newly industrialized countries
NIDDM *see* non-insulin-dependent diabetes mellitus
NIH *see* National Institutes of Health
NIH (U.S. National Institutes of Health) **12**, 40, 68, 142, 198, 241, 291, 461, 512ff, 566ff
– Director **12**, 243, 567
– Office of Recombinant DNA Activities (ORDA) **12**, 243f
nisin **1**, 358
– formation by lactic acid bacteria **9**, 652
– gene **1**, 358ff
– – nucleotide sequence **1**, 360
– prepeptide structure **7**, 345
– primary structure **7**, 330ff
– solution structure, in aqueous solution **7**, 339
– – in lipophilic solvents **7**, 339
– structural gene **7**, 342
– use in foods and beverages **7**, 358
nisin biosynthesis, gene clusters **7**, 343
NITRA process **11A**, 60f
nitrate, effect in microbial growth **1**, 134
– estimation methods **1**, 126
– in groundwater **11C**, 379
– inhibition of phosphate release in activated sludge **11A**, 294
– limit in drinking water **11A**, 178
– monitoring with biosensors **11A**, 128f
nitrated compounds, biodegradation of, by fermenting bacteria **11B**, 286f
– – by sulfate-reducing bacteria **11B**, 286
– remediation techniques **11B**, 290ff
nitrated hydrocarbons, microbial catabolism of **11B**, 275ff
nitrate esters, bacterial denitration of **11B**, 277
– microbial metabolism of **11B**, 277
nitrate reductase **11A**, 39, 59
– immobilization of **11A**, 128f
– in pseudomonads **1**, 419
nitric acid, biogenic corrosion **10**, 281f
– – of concrete **10**, 305
nitric oxide **5A**, 135
nitric oxide reductase **11A**, 59
nitrification **11A**, 38f, 55ff, 117f
– activated sludge process **11A**, 262ff
– – by ammonia **11A**, 263f
– – effect of pH **11A**, 263f
– – inhibition **11A**, 263
– as a monitoring tool, in wastewater treatment plants **11A**, 117
– autotrophic – **11A**, 38
– biofilm modelling **4**, 422f
– following anaerobic wastewater treatment **11A**, 407f
– heterotrophic – **11A**, 38, 58f
– in fixed bed plants **11A**, 359
– in fixed growth reactors **11A**, 335ff

– inhibition by, ammonia **11A**, 263
– – nitrous acid **11A**, 263
– in large scale plants **11A**, 396f
– in wastewater treatment **11A**, 58f
– kinetics **11A**, 396
– microbial fundamentals **11A**, 55ff
– of landfill leachates **11C**, 188
– process scheme **11A**, 117
– rates, in biofilters **11A**, 369
– stoichiometry **11A**, 394f
– use of rotating biological contactors **11A**, 346f
– use of trickling filters **11A**, 343f
– – single-slage units **11A**, 340
– – two-stage plants **11A**, 340f
– yield coefficients **11A**, 394f
nitrification/denitrification process, combined –
  11A, 39f, 60f, 274ff
nitrifiers **11A**, 395ff
– autotrophic – **11A**, 38
– heterotrophic – **11A**, 38f, 59
– in activated sludge **11A**, 262f
– in ASM1 **11A**, 422
nitrifying bacteria, oligonucleotide probes **11A**,
  92
nitrilase, hydrolysis of 3-cyanopyridine **10**, 323f
nitrilase activity, in plants **8A**, 299
– of *Acinetobacter* **8A**, 298
– of *Alcaligenes faecalis* **8A**, 297f
– of *Arthrobacter* **8A**, 293f
– of fungi **8A**, 298f
– – *Torulopsis candida* **8A**, 299
– of *Klebsiella ozaenae* **8A**, 298
– of *Nocardia* **8A**, 293
– of *Pseudomonas* **8A**, 297
– of rhodococci **8A**, 295
nitrilases **8A**, 283, 295ff, **8B**, 287f
– pH optimum **8A**, 296
– purification of **8A**, 296
– temperature optimum **8A**, 296
nitrile containing wastes, bioremediation of, use of
  mixed cultures **8A**, 312
nitrile group, biotransformation of **8A**, 282ff
– chemical properties **8A**, 278
– electronic structure **8A**, 278
nitrile hydratase **1**, 422
– partial hydrolysis of 3-cyanopyridine **10**, 324
nitrile hydratase activity, bacterial, growth medium design **8A**, 288
– in *Rhodococcus erythropolis* **8A**, 285
– in *Rhodococcus rhodochrous* **8A**, 285ff
– of microbes **8A**, 283ff
– – agrobacteria **8A**, 294f
– – *Arthrobacter* **8A**, 293f
– – brevibacteria **8A**, 292
– – corynebacteria **8A**, 290f
– – *Myrothecium verrucaria* **8A**, 295
– – *Nocardia* **8A**, 293
– – pseudomonads **8A**, 294

– – rhodococci **8A**, 284ff
nitrile hydratases **8A**, 283ff
– cobalt content **8A**, 288
– induction by -caprolactam **8A**, 289f
– induction by urea **8A**, 288
– iron containing – **8A**, 289
– of *Rhodococcus*, characteristics **8A**, 286ff
– – primary sequence **8A**, 286
– – structure **8A**, 286ff
– photoactivated preparations **8A**, 289
– production of, acrylamide **8B**, 287f
– – nicotinamide **8B**, 287f
– rhodococci **8A**, 284ff
– – *Rhodococcus rhodochrous* **8A**, 285ff
– stereoselective activity **8A**, 308
nitrile hydrolysis, by hydratases **8A**, 283
– by nitrilases **8A**, 283
nitrile pollutants, bioremediation of **8A**, 307
nitriles **8A**, 277ff
– chemical synthesis of **8A**, 278
– chemical transformations of **8A**, 278f
– enantioselective hydrolysis of **8A**, 309
– heterocyclic –, hydrolysis of **8A**, 287
– isolated from animals **8A**, 282
– naturally occurring- **8A**, 279ff
– pathways of hydrolysis of **8B**, 287
– stereoselective biotransformation of **8A**, 308ff
nitrile transformations, biotechnology of **8A**,
  302ff
– by microorganisms, listing of **8A**, 303ff
– commercial processes, acrylic acid production
  8A, 311f
– redesign of enzymes for, protein engineering
  8A, 314
– search for novel activities **8A**, 313
– use of metabolic engineering **8A**, 314
nitrilotriacetate (NTA), degradation of **11B**, 174
nitrilotriacetic acid (NTA) degradation, in Gram-
  negative bacteria **1**, 417f
nitrite, monitoring with biosensors **11A**, 128f
nitrite oxidation **11A**, 58
– physiology of **11A**, 58
nitrite oxidizers, denitrification by **11A**, 58
– growth characteristics of **11A**, 58
nitrite oxidoreductase **11A**, 58
nitrite reductase **11A**, 39, 59
– immobilization of **11A**, 128f
– in pseudomonads **1**, 419
nitroalkane oxidase **11B**, 275f
– reaction mechanism **11B**, 277
nitroalkanes, biodegradation of **11B**, 275ff
4-nitroanisole, biodegradation of **11B**, 287
nitroaromatic compounds, abiotic reduction **11B**,
  284f
– application of composting **11B**, 290f
– biodegradability of **11B**, 468f
– biodegradation of, by aerobic bacteria **11B**,
  278ff

– – by anaerobic bacteria **11B**, 284ff
– – dioxygenase reaction **11B**, 279
– – monooxygenase reaction **11B**, 278ff
– – pathways **11B**, 278f
– biotic reduction **11B**, 284f
– degradation by genetically engineered microorganisms **11B**, 452
– monooxygenase reaction **11B**, 278
– transformation by plant enzymes **11B**, 377
nitroaromatic hydrocarbons, polycyclic –, carcinogenic potential **11B**, 287
– – metabolism **11B**, 287f
– – mutagenic potential **11B**, 287
nitroaromatics, properties **11A**, 384
– use **11A**, 384
Nitrobacter sp. **11A**, 117, 395ff
– inhibition by ammonia **11A**, 263
– inhibition by nitrous acid **11A**, 263
– net maximum growth rates **11A**, 262
– nitrite oxidation **11A**, 58
– NO₂ oxidation **11A**, 397
nitrobenzene, as the sole source of nitrogen, carbon and energy **11B**, 280
– degradation of, in biological waste air filtration **11C**, 286f
nitrobenzene reductase **11B**, 280
nitrobenzenes, biodegradation of, by dioxygenases **11B**, 280
m-nitrobenzoate, bacterial degradation of **11B**, 153
Nitrococcus, nitrite oxidation **11A**, 58
nitro compounds, as carbon source **11B**, 278
– as nitrogen source **11B**, 278
– biodegradation of **11B**, 186, 273ff
– non-aromatic –, bioconversion of **11B**, 285
– sources **11B**, 275
– synthesized by organisms **11B**, 275f
nitrodiphenylamines, biodegradation of, by anaerobic bacteria **11B**, 288
nitrogen, effect on microbial growth **1**, 133
– supplementation, in contaminated soils **11B**, 431f
nitrogenase **10**, 279
nitrogen assimilation, *in vivo* NMR studies **4**, 213
nitrogen compounds, corrosion by **10**, 278
nitrogen-containing compounds, in municipal wastewater **11A**, 178f
nitrogen content, of humic substances **11B**, 103
nitrogen cycle **1**, 90, **10**, 278f
– in streptomycetes **1**, 443
nitrogen elimination, in sludge liquor **11A**, 60f
– – flow chart **11A**, 60
nitrogen limitation, metabolic effect **1**, 172ff
nitrogen metabolism **1**, 90f
– of bacilli **1**, 387f
nitrogen metabolite regulation **4**, 151
– need for on-line process analysis **4**, 151

nitrogen monoxide, inhibition of phosphate release in activated sludge **11A**, 294
nitrogen mustards, cytotoxicity **8B**, 443
– prodrug activation **8B**, 443f
nitrogen removal **11A**, 270ff
– activated sludge process **11A**, 256f, 269ff
– – alternating nitrification and denitrification process **11A**, 276f
– – intermittent nitrification and denitrification **11A**, 274ff
– – plant configurations **11A**, 269ff
– – post-denitrification **11A**, 277f
– – pre-anoxic zone denitrification **11A**, 270f
– – process for low COD/TKN ratio **11A**, 277
– – simultaneous nitrification and denitrification **11A**, 273f
– – step-feed denitrification process **11A**, 271ff
– by the sequencing batch reactor **11A**, 325f
– – cycle design **11A**, 325f
– efficiency in wastewater treatment plants **11A**, 118
– in fixed-bed plants **11A**, 359
– in wastewater treatment **11A**, 13, 37f, 57
– – ammonia oxidation **11A**, 40, 56
– – ammonification **11A**, 37f
– – combined nitrification/denitrification **11A**, 39f, 60f
– – denitrification **11A**, 39, 59f
– – nitrification **11A**, 38f
nitrogen sources, in fermentation media **3**, 129
nitrogen supply, in plant cell culture **1**, 581
nitrogen uptake rate (*NUR*), measurement **4**, 235
nitroglycerine, bacterial denitration **11B**, 277
– glycerol trinitrate reductase **11B**, 277
nitroorganics, degradation of **11B**, 186
2-nitrophenol monooxygenase **11B**, 280f
4-nitrophenol oxygenase **11B**, 281
nitrophenols, biodegradation of, oxygenolytic removal of nitrite **11B**, 280
– toxicity of **11B**, 280
4-nitrophenol 2-hydroxylase **11B**, 281
4-nitrophenol (4-NP), biodegradation of **11A**, 383ff
2-nitropropane dioxygenase, reaction mechanism **11B**, 277
nitroreductases **11B**, 284f
Nitrosococcus, ammonia oxidation **11A**, 56
Nitrosolobus, ammonia oxidation **11A**, 56
Nitrosomonas sp. **11A**, 56f, 117, 395ff
– ammonia oxidation **11A**, 56f
– effect of nitrogenous oxides on ammonia oxidation **11A**, 57
– inhibition by ammonia **11A**, 263
– inhibition by nitrous acid **11A**, 263
– in nitrifying biofilms **11A**, 98
– net maximum growth rates **11A**, 262
– NH₄ oxidation **11A**, 397
Nitrospira sp., in wastewater treatment **11A**, 102

– nitrite oxidation **11A**, 58

nitro-substituted compounds, biodegradation by
  fungi **11B**, 288ff

nitrotoluenes, aerobic biodegradation **11B**, 282ff

nitrous oxide reductase **11A**, 59

NMDA receptor antagonists **7**, 114

NMR *see* Nuclar magnetic resonance spectroscopy

NMR spectroscopy **5A**, 96

– of protein complexes **5A**, 92f

NMR structures, refinement of **5A**, 94

N_2O reductase, in pseudomonads **1**, 419

NO, corrosion by **10**, 284

noble rot, of wines **9**, 497ff, 727

Nocardia, Baeyer–Villiger biotransformations
  8A, 566

– chemotaxonomic marker **1**, 435

– G + C content **1**, 435

– morphology **1**, 435

– nitrilase activity **8A**, 293

– production of antibiotics **1**, 445

Nocardia amarae, bio-de-emulsifying activity **6**,
  704f

Nocardia lactamdurans, β-lactam biosynthesis **7**,
  251

– chemically defined medium **3**, 131

– clustering of β-lactam biosynthetic genes **7**,
  259ff

Nocardia sp., bio-de-emulsification **6**, 702

– siderochelin A **7**, 208

Noell thermal conversion process **11C**, 26ff

– energy balance **11C**, 30

– input and output data **11C**, 29

– scheme **11C**, 27

nogalamycin, antitumor agent, structure **7**, 658

noggin **5A**, 160

nojirimycin **8B**, 296ff

– α-glucosidase inhibitor **8B**, 297

– biosynthetic pathway **7**, 437

– structure of **8B**, 296

nomenclature, of enzymes **9**, 7ff

non-coding regions **5B**, 366

non-extractable residues *see* bound residues

non-governmental organizations (NGOs) **12**, 627

non-insulin-dependent diabetes mellitus
  (NIDDM) **5A**, 134f

– medical applications of recombinant proteins
  5A, 134

non-maleficence **12**, 119

non-reactor composting **11C**, 117ff, 138ff

non-redundant gene set **5B**, 384

non-redundant sequence **5B**, 384

non-secreted products, purification **3**, 505ff

norisoprenoids, biotransformation of **10**, 407ff

normalization techniques **5B**, 149

normalized libraries **5B**, 372

North American Plant Protection Organization
  12, 208

Norway, contaminated site management, financing
  schemes **11B**, 15

– – legal development **11B**, 10

– – principles **11B**, 12

– – registration **11B**, 17

– – risk assessment **11B**, 19

– – technical development **11B**, 10

– identification of contaminated sites **11B**, 17

– regulatory concerns, action thresholds **11B**, 21

– – conflict management **11B**, 28

– – enforcement **11B**, 27

– – in soil contamination **11B**, 7ff

– – liability rules **11B**, 14

– – organization **11B**, 27

– – priority setting **11B**, 23

– – remediation targets **11B**, 24

– – verification of clean-up **11B**, 24

NOTA (Netherlands Organization for Technology
  Assessment) **12**, 662

notification **12**, 547f

novel gene discovery **5B**, 125ff

– computational strategy for **5B**, 129

novobiocin resistance, in streptomycetes, induction
  of – **7**, 93

NO_2, corrosion by **10**, 284

NRC (U.S. National Research Council) **12**, 47,
  200, 242, 463, 468f, 512

NSC (National Science Council of Taiwan) **12**,
  404ff

NSO *see* myeloma cells

NSTB (National Science and Technology Board
  of Singapore) **12**, 419

NSTDA (National Science and Technology Development Agency of Thailand) **12**, 420f

NTA *see* nitrilotriacetate

NTP *see* nucleoside triphosphates

nuclear fusion **2**, 84, 95

nuclear magnetic resonance (NMR) spectroscopy
  1, 192ff

– bioreactor **1**, 193

– cell spectrum **1**, 192f

– *in vivo* application **1**, 192

– membrane transport studies **1**, 193

– ^{31}P **1**, 192

nuclear magnetic resonance spectroscopy (NMR),
  determination of cell parameters **4**, 198

– measurable cellular parameters **4**, 198

– monitoring of metabolic intermediates **4**, 212f

– pH_i measurement **4**, 203

– ^{31}P-NMR **4**, 198, 203

nuclear Overhauser effect (NOE) **5A**, 93

nuclease **2**, 378

– S1 mapping **2**, 404

nuclease P_1 **6**, 568ff

– of *Penicillium citrinum* **6**, 568f

– properties **6**, 568

– use as a biochemical tool **6**, 569f

nucleic acid **2**, 193ff

– chemical stability **2**, 320
– electron microscopy **1**, 11
nucleic acid-based techniques, detection of functional genes **11A**, 104
– in wastewater treatment, analysis of microbial communities **11A**, 85ff
nucleic acid derivatives, hydrolysis of **8B**, 235f
nucleic acid hybridization, application of DNA probes **2**, 332
– *in situ* **2**, 403
nucleic acid precipitation, with hexadecyl trimethylammonium bromide (CTAB) **9**, 93f
– with polyethylenimine (PEI) **9**, 93
nucleic acids, chlorinated – **8B**, 187
– in microbial biomass, limitations for human food **9**, 233f
– properties **3**, 696
nucleic acid synthesis **2**, 234ff
– direction **2**, 234
nucleic acid therapeutics **2**, 352ff
– anti-sense oligonucleotide **2**, 352, 355
– – action mechanism **2**, 352
– – antiviral effect **2**, 355
– – binding affinity **2**, 355
– membrane transport **2**, 355
– potential intervention points **2**, 353
– requirements for application **2**, 352
– sense oligonucleotide **2**, 352, 357
– – action mechanism **2**, 352
– – application **2**, 357
– – strategic sequence **2**, 357
– stability **2**, 354
– triplex forming oligonucleotide (TFO) **2**, 352, 356f
– – action mechanism **2**, 352
– – application **2**, 356
– – binding conditions **2**, 356
nucleobase *see* base
nucleobase analogs **7**, 647f
– structure **7**, 648
nucleocidin **8B**, 190
nucleolus organizer region (NOR) **2**, 156
nucleophilic substitution, use of catalytic antibodies **8B**, 477
nucleoside antibiotics **7**, 647f
– biosynthetic strategies of formation **7**, 32
– structural class **7**, 40
– structure **7**, 648
nucleoside kinase, phosphorylation by **8B**, 226
nucleoside phosphorylase **6**, 597
nucleoside polyphosphates, fermentative production **6**, 599ff
– – ATP **6**, 599ff
nucleosides **6**, 582ff
– – *see also* purine nucleosides
– fermentative production, adenine arabinoside **6**, 595f
– – adenosine **6**, 591f

– – cytidine **6**, 593f
– – guanosine **6**, 586ff
– – inosine **6**, 582ff
– – orotic acid **6**, 594f
– – orotidine **6**, 594f
– – pseudouridine **6**, 594
– – ribavirin **6**, 596ff
– – thymidine **6**, 594
– – uridine **6**, 592f
– – xanthosine **6**, 585f
– – 3′-deoxyadenosine **6**, 596
– formation by nucleoside phosphorylase **6**, 597
– novel compounds **6**, 598
nucleoside triphosphates (NTP), as phosphate source **8B**, 228ff
nucleosome **2**, 143, 674
nucleotide **2**, 259
nucleotide phosphate metabolism **8B**, 222
nucleotide production, by corynebacteria **1**, 458
nucleotides *see also* purine nucleotides
– as flavor enhancers **6**, 563
– as pharmaceuticals, antimetabolites **6**, 564ff
– – antitumor agents **6**, 567
– – antiviral agents **6**, 566f
– – hypotensive action **6**, 564f
– production **6**, 567ff
– – adenosine monophosphate **6**, 578
– – cytidine **6**, 579
– – from yeast RNA **6**, 568ff
– – guanylate **6**, 577f
– – inosinate **6**, 570
– – uridine monophosphate **6**, 579
– – xanthylate **6**, 576f
nucleotide sequence databanks **5B**, 132, 283f
nucleotide sequences, databases **12**, 325
nucleotide synthesis **2**, 34
– regulation **2**, 34
nucleotides (5′) **6**, 561
nuoc-mam, production of **9**, 536f
nutraceuticals, from microalgae **10**, 127
– of marine origin **10**, 462ff
– – PUFA **10**, 462ff
nutrient availability, effect in microbial growth **1**, 131ff
– – *see also* media
nutrient content, of bread **9**, 246
– of stillage, from barley **6**, 107
– – from corn **6**, 107
– – from manioc **6**, 108
– – from potatoes **6**, 105
– – from rye **6**, 106
– – from sorghum **6**, 108
– – from wheat **6**, 106
nutrient limitation, metabolic effect **1**, 170ff
– pH effect **1**, 176
– polysaccharide production **1**, 172
nutrient proteins, functions **5A**, 33
– of animals **5A**, 33

– of plants **5A**, 33
nutrient requirements, of microorganisms **1**, 143
nutrients, for mammalian cell cultures **3**, 145f
– low molecular weight – **3**, 145
– sterilization effects **3**, 166ff
– temperature sensitivity **3**, 164
– thermal destruction, activation energy **3**, 164
– – scale-up effects **3**, 166
– – thiamine **3**, 169f
nutrient supplementation, in soil bioremediation **11B**, 431f
nutrient supply, plant cell culture **1**, 591
nutrition, contribution of bread **9**, 244ff
nutritional requirements, microbial growth **9**, 199ff
– – growth factors **9**, 200
– – macro elements **9**, 199f
– – trace elements **9**, 200
nutritive value, of crop residue, enhancement **9**, 779f
Nutropin **5A**, 164

O

OAD *see* oxaloacetat decarboxylase
OAPI (Organisation Africaine de la Propriété Intellectuelle) **12**, 314
oat malt, as brewing adjunct **9**, 432
object-oriented programming **4**, 596f
observability analysis, of bioprocesses **3**, 382
observer *see* state observer
OCD *see* oxalyl-CoA decarboxylase
Ochrobactrum anthropi amidase **8A**, 246f
octachloro dibenzodioxin, anaerobic biodegradation **11B**, 263f
octynyladenosine, cardiovascular effects **6**, 564f
– structure **6**, 564
OC value *see* oxygenation capacity
ODC *see* ornithine decarboxylase
odor, from composting, appearance of **11C**, 88
– – control of **11C**, 87ff
– – formation of **11C**, 87ff
– – relation to microbial populations **11C**, 89
– in waste gas **11C**, 264
odor control, in livestock factilities **11C**, 363
– in wastewater treatment pants **11C**, 364
OECD (Organisation for Economic Cooperation and Development) **12**, 29ff, 48, 58, 204, 285, 359, 372ff, 469, 510, 533, 586ff, 611ff
– Blue Book **12**, 614ff, 650
– chemicals control **12**, 623ff
– Good Developmental Principles (GDP) **12**, 208f
– history **12**, 612f
– work on biotechnology safety **12**, 613ff

OEEC (Organisation for European Economic Co-operation) **12**, 29
ogi, production of **9**, 545f
OHHL, of *Vibrio fischeri* **7**, 65f
– – structure **7**, 65
– regulators **7**, 65f
Oidium lactis, in idli fermentation **9**, 539
oil, crude, elimination from tidal environments **6**, 697
– decontamination, use of fertilizers **11B**, 432
– degradation of **11B**, 327f
oil extraction, enzymatic **9**, 719ff
– – coconut oil **9**, 726
– – olive oil **9**, 719f
– – rape seed oil **9**, 721ff
– – soybean oil **9**, 726
– – sunflower oil **9**, 726
oil field emulsions, use of bio-de-emulsifiers, *Corynebacterium petrophilum* **6**, 707f
oil industry, use of polysaccharides **6**, 629f
– – enhanced oil recovery **6**, 629
– – oil well development **6**, 629
– – selective pore blockage **6**, 630
oil production, wastewater of **11A**, 203ff
oil production modification **2**, 651
oil recovery *see* enhanced oil recovery
– microbial enhanced – **10**, 313
oil spills, marine –, biodegradation of **10**, 470
okadaic acid, inhibition of protein kinase C **7**, 677
OKT3 **5A**, 317
oleaginous microalgae **7**, 164ff
– as sources of eicosapentaenoic acid **7**, 169
– cultivation of – **7**, 164
– fatty acid profiles **7**, 165f
– fatty acids, arachidonic acid (ARA) **7**, 167f
– – docosahexaenoic acid (DHA) **7**, 169f
– – eicosapentaenoic acid (EPA) **7**, 168f
– – γ-linolenic acid (GLA) **7**, 167
– lipid contents **7**, 165
oleaginous microorganisms **7**, 140ff
– filamentous fungi **1**, 523
– pathway of triacylglycerol formation **7**, 143ff
– – scheme **7**, 144
oleaginous molds **7**, 155ff
– fatty acid profiles **7**, 156ff
– fatty acids, arachidonic acid **7**, 161f
– – dihomo-γ-linolenic acid (DHGLA) **7**, 160
– – docosahexaenoic acid (DHA) **7**, 162f
– – eicosapentaenoic acid (EPA) **7**, 162
– – eicosatrienoic acid (ETA) **7**, 163
– – γ-linolenic acid (GLA) **7**, 159f
– formation of PUFA **7**, 158
– lipid accumulation **7**, 140
– – efficiency **7**, 142f
– lipid analyses **7**, 147
– lipid contents **7**, 156
– triacylglycerol storage **7**, 146

oleaginous yeasts, cocoa butter equivalent (CBE) **7**, 148, 150
– – process costs **7**, 155
– conversions of fatty acids **7**, 150
– fatty acid profile of lipids **7**, 148f
– increase of stearic acid content **7**, 150ff
– – by direct feeding **7**, 150
– – by inhibition of stearoyl desaturase **7**, 151f
– – by metabolic manipulation **7**, 154
– – by mutation **7**, 152ff
– lipid accumulation **7**, 140ff
– – efficiency **7**, 142f
– lipid analyses **7**, 147
– lipid contents **7**, 149
– triacylglycerol storage **7**, 146
oleandomycin formation, by *Streptomyces antibioticus*, growth-phase dependency **7**, 72
oleuropein, bitterness of olive fruits **9**, 597
oligodeoxynucleoside methylphosphonate **2**, 326
– synthesis **2**, 327f
oligodeoxynucleoside phosphate triester **2**, 328
– alkyl triester **2**, 328
oligogenic traits **5B**, 68
oligoglycosides, structural class **7**, 40
oligonucleotide **2**, 319
– 5′-amino-oligonucleotide structure **2**, 272
– anti-sense **2**, 326
– application **2**, 332ff
– for random mutagenesis **2**, 350
– modified oligonucleotides, application **2**, 324
– selfpriming oligonucleotide **2**, 344
– – for gene construction **2**, 344
oligonucleotide microassays **5B**, 116
oligonucleotide modification **2**, 326ff, 354
– phosphate methylation **2**, 326
– sulfurization **2**, 326
oligonucleotide mutagenesis **2**, 517
oligonucleotide probes, specific for, bacteria involved in floc formation, bulking, foaming **11A**, 94f
– – denitrifying bacteria **11A**, 92
– – methanogens **11A**, 94
– – methanotrophs **11A**, 93
– – nitrifying bacteria **11A**, 92
– – sulfate reducing bacteria **11A**, 93
oligonucleotides, for microarray fabrication **5B**, 230f
– – quality control **5B**, 233
– specific for higher taxa **11A**, 96
oligonucleotide sequence, *in vitro* selection **2**, 350
oligonucleotide synthesis **2**, 319ff
– chain length **2**, 321
– characterization **2**, 330
– criteria for purity **2**, 331
– diester method **2**, 323
– direction **2**, 321
– Merrifield strategy **2**, 323
– methods **2**, 321ff

– modified oligonucleotides **2**, 324ff
– – non-oxygen bridged **2**, 328
– nucleoside protecting group **2**, 321f
– nucleoside protection **2**, 329f
– phosphonate method **2**, 325
– phosphoramidite method **2**, 325, 329f,
– – cycle **2**, 331
– phosphorus protecting group **2**, 323
– purification **2**, 330f
– sequence verification methods **2**, 332
– solid phase **2**, 329f
– – nucleoside attachment **2**, 329
– – steps **2**, 331
– – supports **2**, 329
– strategy **2**, 321
– triester method **2**, 323
– – high efficiency liquid phase (HELP) synthesis **2**, 324
– unmodified oligonucleotide **2**, 319ff
– yield **2**, 321, 330
oligopeptides, biosynthesis of – **7**, 36
– farnesamide *N*-oligopeptides, regioand enantioselective epoxidation of **8B**, 507
– rhodium polypeptide complex, stereoselective hydrogenation by **8B**, 507f
oligosaccharides, complex, assembly of **8B**, 250f
– *in vitro* enzymatic synthesis **8B**, 245
oligosaccharide synthesis, use of glycosidases **8B**, 249
– use of glycosyltransferases **8B**, 258ff
olive fermentation **9**, 593ff, 601ff
– acid formation **9**, 608
– black ripe olives in brine **9**, 614ff
– – microbial growth **9**, 615
– green olives in brine **9**, 602ff
– – brining **9**, 603f
– – control **9**, 608ff
– – fermentation media **9**, 604
– – lye treatment **9**, 603
– – microbial growth **9**, 604ff
– – washing **9**, 603
– history **9**, 595
– lactic acid bacteria in – **9**, 617ff
– natural black olives in brine **9**, 611ff
– – aerobic fermentation **9**, 612
– – anaerobic fermentation **9**, 611f
– – control **9**, 613f
– processing schemes **9**, 602
– waste water treatment **9**, 621f
olive oil, enzymatic extraction **9**, 719f
olives, bitterness **9**, 597
– composition of final products **9**, 620f
– composition of fresh fruits **9**, 596ff
– – fiber **9**, 598
– – oil **9**, 597
– – oleuropein **9**, 597
– – organic acids **9**, 597
– – pectic substances **9**, 598

– – pigments **9**, 598
– – protein **9**, 598
– fermentable substrate in – **9**, 616
– fermented –, definition **9**, 600
– harvesting **9**, 601
– microbial spoilage **9**, 619f
– – malodorous fermentations **9**, 619f
– producing countries **9**, 596
– regulatory affairs **9**, 623
– storage of fermented – **9**, 607f
– trade preparations **9**, 599ff
– transportation **9**, 601
– varieties **9**, 598ff
– – Gordal **9**, 599
– – major producing countries **9**, 600
– – Manzanilla **9**, 599
olive trees, geographical distribution **9**, 595
OMH process **11B**, 338
OMIM *see* On-Line Mendelian Inheritance in
  Man
omphalotin, nematicidal activity **7**, 522
– structure **7**, 522
oncogene **2**, 786ff
– c-neu **2**, 789
– – tissue specificity **2**, 789
– *myc* gene effects **2**, 788
– – suppression **2**, 788
– – tissue specificity **2**, 788
– synergy **2**, 788f
– transgenic animal, ras oncogene **2**, 789
– transgenic mice **2**, 786ff
– viral T genes **2**, 786ff
oncogenes **7**, 645
– classification **7**, 676
oncogene-transformed cell lines, availability **7**,
  682
– substances changing the morphology of – **7**,
  682ff
oncom **9**, 533ff
– nutritional characteristics **9**, 535
– production of **9**, 534ff
– – preparation of peanuts **9**, 534
Oncostatin M **5A**, 159
one-dye/four-lane approach **5B**, 192
on-line analysis, in non-sterile areas **4**, 152
On-Line Mendelian Inheritance in Man (OMIM)
  5B, 63, 89, 111, 141, 285
ooTFD (Object Oriented Transcription Factors
  Database) **5B**, 409
OPD *see* orotidine-5′-phosphate decarboxylase
open-compost biofilter, for treating odors from a
  livestock facility **11C**, 363
open reading frames (ORFs) **5B**, 22, 364, 380
– close-up **5B**, 389
– plotting of **5B**, 319
– statistics **5B**, 322f
– traits **5B**, 388
open source **5B**, 323

operational qualification (OQ) **3**, 741
operon **2**, 250, **5B**, 366
opiate analgesics **8A**, 340ff
– semisynthetic – **8A**, 341
opiate drugs, chemical synthesis of **8A**, 342
– structures **8A**, 341
opiates, semisynthetic –, hydrocodone **8A**, 347
– – hydromorphone **8A**, 347
opine **2**, 553, 557
opinion polls *see* public opinion surveys
opium, of *Papaver somniferum* **8A**, 342
optical biomass sensor **4**, 182ff
– Aquasant retroreflective turbidity probe **4**, 185
– fluorosensor **4**, 185ff
– light scattering analysis **4**, 182ff
– Monitek turbidity probe **4**, 185
– transmission sensor **4**, 182ff
– – flow-cell **4**, 183
– – four-beam infrared sensor **4**, 184
– – *in situ* **4**, 184
optimization of drug candidates **5B**, 354
ORDA (Office of Recombinant DNA Activities)
  12, 244f
ORDB (Olfactory Receptor Database) **5B**, 409
ORFs *see* open reading frames
organelles, contruction of biosensors **4**, 79f
organelle transformation **2**, 579, 587
– application **2**, 587
– direct gene transfer **2**, 587
– vector construction **2**, 587
Organibio (Organisation Nationale Interprofes-
  sionnelle des Bio-industries) **12**, 591, 635
organic acids, conversion, by *Pseudomonas putida* **1**, 422
– effect on microbial growth **1**, 136
– metabolization, in clostridia **1**, 300
– microbial production **6**, 363ff
– – biosynthetic routes **6**, 364ff
– – fruit acids **6**, 368ff
– – itaconic acid **6**, 366f
– production, by filamentous fungi **1**, 523
organic carbon removal **11A**, 115ff
– process scheme **11A**, 116
organic chemicals, synthesis by biological catalysis **8A**, 19
organic compounds **11B**
– anaerobic degradation of **11B**, 169ff
– – comparison to aerobic degradation **11B**, 170
– – electron acceptors **11B**, 170f
– – key reactions **11B**, 17ff
– chlorinated –, reductive dehalogenation of
  11B, 185
– degradation of, biogenic macromolecules **11B**,
  97f
– – halogenated – **11B**, 185
– – *N*-alkyl compounds **11B**, 174
– – *S*-alkyl compounds **11B**, 174
– interaction with soil matrix **11B**, 101

– metabolism by microfungi **11B**, 157ff
– microbial degradation in soil, carbon flow **11B**, 99
– microbial degradation of **11B**, 96, 145ff, 152ff
– – in soil **11B**, 96ff
– – partitioning of carbon **11B**, 111
– persistence of, effect of added substituents **11B**, 274
– – in soil **11B**, 96
– phytoremediation of **11B**, 371ff
– stability in soil **11B**, 105
– with nitro functions, biodegradation of **11B**, 273ff
organic materials, biodeterioration of **10**, 305ff
organic matter *see also* biodegradable organic matter (BOM)
– degradation of, energy content **11C**, 83
– – heat formation **11C**, 83
– removal of, by bilfiltration processes **11C**, 457ff
organic solvents, alcohol dehydrogenases in **8A**, 420ff
– in yeast biotransformations **8A**, 369f
– lipases in **8A**, 50ff
– – increase of catalytic activity **8A**, 51ff
– – increase of enantioselectivity **8A**, 54ff
– – water content of reaction mixture **8A**, 58ff
– log *P* values of **8A**, 420f
– oxidoreductions in **8A**, 423ff
– use in formation of glycosidic linkages **8B**, 254
organic substances, extraction of **11B**, 498f
– in municipal wastewater **11A**, 177, 179f
– in sewage sludge, prescribed limits **11A**, 183
– in wastewater **11A**, 197
organofluorine compounds **8B**, 184
– naturally occuring – **8B**, 189
organohalogens, natural – **8B**, 184ff
– – listing of **8B**, 185ff
organoiodine compounds, isolated from marine organisms **8B**, 190f
organometals, microbial dealkylation **10**, 248f
organophosphorus insecticides, monitoring with biosensors **11A**, 132
organosulfur compounds, in coal **10**, 174
organotrophs **3**, 128
organotrophy **1**, 50
– definition of **10**, 275
organ transplantation, medical applications of recombinant proteins **5A**, 141
oriental fermented foods **9**, 507ff
– ang-kak **9**, 525ff
– bagoong **9**, 537
– katsuobushi **9**, 537
– kimchi **9**, 537f, 645ff
– lao-chao **9**, 525
– meitauza **9**, 525
– miso **9**, 518ff
– natto products **9**, 522f
– nuoc-mam **9**, 536f

– oncom **9**, 533ff
– phaak **9**, 537
– prahoc **9**, 537
– production of **9**, 528f
– puto **9**, 527f
– ragi **9**, 528
– soy sauce **9**, 507ff
– sufu **9**, 523ff
– tempeh **9**, 529ff
ornibactins, structure **7**, 216
– tetrapeptide siderophores of *Pseudomonas cepacia*-like strains **7**, 216f
ornithine-containing lipids, from *Pseudomonas rubescens* **6**, 664
– from *Thiobacillus thiooxidans* **6**, 664
ornithine decarboxylase (ODC) **8B**, 78f
– of *Trypanosoma brucei* **8B**, 78
L-ornithine, decarboxylation of, by ornithine decarboxylase **8B**, 79
L-ornithine production **3**, 451, **6**, 513
orotic acid, production by *Bacillus subtilis* mutants **6**, 595
orotidine, production by *Bacillus subtilis* mutants **6**, 595
orotidine 5'-phosphate, decarboxylation of, by protidine 5'-phosphate decarboxylase **8B**, 82
orotidine-5'-phosphate-decarboxylase (OPD) **8B**, 82
orphan crops **12**, 361, 364
orphan drug, definition **5A**, 505
Orphan Drug Act **5A**, 523, 528, **12**, 145
orphan drug legislation **5A**, 505ff, 514
orphan proteins **5B**, 138
orthogonal separation mechanism **5B**, 276
OSHA (Occupational Safety and Health Administration) **12**, 243, 570
osmolality, effect in microbial growth **1**, 130
osmoprotectants *see* compatible solutes
osmotolerant yeasts, glycerol production **6**, 216f
osteoblastic cells, synthesis of phosphoproteins **5A**, 17
osteogenesis imperfecta **5B**, 69
osteoporosis, medical applications of recombinant proteins **5A**, 133
OSTP (U.S. Office of Science and Technology Policy) **12**, 199, 568ff
OTA (Office for Technology Assessment) **12**, 12ff, 55f, 473, 482, 529f, 534, 544, 569, 658
– Management Board **12**, 21
– view of assessment, objective overall assessment **12**, 18ff
– – subjective overall assessment **12**, 20ff
– view of technology **12**, 16ff
OTR see Oxygen transfer rate
OTS (Organization for Tropial Studies) **12**, 440
oudemansins **7**, 496ff
– blockade of respiration in eukaryotes **7**, 497
– fungistatic activity **7**, 497

– mode of action **7**, 497ff
– possible functions **7**, 501
– producing organisms **7**, 499
– structures **7**, 498
OUR see Oxygen uptake rate
ovalbumin **5A**, 4f
overexpression, of fructose-1-phosphate aldolase **8B**, 14
– of rhamnose-1-phosphate aldolase **8B**, 14
– of tagatose-1,6-diphosphate aldolase **8B**, 14
overproduction, of enzymes **9**, 21, 75ff
overproduction of amino acids **8B**, 315ff
over-the-counter drugs, market share **5A**, 535
oxalate decarboxylase **8B**, 55f
oxaloacetate, decarboxylation of, by oxaloacetate decarboxylase **8B**, 56ff
oxaloacetate decarboxylase (OAD) **8B**, 55ff
– decarboxylation of oxaloacetate **8B**, 57f
– – mechanism **8B**, 57
– of *Pseudomonas putida* **8B**, 57f
Oxalobacter formigenes, oxalyl-CoA decarboxylase **8B**, 67f
oxalyl-CoA decarboxylase (OCD), of *Oxalobacter formigenes* **8B**, 67f
oxamniquine, manufacture of **8A**, 14
– – by benzylic hydroxylation **8A**, 19
oxanosin, antitumor activity **7**, 647
oxidants, use in baked goods **9**, 262
oxidation, in drinking water treatment **11C**, 400ff
– thermodynamics **1**, 55
β-oxidation **1**, 87
oxidation pond, wastewater treatment **4**, 430
oxidation reactions **8B**, 386ff
– enzymatic – **8B**, 386
– – synthesis of heterocyclic compounds **8B**, 386ff
– use of catalytic antibodies **8B**, 477f
oxidative degradation **8B**, 391f
– enzymic –, aromatic compounds **8B**, 391f
oxidative phosphorylation **9**, 188
oxidoreductase **1**, 66
oxidoreductase-catalyzed reactions, examples **8A**, 29
oxidoreductases **5A**, 24ff, 209ff, **9**, 8
– – *see also* dehydrogenases, reducing enzymes
– industrial applications **5A**, 209
– in industrial biotransformations **3**, 454ff
– NAD(P) dependent –, classification **8A**, 394f
– NADPH-dependent – **8A**, 412
– nomenclature **8A**, 478
– noncommercial –, future potential in organic synthesis **8A**, 457f
– of yeast, listing of **8A**, 371
oxidoreductions, in the presence of cyclodextrins **8A**, 425ff
– using organic solvents **8A**, 423ff
2,3-oxidosqualene cyclase **8B**, 28ff
– from pig liver **8B**, 28ff

oxinitrilase, in industrial biotransformations **3**, 460
oxo-acid-lyases **8B**, 45, 90f
2-oxoglutarate, formation in wine making **9**, 475f
oxygen **3**, 13, 328
– – *see also* dissolved oxygen
– as limiting factor in yeast propagation **9**, 332f
– concentration, in bioreactors **3**, 366ff
– – in fermentation broths **3**, 192
– diffusivity **3**, 235
– effect on microbial growth **1**, 134
– estimation methods **1**, 126
– exclusion techniques **1**, 259, 296
– in fermentation media **3**, 130
– influence on microbial growth **10**, 274
– mass balance, in fermentation broths **3**, 192
– – in the gas phase **3**, 193
– – in the liquid phase **3**, 194
– mass transfer resistance **4**, 393
– mole fractions, in bioreactors **3**, 405
– partial pressure, mean driving force **3**, 404
– profile, gas-liquid interface **3**, 190
– – in alginate beads **1**, 229
– requirements, of bacilli **1**, 381
– solubility **3**, 235
– supply, of plant cell cultures **1**, 587f
– toxicity **1**, 135, 293
oxygen analyzer **4**, 60ff
– infrared **4**, 56f
– paramagnetic **4**, 56f, 60
– – characteristics **4**, 61
– – magnetomechanical instruments **4**, 61f
– – practical problems **4**, 62
– – thermomagnetic instruments **4**, 61
oxygenase reactions, oxidation of alkanes **11B**, 196ff
oxygenation, of the rhizosphere, in wetland systems **11A**, 245f
oxygenation capacity (OC value) **11A**, 11
oxygen balance **4**, 37
– around the micromixer, single-impeller system **4**, 335
– – two-impeller system **4**, 336
oxygen concentration, effect in microbial activity **11B**, 430
oxygen consumption, carbon-limited **1**, 168
– energy balance **1**, 142
– fungal **1**, 228
– microbial **1**, 168
– nitrogen-limited **1**, 168
– of pseudomonads **1**, 420
– phosphate-limited **1**, 168
– sulfate-limited **1**, 168
oxygen demand, aerobic metabolism **9**, 195f
oxygen detoxification **1**, 293
– enzymes **1**, 352
– in clostridia **1**, 293
– in lactic acid bacteria **1**, 352

oxygen diffusion, in mycelial filaments **4**, 332f
– in mycelial pellets **4**, 331ff
oxygen exit, measurement, accuracy of **4**, 21
oxygen gas analysis, errors in **4**, 50f
oxygen metabolism, energetics **1**, 352
oxygen partial pressure *see p*$_{O_2}$
oxygen profile, comparison with predictions by different models **4**, 371
– effect of tower height **4**, 375
– in stirred tank models **4**, 322ff
oxygen sensor, position in tower loop reactors **4**, 261f
oxygen supply *see also* bioreactor aeration, membrane aeration, sparged aeration
– bubble-free – **3**, 269ff
– external – **3**, 254
– in stirred tank bioreactors **3**, 113
– limits **3**, 270
– mass transfer coefficients **3**, 269
– of animal cell cultures, methods **3**, 237ff
– of bioreactors **3**, 185ff, 224
– oxygen mass balance **3**, 224
– sparged – **3**, 270ff
oxygen transfer **4**, 38ff, 110f
– across gas/liquid interfacial areas **4**, 110ff
– by silicone rubber **3**, 248
– comparison of single- and two-impeller systems **4**, 337
– correlations in Newtonian fluids **3**, 197
– in air lift fermenters **4**, 374f
– in bioreactors **3**, 190, 412
– – effectiveness factor **1**, 228f
– – in alginate **1**, 228
– – influence of microbial growth **1**, 227f
– – in gels **1**, 227
– – into microfilms **1**, 227
– – to immobilized cells **1**, 226ff
– in large-scale bioreactors **4**, 38
– steady-state model **4**, 39
– to fungi **4**, 331
– well-mixed and plug flow gas model **4**, 39ff
– well-mixed gas and liquid model **4**, 39ff
oxygen transfer coefficient **3**, 240ff, 405ff, 412f
– – *see also* mass transfer coefficient
– volumetric – **3**, 189ff, 324
oxygen transfer rate **9**, 198
oxygen transfer rate (*OTR*) **3**, 190, 366ff, 404ff, **4**, 8, 38f
– determination of **4**, 35
– effects of linear gas velocity **3**, 408
– in Newtonian fluids **3**, 197
– overall **4**, 261ff
– – measurement **4**, 261ff
oxygen uptake, as a function of power input **4**, 327, 333
– *Aspergillus niger* **4**, 332ff
– characteristic time **3**, 212

– kinetics, as a function of circulation-time distribution **4**, 338f
– of cell cultures, during aerobic growth **3**, 192
oxygen uptake rate (*OUR*) **3**, 226, 324, **4**, 8, 545f
– control in bioreactors **3**, 330
– determination of **4**, 33, 36f
– – carbon dioxide analysis **4**, 37
– – dynamic method **4**, 36
– – fed-batch oxygen control **4**, 38
– – in loop reactors **4**, 36f
– – recycling gas phase **4**, 38
– in bioprocesses **3**, 366ff
– measurement **4**, 235, 545
– measurement errors **4**, 50f
– specific – **3**, 385
oxymorphone, transformation by *Pseudomonas putida* **8A**, 347f
oxynitrilases **8B**, 282
– enannoselectivity of **8A**, 301
– *Prunus amygdalus* **8A**, 301
oxyrase **3**, 298
oxytetracycline, chemical structure **10**, 342
oxytocin receptor antagonists **7**, 121
ozonation, effect of the ozone dose, on AOC **11C**, 463f
– – on the regrowth factor of bank filtrate **11C**, 463f
– effect on BOM **11C**, 463f
– of drinking water **11C**, 401f, 411, 421, 451

P

PAC *see* powdered activated carbon
paclitaxel **7**, 665, 667
– inhibition of the microtubular system **7**, 668
– structure **7**, 668
PADH *see Pseudomonas* sp. alcohol dehydrogenase
PAH *see* polycyclic aromatic hydrocarbons
paint, biodeterioration of **10**, 310f
pairwise interactions **5B**, 401
PAL *see* phenylalanine ammonia-lyase
palm-kernel meal, cell wall polysaccharides **9**, 707
Panax ginseng, commercial gisenoside production **1**, 589, 604
– saponin aglycone, structure **7**, 611
pancreatic islets (Langerhans islets), immunoisolated by Ba^{2+} alginate capsules **10**, 556
– implantation in diabetes patients **10**, 555f
pancreatic ribinuclease, unfolding in urea **5A**, 112
panepoxydone, inhibition of signal transduction **7**, 524
D-pantoic acid, direct fermentation process **10**, 327

pantolactone, optical resolution by fungal lactonase **10**, 326
D-pantolactone, synthesis of **10**, 326f
– – enzymatic routes **10**, 327
panudial, antithrombotic compounds, structure **7**,
512
papadam, production of **9**, 540
papain **6**, 544, **8A**, 233ff
– ester hydrolysis **8A**, 233f
– esterification of *N*-protected amino acids **8A**,
235
– formation of amides **8A**, 244f
– hydrolytic resolution of unnatural amino acid
derivatives **8A**, 234
– preferred cleavage sites **8A**, 249
Papaver somniferum, accumulation of benzophenanthridine alkaloids **7**, 605f
– alkaloid yield **1**, 593
– elicitor effect **1**, 593
– opium **8A**, 342
paper, production of, microbial problems **10**, 307
– recycling of **11C**, 8ff
paper industry, biotechnology in **10**, 523ff
paper manufacture, effluent discharge **10**, 538
– enzymes used in **10**, 538ff
paper processing, application of biotechnology
10, 525
paper production, use of α-amylases **5A**, 195f
paper pulp, chlorine bleaching **10**, 536
– recycled –, de-inking of **10**, 539f
papilloma virus, monoclonal antibody therapy
5A, 317
parabactin **7**, 204f
– catecholate siderophore of *Micrococcus denitrificans* **7**, 204f
– structure **7**, 204
Paracoccus denitrificans, denitrification by **11A**,
59
– parabactin **7**, 204f
Paracoccus sp., in wastewater treatment **11A**, 102
paralogs **5B**, 376
parameter estimation **4**, 263
– adaptive optimization, of a baker's yeast culture **4**, 553f
– recursive least square error algorithm, sensitivity matrix **4**, 240
– techniques **4**, 542ff
parameter identification, by the output error
method **4**, 273
paramylum, production by *Euglena gracilis* **10**,
145f
parasexual cycle, in fungi **1**, 519
parasexual recombination **2**, 84
parasites, classification **12**, 76f
– in sewage **11C**, 217
– in sewage sludge **11C**, 217
parathyroid hormone **5A**, 165
parathyroid tissue, allotransplantation of **10**, 557

– encapsulation of **10**, 556f
– xenotransplantation of **10**, 556
parchment paper, biodegradation of **10**, 307
parenteral emulsions **6**, 677f
– particle size **6**, 678
parenteral nutrition, use of amino acids **6**, 468
Paris Convention for the Protection of Industrial
Property **12**, 294
Parkinson's disease, synthesis of L-DOPA **8B**, 95
Parmesan cheese **9**, 378
paromomycins, biosynthetic pathway **7**, 433
particle bombardment **5A**, 432, **5B**, 156
particle separation **3**, 478f
– comparative economics **3**, 500f
– from fermentation broths **3**, 469ff
partition, in aqueous two-phase systems, modelling **4**, 613f
partition coefficients, in liquid-liquid extractions
3, 558ff
– – ethanol **3**, 560
– – lactic acid **3**, 561
– – methanol **3**, 560
– – *n*-butanol **3**, 560
– – *n*-propanol **3**, 560
– in reversed micelle extractions, amino acids **3**,
583
paspalic acid, biosynthesis **1**, 530
passage method, for strain improvement **6**, 316
passive diffusion, substrate uptake **9**, 190
PASTEUR, large-scale fermenter **8A**, 7
– methods for resolving the enantiomers in a ra-
cemate **8A**, 6ff
Pasteur effect, during yeast growth **4**, 388
– in vivo NMR studies **4**, 213
– in yeast **1**, 483, 486
pasteurization **3**, 779, **10**, 287
– of cucumbers **9**, 632f
– of milk **9**, 450f
– of molasses **9**, 328f
– of soy sauce **9**, 517f
– of vinegar **9**, 582
pasteurization of sludge **11C**, 221
PASTEUR Memorial Lecture **8A**, 6f
Patau syndrome **5B**, 76
patchoulol **10**, 382f
– bioconversion of **10**, 410
– biotransformation of **10**, 382f
– preferred sites of oxygen attack on **10**, 410
patentability **12**, 287f, 531
– basic requirements, industrial applicability **12**,
303f
– – non-obviousness **12**, 302f
– – novelty **12**, 302
– – utility **12**, 303f
– exclusions **12**, 305ff
– – by the EPC **12**, 287f
– in public opinion **12**, 143
– of biological subject matter **12**, 306

– – restrictions **12**, 307
– of discoveries **12**, 304f
– of inventions **12**, 302ff
– of living organisms, under the EPC **12**, 291ff
– of processes **12**, 308f
– of products **12**, 308
– of products of nature **12**, 304f
– of use **12**, 309
patentable inventions, characteristics **12**, 302ff
– definitions **12**, 306
– exclusions **12**, 305ff
– – by the EPC **12**, 306
– license contracts, lump sum payments **12**, 320
– – royalties on sales **12**, 320
– selling **12**, 319ff
patent applications **12**, 309ff
– CIP **12**, 315
– deposition of microorganisms **12**, 310f
– – depository authorities **12**, 310f
– filing **12**, 313ff
– for biomedicinal products **12**, 299ff
– patent description **12**, 309f
– publication **12**, 315
patent claims **12**, 311ff
patent databases **12**, 318, 327
patenting **12**, 142f
– of genetic material **5B**, 96, 425
– of living organisms **12**, 142
– – history **12**, 283f
patent laws, WIPO efforts **12**, 286
patent offices, practice, EPO **12**, 291ff
– – U.S. Patent and Trademark Office (PTO)
  12, 290f
patent protection, duration **12**, 316
– of biomedicinal products **12**, 215
patents, costs **12**, 317f
– definition **12**, 282
– dependent – **12**, 309
– description **12**, 309f
– equivalents **12**, 312
– information **12**, 318f
– opposition **12**, 317
– priority **12**, 315f
– process – **12**, 308f
– product – **12**, 308
– purpose **12**, 300f
– use **12**, 309
pathogen-derived resistance **5B**, 154
pathogenicity, characteristics **12**, 44f
– classification **12**, 66ff
– *E. coli* **12**, 84f
– *Saccharomyces cerevisiae* **12**, 89f
pathogenic microorganisms **12**, 89
– bacilli **1**, 376
– – for insects **1**, 376
– – for man and mammals **1**, 376
– bacteria, *E. coli* **12**, 84f
– causing waterborne diseases **11C**, 414ff

– classification **12**, 67ff
– clostridia **1**, 286ff
– contained use **12**, 71ff
– lactic acid bacteria **1**, 331, 344ff
– risk evaluation **12**, 66
– risk groups, bacteria **12**, 73f
– – fungi **12**, 76f
– – parasites **12**, 76f
– – viruses **12**, 75f
pathogenic organism **2**, 839
– risk group **2**, 839f
– – classification **2**, 841
pathogenic yeasts, in sewage **11C**, 217
– in sewage sludge **11C**, 217
pathogens, in biofilters, behavior of **11C**, 467f
– in composting, inactivation of **11C**, 70f
– removal of, during artificial groundwater recharge **11C**, 438
pathothenic acid, microbial production of **10**,
  325ff
4s-pathway **10**, 175
pattern recognition, of bioprocesses **3**, 388ff
– – foaming detection **3**, 396f
– – principal component analysis (PCA) **3**, 391
pBACwich vector **5B**, 174
PCASTRD (Philippine Council for Advanced
  Science and Technology Research and Development) **12**, 416
PCB *see* polychlorinated biphenyls
PCDD *see* polychlorinated dibenzodioxins
PCDF *see* polychlorinated dibenzofurans
PCE *see* tetrachloroethene
PCL *see* *Pseudomonas cepacia* lipase
p_{CO_2} (CO_2 partial pressure), determination of **4**,
  19f
– use in process control **4**, 9
p_{CO_2} electrode, sterilizable **4**, 19
p_{CO_2} meter **4**, 19
PCP *see* pentachlorophenol
PCR *see* polymerase chain reaction
PDB **5B**, 291
PDC *see* pyruvate decarboxylase
peanut milk, fermentation of **9**, 550f
pear juice, enzymatic treatment **9**, 713
– production **9**, 713f
Peclet number **3**, 171ff, **4**, 364
pectate lyase, mode of action **9**, 709
pectate lyases **8B**, 124f
pectic substances, cleavage of **8B**, 125
– structure **8B**, 124
pectinases **9**, 707f
– mode of action **9**, 709
– use in fruit juice production **9**, 712
pectin degradation, in clostridia **1**, 297
pectin esterase, mode of action **9**, 709
pectin lyase, mode of action **9**, 709
pectin lyases **8B**, 125
pectin methylesterase **9**, 718f

pectins, action of enzymes on **10**, 532
pedigree analysis **5B**, 77, 108
pediocin, formation by *Pediococcus* strains **9**, 652
Pediococci, isolation media **1**, 348
Pediococcus **1**, 329, 339f
– growth conditions **1**, 340
– habitats **1**, 339
– in food and feed **1**, 339
– in olive fermentation **9**, 606
– pediocin formation **9**, 652
– species characteristics **1**, 340
– substrates **1**, 340
Pediococcus cerevisiae, in idli fermentation **9**, 539
– in sausage fermentation **3**, 311
Pediococcus damnosus, as beer contaminant **9**, 447
– malolactic fermentation **9**, 491
Pediococcus damnosus (*cerevisiae*) **1**, 339
– bee spoilage **1**, 339
Pediococcus dextrinicus **1**, 339
Pediococcus halophilus, in miso fermentation **9**, 520
PEG *see* polyethylene glycol
Peganum harmala, serotonin biosynthesis **7**, 624
penetrance, of genetic diseases **5B**, 70f
– – definition **5B**, 70
– – incomplete – **5B**, 70
penicillia, infection of wine grapes **9**, 500
Penicillin **2**, 115
penicillin **3**, 449
Penicillin, colony phenotype **2**, 117
penicillin, distribution coefficients, in liquid-liquid extractions **3**, 563
– extraction **3**, 563, 577
– – by liquid membrane techniques **3**, 582
– – carriers **3**, 577
– – concentration profiles **3**, 579f
– – degrees **3**, 577f
– – *in situ* **3**, 564, 586
– – kinetics **3**, 579
– – process flowsheet **3**, 587
– – re-extraction **3**, 577f
– *in situ* conversion, flowsheet **3**, 587
Penicillin, penicillin-producing recombinant **2**, 115ff
penicillin, stereocontrolled synthesis, using kidney acylase **8B**, 356f
penicillin acylase **8A**, 229ff
– amidase activity **8A**, 229
– esterase activity **8A**, 229
– immobilization **3**, 449
– in industrial biotransformations **3**, 449
– kinetic resolution of, phenyl acetyl esters of secondary alcohols **8A**, 231
– – primary and secondary alcohols **8A**, 230
– – pyridyl acetate esters of secondary alcohols **8A**, 231f
– of *Escherichia coli* **8A**, 15

– substrate tuning **8A**, 231
– use in selective protecting group strategy **8A**, 232
– use in transacylations **8B**, 386
penicillin biosynthesis **7**, 119f
– enzymes **7**, 253ff
– – ACVS **7**, 254
– – AT **7**, 257ff
– – IPNS **7**, 256
– gene dosage **7**, 264
– genes **7**, 251
– – cloning of – **7**, 259
– – clustering of – **7**, 259ff
– pathway **7**, 252
– – compartmentalization of – **7**, 262
– regulation of gene expression **7**, 264
– side chain exchange **7**, 257
penicillin fermentation **7**, 292
– by *Penicillium chrysogenum* **4**, 539
– CO_2 measurement **4**, 70f
– optimal glucose feed-rate profiles **4**, 539
– optimization **4**, 528ff
penicillin G, hydrolysis by penicillin acylase **8B**, 386f
– separation and purification system **4**, 620
penicillin G acylase (PGA), formation of ampicillin **8A**, 259
– from *Escherichia coli* **8A**, 258f
– immobilized, formation of loracarbef **8A**, 259f
– production of 6-aminopenicillanic acid **8B**, 280
penicillin production **3**, 449, **4**, 150f
– amount **1**, 156
– by filamentous fungi **1**, 156, 527
– kinetics **3**, 332
– models **3**, 332
– optimization **3**, 10, 344
penicillins, conversion to cephalosporin derivatives **8A**, 16f
– obtained from fermentation **8A**, 15
– semisynthetic – **8A**, 15
– synthesis **8A**, 258ff
Penicillium **1**, 525ff
– bread spoilage **9**, 312
– in ogi fermentation **9**, 545
– penicillin production **1**, 156
– segregated models **3**, 332
Penicillium camemberti, in cheese production **9**, 377, 681
Penicillium chrysogenum **2**, 36, 115f, 531, 538, **4**, 157f, 539, **7**, 249ff
– clustering of β-lactam biosynthetic genes **7**, 259ff
– genealogy **2**, 36
– gene expression, regulation of – **7**, 266
– growth kinetics **1**, 133
– instruments for determination of **4**, 15f
– intracellular **4**, 202ff
– measurement **4**, 15f

– – accuracy of **4**, 21
– – in anaerobic wastewater purification **4**, 451
– mutagen treatment **2**, 36
– on-line analysis of cultivation broth **4**, 157
– optimization **4**, 526ff
– optimum **4**, 6f
– pellet **1**, 118
– penicillin biosynthetic enzymes, AT **7**, 257ff
– penicillin production **1**, 527
– pH-sensitive dyes, BCECF **4**, 203f
– – DCH **4**, 203
– – fluorescein **4**, 203
– use in process control **4**, 6f
Penicillium citrinum, nuclease P₁ **6**, 568f
Penicillium digitatum, stereospecific modification of terpenoids **8B**, 374
Penicillium frequentans, esterase of **8A**, 221
– immobilization effect **1**, 231
Penicillium roqueforti, blue veined cheese flavor compounds **8B**, 347
– in cheese production **9**, 377, 681
penochalasin A, structure **10**, 454
pentachloroethane, degradation by genetically engineered microorganisms **4**, 450
pentachlorophenol, degradation under aerobic conditions **8B**, 210
– reductive dehalogenation of **8B**, 207
pentachlorophenol (PCP), biodegradation **4**, 429
– – aerobic – **11B**, 259f
– – use of bioaugmentation **11B**, 434
– degradation by immobilized cells **11B**, 420
pentadin, proteinaceous sweetener **10**, 577f
pentane-2,3-dione, in beer **9**, 443f
pentosamines, as moieties of antibiotics **7**, 456
pentosanases, use in baking **9**, 675, 678f
pentosans **9**, 271ff
pentose, fermentation pathway **1**, 204
– metabolization, in bacilli **1**, 385
– – in clostridia **1**, 298
pentose phosphate pathway (PPP) **1**, 83, 479ff, 483
– reductive **1**, 84
pentoses, amination of – **7**, 415
pentostatin, inhibition of adenosine deaminase **7**, 648
People's Republic of China *see* China
PEP:carbohydrate phosphotransferase, PTS system **1**, 385
PEP-CL *see* phosphoenolpyruvate carboxylases
pepsin, preferred cleavages sites **8A**, 249
peptidase **2**, 417
– *de novo* formation in malt **9**, 427
– in *Lactococcus* **1**, 355f
peptidases *see* proteases
– classification **5A**, 206
peptide, metabolism in lactic acid bacteria **1**, 354
– transport in *Lactococcus* **1**, 358

peptide amidase, in industrial biotransformations **3**, 461f
– substrate specificity **3**, 461
peptide antibiotics **7**, 67ff, 277ff
– biosynthesis **7**, 284ff
– biotechnology of – **7**, 284ff
– directed biosynthesis **7**, 296
– future prospects **7**, 297f
– nonribosomal origin **7**, 279f
– production by *Bacillus* sp. **7**, 67
– reprogramming, genetic approaches **7**, 297
– ribosomal origin **7**, 278f
– side chain modification **7**, 282
– structures **7**, 278ff
peptide biosynthesis **7**, 284ff
– biosynthetic modules **7**, 286ff
– fermentation procedures **7**, 292, 294
– gene clusters **7**, 286f
– – compilation of – **7**, 287f
– mechanisms **7**, 285
– multienzyme systems **7**, 292
– – domains **7**, 290
– nonribosomal system **7**, 285
– ribosomal system **7**, 285
– structural alterations **7**, 294
– tripeptide model system **7**, 291
peptide bond formation, kinetically controlled-, by Alcalase **8A**, 251
– – coupling involving dehydroamino acids **8A**, 253f
– – coupling involving proline **8A**, 253
peptide bonds **9**, 10
– formation of, by catalytic antibodies **8B**, 418f
peptide cleavage, sequence specific –, by an antibody **8B**, 420ff
peptide coupling, with minimal protection **8A**, 255
peptide drugs **3**, 57ff
– adrenocorticotropin (ACTH) **3**, 64
– analogs **3**, 65
– calcitonin **3**, 65
– homogeneity analysis, technique **3**, 66
– impurities **3**, 65
– Merrifield synthesis **3**, 64
– peptidomimetics **3**, 64
– purification **3**, 57ff
– ribonuclease **3**, 64
– side chain protection **3**, 65
– sources **3**, 57ff
peptide families, analogs **7**, 294f
– – aureobasidin **7**, 295
– – cyclosporin **7**, 295
– – gramicidin **7**, 295
– – gramicidin S **7**, 295
peptide formation, by thermolysin **8B**, 355f
– kinetically controlled –, in anhydrous alcohols **8A**, 251
peptide library, sequence selection **2**, 351

peptide ligand receptors, antagonists of – **7**, 116ff
– – structures **7**, 118ff
peptide mass fingerprints **5B**, 253
– for protein identification **5B**, 329
peptide mass matches, for protein identification **5B**, 330
peptide purification **3**, 601
peptides *see also* synthetic enzymes
– as flavor enhancers **9**, 702, 704
– branched cyclic – **7**, 293
– chlorinated – **8B**, 185
– compilation of – **7**, 299ff, 304ff
– cyclic – **7**, 279, 293
– cyclodepsipeptides **7**, 293
– formed by *in vitro* synthesis **7**, 296f
– gene-encoded – **7**, 283f
– lactones **7**, 293
– linear **7**, 283, 293
– moncyclic – **7**, 279f
– multicyclic – **7**, 282f
– of nonribosomal origin **7**, 279ff, 284
– – from actinomycetes **7**, 304ff
– – from ascidians **7**, 313
– – from bacteria **7**, 304, 308f
– – from cyanobacteria **7**, 307f
– – from fungi **7**, 309f
– – from molluscs **7**, 314
– – from plants **7**, 311f
– – from scorpions **7**, 314
– – from sea hares **7**, 314
– – from spiders **7**, 314
– – from sponges **7**, 312f
– – future prospects **7**, 298
– of ribosomal origin **7**, 278ff
– – D-amino acids **7**, 281
– – features **7**, 284
– – from frogs **7**, 301f
– – from insects **7**, 300
– – from mammals **7**, 302f
– – from microbial sources **7**, 299
– – from molluscs **7**, 301
– – from plants **7**, 299
– – from scorpions **7**, 301
– – from snakes **7**, 302
– – from spiders **7**, 301
– – future prospects **7**, 297
– – production of – **7**, 283
– – structures **7**, 278ff
– protease-catalyzed formation **8A**, 247
– protease-catalyzed hydrolysis **8A**, 247ff
peptide sequence databanks **5B**, 131
peptide siderophores **7**, 212ff
– alterobactin **7**, 212
– azotobactin **7**, 216
– ferribactin **7**, 214
– ferrocins **7**, 212f
– ornibactins **7**, 216f
– pseudobactins **7**, 213ff

– pyoverdins **7**, 213f
peptide synthesis **3**, 453
– bioreactor design **3**, 461
– enzymatic – **3**, 452ff
– – continuously operated – **3**, 461
– integrated product recovery **3**, 461
– use of CLECs **8A**, 256
– use of immobilized enzymes **8A**, 255
peptide synthetases **7**, 286ff
– adenylate forming domains **7**, 286
– branched cyclopeptides **7**, 293
– carrier domains **7**, 289
– compilation of – **7**, 293
– condensation domains **7**, 290
– consensus sequences **7**, 289
– cyclodepsipeptides **7**, 293
– cyclopeptides **7**, 293
– domain construction **7**, 288
– epimerization domains **7**, 290
– lactones **7**, 293
– linear peptides **7**, 293
– N-methylation domains **7**, 290f
– thioesterase domains **7**, 291
peptidic drugs, structural class **7**, 39
peptidoglycan **1**, 19f
– in *Bacillus* spores **1**, 391
peptidyl-prolyl isomerase (PPI) **3**, 534
– as catalytic chaperone **3**, 540
peptolide SDZ, 3D-structure, in apolar environment **7**, 562
peptolide SDZ 214-103 synthetase **7**, 580
Pep5, biosynthetic pathway **7**, 334
– prepeptide structure **7**, 345
– primary structure **7**, 332ff
– structural gene **7**, 344
PER *see* tetrachloroethene
peracids, in Baeyer–Villiger reactions **8A**, 538f
percolation **11C**, 209f
Percotrate **11C**, 209f
percutaneous transluminal coronary angioplasty *see* PTCA
perfluorocarbon, in microcapsules **10**, 560
perfusion chromatography **3**, 50
– in enzyme purification **9**, 109
perfusion culture **3**, 28
– high cell densities **3**, 34
perillic acid, product of limonene bioconversion **10**, 393f
perillyl alcohol, product of limonene bioconversion **10**, 393f
pernilase **10**, 85
peroxidase, bleaching of paper pulp **10**, 536
peroxidases, depolymerization of lignite **10**, 165ff
– hydroxylation by **8A**, 504f
peroxisome **1**, 33
persistence of organic compounds, in soil **11B**, 96
personnel, in biotechnological facilities, monitoring **3**, 764

– – validation **3**, 764
pertussis vaccines **5A**, 138
pesticides, application of, formation of bound resi-
    dues **11B**, 109f
– halogenated –, distribution of **8B**, 177
– in wines **9**, 487
– monitoring with biosensors **11A**, 131f
– public opinion **12**, 125
PEST sequence **5B**, 337
Petroff-Hausser slide **1**, 119
petroleum, biodegradation of **11B**, 429
– optimum nitrogen level **11B**, 432
petroleum hydrocarbons, bioremediation, C–N ra-
    tio **11B**, 431
– – optimum pH **11B**, 431
– – with the heap technology **11B**, 325
petroleum oil, bioremediation of **11B**, 427
Peutz–Jeghers syndrome **5B**, 83, 86
Pfam **5B**, 134, 289
PFBP (Protein Function and Biochemical Net-
    works) **5B**, 410
PFR *see* plug flow reactor
PGA *see* penicillin G acylase
pH, influence of CO_2, anaerobic wastewater proc-
    ess **4**, 464
PHA *see* polyhydroxyalkanoic acid
– – *see* polyhydroxyalkanoic acids
phaak, production of **9**, 537
PHA analysis **6**, 434ff
– by chromatographic methods **6**, 436
– by gravimetric methods **6**, 435
– by HPLC **6**, 437
– by pyrolysis **6**, 437
– by spectroscopic methods **6**, 435f
– by turbidometric methods **6**, 435
– staining reactions **6**, 434f
PHA biosynthesis **6**, 412ff
– in eukaryotes **6**, 423f
– pathways in prokaryotes **6**, 414ff
– – *Alcaligenes eutrophus* **6**, 415
– – in recombinant bacteria **6**, 421ff
– – *Pseudomonas aeruginosa* **6**, 415
– – *Pseudomonas oleovorans* **6**, 415
– – *Rhodospirillum rubrum* **6**, 415
– regulation **6**, 423
PHA degradation, by spontaneous hydrolysis **6**,
    432
– extracellular – **6**, 428ff
– in the environment **6**, 430f
– intracellular – **6**, 431
PHA depolymerases, extracellular – **6**, 428ff
– – genetics **6**, 429f
– granule-associated – **6**, 427
Phaenerochaete chrysosporium **1**, 535
Phaeophyta, characteristics of **10**, 115
Phaffia rhodozyma **6**, 25
– astaxanthin production **7**, 175
phage *see* bacteriophage

phage control, in cheese production **9**, 372
– in cultured milk plants **9**, 398
phagemid **2**, 269
phage reproduction, physiology **1**, 552ff
phages, as contaminants, in lactic acid fermenta-
    tion **6**, 296
PHA granules **6**, 425ff
– associated proteins **6**, 427f
– – in *Rhodoccocus ruber* **6**, 428
– – PHA depolymerases **6**, 427
– – phasins **6**, 428
– – PHA synthases **6**, 427
– isolation **6**, 425ff
– – by centrifugation **6**, 426f
– structural organization **6**, 425
Phanerochaete chrysosporium, lignin-degrading
    white-rot fungi **11B**, 101
– oxidation of polycyclic aromatic hydrocarbons
    11B, 230ff
– TNT degradation **11B**, 289f
PHA (polyhydroxyalkanoic acids) **6**, 493ff
– – *see also* poly(3HB)
– applications, combination with other materials
    6, 447f
– – in agriculture **6**, 447
– – in medicine and pharmacy **6**, 446f
– – in the packaging industry **6**, 445f
– chemical synthesis **6**, 452
– commercial outlook **6**, 452
– composition, effect of temperature **6**, 420f
– copolyesters **6**, 432
– degree of polymerization **6**, 413
– extraction from biomass **6**, 444
– functions **6**, 424f
– fungal – **6**, 408
– general properties **6**, 432
– homopolyesters **6**, 432
– isolated from the environment **6**, 409f
– length of carbon chain **6**, 406ff
– microbial production **6**, 403ff
– – history **6**, 405
– occurrence in bacteria **6**, 406ff
– – *Alcaligenes eutrophus* **6**, 413
– – *Escherichia coli* **6**, 413
– occurrence in eukaryotes **6**, 412f
– – animals **6**, 413
– – humans **6**, 413
– – plants **6**, 413
– occurrence in fungi, *Physarum polycephalum* **6**,
    413
– – *Saccharomyces cerevisiae* **6**, 413
– patents **6**, 449ff
– physical properties, thermoplasticity **6**, 433
– piezoelectric properties **6**, 434
– production by fermentation **6**, 437ff
– – by *Alcaligenes eutrophus* **6**, 437ff
– – by *Alcaligenes latus* **6**, 439
– – by *Azotobacter vinelandii* **6**, 439

– – by *Chromobacterium violaceum* **6**, 440f
– – by methylotrophic bacteria **6**, 441
– – by *Pseudomonas oleovorans* **6**, 440
– – by recombinant *E. coli* **6**, 442ff
– production in plants, *Arabidopsis thaliana* **6**, 443
– prokaryotic inclusions **6**, 410ff
– – structure **6**, 411
– recovery **6**, 443ff
– – chemical methods **6**, 445
– – enzyme process **6**, 444
– – physical methods **6**, 445
– solution properties **6**, 433f
– structure **6**, 432f
– 3-hydroxyalkanoic acid constituents **6**, 407
– – branched **6**, 408
– – unsaturated – **6**, 407
– – with ester groups **6**, 409
– – with halogen atoms **6**, 409
pharmaceutical development, chances **12**, 214ff
– risks **12**, 214ff
pharmaceutical intermediates *see* enantiomerically pure pharmaceutical intermediates
– enantiomerically pure, use of lipases **8A**, 146
pharmaceutical precursors, kilogram-scale routes to, involving amidases **8B**, 285
– – involving lipases **8B**, 284
– – involving proteases **8B**, 284f
pharmaceutical products *see also* biomedicinal products
– efficacy **12**, 223f
– – tests **12**, 223f
– extensions of patent duration **12**, 316f
– licensing policy interpretation **3**, 747
– patent restrictions **12**, 307
– process development **12**, 224f
– quality **12**, 222
– safety **12**, 222f
– – regulations **12**, 222
pharmaceuticals, as products of marine biotechnology **10**, 450ff
– regulatory environment **8A**, 393
– world market **8A**, 393
pharmacogenomics **5B**, 92ff, 228, 418
pharmacokinetics, of antibodies, preclinical testing **5A**, 298
– of CDP571 **5A**, 348ff
pharmacology, of antibodies **5A**, 289ff, 297f
– – preclinical testing **5A**, 298
pharmacopoeias **12**, 224
pharmaco-toxicological testing, of antibodies **5A**, 307ff
– – tissues used **5A**, 308
phase I clinical trials, CMA-676 **5A**, 360f
– of HIV gene therapy **5A**, 482ff
phase separation **3**, 595ff
– aqueous polymer systems **3**, 595f
– PEG-dextran systems **3**, 599f

– phase volume ratio **3**, 596
– polymer-polymer system **3**, 595
– polymer-salt system, immiscible – **3**, 595
– ternary phase diagram **3**, 595
– two-phase systems **3**, 595ff
– – electrical potential difference **3**, 598
– – mass transfer **3**, 605
– – PEG concentration **3**, 601
– – phase components **3**, 595
– – polymer effects **3**, 598
– – polymer recycling **3**, 603
– – salt recycling **3**, 603
– – shuttle polymers **3**, 600
phasins **6**, 428
– interaction with surface of PHA granules **10**, 427
phasmid **2**, 260
PHA synthases **6**, 471ff
– catalytic mechanisms **6**, 419
– classification **6**, 417f
– expression in recombinant bacteria **6**, 421ff
– gene structure **6**, 418
– granule-associated – **6**, 427
– preferred substrates **6**, 418
PHB *see* poly-β-hydroxybutyric acid
PHB *see* poly-β-hydroxybutyrate
– – *see* Poly-β-hydroxybutyric acid
– – *see* poly(3HB)
p-HBAH **4**, 157f
pH calculation, model for anaerobic wastewater purification **4**, 463ff
– – biological phase **4**, 463
– – gas phase **4**, 467
– – liquid phase **4**, 464ff
pH control, CO_2 production **4**, 33
– use of expert systems, fault diagnosis **4**, 630f, 633ff
pH controller, logic for finding faults **4**, 631
– parameter-adaptive **4**, 475f
pH dependence, of enzyme catalysis **9**, 49f
pH effect, on microbial growth **9**, 198
pH electrode, assembly of **4**, 15
– Nernst potential, aging **4**, 16
phellodonic acid, structure **7**, 505
phenanthrene, bacterial degradation of **11B**, 153, 222ff
– bioremediation **11B**, 428
– complete metabolization and mineralization **11B**, 222ff
– hydroxylation of **8A**, 501
– oxidation by *Phanerochaete chrysosporium* **11B**, 233
– physical and chemical properties of **11B**, 218
– structure of **11B**, 216
phenocopy **5B**, 68
phenol, anaerobic degradation **6**, 690
phenol degradation, by immobilized microorganisms **1**, 241f

phenolic compounds **11A**, 246f
– formation of **8B**, 346f
– monitoring with biosensors **11A**, 130f
– removal by plants, in westland systems **11A**, 246f
phenol oxidases, depolymerization of coal and coal humic substances **10**, 168f
– modification of lignon **10**, 530
phenols, anaerobic degradation of **11B**, 177f
– bacterial degradation of **11B**, 153
– bioremediation with the heap technology **11B**, 326
– chlorinated – **8B**, 188
– degradation by immobilized cells **11B**, 418
– formation by microbial oxidation of aromatic substrates **8A**, 499ff
– halogenation of **8B**, 196
– in municipal wastewater **11A**, 177
– *o*-ring cleavage, by microfungi **11B**, 159
phenotypic markers **5B**, 167
phenoxazinone synthase, effect of glucose on activity **7**, 73
phenoxyalkanoic acids, bioredegradation of **11B**, 260f
phenylacetate catabolism, pathway **8B**, 66
phenylacetyl, as protecting group, in peptide chemistry **8A**, 261
phenylacetylcarbinol, formation of **8A**, 9f
phenyl acetyl esters of secondary alcohols, kinetic resolution by penicillin acylase **8A**, 231
phenylalanine, biosynthetic pathway, rate limiting steps **8B**, 320
– overproducing organisms **8B**, 315ff
– regulation of biosynthesis **8B**, 315
– secretion from the cell **8B**, 320f
phenylalanine ammonia-lyase (PAL) **8B**, 137ff
– deamination of phenylalanine **8B**, 139f
– inhibitors **8B**, 140
– of *Rhodotorula glutinis* **8B**, 139ff
– substrates **8B**, 140
– – pyridyl cinnamates **8B**, 141
phenylalanine dehydrogenase **6**, 541f
– in industrial biotransformations **3**, 455
– oxidative deamination of L-phenylalanine **6**541
– synthesis of L-amino acids **6**, 542
phenylalanine dehydrogenases, applications of **8A**, 446
– characteristics of **8A**, 446
D-phenylalanine, engineering biochemical pathway, for whole cell biosynthesis **8B**, 327
– novel biosynthetic pathway **8B**, 325ff
phenylalanine hydroxylase (PAH) gene **5B**, 70
L-phenylalanine, bacterial production of **8B**, 314ff
– biosynthesis **6**, 473ff
– – in *Corynebacterium glutamicum* **6**, 474
– biosynthetic pathway **6**, 511
– biosynthetic pathways **8B**, 314f

– degradation **6**, 519f
– feedback inhibition of prephenate dehydratase **8B**, 319
L-phenylalanine ammonia lyase **6**, 519f
L-phenylalanine production **3**, 455, **6**, 510
– enzymatic process **6**, 473ff
– fermentation technology **6**, 474f
– strain improvement **6**, 474
phenylketonuria (PKU) **5B**, 70, 84
phenyl methyl ether, anaerobic demethylation of **11B**, 172f
2-phenyloxazolin-5-ones, lipase-catalyzed ring opening **8A**, 116
phenylpyruvate, decarboxylation of, by phenylpyruvate decarboxylase **8B**, 86
phenylpyruvate decarboxylase **8B**, 85
pheromone, action in bacteria **2**, 55
pheromones, component, synthesis by yeast alcohol dehydrogenase **8B**, 360
– control of pest insects **10**, 498f
– crockroach aggregation, haloorganic – **8B**, 177
– enzymatic synthesis of **8B**, 361f
– receptor-based sensors **4**, 96
– synthesis of **8B**, 22
Philippines, biotechnology **12**, 413ff
pH$_i$ measurement **4**, 202ff
– flow cytometric **4**, 203
– NMR **4**, 203
– radiometric **4**, 203
– sample-flow analysis **4**, 203f
phloem, cells **1**, 580
pH measurement, in production bioreactors **3**, 322
pH meter **4**, 16
– reliability **4**, 20
phopholipids, yeast "lecithin", extraction of spent brewer's yeast **7**, 183
phorbol, structure **7**, 678
phorbol esters, activation of protein kinase C **7**, 677
– tumor promoters **7**, 677
Phoredox process, in enhanced biological phosphorus removal **11A**, 298
phosphatase, in yeast **1**, 490f
phosphate, estimation methods **1**, 126
– pK_a values **8B**, 222
phosphate bond, energy-rich **1**, 54
phosphate ester, hydrolysis of, use of catalytic antibodies **8B**, 463f
phosphate ester hydrolysis, by nuclease **8B**, 236
– by phosphatase enzymes **8B**, 236f
phosphate limitation, metabolic effect **1**, 179
phosphate metabolite regulation, need for on-line process analysis **4**, 151
phosphate monoester, cleavage of, use of catalytic antibodies **8B**, 435f
phosphate precipitation, biologically induced chemical – **11A**, 290f

– in sludge treatment **11A**, 307f
– in the sidestream process **11A**, 299
– physico-chemical background **11A**, 308ff
phosphate release, by activated sludge **11A**, 291ff
– – inhibition of **11A**, 294
phosphate removal **11A**, 41f
– activated sludge process **11A**, 257
– biological – **11A**, 41f
– enhanced biological –, EBPR pilot plant **11A**, 304f
phosphate sources, inorganic phosphate **8B**, 228
– nucleoside triphosphates **8B**, 228ff
phosphate uptake system **1**, 179
phosphatidylglycerol production **3**, 454
phosphines and phosphine oxides, kinetic resolution with cholesterol esterase **8A**, 217f
PhosphoBase **5B**, 410
phosphodiester, cleavage of, use of catalytic antibodies **8B**, 436f
phosphodiesterase, of snake venom **6**, 569
– phosphorylation by **8B**, 226
phosphodiester formation, coupling with adenosine triphosphate **8B**, 232
– coupling with cytidine triphosphate **8B**, 232f
– coupling with uridine triphosphate **8B**, 233
phosphoenolpyruvate **8B**, 229ff
– enzymatic synthesis of **8B**, 229
– phosphorylation potential **8B**, 229
phosphoenolpyruvate carboxylases (PEP-CL) **8B**, 84f
– listing of **8B**, 85
phosphofructokinase, activity, in immobilized microorganisms **1**, 237f
– gene **1**, 481
– regulation **1**, 486
phosphofructokinase (PFK), freeze-dried, stability **3**, 706
phosphoglucose isomerase, genes, (*DCD30*) **1**, 481
– – (*PGI1*) **1**, 481
phosphoglycerate kinase, gene (*PGK1*) **1**, 482
phosphoglycerate kinase (PGK), refolding, temperature sensitivity **3**, 547
phosphoglycerate mutase, gene (*GPM1*) **1**, 482
phosphoinositol kinase inhibitors **7**, 681
phosphoinositol metabolism, inhibitors of – **7**, 680ff
– – structures **7**, 681
phosphoketolase pathway, microbial metabolism **9**, 184
phospholipase D, in industrial biotransformations **3**, 454
phospholipase inhibitors, from basidiomycetes **7**, 518f
phospholipid bilayer **2**, 95
phospholipid fatty acid analysis **11C**, 451f
phospholipids **7**, 183f
– anaerobic degradation of **11A**, 34f

– as nutrients, for mammalian cell cultures **3**, 146
– medicinal uses **7**, 184
– structures **7**, 139
– use in cosmetics **6**, 676
phosphonates, inhibitors of phosphodiester cleavage **8B**, 236
4-phosphopantetheine **9**, 35
phosphoproteins **5A**, 17ff
– secreted into milk **5A**, 17
phosphoramidate esters, hydrolysis by alkaline phosphatase **8B**, 237
phosphoric acid esters, as nonionic surfactants **6**, 673
– use in cosmetics **6**, 673
phosphor limitation, effect on *Escherichia coli* **1**, 173
– – metabolism **1**, 172ff
phosphorothioate oligodeoxynucleotide **2**, 326f
– synthesis of non-oxygen bridged – **2**, 328
phosphorus, in wastewater **11A**, 197
– physico-chemical refixation **11A**, 307
– supplementation, in contaminated soils **11B**, 431f
phosphorus compounds, in municipal wastewater **11A**, 178f
phosphorus feedback, during sludge treatment **11A**, 306ff
phosphorus precipitation, chemicals for – **11A**, 310
– post-precipitation **11A**, 312
– pre-precipitation **11A**, 310f
– process schemes **11A**, 311
– simultaneous – **11A**, 311f
phosphorus release, during sludge treatment **11A**, 306ff
– enhanced-, biochemical models **11A**, 294f
phosphorus removal, by the sequencing batch reactor **11A**, 325f
– – cycle design **11A**, 325f
– chemical – **11A**, 285ff, 308ff
– – basic principle **11A**, 308ff
– – flocculation and filtration **11A**, 312
– – influence on biological wastewater treatment **11A**, 312f
– – metal phosphates **11A**, 309
– – post-precipitation **11A**, 312
– – pre-precipitation **11A**, 310f
– – process design **11A**, 310ff
– – simultaneous precipitation **11A**, 311f
– – slude production **11A**, 313ff
– chemical precipitation **11A**, 119
– enhanced biological – (EBPR) **11A**, 119, 285ff, 306
– – aerobic phosphate uptake by activated sludge **11A**, 291ff
– – anaerobic phosphate release by activated sludge **11A**, 291f

– – dependence on phosphate concentration **11A**, 293f
– – EBPR process **11A**, 303ff
– – historical development **11A**, 288f
– – increase of easily biodegradable substrate **11A**, 301f
– – influence of wastewater quality **11A**, 299ff
– – inhibition of phosphate release **11A**, 294
– – Johannesburg (JHB) process **11A**, 303
– – mainstream process **11A**, 297ff
– – microbiological aspects **11A**, 289ff
– – nitrate input into the anaerobic zone **11A**, 302f
– – oxygen input into the anaerobic zone **11A**, 302f
– – process design **11A**, 297ff
– – role of *Acinetobacter* **11A**, 296
– – sidestream process **11A**, 299f
– – sludge production **11A**, 303ff
– in fixed bed plants **11A**, 359
– in wastewater treatment **11A**, 13
– modeling of **11A**, 422f
phosphorus uptake, by activated sludge **11A**, 290ff
– by microorganisms **11A**, 289f
– enhanced –, biochemical models **11A**, 294f
phosphorylation **8B**, 221ff
– enzymes **8B**, 223ff
– enzymic synthesis of heterocyclic compounds **8B**, 389
phosphoryl transfer **8B**, 222ff
– cofactor regeneration **8B**, 228
– mechanism of **8B**, 224
– phosphate source **8B**, 228
phosphoryl transfer enzymes **8B**, 223ff
– classification of **8B**, 223ff
– listing of **8B**, 225
phosphoryl transfer reactions **8B**, 223ff
– enzyme classes **8B**, 223ff
– two groups of **8B**, 223
phosphoryl transfer reagents, phosphorylation potentials of **8B**, 227
phosphotriester, hydrolysis of, use of catalytic antibodies **8B**, 437
Phostrip process **11A**, 289
– scheme **11A**, 300
photoaffinity labels **9**, 42f
photobioreactors, closed – **10**, 122f
– environments for microalgae **10**, 117
– inclined surface type **10**, 121f
– lab-scale – **10**, 118f
– plate type – **10**, 125f
– semi-closed – **10**, 122f
– tubular – **10**, 123ff
photolithotrophs **3**, 128
photoorganotrops **3**, 128
photophosphorylation **1**, 59f
photosynthesis **1**, 59f

– electron transport chain **1**, 59
– – cyclic electron flow **1**, 60
– – Z-scheme **1**, 60
– energy transduction **1**, 61ff
– photosystems **1**, 59ff
– reaction center **1**, 59
– reducing equivalent **1**, 61
– reversed electron-flow **1**, 62
photosynthetic membrane **1**, 23
photosystems, in plants **1**, 59ff
phototrophic bacteria, intracytoplasmic membranes **1**, 24
– plasmalemma **1**, 24
phototrophic microorganisms, cultivation of **10**, 117
phototrophs **3**, 128
phototrophy **1**, 50
– definition of **10**, 275
PHRAP **5B**, 30
phreatophytes, use in phytoremediation **11B**, 379
PHRED **5B**, 30
pH regulation, in anaerobic wastewater purification **4**, 474ff
phthalate, bacterial degradation of **11B**, 153
phthalic acid, oxidation by *Phanerochaete chrysosporium* **11B**, 233
pH tolerance, of pediococci **1**, 340
pH value, effect on microbial activity **11B**, 430f
– effects on, microbial growth **3**, 130
– – product formation **3**, 130
– influence on microbial growth **10**, 274
– of media **1**, 180f
– optima, for microorganisms **1**, 131
Phycomyces blakesleeanus **1**, 522
– β-carotene production **7**, 175
– life cycle **1**, 523
Phycomycetes **1**, 517, 522ff
phylloplane **5B**, 12
phylloquinone, production of **10**, 335f
– structure **10**, 336
phylogenetic analysis, distance **5B**, 138
– maximum likelihood **5B**, 138
– parsimony **5B**, 138
phylogenetic system, rRNA superfamily **1**, 403
– tree of organisms **1**, 252
– tripartite **1**, 402
phylogenetic tree **10**, 66ff
physical genetic map, of *E. coli* **5B**, 10
physical mapping **5B**, 42
– by sequence tag connectors **5B**, 177
physiological state, of bioprocesses, control **3**, 387f
physiological state variables, in bioprocesses **3**, 362
– – calculation **3**, 384ff
– – cascaded control **3**, 387
physiological variables, in bioprocesses **3**, 361
– – calculation **3**, 364ff

– – direct reconstruction **3**, 372ff
– – estimation **3**, 371
phytanyl diether lipid, of methanogens **1**, 259
phytase **5A**, 210f
– characteristics **5A**, 211
– industrial applications **5A**, 210f
– of *Aspergillus* **5A**, 210f
– – properties **5A**, 210
– reaction catalyzed by – **5A**, 210
– recombinant – **5A**, 212
– substrate specificities **5A**, 211
– use in feed industry **5A**, 210ff
phytic acid, hydrolysis by yeast phosphatases **8A**, 379
– in feeds **5A**, 210
phytoalexin **1**, 593
– induction of synthesis **1**, 593
phytoalexine **2**, 647
phytoalexins **10**, 494
phytochelatins **10**, 235, 238
phytochelatin synthases **10**, 236
phytochemical production, commercial **1**, 604
phytoextraction, definition of **11B**, 386
– of metals, chelate-assisted – **11B**, 388
– – continuous – **11B**, 388ff
– – hyperaccumulator species **11B**, 389ff
– – induced – **11B**, 387f
– – metal bioavailability **11B**, 389
– – use of metal chelates **11B**, 387f
phytohemagglutinin **1**, 35
phytohormones, in plant cell culture **1**, 581, 591
phytoremediation **11B**, 365
– agronomic inputs **11B**, 380f
– applications **11B**, 373
– clean-up time **11B**, 382
– contaminant uptake **11B**, 382
– degradation by rhizospheric bacteria **11B**, 377
– degradation processes **11B**, 374ff
– design **11B**, 379ff
– development of **11B**, 386
– groundwater remediation **11B**, 381
– irrigation **11B**, 380
– mechanisms for organic contaminants **11B**, 378
– of metals **11B**, 385ff
– of organic compounds **11B**, 371ff
– plant density **11B**, 380
– plant species used **11B**, 372f
– root concentration factor **11B**, 381f
– transpiration of water **11B**, 381f
– transpiration stream concentration factor **11B**, 381f
– transport of, chemicals **11B**, 373
– – of water **11B**, 373
– – oxygen **11B**, 373
– treatability studies **11B**, 379f
– use of aquatic plants **11B**, 379
– use of grasses **11B**, 379
– use of phreatophytes **11B**, 379

phytostimulation **11B**, 373
Pichia **1**, 269
Pichia miso, esterase of **8A**, 219
Pichia pastoris **1**, 280, 470ff, **2**, 524
– alcohol conversion to aldehydes **8B**, 345
– expression of heterologous proteins **9**, 85
Pichia pinus **2**, 523ff
Pichia saitoi, in sour dough **9**, 297
– in sour dough starters **9**, 255ff
pickle fermentation **1**, 333
pickles *see also* cucumbers
– production statistics **9**, 632
– sensory properties **9**, 638
– spoilage **9**, 638f
– – gaseous – **9**, 638f
– – softening **9**, 639
picornavirus **2**, 687
– internal ribosomal entry site (IRES) **2**, 687
– ribosome landing pas (RLP) **2**, 687
– translation initiation **2**, 687
PIGBASE **5B**, 285
pig liver, flavin dependent monooxygenases **8A**, 575
pig liver alcohol dehydrogenase (PLADH), applications of **8A**, 452
– characteristics of **8A**, 452
pig liver esterase (PLE) **8A**, 194ff
– active site model **8A**, 207ff
– asymmetrization of, cyclic *meso*-1,2-dicarboxylates **8A**, 196f
– – cyclohexene *meso*-1,2-dicarboxylates **8A**, 197
– – cyclopentane *meso*-1,3-dicarboxylates **8A**, 197f
– – diacylated cyclopentene *meso*-diols **8A**, 198f
– – prochiral glutarates **8A**, 195f
– – prochiral malonates **8A**, 195
– – tricyclic *meso*-diesters **8A**, 197ff
– chemoselective hydrolysis by **8A**, 207
– enantioselective hydrolysis of, *meso*-diesters **8A**, 196ff
– – prochiral diesters **8A**, 195f
– kinetic resolution of, quaternary centers **8A**, 202
– – racemic acyclic and cyclic 1,2-diols **8A**, 203
– – racemic cyclic diesters **8A**, 201
– – racemic -caprolactones **8A**, 201f
– – racemic esters **8A**, 198ff
– – racemic methyl alkylphenylphosphinoylacetates **8A**, 206
– – secondary alcohols **8A**, 202
– – tertiary alcohols **8A**, 202
– polyethylene glycol monomethyl ether (MPEG) linked –, enantioselective acylation of *meso*-diols **8A**, 207
– practical considerations **8A**, 207
– recognition of planar chirality **8A**, 205
– regioselective hydrolysis by **8A**, 206
– specificity of **8B**, 367f

– stereoselective hydrolysis **8B**, 353f
– synthesis of heterocyclic compounds **8B**, 384f
– use in the generation of planar chirality, in tri-
  carboxylchromium compounds **8A**, 198
pig liver 2,3-oxidosqualene cyclase **8B**, 28f
pig lung, flavin dependent monooxygenases **8A**,
  575
pigments, as products of marine biotechnology
  10, 466
pilatin **7**, 504f
– structure **7**, 504
pili **1**, 17
– structure **1**, 18
α-pinene, degradation of, in a biofiltration column **11C**, 287
pinenes **10**, 404ff
– bioconversion products from **10**, 405
– biotransformation of **10**, 380f, 404
– degradation of **10**, 404
pinicoloform, inhibition of differentiation of leukemia cells **7**, 524
piping system, for biotechnological facilities **3**,
  758
PIR **5B**, 286
PIR International Protein Sequence Database **12**,
  326f
PKA *see* porcine kidney aminoacylase
PKS *see* polyketide synthases
PKU *see* phenylketonuria
Plackett-Burman design, in medium optimization
  3, 134f
PLADH *see* pig liver alcohol dehydrogenase
planar chirality, recognition by pig liver esterase
  8A, 205
planned introductions *see* field trials
plant, as *Bacillus* habitat **1**, 375
– biocontrol of pathogens **1**, 423
– cormus organization **1**, 579
– metabolism **1**, 580
– propagation **1**, 603
– – pathogen-free plants **1**, 603
– protoplast **1**, 36
plant biotechnology **2**, 629
plant breeders' rights **12**, 284f
plant breeding, acceleration of traditional – **5B**,
  152f
– by plant tissue culture **1**, 603
– crop breeding plan **12**, 477
– limitations **2**, 629
plant cell **1**, 34f, 577ff
– antisense DNA technology **1**, 602
– as host cell **3**, 285
– compartments **1**, 578
– differentiation **1**, 579
– epidermal **1**, 579
– genetic engineering **1**, 601f, **2**, 547ff
– membrane permeabilization **1**, 598
– meristematic **1**, 579

– mesophyll **1**, 580
– morphology **1**, 578
– organization **1**, 578ff
– recombinants, in fermentation processes **3**, 285
– recovery, from bioreactors **3**, 473
– shear sensitivity **1**, 586
– surface layer **1**, 34
– suspension characteristics **1**, 585
– transformation **2**, 573ff
– types **1**, 579
– vascular **1**, 580
– wall **1**, 35, 578
plant cell culture **1**, 40f, 579ff
– agitation damage **3**, 230
– biomass kinetics **1**, 586
– biosynthetic pathway elucidation **1**, 590
– biotransformation reactions **1**, 595f, 599
– cell density **1**, 584
– cell differentiation **1**, 590
– cell division synchrony **1**, 584
– cell immobilization **1**, 598
– cell line selection **1**, 592
– characteristics **1**, 583f
– CO_2 fixation **1**, 580
– differentiated – **1**, 600f
– doubling time **1**, 585
– enzyme production **1**, 589
– genetic stability **1**, 584
– growth, assessment **1**, 588
– – phases **1**, 585f
– – rate **1**, 592
– in bioreactors **1**, 586f
– *in vitro* **1**, 579, 581ff
– macronutrients **1**, 591
– medium optimization **1**, 590
– metabolism **1**, 580
– nutrient requirements **1**, 580f
– oxygen supply **1**, 587f
– – effects on product formation **1**, 587
– physical conditions **1**, 595
– physiological stability **1**, 584
– phytohormones **1**, 581, 591
– productivity **1**, 592
– requirements **1**, 41
– secondary compound production **1**, 589ff
– stirrer types **1**, 587
– substrate kinetics **1**, 586
– suspension – **1**, 583ff
– – ^{31}P-NMR spectrum **1**, 192f
– two-phase – **1**, 598
– types **1**, 581
– ventilation **1**, 587
– vessel design **1**, 595
plant cell cultures **7**, 604
– biotechnological importance **7**, 596ff
– biotransformations **7**, 621ff
– commercial interest in – **7**, 595
– expression of pathways **7**, 603

– hydroxylation of steroids **8A**, 489
– maintenance of **10**, 27ff
– metabolic engineering **7**, 623ff
– of transgenic plants **7**, 623f
– production of secondary metabolites **7**, 593ff
– regulation of secondary pathways **7**, 623
– technology **7**, 596
plant cells, electrofusion of, hybrid yield **10**, 515f
– – under microgravity conditions **10**, 515
plant cell suspension cultures, secondary product
   formation **7**, 598ff
– – anthocyanins **7**, 609f
– – anthraquinones **7**, 601ff
– – antitumor compounds **7**, 614ff
– – benzophenanthridine alkaloids **7**, 603ff
– – betalains **7**, 611
– – cardiac glycosides **7**, 617
– – cinnamic acid derivatives **7**, 598ff
– – immunologically active polysaccharides **7**,
   612
– – monoterpene indole alkaloids **7**, 606ff
– – morphinan alkaloids **7**, 613
– – naphthoquinones **7**, 601f
– – protoberberine alkaloids **7**, 603ff
– – quinoline alkaloids **7**, 614
– – tropane alkaloids **7**, 613f
– – vanillin **7**, 617f
plant cell wall, content in plant material **9**, 708
– degradation **9**, 706ff
– polysaccharides **9**, 707
– structure **9**, 706
plant defense, against microorganisms **2**, 647
– – chitinase production **2**, 647
– – phytoalexine production **2**, 647
– – *Pseudomonas syringae* pv. *tabaci* **2**, 647f
– – stilbene production **2**, 647
plant genetic engineering, potential for plant
   breeding **2**, 598
plant growth regulators, from secondary metabolites **7**, 9
plantibodies **10**, 493
plant improvement **2**, 638ff
plant material, composting of **11C**, 17
plant material fermentation **1**, 333
– *Lactobacillus mesenteroides* **1**, 341
– microbial succession **1**, 333
plant matter, decaying process **11B**, 99ff
– decaying velocity **11B**, 100
– degradation processes, two-stage exponential
   function **11B**, 100
– degradation rate **11B**, 101
– humification rate **11B**, 101
– microbial degradation of, carbon partitioning
   11B, 99
plant-microorganism interaction **1**, 422
plant oils, commercial markets **7**, 135
– fatty acid composition **7**, 136
– fatty acid profiles **7**, 158

– modification by genetic engineering **7**, 138
plant organ culture **1**, 579, 605
– metabolism **1**, 580
– transformation **1**, 600
– types **1**, 581
plant pathogenicity factors, bacterial – **10**, 491
plant pathogens, bacteria **11C**, 234
– biocontrol of, use of bacteria **10**, 487f
– – use of fungi **10**, 488
– detection tools **10**, 499f
– fungi **11C**, 235
– in composting, inactivation of **11C**, 77
– soilborne –, control by compost **11C**, 77f
– viruses **11C**, 234
Plant Pest Act **12**, 462, 472
plant product modification **2**, 650f
plant protection, biotechnology in **10**, 485ff
– by *Bacillus thuringiensis* **10**, 496ff
– by insect pathogenic microorganisms **10**, 496ff
– by integration of *B.t.* toxin genes into plant genomes **10**, 498
– by pheromones **10**, 498f
– by systemic acquired resistance **10**, 494f
– transgenic plants, herbicide resistance of **10**,
   493f
– – production of plantibodies **10**, 493
– – virus resistant – **10**, 492f
– use of, arthropods **10**, 495f
– – sterile insect release **10**, 495f
– – transgenic plants producing compounds
   against phytopathogens **10**, 491f
plant protection products **12**, 563
– biotechnologically produced – **10**, 501
plant protoplasts, effects of microgravity on, adaptation strategies **10**, 510f
Plant Quarantine Act (PQA) **12**, 201ff
plant regeneration **2**, 124f, 629ff
– callus formation **2**, 124
– chloroplast segregation **2**, 126
– from multicellular explants **2**, 630f
– from protoplasts **2**, 631
– Gramineae **2**, 125
– intergeneric hybrid **2**, 128
– maize **2**, 125
– Solanaceae **2**, 125
– transgenic plants **2**, 629
plant residues availability of **11C**, 104ff
– composition of **11C**, 104ff
– composting of **11C**, 101ff
– for composting **11C**, 102ff
– – amount of **11C**, 102ff
– – bulk density **11C**, 106
– – disintegration of **11C**, 114
– – heavy meal content **11C**, 105
– – mixing of **11C**, 116f
– – optimal water content **11C**, 109f
– – origin of **11C**, 102
– – properties of **11C**, 106

– – screening of **11C**, 115f
– – seasonal availability **11C**, 108
– – sorting **11C**, 116f
– – structure of **11C**, 109
– of agricultural plants **11C**, 104
– reactor composting **11C**, 122f, 123
plants, alkaloids from **8A**, 330f
– – use of cell culture **8A**, 331
– – use of metabolic engineering **8A**, 331
– as sources of nonribosomal peptides **7**, 311f
– as sources of ribosomal peptides **7**, 299
– conditions of reduced gravity **10**, 508ff
– degradation of organic contaminants **11B**, 374ff
– direct uptake of chemicals **11B**, 375
– effects of microgravity on **10**, 509f
– hyperaccumulator species **11B**, 389ff
– nitrilase activity **8A**, 299
– phytoremediation of organic compounds **11B**, 371ff
– secondary metabolites, from cell cultures **7**, 595ff
– – uses **7**, 595
– use in terrestrial ecotoxicity test systems **11B**, 136
– virus resistant – **10**, 492f
plant secondary metabolites, abiotic induction/ stimulation factors **1**, 595
– elicitation by stress **1**, 593f
– extraction with solid adsorbents **1**, 600
– industrial production **1**, 589ff
– isolation **1**, 598f
– precursor feeding **1**, 595
– production induction/stimulation **1**, 593f
– secretion **1**, 598
– two-phase extraction **1**, 600
plant somatic hybrid **2**, 127
– allotetraploid **2**, 127
– amphidiploid **2**, 127
– asymmetric **2**, 128
– incompatibility **2**, 128
– intergeneric **2**, 128
– symmetric **2**, 128
plant tissue **1**, 37
plant tissue culture **1**, 579, 582f
– application in agriculture **1**, 603
– for plant breeding **1**, 603
– metabolism **1**, 580
– types **1**, 581
plant tissue cultures, expression of pathways **7**, 597
– novel pharmacologically active compounds **7**, 621
– secondary metabolites **7**, 595
– technology **7**, 596
plant toxins, ricin **3**, 71
plant transformation **5B**, 152
– by *Agrobacterium* **2**, 549ff, 573ff
– technologies **5B**, 153

plant varieties, protection **12**, 284ff
plant variety rights **12**, 289
plant viral vector **2**, 564ff
– advantages in gene transfer **2**, 565
– design **2**, 569ff
– expression vector **2**, 566f
– limitations **2**, 567
– replacement vector **2**, 568
– transformation vector **2**, 567f
plant virus **2**, 564ff
– classification **2**, 564f
– DNA virus **2**, 564, 567f
– – double-stranded **2**, 567
– – single-stranded **2**, 567
– RNA virus **2**, 565ff
– – capacity for nucleic acid uptake **2**, 567
– – gene expression stability **2**, 567
– – replication fidelity **2**, 567
– – strategies for plant transformation **2**, 565
plasmalemma of bacteria *see* membrane
plasma membrane, lipid-linked proteins **5A**, 17
Plasmaviridae **1**, 550
plasmid **2**, 429
– conjugative **2**, 429ff
– – helper plasmid **2**, 429ff
– *E. coli* R1 **2**, 411
– high-copy **2**, 382
– in eukaryotes **2**, 509
– low-copy **2**, 382, 384
– map **2**, 52
– mobilizable vector **2**, 429ff
– *Phanerochaete* plasmid **2**, 540
– replication system **2**, 382
– rescueing **2**, 461, 535
– self-transmissible **2**, 429
– shuffling **2**, 517
plasmid-containing cells, fermentation of **5A**, 448f
– – high-performance technology **5A**, 448
plasmid copy number, effect on growth rate **4**, 490
plasmid DNA *see also* DNA
– cell lysis **5A**, 450
– clinical-grade – **5A**, 465
– clinical trials **5A**, 460
– drug master file **5A**, 461
– fermentation **5A**, 446ff
– good manufacturing practice **5A**, 445, 452ff
– in vaccination **5A**, 444, 465f
– manufacturing conditions, quality assurance **5A**, 445
– – quality control **5A**, 445
– pharmaceutical requirements **5A**, 460f
– pilot production **5A**, 453
– purification **5A**, 446ff, 450f
– quality control **5A**, 458ff
– – DNA quantity **5A**, 459
– – restriction analysis **5A**, 459

– – sequencing **5A**, 459
– – sterility **5A**, 459
– therapeutic –, future of **5A**, 465f
– therapeutic use, guidelines **5A**, 461
plasmid DNA manufacturing, design **5A**, 449ff
– documentation **5A**, 465
– downstream processing **5A**, 454f
– drug master file **5A**, 457f
– *E. coli* host cells **5A**, 447
– facility requirements **5A**, 461f
– fermentation **5A**, 453f
– fermentation of cells **5A**, 448f
– flow diagram **5A**, 454
– guidelines **5A**, 461
– harvesting **5A**, 453f
– host cells **5A**, 449
– in-process control **5A**, 463
– large scale – **5A**, 443ff
– master cell bank **5A**, 452f
– plasmid isolation **5A**, 447f
– – laboratory scale – **5A**, 447f
– plasmid stability **5A**, 451f
– process types **5A**, 458
– process validation **5A**, 462f
– purification **5A**, 447f, 453, 455f, 458
– – lysis procedures **5A**, 447f
– quality assurance **5A**, 463f
– quality control **5A**, 463f
– safety aspects **5A**, 449
– strain selection **5A**, 452
– upscaling **5A**, 456f
– use of antibiotic resistance genes **5A**, 446
– vector selection **5A**, 452
– yield **5A**, 449, 458
plasmid DNA therapeutics, good manufacturing
  practice **5A**, 445
– guidelines **5A**, 445
– manufacturing conditions, quality assurance
  5A, 445
– – quality control **5A**, 445
– regulatory aspects **5A**, 445f
– safe manufacturing process **5A**, 445f
– system development process **5A**, 445
plasmid preparation, research-grade – **5A**, 448
plasmids **12**, 91f
– in bacilli **1**, 395
– incompatibility group (incP) **1**, 410
– in *Lactococcus* **1**, 361
– in pseudomonads **1**, 409f
– – host range **1**, 410
– in recombinant fermentation **4**, 489ff, 540
– – content **4**, 498f
– – stability **4**, 497ff
– in streptomycetes **1**, 440
– loss, F_{25} value **3**, 286
– – kinetics **3**, 285
– – probability **3**, 285
– naturally occurring, genetic properties **12**, 80f

– pAMβ1 **1**, 362
– pWWO **1**, 201, 411
– – transposon **1**, 411
– – *xyl* gene regulation **1**, 411
– recombinant – **3**, 287
– – pSC101-trpI15 × 14 **3**, 290
– – stability **3**, 285ff, 291
– safety **12**, 80
– segregational instability **1**, 119
– stability **12**, 170f
– – gene expression effects **3**, 286
– stable mutant isolation **3**, 289
– TOL **1**, 201, 411
– types in yeast **1**, 475, 499
plasmid stability **5A**, 451ff
– effect on plasmid yield in manufacturing processes **5A**, 451
– in manufacturing **5A**, 451
– testing of **5A**, 451
plasmids types in yeast *see also* yeast plasmids
plasmid vector **2**, 382ff, 429, 458ff, 557ff
– – *see also* vector
– broad host range **2**, 429ff, 460
– – cloning site **2**, 431
– – pKT vector **2**, 433
– – pSUP vector **2**, 432ff
– – selectable marker **2**, 431
– cassette construction **2**, 560
– cloning vector **2**, 435
– expression vector in animals **2**, 723
– – structure **2**, 723
– for *Bacillus* **2**, 458ff
– IncP-1 derivative **2**, 429ff
– IncQ derivative **2**, 430f
– IncW derivative **2**, 430f
– integrative **2**, 430
– introduction of phage control units **2**, 382f
– narrow host range **2**, 429, 433ff
– – cloning site **2**, 434
– – marker **2**, 434
– – replicon **2**, 434
– pACYC derivatives **2**, 433f
– pAMβ1 **2**, 460
– pBluescript II SK + /–, map **2**, 383
– pBR derivatives **2**, 433f
– pBR322 **2**, 392f
– pGV451 **2**, 288
– pIP404 **2**, 460
– plasmid instability **2**, 459
– pMB1 (Co1E1) **2**, 382
– pRK290 **2**, 430f
– probe vector **2**, 384
– pSP72, map **2**, 383
– pUC **2**, 382f
– – map **2**, 383
– replicative **2**, 430
– RP4 **2**, 429, 446
– – cointegrate formation **2**, 447

– RSF1010 derivatives **2**, 432f
– split-end vector systeme (SEV) **2**, 559
– transfer to original host **2**, 437
– transposon carrier **2**, 435, 441f
– viral replication element **2**, 724
– – simian virus 40 (SV40) **2**, 724
plasmid vectors, use in gene therapy **5A**, 452
– use in vaccination **5A**, 452
plasmin **9**, 368f
plasminogen **9**, 369
plasminogen activators, genetic engineering of
  5A, 540
– in therapy **5A**, 168f
plasmodesmata **1**, 34
plasmogamy **2**, 75
plasmolysis **3**, 517
– of yeast cells **9**, 703
plasters, biodeterioration of **10**, 311
plastic materials, additives for, biodegradability
  of **10**, 310
– biodeterioration of **10**, 308ff
plastics, production by pseudomonads **1**, 420
– recycling of **11C**, 8ff
plastidal genome **2**, 169f
– inheritance **2**, 170
– restriction fragment length **2**, 169
– *Vicia faba* **2**, 169
plate frame filter, baker's yeast dewatering **9**,
  339f
platelet activating factor receptor antagonists **7**,
  114f
platelet aggregation, screening of receptor-active
  compounds **7**, 110
platelet aggregation inhibitors, from basidiomycetes **7**, 512f
– – structures **7**, 512
platelet-derived growth factor **5A**, 161f
– in therapy **5A**, 161
platelet endothelial cell adhesion molecule **5A**,
  145
platelet receptor blockage, with ReoPro **5A**, 367
PLE *see* pig liver esterase
Plectovirus **1**, 550
pleuromutilin, antibiotic action **7**, 496
– production by fermentation of *Pleurotus* **7**, 496
– structure **7**, 495
pleurotellic acid, structure **7**, 505
pleurotellol, cytotoxic activity **7**, 507
– structure **7**, 505
Pleurotus ostreatus **1**, 534f
– degradation of PAH **11B**, 63
– fruit body formation **2**, 121
– strain improvement **2**, 121
Pleurotus sp., pleuromutilin production **7**, 496
plug flow reactor, overall reaction rate **4**, 339
plug flow reactor (PFR), mass balance equation
  11A, 329
plug flow reactors, fixed-bed – **3**, 431

^{31}P-NMR **4**, 209, 211, 213
– spectrum **4**, 198f
P–O bonds, cleavage of **8B**, 235ff
– formation of **8B**, 228ff
– forming reactions, listing of **8B**, 234f
podophyllotoxins **7**, 617
– as inhibitors of topoisomerase II **7**, 655
podoscyphic acid, structure **7**, 510
Podoviridae **1**, 550
poi, production of **9**, 548
point mutation **2**, 9
Poisson–Boltzmann equation **5A**, 71, 94
polarization testing, of raw sugar **6**, 10
pollen exchange **12**, 164
pollen transformation **2**, 584
– *in vitro* matured pollen **2**, 584
pollutant degradation, in trickling filters, kinetics
  11C, 341f
– – limiting steps **11C**, 340ff
pollutant degrading microbes, measurement in
  subsoil **11B**, 403
pollutant diffusion, in soil **11B**, 74
– kinetic data **11B**, 73ff
pollutants *see also* contaminants
– bioavailibility of, increase of the desorption
  process **11B**, 79ff
– – increase of the mass transfer rate **11B**, 79ff
– – in soil **11B**, 67ff
– – in the water phase **11B**, 64ff
– biodegradation in soil **11B**, 71ff
– biodegradation rate **11B**, 74
– cometabolic degradation of **11B**, 152ff
– concentration gradients in biofilms **11C**, 268
– crystallization of **11B**, 64f
– crystal size, effect in biological growth rate
  11B, 65
– degradation of, in biological waste gas purification **11C**, 291
– environmental, monitoring of **11A**, 126f
– estimation of, biological test procedures **11B**,
  496f
– – mobility of **11B**, 491
– humification of **11B**, 68
– hydrophobic liquid compounds, direct uptake
  through the cell membrane **11B**, 66f
– immobilization of **11B**, 68
– incomplete degradation of **11B**, 62
– increase of solubility in the water phase **11B**,
  78f
– in municipal wastewater, concentration limits
  11A, 164
– in non-aqueous phase liquids, sufactant catalyzed microbial uptake **11B**, 67
– in waste gas **11C**, 264
– – treatment in trickling filters **11C**, 333ff
– lack of bioavailability, potential advantages
  11B, 82f
– loads of, in groundwater **11C**, 377ff

– microbial degradation of **11C**, 267ff
– organic –, biodegradation of **11B**, 145ff, 371ff
– recalcitrant – **8B**, 202
– solubility in water **11B**, 64ff
– surfactant-based biodegradation **11B**, 78f
– technical options to improve bioavailability
   11B, 78ff
– treatment of, in waste gas **11C**, 266
– with nitro functions, biodegradation of **11B**,
   273ff
pollutant transport, chemical-physical effects **11B**,
   50f
– effect of geochemical processes in the subsurface **11B**, 49ff
– influence of soil properties **11B**, 50
– influence of subsurface water composition **11B**,
   51
– in groundwater **11B**, 44
– in soil, colloids as medium **11B**, 54f
– – effect of geochemical barriers **11B**, 52
– – effect of hydrochemical barrier **11B**, 52
– – sorption processes **11B**, 52f
pollution levels, of PAH in soil **11B**, 217
polyacrylamide gel electrophoresis **2**, 264f
– capillary gel **2**, 279
– gel characteristics **2**, 265f
– gel handling **2**, 279
– isoelectric focusing gels **9**, 64
– native gels **9**, 62f
– sodium dodecyl sulfate gels **9**, 63
– staining gels **9**, 64f
polyacrylamide gel electrophoresis (PAGE) **3**,
   720f
– buffer pH **3**, 721
– buffer system, continuous – **3**, 721
– – discontinuous – **3**, 721
– Ferguson plot analysis **3**, 722
– gel analysis **3**, 725
– native – **3**, 723
– SDS-PAGE **3**, 723, 725
– – Laemmli system **3**, 724
polyacrylamide gel (PAA) **3**, 720
– gradient gel **3**, 722
– porosity **3**, 721
– resolving gel **3**, 721
– stacking gel **3**, 721
poly-adenylation sites **5B**, 365
poly-alanine, immobilized –, preparation of **8B**,
   500f
polyamines, analysis of **1**, 408
– extraction of **1**, 408
– HPLC chromatography of **1**, 409
– in chemotaxonomy **1**, 408
– in pseudomonads **1**, 405ff
polyampholytes, as phase-forming polymers **3**,
   596
polyaromatic compounds, chemical properties
   11B, 218

– degradation of **11B**, 211ff
– physical properties **11B**, 218
poly-β-hydroxy alkanoates **7**, 177ff
– occurrence in Pseudomonas oleovorans **7**, 180
poly-β-hydroxybutyrate (PHB) **1**, 26
– accumulation in bacteria **7**, 178
– biosynthetic pathway **1**, 212, **7**, 178
– heterologous gene expression **1**, 26
– microbial production **1**, 420
– structure **7**, 177
– synthesis by condensation of acetyl-CoA units
   7, 179
poly-β-hydroxybutyric acid (PHB) **10**, 308
– effect on bioflocculation **4**, 436
– synthesis by *Alcaligenes latus* **4**, 42ff
poly-β-hydroxybutyric acid (PHB) production **4**,
   42ff
– elemental balancing **4**, 42ff
– – instantaneous error analysis **4**, 51ff
– principles of fluxes **4**, 42ff
– reaction rates, dynamic error analysis **4**, 54ff
– stoichiometry **4**, 42
polychlorinated biphenyl (PCB), degradation of,
   by pseudomonads **1**, 415
polychlorinated biphenyls (PCB), biodegradation
   of, aerobic – **11B**, 262f
– – bacterial – **11B**, 155f, 315
– – by genetically engineered microorganisms
   11B, 446ff
– – by plants **11B**, 378f
– – mechanisms **11B**, 261f
– – use of bauaugmentation **11B**, 434
– bioremediation of **11B**, 427ff
– dechlorination of **11A**, 46
– degradation of **8B**, 207
– extraction of **11B**, 499
– monitoring with biosensors **11A**, 133f
– oxygenolytic cleavage of **8B**, 209
– patterns of dechlorination **8B**, 208
– principle of dehalogenation **11B**, 155f
polychlorinated dibenzodioxins (PCDD), biodegradation of **11B**, 262f
polychlorinated dibenzofurans (PCDF), biodegradation of **11B**, 262f
polycistronic transcript **2**, 250
polyclonal antibodies **5A**, 317
polycyclic aromatic hydrocarbons, hydroxylation
   of **8A**, 501
polycyclic aromatic hydrocarbons (PAH), bioavailability of **11B**, 217ff
– biodegradability of **11B**, 468f
– biodegradation of **11A**, 386ff, **11B**, 315
– – by plants **11B**, 378f
– – PAH dissolved in *n*-dodecane standard emulsion **11A**, 389f
– – PAH dissolved in water **11A**, 388f
– – stoichiometric and kinetic coefficients **11A**,
   390

- bioremediation with the heap technology **11B**, 326
- bound residues **11B**, 115f
- – formation in the organic soil matrix **11B**, 111
- – in soil contaminated by tar oil **11B**, 116f
- – mibilization of **11B**, 115f
- – mineralization of **11B**, 115f
- – stability of **11B**, 115
- cometabolic transformation of **11B**, 221, 226ff
- – toxicologically relevant potential **11B**, 229f
- complete metabolization and mineralization **11B**, 220f
- degradation of **11B**, 62f
- – by *Pleurotus ostreatus* **11B**, 64
- – slurry decontamination process **11B**, 340f
- – use of bioaugmentation **11B**, 434f
- distribution coefficient k_{oc} **11B**, 217
- extraction of **11B**, 499
- microbial degradation of **11B**, 219ff
- – initial oxidation reaction **11B**, 220
- – pathways **11B**, 219f
- model standard for analysis of **11B**, 215ff
- monitoring of **11A**, 121
- nitrated –, metabolism of **11B**, 287f
- non-specific oxidation by radicals **11B**, 221, 230ff
- – toxicological effect **11B**, 232f
- occurence **11B**, 215f
- octanol-water distribution coefficient k_{ow} **11B**, 217
- physico-chemical properties **11B**, 217ff
- pollution levels **11B**, 217
- primary oxidation products **11B**, 231
- properties **11A**, 386
- solubility characteristics **11B**, 217
- sorption of, hydrophobic – **11B**, 218
- structures of **11B**, 216
- toxicological relevance **11B**, 219
- transfer from soil of liquid overlay phase **11B**, 80
- use **11A**, 386
polycyclic biotransformation products **8B**, 376ff
- use of alcohol dehydrogenases **8B**, 382f
polydactyly **5B**, 69f
polyenes, cationic cyclization of, use of catalytic antibodies **8B**, 430f
- structures **7**, 34
polyester amides, biodegradable – **10**, 434
polyester resins, Biomax® **10**, 432
- Bionolle® **10**, 432f
- Corterra® **10**, 433
- Eastar Bio® **10**, 433
- Ecoflex® **10**, 433
- Lacea® **10**, 433
- Nature Works™ **10**, 433
- Resomer® **10**, 433f
- semisynthetic – **10**, 432ff
- synthetic – **10**, 432ff

polyesters **7**, 177ff
- lipase-catalyzed degradation **8A**, 145
- lipase-catalyzed polymerization **8A**, 144
- poly-β-hydroxyalkanoates **7**, 177ff
polyethers, structures **7**, 34
polyethylene glycol (PEG) **2**, 97
- anaerobic degradation of **11B**, 173
- as phase-forming polymer **3**, 596, 599
- – affinity ligand binding **3**, 598
- – shuttle polymer **3**, 600
- as protein folding aid **3**, 542
- contaminants with negative effects **2**, 98
- molecular effect **2**, 102
- protein stabilization effects **3**, 705
- synergistic fusogen contaminants **2**, 97f
polyethylene oxide **5B**, 274
polyethylenimine (PEI), for lysate clarification **9**, 93
polygalacturonase, mode of action **9**, 709
polygenic traits **5B**, 65ff
- continuous – **5B**, 65
- discontinuous – **5B**, 65
polyhydroxy alcohols, formation by yeasts **6**, 220ff
- – arabitol **6**, 220
- – erythritol **6**, 220
- – mannitol **6**, 222
- – ribitol **6**, 222
- – sorbitol **6**, 222
- – xylitol **6**, 223
- formation from *n*-alkanes **6**, 222
polyhydroxyalkanoate production **1**, 212
polyhydroxyalkanoic acid (PHA) **1**, 26
- production by pseudomonads **1**, 420
polyhydroxyalkanoic acids *see* PHA
polyhydroxyalkanoic acids (PHA) **10**, 426ff
- applications of **10**, 170, 431
- biosynthesis of **10**, 170f
- biosynthetic pathways **10**, 427f
- – *in vivo* **10**, 428f
- commercialization of **10**, 431
- fermentative production of **10**, 429f
- PHA granules, interaction with phasins **10**, 427
- PHA synthases, biochemistry of **10**, 426f
- – *in vitro* PHA biosynthesis **10**, 429
- synthesis, from lignite depolymerization products **10**, 171f
- – in agricultural crops **10**, 430
- – in recombinant *E.coli* strains **10**, 430
- – in transgenic plants **10**, 430f
polyhydroxybutyric acid *see* poly(3HB)
(poly)-isoprenoid, biosynthesis **10**, 581
polyketide chain assembly **10**, 354f
polyketide drugs, structural class **7**, 39
polyketides **7**, 32ff
- assembly-line biosynthesis of **10**, 359
- biosynthesis of, actinorhodin pathway **10**, 349
- – lovastatin **10**, 345ff

– – tetracenomycin pathway **10**, 349
– biosynthetic pathways **10**, 343
– chemical structures **10**, 342
– macrocyclic – **10**, 354
– structures **7**, 34
polyketide synthases **7**, 32ff
– scheme **7**, 33
polyketide synthases (PKS) **10**, 341ff
– acyl-carrier protein domain **10**, 361f
– – specificity of **10**, 361f
– acyl transferase domains, specificity of **10**, 360f
– aromatase/cyclase **10**, 351
– aromatic –, models of **10**, 350
– auxiliary units **10**, 351
– catalytic sites of **10**, 345
– chain assembly **10**, 345ff
– chain transfer **10**, 358
– chalcone-like – **10**, 352ff
– – reactions **10**, 352f
– crosstalk to fatty acid synthases, in *Mycobacterium* **10**, 366
– – in *Streptomyces* **10**, 365
– erythromycin PKS **10**, 357
– expression of modules **10**, 357
– iterative –, type I **10**, 344ff
– – type II **10**, 348ff
– ketoreductase **10**, 351
– ketoreductase domain, specificity of **10**, 361
– ketosynthase domain, specificity of **10**, 359f
– linker-mediated chain transfer **10**, 363
– linkers, coiled-coil nature of **10**, 364
– – inter-polypeptide – **10**, 363f
– – intra-modular – **10**, 364
– – intra-polypeptide – **10**, 363
– – properties of **10**, 362f
– minimal PKS **10**, 348ff
– modular – **10**, 354ff
– – architecture **10**, 344, 356
– – hierarchial organization of **10**, 355ff
– – properties of modules **10**, 356
– module swaps **10**, 358
– properties of domains **10**, 358f
– substrate specificity **10**, 357f
– thioesterase domain, specificity of **10**, 362
– types of **10**, 343f
poly-L-alanine, epoxidation of, chalcone **8B**, 499f
– – "chalcone-type" substrates **8B**, 499ff
poly-leucine, epoxidation by **8B**, 499ff
– – preparation of natural products and prodrugs **8B**, 501f
– immobilized –, preparation of **8B**, 501
poly-L-leucine, preparation of **8B**, 503f
– stereoselective carbonylation by, palladium catalyzed – **8B**, 505f
poly-L-valine, stereoselective electrochemical reduction by **8B**, 505f
polymerase chain reaction (PCR) **2**, 263, 333ff, 393f, **5B**, 44, 83

– amplification scheme **2**, 334f
– anchor **2**, 294f
– applications **2**, 340ff
– – for gene cloning **2**, 394
– – in diagnostics **2**, 340f
– asymmetric PCR **2**, 341f
– blocking primer PCR **2**, 287
– contamination avoidance **2**, 339
– detection of plant pathogens **10**, 499f
– DNA template **2**, 337
– – purity **2**, 338
– for heterologous gene detection **2**, 759f
– history **2**, 334
– impact in sequencing **2**, 269f
– inverse **2**, 293ff
– nested priming **2**, 338
– primers **2**, 338
– product analysis **2**, 339
– product carry-over avoidance **2**, 339
– product detection **2**, 338f
– product indentification **2**, 339
– product purification **2**, 288
– products **5B**, 196
– properties **2**, 269
– reactants **2**, 336f
– semi-random primer approach **2**, 295f
– specificity **2**, 338
– – hot start **2**, 338
– template copy yield **2**, 334ff
– thermal asymmetric PCR **2**, 287
– two-step reaction **2**, 338f
– use of thermostable DNA polymerases **10**, 87f
polymerizations, lipase-catalyzed – **8A**, 144f
polymer production **1**, 174, 212
polymers, biodegradability of **10**, 309f
– biodeterioration of **10**, 308ff
– containing 3-mercaptobutyric acid **10**, 432
– containing 3-mercaptopropionic acid **10**, 431f
– hydrolysis, by immobilized enzymes **3**, 454
– in two-phase systems **3**, 596
– – concentration effects **3**, 598
– – molecular weight effects **3**, 598
– – recycling **3**, 603
polymorphic DNA markers **5B**, 42
polynucleotide **2**, 319
polynucleotide phosphorylase, recombinant – **6**, 606
polynucleotide synthesis, enzyme catalyzed phosphate bond formations **8B**, 389
polyol dehydrogenases **8B**, 306
polyols *see* polyhydroxy alcohols
– oxidation of **8B**, 300
– – according to the Bertrand-Hudson rule **8B**, 301
– protein stabilization effects **3**, 705
polyoses, in wood **10**, 306
polyoxyethylene octyl phenols *see* Triton
polypeptide backbone **5A**, 60

polypeptide drugs **3**, 58
– conjugated – **3**, 61, 69f
– cyclic – **3**, 58
– from animal tissue **3**, 58f, 62
– from bacteria **3**, 60
– from blood **3**, 62, 59
– from fungi **3**, 60
– from mammalian cell cultures **3**, 60
– from plant tissue **3**, 58f
– human – **3**, 58ff
– – human growth hormone (hGH) **3**, 61
– microbial contamination **3**, 62
– recombinant – **3**, 60
– – host cells **3**, 60
– semisynthetic – **3**, 61, 69f
– synthetic – **3**, 60, 64
– virus contamination, inactivation **3**, 61
– – removal **3**, 61
polypeptides *see* proteins
– as epoxidation catalysts **8B**, 499ff
– conformational properties **5A**, 60
polyphenol oxidase, in cocoa fermenttion **9**, 570
polyphosphate accumulating bacteria, aerobic phosphate uptake **11A**, 292f
– aerobic polysaccharide storage **11A**, 293
– anaerobic phosphate release **11A**, 292f
– in activated sludge **11A**, 296f
– synthesis of storage material **11A**, 292f
– TEM electron micrographs **11A**, 291
polyphosphate accumulation **11A**, 41f
– by *Acinetobacter* sp. **11A**, 41
polyphosphate bodies, metal accumulation in **10**, 238
polyphosphate granules **11A**, 290f
polyphosphates, characteristics of **11A**, 289f
– degradation under anaerobic conditions **11A**, 292
– structure of **11A**, 289f
polyploidy **2**, 8
– role in crop plant evolution **5B**, 147
Polyporus versicolor, laccase **9**, 713
polyprene hydrocarbons, as components of the alga *Botryococcus braunii* **7**, 175f
polyprenoids **7**, 175f
polyribonucleotide nucleotidyltransferase, bacterial – **6**, 604
– immobilization **6**, 605
polyribonucleotides, formation from ribonucleoside diphosphates **6**, 604ff
polysaccharide production, batch culture **6**, 623
– continuous culture **6**, 623f
– industrial – **6**, 624
– product recovery **6**, 624f
– – precipitation with polar solvents **6**, 625f
polysaccharides **1**, 413, **6**, 613ff
– – *see also* biopolymers
– – *see also* exopolysaccharides
– acceptability **6**, 630ff

– as bioemulsifiers **6**, 664f
– degradation, by bacilli **1**, 384
– – by clostridia **1**, 297
– enhancement of characteristics **6**, 622f
– extracellular –, metal binding by **10**, 234ff
– food uses **6**, 627f
– – properties **6**, 627
– heteropolysaccharides **6**, 615
– hompolysaccharides **6**, 615
– immunologically active –, from plant cell cultures **7**, 612
– non-food uses **6**, 628f
– oil field usage **6**, 629f
– producing microorganisms **6**, 616ff
– – strain development **6**, 619
– production, by *Agrobacterium radiobacter* **1**, 173
– – nutrient limitation effect **1**, 172
– production by algae **10**, 130
– specifications **6**, 632
– viscosity enhancement **6**, 623
polysaccharide synthesis, derepression **6**, 620f
– genetics **6**, 621
polysaccharide yield, improvement **6**, 621f
polysome **1**, 14, **2**, 248
polysporic acid, iron-binding pigments **7**, 231
– structure **7**, 231
polythioesters, containing 3-mercaptobutyric acid **10**, 432
– containing 3-mercaptopropionic acid **10**, 431f
polyunsaturated fatty acids (PUFA) **7**, 138f
– biosynthetic pathways **10**, 335
– dietary uses **7**, 158
– formation by protozoa **10**, 141f
– from microalgae **10**, 109
– in bacteria **7**, 147ff
– in fungi **7**, 151, 158
– in microalgae **7**, 165ff
– nutritional roles **7**, 164
– optimization of production **10**, 464f
– producing organisms **10**, 462ff
– – aquaculture of **10**, 465
– – molecular biology of microbial PUFA synthesis **10**, 465
– production by algae **10**, 129f
polyunsaturated fatty acids (PUFAs) **8A**, 137
– lipase selectivities **8A**, 143f
poly(3HB), in *Saccharomyces cerevisiae* **6**, 412f
– medical uses **6**, 446f
– patent applications **6**, 448ff
poly(3HB) (polyhydroxybutyric acid, PHB) **6**, 403ff
– – *see also* PHA
– complexes, in prokaryotes **6**, 410ff
– microbial production **6**, 403ff
pooled progeny techniques **5B**, 171
population balance equation (PBE) **3**, 88
population genetics **5B**, 418

population models, for bioprocesses **3**, 87
populations, genetically isolated – **5B**, 110
porcine islets, in microencapsulated-cell therapy,
 isolation procedure **10**, 565f
porcine kidney aminoacylase (PKA) **8A**, 245ff
 – hydrolysis of *N*-terminal amides **8A**, 246ff
 – resolution of *α*-amino acids **8A**, 246f
porcine lipoxygenase, hydroperoxide formation
 8A, 503f
porcine pancreatic lipase (PPL) **8A**, 263f, **8B**,
 368ff
 – asymmetric hydrolysis **8B**, 368f
 – catalysis of asymmetric syntheses **8A**, 67f
 – catalyzed hydrolysis **8B**, 358f
 – dipeptide formation **8A**, 262
 – enantioselective reactions catalyzed by primary
 alcohols **8A**, 91f
 – general features **8A**, 44
 – in industrial biotransformations **3**, 458
 – resolution of, acyclic secondary alcohols **8A**, 76
 – – 2-alkanols **8A**, 75
 – – cyclic secondary alcohols **8A**, 77
 – selective acylation of glucose **8A**, 119
Poriales **1**, 534
Poria monticola, lignin degradation **6**, 147
porin **1**, 76
porins **5A**, 3
Porphyra columbina, edible seaweed **10**, 462
Porphyridium curentum, formation of polyunsaturated fatty acids **7**, 168
porphyrin metalation, use of catalytic antibodies
 8B, 477
portals **5B**, 341
portland cement, composition of **10**, 301
 – corrosion tests **10**, 302f
ports, for bioreactors **3**, 421
positional candidate method **5B**, 109
positional cloning **5B**, 54, 109, 139, 170ff, 177
 – of plants **5B**, 165ff
Position-Specific-Iterated BLAST **5B**, 130
post-denitrification, activated sludge process **11A**,
 345
 – biofilm reactors **11A**, 345
postgenome research, *Arabidopsis thaliana* **5B**,
 28f
 – *Archaeoglobus fulgidus* **5B**, 20
 – *Caenorhabditis elegans* **5B**, 32f
 – *Drosophila melanogaster* **5B**, 35
 – *E. coli* **5B**, 12ff
 – yeast **5B**, 24f
post-harvest losses, of corps **5B**, 146
post-source decay technique **5B**, 254
post-translational modification, glycosylation **12**,
 166
post-translational modifications **5B**, 335
potable water *see* drinking water
potao starch, wastewater from – **11A**, 202

potassium, determination with pyruvate kinase **9**,
 144f
potassium efflux system, *Escherichia coli* **1**, 177
potassium limitation, effect of ammonium **1**, 177
 – effect on product formation **1**, 178
 – energetic effects **1**, 177
 – metabolic effect **1**, 172, 176
potassium phosphate, in two-phase systems **3**,
 596
potassium uptake system, *Escherichia coli* **1**, 176
potato, micropropagation **1**, 604
potatoes, analysis of **6**, 62
 – as raw material for ethanol production **6**, 62
 – in industrial ethanol production **6**, 158f
 – – flow diagram **6**, 159
 – nutrient content of stillage **6**, 105
 – processing in ethanol production **6**, 90
 – starch content **6**, 109
 – wet cleaning **6**, 77
potato lipoxygenase **8B**, 365
potato processing, wastewater, flow fractions
 11A, 205f
 – – specific amounts **11A**, 205f
 – – treatment **11A**, 206
potato starch, composition **6**, 32, 34f
 – production **6**, 34
potato tuber lipoxygenase, hydroperoxide formation **8A**, 503
Potentiometric sensor, for biomass determination
 4, 193
pottash **9**, 258
powdered activated carbon (PAC) **11C**, 396f
Power input, calculation of **4**, 13f
 – in bioreactors **4**, 126f
 – instruments for determination of **4**, 13f
 – to bioreactors **3**, 405ff, 416
power law model **4**, 369
pozol, production of **9**, 549
p_{O_2} electrode, calibration **4**, 18
 – galvanic **4**, 16f
 – polarogram **4**, 16f
 – polarographic **4**, 16f
 – sterilizable **4**, 17
p_{O_2} (oxygen partial pressure) **4**, 7f
 – determination of **4**, 16ff
 – measurement, accuracy of **4**, 21
 – use in process control **4**, 7f
PPD (Primary Production Department of Singapore) **12**, 420
ppGpp **7**, 61ff, 75ff
ppGpp synthesis, activation of – **7**, 63
PPL *see* porcine pancreatic lipase
PPL (porcine pancreatic lipase) **8A**, 42f
PQQ-enyzmes **4**, 88
Prader–Willi syndrome (PWS) **5B**, 73f
prahoc, production of **9**, 537
Prasinophyceae, characteristics of **10**, 111
prebiotic reactions **5A**, 2

precipitation, effect on BOM **11C**, 463
– in drinking water treatment **11C**, 383ff
– of metal ions **11A**, 42
– processes during **11C**, 386
Precision, definition of **4**, 45f
preclinical testing, of antibodies **5A**, 289ff, 303ff
– – assays **5A**, 291ff
– of CDP571 **5A**, 345f
precocene II, dihydroxylation of **8A**, 515
precursor metabolite **1**, 81
pre-denitrification, use of trickling filters **11A**, 341ff
– – flow chart **11A**, 342
prediction, of antibody conformations **5A**, 223
– of protein interactions **5A**, 93ff
– of protein structure and function **5A**, 55ff, 65ff, 88
– – antigenic sites **5A**, 67f
– – coiled-coil segments **5A**, 67f
– – quaternary structure **5A**, 77f
– – secondary structure **5A**, 65ff
– – solvent accessibility **5A**, 76f
– – tertiary structure **5A**, 73ff
– – transmembrane regions **5A**, 67f
predictor-corrector method, for bioprocess control **3**, 380
prednisolone, multienzyme synthesis **8B**, 382f
pre-epidermin **7**, 342
pre-ferments **9**, 274ff
– formulation, lactobacilli in – **9**, 277
– preparation of **9**, 275
preimplantation diagnosis **5B**, 84ff
pre-lantibiotics *see* lantibiotic prepeptides
Prelog's rule **8A**, 427f
– asymmetric reduction of carbonyl compounds **8A**, 428
– for carbonyl reduction **8A**, 380
pre-nisin **7**, 341
prephenate dehydratase, deregulation of **8B**, 317ff
– feedback inhibition, by L-phenylalanine **8B**, 319
pre-precipitation, of phosphate **11A**, 310f
prescription drugs, market share **5A**, 535
– side effects **5B**, 92
Prescription Drug User Fee Act **5A**, 500ff, 528
press coverage, of agricultural biotechnology **12**, 482
– of genetic engineering, articles **12**, 497
– – authors **12**, 498
– – in Germany **12**, 495ff
– – shortcomings **12**, 502f
– – tendencies **12**, 500f
– – topics **12**, 498f
– of science and technology **12**, 481
pressure, instruments for determination of **4**, 10
– measurement, accuracy of **4**, 21
– SI unit **4**, 10

primary alcohols, acylation in, modified sugars **8A**, 121ff
– – unmodified sugars **8A**, 119ff
– enantioselective reactions, catalyzed by PCL **8A**, 89
– – catalyzed by PPL **8A**, 91f
– kinetic resolution by penicillin acylase **8A**, 230ff
– lipase-catalyzed enantioselective reactions, alcohols with quaternary stereocenters **8A**, 95
– lipase enantioselectivity **8A**, 87ff
primary cell cultures **10**, 28
primary metabolism **1**, 15, 442ff
primary structure *see* protein structure
prime plasmid **2**, 57f
– F prime **2**, 57
– isolation **2**, 58
– R prime **2**, 57f
primer identification **5B**, 229
principle component analysis (PCA), of bioprocesses **3**, 391
PRINTS **5B**, 289
problem of the commons **12**, 142
process analysis, instruments **4**, 6ff
process cheese products **9**, 363f
process combinations, wastewater treatment, schemes **11A**, 408
process control, *see also* Control of bioreactor systems
– – *see* bioreactor control, fermentation control
– control performance **4**, 263
– control system design **4**, 263
– in anaerobic wastewater purification **4**, 474ff
– – hydrogen regulation **4**, 478f
– – pH regulation **4**, 474ff
– – regulation of gas production **4**, 477
– – substrate regulation **4**, 477f
– – temperature optimization **4**, 479f
– instruments **4**, 6ff
– – reliability of **4**, 20f
– pH effect **4**, 7
process estimation, adaptive linear –, structure **3**, 337
– extended Kalman Filter (EKF) **3**, 335
– methods **3**, 334ff
– – filtering **3**, 336
– model-based – **3**, 335ff
– – linear black-box – **3**, 337ff
– – non-linear black-box – **3**, 338
– number of parameters **3**, 336
process model, baker's yeast production, application for optimization **4**, 400ff
– – verification **4**, 399
– of baker's yeast production **4**, 390ff
process modelling, in bubble column reactors **4**, 372ff
process patents **12**, 308f
process qualification (PQ) **3**, 741

process validation   **3**, 739ff, 760ff
– – *see also* facility validation
– benefits   **3**, 741
– master validation plan (MVP)   **3**, 741
– program   **3**, 742
– report   **3**, 762
– sequence   **3**, 761
process versus product controversy   **12**, 39ff, 460
ProDom   **5B**, 289
prodrugs, activation of, antibody mediated   **8B**, 442ff
– carbamate, cleavage by abzymes   **8B**, 444f
– of chloramphenicol, hydrolysis of   **8B**, 442f
– of 5-fluorodeoxyuridine, activation of   **8B**, 442f
product characteristics, effects on process validation   **3**, 741
product clarification   **3**, 42ff
– insoluble products   **3**, 42
product compost, hygienic safety of   **11C**, 234ff
– – indicator organisms   **11C**, 234f
product concentration, kinetics   **3**, 332
product consistency, of therapeutic antibodies   **5A**, 254f
product development, biomedicinal   **12**, 312ff
– cost   **12**, 215
– decision making   **12**, 232f
– managing   **12**, 228ff
– motivation   **12**, 235f
– organizational structures   **12**, 235f
– – matrix organization   **12**, 235
– – project steering committee   **12**, 235
– phases   **12**, 230ff
– planning   **12**, 228ff
– – risk-oriented   **12**, 229ff
– pre-development phase   **12**, 232
– rules   **12**, 214
– technical aspects   **12**, 224ff
– time   **12**, 215
product efficacy   **12**, 223
product formation, anaerobic –, electron balance   **3**, 299
– – extracellular –   **3**, 299
– kinetics   **3**, 86ff, 299ff
– models   **3**, 87
– net rates, volumetric   **3**, 86
product inhibition, of alcohol dehydrogenases   **8A**, 429f
production, of enzymes, as fine chemical   **9**, 73ff
production-integrated environmental protection   **11A**, 562ff
production media *see* fermentation media
production variables   **3**, 743
production water system   **3**, 755ff
– monitoring   **3**, 757
– validation   **3**, 759
product isolation   **3**, 40ff
productivity number, definition of   **8A**, 366
product patents   **12**, 308

product profile   **12**, 218ff
– benefits   **12**, 220
– contents   **12**, 220
product purification   **3**, 17, 40ff, 45
– amino acid clipping   **3**, 17
– antibiotics   **3**, 45
– by liquid-liquid extraction   **3**, 558ff
– – fat-soluble –   **3**, 565
– contamination, by ingredients   **3**, 17
– electric field applications   **3**, 50
– extracellular products   **3**, 594
– final steps   **3**, 695ff
– fractionation   **3**, 47
– from fermentation boths   **3**, 472
– intracellular products   **3**, 594
– peptides   **3**, 45
– principal techniques   **3**, 558
– small molecules   **3**, 557ff
– yield   **3**, 41
product purification proteins *see* protein purification
product quality   **2**, 838, **12**, 222, 224ff
– alteration   **2**, 798
product recovery, acetone/butanol fermentation   **6**, 253ff
– – distillation   **6**, 253
– – solvent extraction   **6**, 254f
– from reverse micelles   **8A**, 49
– glycerol   **6**, 219
– in citric acid fermentation   **6**, 327
– – by ion-exchange procedures   **6**, 330
– – by precipitation   **6**, 328f
– – by solvent extraction   **6**, 329f
– in gluconic acid production   **6**, 356f
– in industrial polysaccharide production   **6**, 624f
– – use of solvents   **6**, 625f
– in itaconic acid production   **6**, 367
– in lactic acid fermentation   **6**, 299f
– in manufacture of therapeutic antibodies   **5A**, 262f
– in PHA production   **6**, 443ff
– – chemical methods   **6**, 445
– – enzyme process   **6**, 444
– – physical methods   **6**, 445
– yeast biotransformations   **8A**, 383f
– 2,3-butanediol, membrane processes   **6**, 287
– – solvent extraction   **6**, 287
products *see* biotechnological products
product safety   **12**, 218, 222f
– regulations   **12**, 149f
product separation, from fermentation broth, clarification   **3**, 479
– – concentration   **3**, 479
– methods   **3**, 305
– non-secreted products   **3**, 505ff
product solubilization, assay procedures   **3**, 507
product stability, improvement, by excipients   **3**, 703ff

proferrorosamine A, structure **7**, 208
profile optimization, for batch bioreactors **4**, 528
– for fed-batch bioreactors **4**, 530
– optimal feed rate profiles, for type I fermentation processes **4**, 533
– – for type II fermentation processes **4**, 534
– – for type III fermentation processes **4**, 536ff
– optimal feed rates-profiles, computational scheme for determination **4**, 536ff
– pH **4**, 526ff
– temperature **4**, 526ff
proflavine **2**, 26
progeny type **2**, 152
– double haploid line (DH) **2**, 152f
– near-isogenic line (NIL) **2**, 153
– recombinant inbred line (RIL) **2**, 152f
progesterone, biotransformation of **8B**, 381
– conversion to testosterone **8A**, 542
– hydroxylation of **8A**, 488, 505
– – by *Rhizopus arrhizus* **8A**, 477
– 11-α-hydroxylation **8A**, 12f
programmed cell death *see* apoptosis
project manager, qualities **12**, 234f
– responsibilities **12**, 234
prokaryotes *see* bacteria
– DNA transfer to eukaryotes **7**, 263
– effects of microgravity on **10**, 509
– transformation systems, for β-lactam production **7**, 253
prokaryotic cell **1**, 13ff
– – *see also* microorganisms
– adhesion **1**, 18
– macromolecular composition **1**, 15
– metabolism **1**, 15
– structural organization **1**, 13ff
prokaryotic cell culture, products **3**, 106
prokaryotic cells, DNA content **5A**, 4
– lack of Golgi apparatus **5A**, 14
– protein compartmentalization **5A**, 6
– protein content **5A**, 4
Proleukin **5A**, 151
proline isomerases **9**, 23
L-proline production **6**, 492f, 513
– by direct fermentation **6**, 492
prolyl dipeptidyl-aminopeptidase, from *Lactococcus lactis* **8A**, 256
promoter **2**, 234
– – *see also* transcription promotor
– – *see* transcription promoter
– in filamentous fungi **2**, 538
– – for expression systems **2**, 538
– – strong promoters **2**, 538
– in yeast **2**, 519
– – conditional promoter **2**, 520
– – hybrid promoter **2**, 520
– – limited regulation **2**, 519
– – regulated promoter GAL1 **2**, 520
– – regulatory promoter PHO5 **2**, 520

– – upstream activating sequence (UAS) **2**, 519
– – upstream repressing sequence (URS) **2**, 519
promoter prediction tools **5B**, 369
promoters, of baculovirus **9**, 87
– of *E. coli* **9**, 76ff
– – bacteriophage λ promoters **9**, 77
– – bacteriophage T7 promoters **9**, 77f
– – lac promoter **9**, 76
– – lpp promoter **9**, 77
– – trp promoter **9**, 77
– of *Saccharomyces cerevisiae* **9**, 82ff
– – constitutive – **9**, 83
– – hybrid – **9**, 84
– – regulated – **9**, 83f
ProNet (Myriad) **5B**, 410
1,3-propanediol, fermentative production of **10**, 425
– formation by *Klebsiella pneumoniae* **10**, 582
1,3-propanediol production **1**, 206
n-propanol, partition coefficients, in liquid-liquid extraction **3**, 560
prophage **1**, 558f
– detection **1**, 561
– homoimmunity **1**, 558
propionate, in syntrophic metablism **11A**, 463f
propionates, as bread preservatives **9**, 315
Propionibacterium, in Swiss cheese production **9**, 376f
propionic acid, production of, by *Propionibacterium* sp. **8B**, 345
propranolol, enantiomeric purity of **8A**, 84
– lipase-catalyzed routes to **8A**, 88
proprietary medicinal products, committee for – **5A**, 509
PROSAMO study **12**, 138
PROSITE **5B**, 289
– pattern **5B**, 138
– profiles **5B**, 138
prostaglandins, chlorinated – **8B**, 186
prostaglandin synthesis, use of cyclopentane monoester **8A**, 198
– use of porcine pancreatic lipase **8B**, 369f
prostaglandin synthons, use of horse liver alcohol dehydrogenase **8B**, 372f
prosthetic group *see* coenzymes
prosthetic groups, of enzymes **9**, 28ff
protease-catalyzed reactions **8A**, 247ff
protease inhibitors **9**, 687f
– biocompatible, for fermentation media **3**, 152
– microbial –, antitumor effects **7**, 675
proteases **5A**, 205ff, **8A**, 244ff
– actions **5A**, 205
– aspartyl proteases **8A**, 248
– bone cleaning **9**, 697f
– catalytic mechanism **9**, 691
– cell wall proteinas, of *Lactococcus lactis* **1**, 354f
– classification **5A**, 206ff, **8A**, 247
– – aspartyl proteases **5A**, 207

– – cysteine proteases **5A**, 207
– – detergent proteases **5A**, 208
– – metalloproteases **5A**, 207
– – serine proteases **5A**, 206f
– commercial – **9**, 687
– for food protein conversion **9**, 686ff
– from hyperthermophiles **10**, 86
– from pseudomonads **1**, 421
– from thermophiles **10**, 86
– heat-stable – **10**, 85ff
– in detergents **5A**, 208
– industrial applications **5A**, 205ff
– in silage **9**, 777
– metalloproteases **8A**, 248f
– modificaiton of wheat gluten **9**, 698
– nomenclature **8A**, 244
– – for the cleavage of **8A**, 249
– preferred cleavage sites **8A**, 249
– production by genetically modified microorganisms **5A**, 208
– serine proteases **8A**, 248
– substrate specificity **5A**, 206
– synthesis of L,L-diastereomers **8A**, 253
– synthesis of pharmaceutical precursors, kilogram-scale routes **8B**, 284f
– tenderization of meat **9**, 698
– thioproteases **8A**, 248
– types **5A**, 205
– use in baking **9**, 675, 679
– with esterase activity **8A**, 221ff
protein, anaerobic digestion of **11C**, 19ff
– hydrolyzed animal – **9**, 704
– hydrolyzed vegetable – **9**, 704
– in bread, biological value **9**, 245f
protein activity **3**, 718
– measurement **3**, 718, 730ff
protein aggregation, denaturant effects **3**, 547
– during refolding, *in vivo* – **3**, 547ff
– hydrophobic interactions **3**, 549
– inhibitors **3**, 550
– irreversible – **3**, 547
protein analysis, capillary electrophoresis **5B**, 272ff
– Folin-Lowry assay **3**, 732
– Kjeldahl nitrogen assay **3**, 732
– methods **3**, 529, 719ff
– quantitative amino acid analysis **3**, 732
– single cell – **5B**, 275
– tools **5B**, 340
– using mass spectrometry **5B**, 251ff
proteinase, of *Lactococcus lactis* **1**, 354f
protein assays **9**, 60ff
– biuret procedure **9**, 61
– Kjeldahl's method **9**, 61
– UV absorbance **9**, 61
protein assembly **2**, 690
protein–carbohydrate interaction **5A**, 98
protein chemistry, analytical – **3**, 717ff

protein chips **5B**, 437
protein classification **5B**, 135ff
– chemical constituents **5A**, 6
– compartilization **5A**, 6ff
– – in eukaryotic cells **5A**, 6
– – in prokaryotic cells **5A**, 6
– functional – **5A**, 4f
– structure **5A**, 6
protein cleavage, chemical methods **2**, 417
– enzymatic methods **2**, 417
protein complexes **5A**, 96ff, **5B**, 354
– enzyme–inhibitor complexes **5A**, 96
– recognition of targets **5A**, 96
– with HIV protease inhibitors **5A**, 97f
protein composition, analysis **3**, 718, 733
– – peptide mapping **3**, 734
– – protein fragmentation **3**, 734
protein condensates, use in cosmetics **6**, 674
protein conformation, torsion angles **9**, 11f
protein crystallography **5A**, 96
– Fourier maps **5A**, 92
– of protein complexes **5A**, 92
protein database searches **5A**, 49, 53, 57ff, 66
protein degradation **2**, 701f
– ubiquitin-dependent pathway **2**, 701
protein design **5B**, 355ff
– for purification **3**, 601
– – tryptophan-rich fusions **3**, 601
– use of databases **12**, 335
protein detection, by laser induced fluorescene **5B**, 272
– by mass spectrometry **5B**, 272
– by UV absorbance **5B**, 272
protein digests, MALDI-TOF analysis of **5B**, 252
protein disulfide isomerase (PDI), as catalytic chaperone **3**, 540
protein disulfide isomerases **9**, 23
protein–DNA docking **5A**, 96
protein–DNA interactions **5A**, 100
protein docking, binding sites for small ligands **5A**, 95
– *de novo* design of enzyme inhibitors **5A**, 95
– docking algorithms **5A**, 94
– simulation **5A**, 94
– structure predictions **5A**, 94f
– with macromolecules **5A**, 95
protein domains, prediction of **5B**, 337
protein drugs **3**, 57ff
– expression system **3**, 67
– – *Escherichia coli* **3**, 67f
– – immortal cell line **3**, 68
– – mammalian – **3**, 68
– – methionine removal **3**, 67
– – substrate-assisted catalysis **3**, 67
– for AIDS therapy, CD4 **3**, 69
– fusion **3**, 71
– immune complications, asparaginase **3**, 63
– – streptokinase **3**, 63

– lifetime **3**, 70
– purification **3**, 62
– – partition chromatography **3**, 62
– – ultrafiltration **3**, 62
– purity **3**, 66f
– recombinant fermentation **3**, 66ff
– recombinant therapeutics **3**, 67
– sources **3**, 57ff
– – cell culture **3**, 60ff
– – conventional fermentation **3**, 60ff
protein electrophoresis **3**, 719ff
– supports **3**, 720
protein engineering, for folding studies **3**, 535
– of nitrile transforming enzymes **8A**, 314
protein expression, in eukaryotes **3**, 531f
– in prokaryotes **3**, 533
protein folding **2**, 690f, **3**, 531ff, **5A**, 111ff, **9**, 19ff,
   12, 167f
– – *see also* protein refolding
– catalysis **9**, 22f
– – molecular chaperones **9**, 22f
– chaperone **2**, 691
– conditons, optimization **3**, 550
– cytoplasmic protein **9**, 19ff
– disulfide isomerases **9**, 23
– helper proteins **2**, 691
– in eukaryotes **3**, 531ff
– in mutants, effects **3**, 535f
– – pathways **3**, 536
– – temperature-sensitive – **3**, 535
– in prokaryotes **3**, 531ff, 534f
– *in vitro* **5A**, 111, 115ff
– *in vivo* – **3**, 531ff, 535ff
– – folding studies **3**, 535
– molecular chaperones **3**, 534, 537
– native conformation **3**, 534
– of inclusion bodies **5A**, 111ff
– processes **3**, 534
– proline isomerases **9**, 23
– structural model **3**, 545
– studies, protein engineering **3**, 535
protein folding aids, antibodies **3**, 541
– cofactors **3**, 541
– cosolvents **3**, 542
– folding pathways **3**, 541
– inhibitors **3**, 541
– synthetic – **3**, 540f
protein folding space **5B**, 348
protein folds **5B**, 347f
– comparison **5B**, 350
– number of **5B**, 348
– recognition by threading **5B**, 352f
– stability **5B**, 350
– superfolds **5B**, 348
protein functionality **5A**, 47ff
protein functions **5A**, 1f, 5, 22ff, 52ff
– definition **5A**, 54
– enzymes **5A**, 22ff

– knowledge-based prediction **5A**, 55ff
– multiplicity **5A**, 4
protein glycosylation **2**, 697ff, **3**, 534, **5A**, 9
– effects on protein stability **2**, 701
– glycoprotein processing **2**, 700f
– glycosylation site binding protein (GSBP) **3**,
   534
– N-glycosylation **2**, 699
– O-glycosylation **2**, 699
– physiological function **2**, 699f
– sialylation **2**, 701ff
– trimming **2**, 698ff
protein hydrolysates, bitterness **9**, 693f
– blood cell – **9**, 700
– soy – **9**, 701
– – whipping properties **9**, 695f
– taste **9**, 705
protein hydrolysis, control **9**, 688ff
– – osmometry **9**, 690
– – pH-stat technique **9**, 689f
– effect **9**, 692
– enzymatic – **9**, 687ff
– key indices **9**, 692
– kinetic aspects **9**, 690ff
– membrane processes **9**, 698f
– with hydrochloric acid **9**, 702
protein identification **5B**, 327ff
– by sequence similarity **5B**, 332
– from mass spectrometry **5B**, 329ff
– from sequence data **5B**, 331ff
– from 2D gels **5B**, 327, **5B**, 327ff
– *p*I/MW measurements **5B**, 328
– tools **5B**, 326ff
– – Web links **5B**, 327
– Western blotting **5B**, 328
– ^{32}P detection **5B**, 328
protein inactivation, by chemical events **3**, 702
– irreversible – **3**, 702
– reversible – **3**, 702
protein indentification, immunodetection **2**, 399
protein inhibitors, in HIV gene therapy **5A**, 476ff
– – dominant-negative proteins **5A**, 476f
– – intracellular antibodies **5A**, 477f
protein interactions **5A**, 87ff, **5B**, 335f
– analytical ultracentrifugation **5A**, 91
– association constant **5A**, 88
– calorimetric methods **5A**, 90f
– continuum models for – **5A**, 94
– crystallography **5A**, 92
– electrostatic – **5A**, 89
– hydrogen bonds **5A**, 89
– hydrophobic – **5A**, 89
– NMR spectroscopy **5A**, 92f
– prediction, force field methods **5A**, 93f
– quantitative description **5A**, 88f
– strengths **5A**, 88
– surface–plasmon resonance **5A**, 92
– theoretical methods **5A**, 93ff

protein isolation **3**, 506, **5B**, 244ff
– heterogenous sample **5B**, 245f
– homogenous sample **5B**, 245
– inclusion bodies **3**, 530
protein kinase inhibitors, antitumor agents **7**, 677ff
– structures **7**, 678
protein–ligand complexes **5A**, 94, 96f
– enzyme–inhibitor complexes **5A**, 96
protein–ligand docking, ligand design **5A**, 95
– ligand optimization **5A**, 95
protein ligands, as protein folding aids **3**, 541
protein localization **2**, 693ff, **5A**, 5ff
– membrane anchor **2**, 693ff
– signal recognition particle (SPR) **2**, 691
protein–macromolecule docking **5A**, 95f
protein-metal interaction, in affinity partitioning **3**, 601
protein mobilization **5B**, 273
protein modeling **5A**, 73ff
– homology modeling **5A**, 75ff
protein modification **2**, 697ff
– acylation **2**, 697f
– co-/posttranslational **2**, 697ff
– posttranscriptional – **5A**, 49
– posttranslational **2**, 249, **5A**, 49ff
– subcellular localization **2**, 698
protein monitoring **4**, 205ff
– flow cytometry **4**, 206f
– penicillin-G acylase **4**, 207
– sample-flow analysis **4**, 207f
protein–nucleic acid complexes **5A**, 100f
protein overexpression **2**, 410
– – *see also* heterologous protein production
– in *E. coli* **2**, 410
protein partitioning, affinity – **3**, 598f
– – ligands **3**, 598
– β-galactosidase **3**, 601
– by reversed micellar extraction **3**, 607ff
– – affinity-based – **3**, 609
– – equipment **3**, 611
– – surfactant concentration effects **3**, 608
– effects of, pH value **3**, 598, 602
– – protein size **3**, 597, 608f
– – salt concentration **3**, 598
– – salt type **3**, 597f
– influencing factors **3**, 597ff, 607ff
– *in situ* – **3**, 604
– in two-phase systems, equipment **3**, 605
– – IDA-PEG system **3**, 602
– lactate dehydrogenase **3**, 600
– partition coefficients **3**, 596ff
– peptides **3**, 601
protein phosphatase inhibitors, antitumor agents **7**, 677ff
– structures **7**, 678
protein phosphorylation, role in antibiotic production **7**, 81f

protein processing, in eukaryotes **3**, 532
– in prokaryotes **3**, 533ff
protein production **1**, 380ff, **2**, 666, **3**, 24, 26
– – *see also* enzyme production, heterologous gene expression
– by clostridia **1**, 310
– by plant cell cultures **1**, 589
– by pseudomonads **1**, 421
– costs **3**, 20
– diagnostic reagents **3**, 26
– glycoproteins **3**, 32
– hetrologous –, by bacilli **1**, 396ff
– – human interleucin **1**, 396
– – in yeast **1**, 496ff
– host, cell development **3**, 30
– – organisms **3**, 24, 530
– human reagents, *in vivo* – **3**, 26
– in insect cells, mammalian proteins **2**, 729f
– in plants, amino acid composition improvement **2**, 650
– – protein drug production **2**, 650
– – seed proteins **2**, 650
– parameters **2**, 666
– recombinant DNA proteins **3**, 11
– regulations **3**, 20, 29
– safety **3**, 29
– single cell protein **1**, 278
– therapeutics **3**, 10, 26
protein property prediction **5B**, 333ff
– formulas for **5B**, 334
– tools, Web links **5B**, 340
– 3D structure, tools **5B**, 368
ProteinProspector **5B**, 262
protein–protein complexes, antibody–antigen complexes **5A**, 99f
– antigen–antibody complexes **5A**, 99
– protease inhibitors **5A**, 99
protein–protein docking **5A**, 77, 94ff
protein purification **3**, 29, 57ff, 593ff, **4**, 620
– analysis of purity **3**, 718ff
– chromatography **3**, 48ff
– extracellular enzymes **3**, 610
– intracellular enzymes **3**, 610f
– steps **3**, 48f
– virus removal **3**, 29
– with two-phase systems, polymer recycling **3**, 603
– – salt recycling **3**, 603
protein purification tag **2**, 417f
– affinity purification **2**, 417
– tag peptides **2**, 418
protein purity **3**, 718ff, 735
– contaminants **3**, 719
– impurities **3**, 718f
– microheterogeneity **3**, 724
protein quantitation **3**, 726
protein refolding **3**, 527ff
– – *see also* protein folding

– binding protein (BiP) function **3**, 532
– binding protein (BiP) level, genetic manipulation **3**, 32
– diffusion-collision-adhesion model **3**, 544
– effects of, denaturant concentration **3**, 545
– – environmental parameters **3**, 545
– – temperature **3**, 547
– folding pathways **3**, 541
– *in vitro* – **3**, 527ff
– – analysis **3**, 536
– models **3**, 543, 544f
– rates **3**, 541
– specific ligands **3**, 541
protein–RNA interactions **5A**, 101
proteins **3**, 718
– – *see also* enzymes
– adhesive –, of marine mussels **10**, 467
– anaerobic degradation of **11A**, 32ff
– antifreeze – **10**, 467f
– conjugated – **5A**, 9ff
– content of various cells **5A**, 4
– contractile – **5A**, 31ff
– deamidation **9**, 705
– *de novo* sequencing of **5B**, 255, 263f
– detection of **5B**, 248
– electron microscopy **1**, 11
– enzymatic modification **9**, 685ff
– evolution of **5A**, 2f, **5B**, 349f
– export of heterologous –, *E. coli* **9**, 78
– – *Saccharomyces cerevisiae* **9**, 84
– fatty acrylated – **5A**, 17
– fermentation of **11A**, 36f
– – biogas amounts and composition **11A**, 36f
– homology modeling **5B**, 351f
– hydrolysis **3**, 733
– identification based on MS/MS **5B**, 260ff
– import **1**, 79
– in microbial biomass **9**, 223f
– in milk **9**, 364ff
– *in vitro* folding **5A**, 112
– isoelectric pH value (p*I*) **3**, 724
– isoprenylated – **5A**, 18
– Kjeldahl conversion factors **9**, 688
– low-abundance **5B**, 250
– metabolism, in lactic acid bacteria **1**, 354ff
– metal-binding – **10**, 234ff
– multimodular –, properties of linkers **10**, 362f
– overproduction, effect on polysomes **1**, 495
– – in *Aspergillus* **1**, 536f
– post-labeling **5B**, 249
– post-translational modification, medium effects **3**, 155
– prediction of, absorptivity **5B**, 333
– – active sites **5B**, 333f
– – domains **5B**, 337
– – isoelectric point **5B**, 333
– – location in cells **5B**, 336
– – modification sites **5B**, 335

– – molecular weight **5B**, 333
– – secondary structure **5B**, 337f
– – shape **5B**, 336f
– – stability **5B**, 336f
– – 3D folds **5B**, 338f
– pre-labeling **5B**, 248
– properties **3**, 696
– recombinant –, large-scale production **5A**, 112
– – market **5A**, 533
– – medical applications **5A**, 125ff
– – regulations **5A**, 495
– – sales **5A**, 533
– regulatory – **5A**, 38ff
– release of intracellular – **3**, 508ff, 513, 519
– – chemical cell permeabilization **3**, 516f
– – from *Escherichia coli* **3**, 516f
– structural elements **3**, 697
– structural signatures **5B**, 350f
– structure information **5B**, 345ff
– – Web sites **5B**, 356
– sweet-tasting –, production of **10**, 577
– translocation **1**, 79
protein secretion **1**, 79, **2**, 414f, 521, 691ff
– bacteriocin release protein (BRP) **2**, 415
– heterologous proteins **2**, 347
– in eukaryotes **3**, 531f
– in prokaryotes **3**, 534f
– invertase **2**, 521
– in yeast **1**, 500
– leader sequence **1**, 500
– limitations in *E. coli* **2**, 415
– protein processing **2**, 521
– regulated **2**, 694ff
– secretory pathway **2**, 414
– signal sequence **2**, 414, **3**, 531
– yeast σ-factor **2**, 521
protein secretory pathway **2**, 691ff
– secretory granules **2**, 694
– signal peptide **2**, 691
– – looped insertion **2**, 693
protein separation **5B**, 247ff
– by 2D gel electrophoresis **5B**, 247ff
protein sequence **5A**, 43ff
– analysis **5A**, 50ff, 55ff
– cDNA **5A**, 50f
– database-aided homology search **5A**, 57ff
– databases **5A**, 49
– – errors **5A**, 54ff
– – identification algorithm **5A**, 54f
– – quality **5A**, 53ff
– motif searches **5A**, 58
– profile searches **5A**, 58
– significance of weak homologies **5A**, 58
– similarity searches **5A**, 56
– WWW pointers **5A**, 56
protein sequence databases **5B**, 286ff
protein sequences, databases **12**, 325
protein sequence–structure relationships **5A**, 59

protein sequencing **3**, 733
– Edman sequencing **3**, 733
protein signature databases **5B**, 289
protein size **5A**, 49
– effect on protein partitioning **3**, 597
protein solubility **9**, 18f
protein solubilization **3**, 513ff
– isoelectric point dependence **3**, 609
– pH dependence **3**, 609
– – molecular weight effects **3**, 609
protein sorting **2**, 691ff
– membrane isertion sequence **2**, 691
– polarized transport **2**, 695ff
– retention sequence **2**, 696
– – KDEL **2**, 696
– signals **2**, 694ff
– signal sequence **2**, 691ff
– stop transfer sequence **2**, 693f
– to ER **2**, 696
– to Golgi apparatus **2**, 696
– to lysosomes **2**, 695
– to nucleus **2**, 695f
– – nuclear localization sequence (NLS) **2**, 695f
protein splicing, posttranslational – **5A**, 50f
protein stability, in freeze-drying **3**, 703ff
– storage, after freeze-drying **3**, 704
– trehalose effects **3**, 706
protein stabilization **9**, 16ff
protein–steroid complexes **5A**, 98f
protein structure **3**, 528, **5A**, 43ff, **9**, 9ff
– *alpha*-helices **3**, 528
– analytical techniques **3**, 529
– *beta*-sheet **3**, 528
– Brookhaven Protein Database **5A**, 66, 92
– chemical – **5A**, 50
– conformational searches **5A**, 74
– crystallographic **5A**, 62
– determination, by NMR spectroscopy **9**, 23
– – by X-ray crystallography **9**, 23
– – force field methods **5A**, 93f
– domains **9**, 15ff
– helix types **5A**, 60ff
– hierarchial description **5A**, 47f
– investigation of primary – **5A**, 50ff
– knowledge-based prediction **5A**, 55ff
– motifs **9**, 13ff
– polypeptide backbone **5A**, 60
– prediction of **5A**, 88
– quaternary – **5A**, 77ff
– – phenomenology **5A**, 77
– – prediction **5A**, 77f
– Ramachandran plot **5A**, 59
– random coils **3**, 528
– secondary – **3**, 528, **5A**, 59ff
– – computer algorithm **5A**, 63
– – extended – **5A**, 61f
– – folding types **5A**, 64
– – loop segments **5A**, 62f

– – polypeptide segments **5A**, 62
– – predictions **5A**, 65ff
– – repetitive – **5A**, 60ff
– – structural classes **5A**, 63ff
– – types **5A**, 59ff
– solvation energy **5A**, 71
– structural comparisons, techniques **5A**, 72
– tertiary – **3**, 529, **5A**, 68ff
– – construction principles **5A**, 69ff
– – modeling **5A**, 73
– – phenomenology **5A**, 68ff
– – predictions **5A**, 73ff
– – solvent accessibility **5A**, 76f
– – structural families **5A**, 71ff
– three-dimensional – **5A**, 72
– torsional angle **5A**, 59
protein superfamilies **5B**, 138
protein surfaces **5A**, 70f
protein synthesis **1**, 99
– direction **2**, 234
– elongation **1**, 99
– initiation **1**, 99
– termination **1**, 99
protein targeting **2**, 521, 589
– heterologous proteins **2**, 635
– in plants **2**, 635f
– signal peptide **2**, 635
– transit peptide **2**, 635
– vacuolar **2**, 635
protein technologies **5B**, 325ff
protein topologies **5A**, 74
protein transport **2**, 691ff
protein truncation test **5B**, 115f
protein turnover **2**, 701f
– (de-)stabilizing amino acid sequences **2**, 702
– PEST sequences **2**, 702
protein types **5A**, 1ff, 5ff
protein variability **5A**, 5
protein visualization **3**, 726
Proteobacteria **1**, 402
– in wastewater treatment **11A**, 102
proteobacterial rRNA, in activated sludge samples **11A**, 97
proteolytic enzymes **10**, 86
proteome, of *Drosophila* **5B**, 34
Proteome Analysis Database **5B**, 291
proteometrics, definition of **5B**, 264
proteomics **5B**, 150f, 437
– capillary electrophoresis **5B**, 269ff
– cell and tissue samples **5B**, 245ff
– – sub-population of cells **5B**, 246f
– cell and tissue sampling, subcellular components **5B**, 247
– databases **5B**, 290f
– definition **5B**, 243
– mass spectrometry **5B**, 241ff
– 2D gel electrophoresis **5B**, 241ff

Proteus vulgaris, use in bioelectrochemical fuel cells **10**, 15ff
protists *see also* protozoa, microalgae
– biotechnology with **10**, 137ff
– classification of **10**, 138ff
– evolution of **10**, 138f
– expression of foreign genes **10**, 146f
– mass cultivation of **10**, 144ff
– products from **10**, 140ff
– – astaxanthin **10**, 145
– – β-carotene **10**, 145
– – bioactive compounds **10**, 140f
– – PUFAs **10**, 141f
– the eukaryote tree **10**, 139
protoberberine alkaloids, from plant cell cultures **7**, 603ff
– structures **7**, 604
protocatechuate, central intermediates, in xenobiotic degradation **11B**, 153
protochelin, catecholate siderophore of *Azotobacter vinelandii* **7**, 203
– structure **7**, 203
protocol process **6**, 210
– for glycerol production **6**, 219
proton-motive force **1**, 64
– in *Lactococcus* **1**, 351
protooncogene induction, inhibition of – **7**, 682
protooncogenes, in tissue culture cells **12**, 95
protophyta *see* microalgae
protoplast **2**, 103
– L-form **2**, 104
protoplast formation **2**, 103ff
– antibiotic-mediated **2**, 104
– bacterial **2**, 103
– enucleate **2**, 114
– from plants **2**, 124ff
– fungal **2**, 109f
– glycine-mediated **2**, 103f
– Gram negative bacteria **2**, 104
– Gram positive bacteria **2**, 103
– plant tissue sources **2**, 124
protoplast fusion **2**, 85, 95, 103
– bacterial **2**, 103, 105
– – diploid **2**, 106
– – genetic mapping **2**, 106f
– – hybrid antibiotic **2**, 108f
– – interspecies **2**, 108f
– – intraspecies **2**, 107
– – recombinant frequency **2**, 108
– – strain improvement **2**, 107
– – versus conjugation **2**, 106f
– – with killed donor **2**, 105
– in Corynebacteria **2**, 483
– in fungi **2**, 109ff
– – frequency **2**, 113
– – heterokaryon sporulation **2**, 116f
– – incompatibility **2**, 117
– – intergeneric **2**, 119

– – interspecies **2**, 116f
– – intraspecies **2**, 115
– – limitations **2**, 116
– – methodology **2**, 113
– – mitochondrial incompatibility **2**, 118f
– plant protoplasts **2**, 123ff
– recombinant recovery **2**, 107
– techniques **2**, 126
protoplast-organelle fusion *see* transfusion
protoplast regeneration **2**, 103ff, 576f
– bacterial **2**, 103
– cell wall formation **2**, 110
– fungal **2**, 109f
– – conditions **2**, 111
– plant **2**, 124ff
– recombinant selection **2**, 113
– reversion frequency **2**, 104
protoplast transformation **2**, 576f
– polyethylene glycol (PEG)-mediated **2**, 577ff
protozoa *see also* protists
– biotechnology with **10**, 137ff
– generation times **10**, 144
– in artificial groundwater recharge **11C**, 434
– in biofilms **11A**, 66
– in biofiltes **11C**, 468
– occuring in biofilms **11A**, 357
– products from, enzymes **10**, 142ff
– use in bioassays **10**, 140f
prourokinase, use in therapy **5A**, 131
Provolone cheese **9**, 378
Prunus amygdalus, oxynitrilase of **8A**, 301
Prymnesiophyta, characteristics of **10**, 113
pseudoaffinity chromatography, in enzyme purification **9**, 58
pseudobactins **7**, 213ff
– peptide siderophore of *Pseudomonas fluorescens* **7**, 213
– structures **7**, 213, 215
pseudo-enzymes *see* synthetic enzymes
pseudoephedrine, industrial syntheis of **8B**, 24
Pseudomonadaceae **1**, 401ff
– aromatic hydrocarbon hydroxylation **1**, 422
– as biocontrol agents **1**, 423
– α subclass **1**, 407
– β-subclass **1**, 407
– catabolic transposons **1**, 411
– characteristics **1**, 402
– chemotaxonomical identification **1**, 406ff
– degradation of, aromatics **1**, 414
– – halogenated hydrocarbons **1**, 415
– – hydrocarbons **1**, 413
– – polychlorinated biphenyls (PCBS) **1**, 415
– – xenobiotics **1**, 414
– DNA hybridization **1**, 412
– ecology **1**, 410ff
– electron tansfer pathway **1**, 417
– flocculation **1**, 413
– gene probing **1**, 411f

– genome organization **1**, 409f
– – integrons **1**, 410
– γ-subclass **1**, 407
– habitats **1**, 411
– herbicide detoxification **1**, 423
– identification techniques **1**, 411f
– *in situ* detection **1**, 412
– interaction with plants **1**, 422
– markers for monitoring **1**, 414
– metabolic properties **1**, 414
– – genetic organization **1**, 414
– monoterpene metabolism **1**, 415
– natural classification **1**, 404
– oxygen limitation response **1**, 420
– phylogenetic status **1**, 402
– plasmids **1**, 409f
– production of, α-galactosidases **1**, 421
– – amylase **1**, 421
– – lipases **1**, 421
– – proteases **1**, 421
– – reserve polyester (PHB) **1**, 420
– – siderophores **1**, 423
– resistance to antibiotics **1**, 410
– serological identification **1**, 412
– tartrate metabolization **1**, 422
pseudomonads, amidase activity **8A**, 294
– nitrile hydratase activity **8A**, 294
Pseudomonas **1**, 401ff
– anaerobic wastewater process **4**, 447
– atropine metabolism **8A**, 335f
– authentic genus **1**, 403f
– bacteriophages **1**, 568
– cocaine metabolism **8A**, 337
– degradation of xenobiotics **1**, 201
– diagnostic fatty acid content **1**, 405
– formate dehydrogenase **8A**, 403
– genus structure **1**, 402
– in industrial biotransformations **3**, 459
– nitrilase activity **8A**, 297
– *ortho*-pathway for 3-chlorobenzoate **1**, 203
– – expansions **1**, 203
– polyamine content **1**, 405
– species **1**, 405
– substrate spectrum expansion **1**, 201
– TOL-plasmid pWWO **1**, 201
– – gene regulation **1**, 202
Pseudomonas acidovorans M3GY, degradation of
  chlorobiphenyls **11B**, 447f
Pseudomonas aeruginosa, arginine deiminase
  pathway **1**, 420
– genome **1**, 409
– in biofilm **11C**, 448ff
– pathways of PHA biosynthesis **6**, 415
– pyochelin **7**, 207f
– rhamnolipids **7**, 180f
Pseudomonas aeruginosa AC869(pAC31), degradation of chlorobenzoate **11B**, 446

Pseudomonas aeruginosa EGSOX, degradation of
  dibenzothiophene **11B**, 452
Pseudomonas aeruginosa FRD1(pUTK50), biore-
  porter strains **11B**, 456
Pseudomonas alcaligenes DM201(pC1), degrada-
  tion of dibenzothiophene **11B**, 452
Pseudomonas amylodermatosa, isoamylase **9**, 746
Pseudomonas capacia, growth kinetics **1**, 133
*Pseudomonas cepacia see also Burkholderia cepa-
  cia*
Pseudomonas cepacia, cepabactin **7**, 224
Pseudomonas cepacia JHR22, degradation of chlorobiphenyls **11B**, 447
Pseudomonas cepacia-like strains, ornibactins **7**,
  216f
Pseudomonas cepacia lipase (PCL) **8A**, 41ff
– Baeyer–Villiger biotransformations **8A**, 568
– catalysis of asymmetric syntheses **8A**, 67f
– enantiopreference of **8A**, 87
– – empirical rules **8A**, 88
– enantioselective reactions catalyzed by primary
  alcohols **8A**, 89f
– enantioselective reactions of carboxylic acids,
  with a stereocenter at the α-carbon **8A**, 106
– general features **8A**, 45
Pseudomonas cepacia RHJ1, degradation of 2,4-
  diophenoxyacetic acid **11B**, 451
Pseudomonas fluorescens **10**, 157
– adhesion-deficient variant **11B**, 411
– ferribactin **7**, 214
– ferrocins **7**, 213
– pseudobactins **7**, 213ff
– pyoverdins **7**, 213f
Pseudomonas fluorescens HK44, bioreporter
  strains, naphtalene **11B**, 454f
Pseudomonas fluorescens spp. *cellulosa* **1**, 405
– cellulolytic system **1**, 413
Pseudomonas fluorescens 10586/pUCD607, *lux*
  biosensor **11B**, 454
Pseudomonas fluorescens 10586s/FAC510, *lux* biosensor **11B**, 454
Pseudomonas fluorescens 410PR(pWWOΔpm),
  metabolism of nitrotoluene **11B**, 452
Pseudomonas lipases **8A**, 76f
– enantiopreference of **8A**, 87
– enantioselective reactions of cyclic secondary alcohols **8A**, 82ff
– enantioselective reactions with racemic acyclic
  alcohols **8A**, 79ff
– resolution of 2-alkanols **8A**, 78
– synthesis of β-blockers **8B**, 385
Pseudomonas mendocina, trichloroethylene degradation **11B**, 410
Pseudomonas oleovorans **6**, 406
– alkane utilization by **11B**, 203f
– biotechnological production of PHA **6**, 440
– pathways of PHA biosynthesis **6**, 415
– PHA synthase, gene structure **6**, 418

– PHA synthesis from lignite depolymerization products **10**, 171f
– poly-*β*-hydroxyalkanoates **7**, 180
Pseudomonas ovalis, use in bioelectrochemical fuel cells **10**, 18f
Pseudomonas pseudoalcaligenes KF/07-D2, degradation of trichloroehtylene **11B**, 449
Pseudomonas pseudoalcalignes, maleate hydratase **8B**, 114
Pseudomonas putida **1**, 405, **8B**, 363f
– aminopeptidase, asymmetric hydrolysis of lupinic acid amide **8A**, 245f
– – resolution of racemic *α*-amino acids **8A**, 245
– Baeyer–Villiger biotransformations **8A**, 555ff
– benzoic acid catabolism **2**, 66
– – *meta*-pathway **2**, 66
– – *ortho*-pathway **2**, 66
– benzoylformate decarboxylase **8B**, 65ff
– camphor metabolism **8A**, 543f
– *cis-cis*-muconic acid production **2**, 64ff
– dihydroxylation of substituted monocyclic arenes **8A**, 506f
– dioxygenase reactions **8A**, 506
– formation of alcyclic compounds by oxidation reactions **8B**, 374ff
– genome **1**, 409
– hydroxylation of carboxylic acids **8B**, 363
– immobilization effect **1**, 242f
– monosaccharide carboxylic acid dehydratases **8B**, 115
– morphine dehydrogenase **8A**, 348f
– organic acid conversion **1**, 422
– oxaloacetate decarboxylase **8B**, 57f
– oxidation of styrene **8B**, 375
– phenol degradation **1**, 242
– TOL plasmid, pWWO **2**, 65
– transformations of morphine alkaloids **8A**, 346ff
– use in ecotoxicological testing **11B**, 137
Pseudomonas putida EGSOX, degradation of dibenzothiophene **11B**, 452
Pseudomonas putida F1/pSMMO20, degradation of trichloroethylene **11B**, 450
Pseudomonas putida G786(pHG-2), degradation of pentachloroethane **11B**, 450
– degradation of trichloroethylene **11B**, 450
Pseudomonas putida IPL5, degradation of polychlorinated biphenyls **11B**, 449
– field application vectors **11B**, 449
Pseudomonas putida KF715-D5, degradation of trichloroethylene **11B**, 449
Pseudomonas putida (pDA261, degradation of biphenyls **11B**, 448
Pseudomonas putida PPO300(pRO101), degradation of 2,4-diphenoxyacetic acid **11B**, 451
Pseudomonas putida PPO301(pRO103), degradation of 2,4-diphenoxyacetic acid **11B**, 451

Pseudomonas putida TVA8, bioreporter strains, BTEX compounds **11B**, 454
Pseudomonas rubescens, ornithine-containing lipid formation **6**, 664
Pseudomonas sp., alginate production **6**, 620
– biopolymer formation **6**, 617
– degradation of chlorobiphenyls **11B**, 447
– degradation of toluene **11B**, 445
– eugenol degradation pathway **8B**, 338
– production of ferulic acid **8B**, 337
– production of pseudouridine **6**, 594
– rhamnolipid formation **6**, 663
– sulfoxidations by **8A**, 574
Pseudomonas sp. alcohol dehydrogenase (PADH), applications of **8A**, 453
– preparation of chiral hydroxy compounds **8A**, 453
– synthesis of L-carnitine **8A**, 453
Pseudomonas sp. B13, degradation of chlorobenzoates **11B**, 445f
– simultaneous degradation of chloro- and methyl aromatic compounds, hybrid pathway **11B**, 445
Pseudomonas sp. clone A (pWWO-Km), degradation of trinitrotoluene **11B**, 452
Pseudomonas sp. HL7bR(pC1), degradation of dibenzothiophene **11B**, 452
Pseudomonas sp. strain JR1A::ipb, degradation of, isopropylbenzene **11B**, 449f
– – trichloroethylene **11B**, 449f
Pseudomonas sp. UG14Lr, bioreporter strains **11B**, 455
Pseudomonas sp. US1 ex., degradation of chlorobenzoate **11B**, 446
Pseudomonas strain Shk1, toxicity monitoring bioreporter **11B**, 456
Pseudomonas stutzeri, denitrification pathway **1**, 419
– maltose production **9**, 747
Pseudomonas syringae, frost damage in plants **1**, 424
pseudomurein **1**, 258
– of methanogenic bacteria **1**, 21
pseudomycelia, of yeasts **1**, 115
pseudopterosin, antiinflammatory compound of marine origin **10**, 455ff
pseudopteroson E, structure **10**, 457
pseudouridine, production by a *Pseudomonas* strain **6**, 594
pseudovirion *see* viral vector
psoriasis **5B**, 117
– medical applications of recombinant proteins **5A**, 143
– monoclonal antibody therapy **5A**, 319
– vaccination **5A**, 143
Psychoda **11A**, 9
psychrophiles **10**, 273
– characteristics of **10**, 63ff
– definition of **10**, 63

– enzymes of **10**, 460
psychrophilic bacilli *see* bacilli
psychrophilic microorganisms, specific growth
   rate **1**, 128
psychrotolerants, characteristics of **10**, 63ff
– definition of **10**, 63f
psychrotrophs **10**, 273
PTCA (balloon angioplasty) **5A**, 315, 375ff
– monoclonal antibody therapy **5A**, 316
– therapy with ReoPro **5A**, 375
– use of ReoPro **5A**, 376
PTO (U.S. Patent and Trademark Office) **12**, 350
– practice **12**, 290f
public health, impacts, from biological waste **3**,
   784f
Public Health Services Act **5A**, 498
public opinion surveys, on biotechnology **12**,
   123ff
– – in Australia **12**, 126ff
– – in China **12**, 127
– – in Denmark **12**, 660f
– – in Europe **12**, 133f, 658
– – in France **12**, 449ff
– – in Germany **12**, 660
– – in Japan **12**, 124ff, 146ff
– – in New Zealand **12**, 124ff, 146f
– – in The Netherlands **12**, 192, 602
– – in the U.K. **12**, 124, 126, 192, 662
– – in the USA **12**, 133, 146ff, 473, 658ff
– on genetic manipulation **12**, 128ff
– – in France **12**, 451ff
– – in Japan **12**, 128ff
– – in New Zealand **12**, 128ff
PUFA *see* polyunsaturated fatty acids
PUFAs
pulchellalactam, antiinflammatory compound of
   marine origin **10**, 456f
– structure **10**, 457
pulcherrimic acid, iron-binding compounds **7**, 233
– structure **7**, 233
pullulan **6**, 620
– fermentation parameters **6**, 645
– film formation **6**, 645
– food industry uses **6**, 645
– formation by *Aureobasidium pullulans* **6**, 644
– structure **6**, 644
pullulanase **1**, 27, **9**, 669f, 746
– in clostridia **1**, 310
– purification **3**, 602
pullulanases **10**, 76ff
– action of **10**, 75
– classification of **10**, 76
– thermoactive – **10**, 77f
pulmonary inflammatory disorders, monoclonal
   antibody therapy **5A**, 315
Pulmozyme **5A**, 167
pulp and paper processing, applications of biotechnology **10**, 523ff

– use of enzymes **10**, 525ff, 532ff
– use of microorganisms **10**, 532ff
pulping *see also* biopulping
– biomechanical – **10**, 533f
– bleaching, use of xylanase **10**, 535ff
– chemical – **10**, 534f
– – enzymatic treatment **10**, 535
– – microbial treatment **10**, 535
– mechanical – **10**, 533ff
– – enzymatic treatment **10**, 534
– – microbial treatment **10**, 533f
pulsed-field electrophoresis **2**, 162
pulsed field gel electrophoresis **5B**, 44
pumice, as carrier materials of biofilters **11C**, 466
pump and treat processes **11B**, 312
Pure Food and Drugs Act **5A**, 498
Pure Food Campaign **12**, 479
purification, of enzymes **9**, 50ff, 88ff
– – affinity chromatography **9**, 56ff
– – cell rupture **9**, 91f
– – covalent chromatography **9**, 59
– – desalting **9**, 52f
– – extraction **9**, 50f
– – high pressure liquid chromatography **9**, 56
– – hydrophobic interaction chromatography **9**,
   58f
– – hydroxyapatite chromatography **9**, 55
– – ion exchange chromatography **9**, 53ff
– – lysate clarification **9**, 92ff
– – precipitation **9**, 51f
– – size exclusion chromatography **9**, 56
purification procedures, in citric acid fermentation **6**, 327
– in lactic acid fermentation **6**, 299f
purine composition graph **5B**, 391
purine nucleosides **6**, 587
– – *see also* nucleosides
purine nucleotides **6**, 571ff
– – *see also* nucleotides (5′)
– biosynthetic pathway **6**, 571
purine nucleotide synthesis, by *Brevibacterium
   ammoniagenes*, mutant strains **6**, 574
– – regulation **6**, 574
– by *Brevibacterium ammoniagenes*, transformed
   strains **6**, 575f
purinolysis **1**, 306f
– biotechnological application **1**, 306
purity assessment, of enzymes **9**, 59ff
– – by high pressure liquid chromatography **9**,
   65
– – by polyacrylamide gel electrophoresis **9**, 62ff
– – specific activity **9**, 59ff
purpureaglycoside A **1**, 597
purpurin production, by cell cultures **1**, 589
puto, production of **9**, 527f
putrebactin, structure **7**, 211
putrescine, in pseudomonads **1**, 405ff
pyloric stenosis **5B**, 66, 67

pyochelin, catecholate siderophore of *Pseudomonas aeruginosa* **7**, 207f
– structure **7**, 207
pyoverdine **1**, 423
pyoverdins **7**, 213ff
– defective mutants **7**, 214
– peptide siderophore of *Pseudomonas fluorescens* **7**, 213
– structures **7**, 213, 215
pyrazines, hydroxylation of **8A**, 502
pyrene, complete metabolization and mineralization **11B**, 226
– degradation by bacteria **11B**, 226
– – *Mycobacterium* sp. **11B**, 228ff
– hydroxylation of **8A**, 501
– oxidation by *Phanerochaete chrysosporium* **11B**, 233
– physical and chemical properties of **11B**, 218
– structure of **11B**, 216
Pyrenomycetidae, biotechnologically important species **1**, 529
pyrethroid insecticides, using *Arthrobacter* lipase resolution **8B**, 369
pyrethroid precursors, synthesis of, use of a monooxygenase **8B**, 282
– – use of an esterase **8B**, 283
– – use of oxynitrilases **8B**, 282
pyridine nucleotide coenzymes **1**, 71
pyridines, hydroxylation of **8A**, 501f
pyridoxalphosphate **1**, 72
pyridoxal-5′-phosphate **9**, 35
pyridoxal 5′-phosphate enzymes **6**, 520ff
– reactions catalyzed by – **6**, 521
pyridoxine (vitamin B₆), chemical synthesis of **10**, 327
– producing microorganisms **10**, 327f
pyridyl acetyl esters of secondary alcohols, kinetic resolution of **8A**, 232
pyrimidine degradation, in clostridia **1**, 308
pyrimidine dimer **2**, 12
pyrimidine nucleosides, regioselective hydrolysis, by subtilisin **8A**, 223
pyrite, microbial removal from coal **10**, 173f
Pyrococcus archaebacteria, amylases **9**, 745
Pyrococcus furiosus **10**, 65
pyrolysin **10**, 85
pyrolysis, of waste material **11C**, 25ff
pyrophosphate, pK_a values **8B**, 222
pyrosequencing **5B**, 210
pyrotechnology, of solid waste treatment **11C**, 5ff
pyrroles, chlorinated – **8B**, 186
pyrroloquinoline quinone **9**, 35
– microbial production of **10**, 332
pyrroloquinoline quinone (PPQ) **1**, 71, 173ff, 178f, 274
– metabolic function **1**, 173
pyruvate **9**, 35

– anaerobic dissimilation, in lactic acid bacteria **1**, 355
– degradation by clostridial enzymes **6**, 241
– fermentation **1**, 182
– – pH dependence **1**, 182
– – product quantity **1**, 182
– formation by *Clostridium acetobutylicum* **6**, 239
– formation in wine making **9**, 475f
pyruvate carboxylase, fungal –, in mitochondria **6**, 312
– – role in citric acid formation **6**, 312
pyruvate decarboxylase, gene (*PDC1*) **1**, 482
pyruvate decarboxylase (PDC) **8B**, 23ff, 49ff
– activation of **8B**, 53
– acyloin condensation **8B**, 23ff
– biotransformations **8B**, 53ff
– decarboxylation of fluoropyruvate **8B**, 54, 56
– mechanism **8B**, 49ff
– of *Saccharomyces cerevisiae* **8B**, 50ff
– of yeast **8B**, 54f
– of *Zymomonas mobilis* **8B**, 49f, 53ff
– structure **8B**, 49ff
pyruvate kinase, determination of K⁺ **9**, 144f
– gene (*PYK*) **1**, 482
– in industrial biotransformations **3**, 459
– phosphorylation by **8B**, 224
– regulation **1**, 486
pyruvate metabolism, in clostridia **1**, 298
P1-derived artificial chromosomes (PAC) **5B**, 195

Q

quality assurance, in yeast production **9**, 344ff
quality control, of cell cultures **10**, 33ff
– of plasmid DNA **5A**, 458ff
– – DNA quantity **5A**, 459
– – restriction analysis **5A**, 459
– – sequencing **5A**, 459
– – sterility **5A**, 459
quality traits **5B**, 146
quantitative trait loci **5B**, 80
quantitative trait locus (QTL) **2**, 150
– detection **2**, 151
– localization methods **2**, 150
– phenotypical effect **2**, 151
– type frequencies **2**, 151
quaternary stereocenters, of alcohols **8A**, 94f
– of carboxylic acids **8A**, 109
quaternary structure *see* protein structure
questin, Baeyer–Villiger oxidation **8A**, 548
– ring cleavage of **8A**, 543
quinoline, hydroxylation of **8A**, 502
quinoline alkaloid, from plant cell cultures **7**, 614
quinolines, chlorinated – **8B**, 187
quinone coenzyme **1**, 71
quinones, in chemotaxonomy **1**, 406

– in pseudomonads **1**, 404f
– isolation **1**, 406
p-quinones, antitumor agents **7**, 652
quinoxaline antibiotics, antitumor agents **7**, 652
quorum filters, in biofilms **11C**, 450
quorum sensing, in flocs and biofilms **11A**, 78f
quorum sensors **7**, 65

R

RA *see* rheumatoid arthritis
rabbit muscle aldolase (RAMA) **8B**, 7ff
– synthesis of azasugars **8B**, 10ff
– synthesis of carbohydrates **8B**, 10f
– synthesis of cyclitols **8B**, 13
– use in synthesis of heterocyclic compounds **8B**, 388f, 391
racemases *see* amino acid racemases
RAC (rDNA Advisory Committee) **12**, 68, 150, 243ff, 471, 514, 524, 566ff
radiation biology, in space **10**, 516f
radiation hybrid map **5B**, 44, 174
radioactive metabolites **5B**, 248
radioimmunoassays, use in drug screening **10**, 44f
radionuclides, biosorption of **10**, 230ff
– – by algae **10**, 230f
– – by fungi **10**, 229f
– precipitation as phosphates **10**, 244
radiotherapy, monoclonal antibody therapy **5A**, 320
Rad51, protein **5B**, 71
RAFI (Rural Advancement Fund International) **12**, 365
ragi, mycological profile **9**, 528
rain water, in industrial wastewater **11A**, 194f
Ralstonia elutropha AE1216, degradation of chlorobiphenyls **11B**, 446f
Ralstonia eutropha AEK301/pYK3021, degradation of trichloroethylene **11B**, 451
Ralstonia eutropha AE707, degradation of chlorobiphenyls **11B**, 446f
Ralstonia eutropha ENV307(pUTK60), biphenyl/ PCB bioreporter **11B**, 453
Ramachandran plot **5A**, 59, 62, **9**, 11, 16
ranalexin, structural gene **7**, 280
random amplified polymorphic DNA, detection of plant pathogens **10**, 500
random amplified polymorphic DNA (RADP) **5B**, 167
random mutagenesis, of lipases **8A**, 149f
random shotgun sequencing **5B**, 193f
rapamycin **7**, 546ff
– biosynthetic clusters **7**, 21
– chemical structure **10**, 342
– immunosuppressive properties **7**, 546f
– structure **7**, 546

rapamycin derivatives **8A**, 133
rape seed, cell wall polysaccharides **9**, 707
– feed products **9**, 721
– – protein meals **9**, 725
rapeseed oi, genetic engineering of – **7**, 138
rape seed oil, enzymatic extraction **9**, 721ff
– – process scheme **9**, 722
– – yield **9**, 722
Raphidophyta, characteristics of **10**, 114
rapid filters, applications of **11C**, 392ff
– closed – **11C**, 391f
– – types of **11C**, 392
– design of **11C**, 390
– dual-media – **11C**, 391
– filter running times **11C**, 394
– mono-layer – **11C**, 390f
– multi-media – **11C**, 391
– operation of **11C**, 393
– regeneration of **11C**, 394
– scheme **11C**, 390
rapid filtration, in drinking water treatment **11C**, 389ff
rare transcripts **5B**, 372
RAS/cAMP signaling pathway **1**, 487f
rasemic amino acids, optica resolution, chemical methods **6**, 543
– – enzymatic methods **6**, 543ff
– – physiochemical methods **6**, 543
– resolution, enzymatic methods **6**, 544ff
ras-farnesyltransferase inhibitors **7**, 685
– structures **7**, 686
RAS gene, in yeast **1**, 487f
raspberry ketone, Baeyer–Villiger degradation by *Beauveria bassiana* **8A**, 548
– formation of **8A**, 377
rate equations, assumptions for modelling **7**, 410f
rational entropy, in enzyme catalysis **8B**, 415ff
Rauwolfia serpentina, biosynthetic pathway of aj-maline **8A**, 353f
– strictosidine synthase **8A**, 352
raw materials, for fermentation processes **6**, 49f
– grinding **6**, 77
– – use of mills **6**, 78
– in acetic acid fermentation **6**, 383f
– in citric acid production, glucose **6**, 324
– – molasses **6**, 318f
– – sucrose **6**, 324
– in classical ethanol production, analysis **6**, 109ff
– – starch-based – **6**, 62ff
– in industrial ethanol production **6**, 136ff
– – cassava **6**, 140
– – corn **6**, 140
– – food industry wastes **6**, 139f
– – fruit crops **6**, 138
– – Jerusalem artichoke **6**, 141
– – lignocellulosic materials **6**, 143ff
– – sorghum **6**, 141
– – starch **6**, 140

– – sugar beet **6**, 137
– – sugarcane **6**, 136f
– – sweet potato **6**, 141
– – waste sulfite liquor **6**, 138f
– – whey **6**, 139
– in lactic acid fermentation **6**, 297f
– – glucose **6**, 297
– – maltose **6**, 297
– – sucrose **6**, 297
– – whey **6**, 297
– prices, for sugar feedstocks **6**, 52
– starch-based **6**, 31ff
– starch-containing **6**, 62ff
– – composition of **6**, 62ff
– – potatoes **6**, 62
– – wheat **6**, 62f
– sugar-based **6**, 5ff
– – composition of **6**, 12
– – provenance **6**, 6ff
– – uses **6**, 24ff
raw material strategies, economical problems **6**, 47ff
– non-renewable resources **6**, 54ff
– – environmental certificates **6**, 55f
– – environmental taxes **6**, 55f
– renewable resources **6**, 54ff
raw spirit, distillation from mashes **6**, 100ff
raw sugar *see* sugar
RCEP (Royal Commission in Environmental Pollution) **12**, 576f
RDBMS **5B**, 292
R&D Center for Biotechnology of Indonesia *see* LIPI
rDNA, potential biohazards **12**, 511f
– protection of workers **12**, 271
– transformation **12**, 162
rDNA containing organisms, large-scale uses **12**, 246ff, 614
– shipment **12**, 249f
– transport **12**, 249f
rDNA experiments, classification **12**, 69, 243ff, 256f, 512, 517
– Japanese regulations **12**, 595
– notification **12**, 268
– risks **12**, 512f
rDNA operations, biosafety **12**, 63ff
– classification **12**, 267ff
– construction of facilities, for production **12**, 259f
– – for research **12**, 259
– – permits **12**, 248
– permits **12**, 267f
– regulations **12**, 239ff
– supervision in Japan **12**, 253ff
– – institutions **12**, 253f
– supervision in the European Union **12**, 266ff
– supervision in the USA, NIH guidelines **12**, 241ff

rDNA products, applications **12**, 175
– consumer acceptance **12**, 135, 157ff, 192f
– decision trees **12**, 175ff, 189f
– development, phases **12**, 173f
– EC regulations **12**, 533
– free circulation **12**, 533
– hazard analysis **12**, 159ff
– labeling **12**, 465f
– market approval **12**, 248f, 270
– of Japan **12**, 377
– of Korea **12**, 388
– risk assessment **12**, 157ff, 563ff
– types **12**, 175
rDNA technology **12**, 159ff
– dangers **12**, 44ff
– ethical aspects **12**, 135ff, 189ff
– metabolic reprogramming **12**, 172
– public controversy **12**, 513ff
– – periods **12**, 513f
– technical aspects **12**, 159ff
– use **12**, 160
rDNA work, Council Diretive **12**, 518f
reaction equilibrium **1**, 53
– constant **1**, 53
Reaction rates, dynamic error analysis **7**, 54ff
reactive extraction *see* carrier extraction
reactive extractions, in wastewater treatment **11A**, 566
reactive immunization, induction of catalysis in antibody binding sites **8B**, 439ff
reactor composting **11C**, 122f
reactor processes *see also* bioreactors
– slurry state processes **11B**, 472
– solid state processes **11B**, 472
reactor systems, choice of, for enzymatic processes **8A**, 431ff
readout methods, in drug screening **10**, 44ff
– – fluorescence detection methods **10**, 45ff
– – microvolume techniques **10**, 47
– – radiolabeling **10**, 44
– – single molecule detection **10**, 47
real-time analysis environments **5B**, 435
rearrangements, use of catalytic antibodies **8B**, 469f
REBASE **5B**, 284, 410
receptor, of low molecular weight ligans, screening of antagonists **7**, 112ff
– of peptide ligands, screening of antagonists **7**, 116ff
receptor-active compounds *see also* receptor agonists, receptor antagonists
– new screening methods **7**, 107ff
– screening assays **7**, 109ff
– – Magnus test **7**, 109f
– – using radio-labeled ligands **7**, 110
– – using recombinant cells **7**, 110f
– screening of, agonists **7**, 122ff
– – antagonists **7**, 111ff

receptor agonists *see also* receptor-active compounds
- erythromycin A **7**, 122ff
- motilides **7**, 122f
- muscarine **7**, 123
- zearalenone **7**, 123
receptor antagonists **7**, 111ff
- – *see also* receptor-active compounds
- androgen receptor **7**, 116
- ANP receptor **7**, 120f
- arginine-vasopressin receptor **7**, 121
- CCK receptors **7**, 116, 121
- C5a receptor **7**, 121
- depamine receptor **7**, 114
- devazepide receptor **7**, 121
- endothelin receptor **7**, 116ff
- estrogen receptor **7**, 115f
- fibrinogen receptor **7**, 115
- ligand receptors **7**, 112
- LTB4 receptor **7**, 114
- muscarinic acethycholine receptor **7**, 112f
- neurokinin receptors **7**, 120
- NMDA receptor **7**, 114
- oxytocin receptor **7**, 121
- PAF receptor **7**, 114f
- peptide ligands **7**, 117ff
- structures **7**, 113
- substances P receptor **7**, 120
receptor binding, of viruses, inhibitors **7**, 124ff
receptor genes, of recombinant cells, use in screens of receptor-active compounds **7**, 110
receptor-mediated endocytosis **5A**, 30f
recombinant antibodies *see* antibodies
recombinant cells, host cell selection **3**, 284
- in fermentation processes **3**, 283ff
recombinant DNA, safety considerations of OECD **2**, 847
- safety debate **2**, 837
Recombinant DNA Advisory Committee (RAC) **5A**, 519f, 525f
- gene therapy regulation **5A**, 520, 523f
- guidelines **5A**, 523f
- regulation of biotechnology products **5A**, 519f
recombinant-DNA organism **2**, 837, 843f
- risk potential classification **2**, 837ff
recombinant DNA technology, brewer's yeast improvement **9**, 439f
recombinant fermentation, application of parameter estimation methods **7**, 495
- bioprocess kinetics **7**, 487ff
- bioprocess modelling **7**, 487ff
- determination of genetic parameters **7**, 491ff
- – gene expression efficiency **7**, 491
- – growth ratio **7**, 491ff
- – kinetic model **7**, 492ff
- – method **7**, 492ff
- – plasmid loss rate **7**, 491ff
- determination of kinetic parameters **7**, 491ff

- effect of specific growth rate, on gene expression rate **7**, 499f
- – on plasmid content **7**, 498f
- – on plasmid stability **7**, 500f
- effect of temperature, on gene expression rate **7**, 501f
- – on plasmid-harboring cell fraction **7**, 502f
- instability parameter **7**, 494
- microbiological process design parameters **7**, 496
- optimization **7**, 496ff
- – dilution rate **7**, 497ff
- ratio of plasmid-free to plasmid-harboring cells **7**, 492ff
recombinant gene *see* heterologous gene
recombinant inbred lines **5B**, 166
recombinant microorganisms, in fermentation processes **3**, 283ff
recombinant organisms **3**, 283ff
- – *see* GMOs (genetically modified organisms)
- amino acid production **3**, 287ff
- applications, safety precautions **3**, 66
- as biological waste **3**, 774
- in fermentation processes **3**, 287ff
- public risk perception **3**, 775
recombinant plants *see* transgenic plants
recombinant proteins *see* proteins
recombinants, productivity, genetic parameters **7**, 491ff
- – kinetic parameter **7**, 491ff
recombinant selection **2**, 85, 105, 394f, 434ff, 707ff, 763
- combined selection procedure **2**, 710
- DNA identification technique **2**, 757
- filter immunoassay **2**, 709
- fluorescence activated cell sorting (FACS) **2**, 709
- for homologous recombination **2**, 714f
- high level expression **2**, 709f
- hypoxanthine-aminopterin-thymidine system **2**, 129
- indirect screening procedures **2**, 710
- mammal cell **2**, 129
- methods **2**, 709
- plants **2**, 126
- – organelle-encoded marker **2**, 126
- – selectable marker **2**, 126
- techniques in mammals **2**, 757
- time in mammals **2**, 756
recombinant stability **2**, 59
recombinant technology, potential **5A**, 533
recombination **1**, 559, **2**, 143, 434ff, 472
- chromosomal integration **2**, 785
- effect of gene integration **2**, 776f, 781
- free recombination **2**, 148ff
- frequency **2**, 143ff
- homologous **2**, 714
- – screening methods **2**, 714f

– homologous integration **2**, 785f
– illegitimate (non-homologous) **2**, 714
– in fungal meiosis **2**, 80
– – practical application **2**, 80ff
– – strain improvement **2**, 80
– in fungal mitosis **2**, 84
– – hybrid conidia **2**, 85
– – industrial application **2**, 86
– – intergeneric **2**, 87
– – interspecific **2**, 87
– – intraspecific **2**, 87
– – restrictions **2**, 85
– in fungi **2**, 75ff, 531
– – homologous **2**, 535
– – non-homologous **2**, 531
– in yeast **2**, 510ff, 522f
– – homologous **2**, 510
– – multiple tandem repeat **2**, 510
– site-specific **2**, 472, 587
– – *see also* homologous recombination
– – with phage recombinase **2**, 588
recombination process **2**, 95
recording, of mRNA **12**, 165f
– – frameshifting **12**, 165f
– – hopping **12**, 166
– – selenocysteine incorporation **12**, 166
rectification, of product spirit **6**, 102ff
– – apparatus **6**, 102f
recurrence risk **5B**, 66f
recycle bioreactors **3**, 432, 441
recycling, from garbage **11C**, 10f
– of solid wastes **11C**, 8ff
– of solid waste streams **11B**, 330
red algae **10**, 111
redox mediators, interaction with the respiratory
    chain **10**, 14
– structure of **10**, 14
– use in bioelectrochemical fuel cells **10**, 13f
– – mode of action **10**, 13f
redox meter, sterilizable **7**, 19
redox potential, definition **7**, 9
– determination of **7**, 18f
– – in anaerobic wastewater purification **7**, 451
– influence on microbial growth **10**, 274
– rH value **7**, 9
– use in process control **7**, 9
redox reactions, use of catalytic antibodies **8B**,
    477ff
red rice *see* ang-kak
reducing agents, use in baked goods **9**, 262f
reductases *see also* dehydrogenases, oxidoreducta-
    ses
– substrate spectra **8A**, 437
reduction reaction, thermodynamics **1**, 55
reduction reactions, catalyzed by yeast enzymes
    8B, 370ff
– use of catalytic antibodies **8B**, 478f
redundancy, in automation systems **7**, 569ff

reference substances, coal derived –, testing of
    10, 177ff
refinement plants, wastewater concentrations
    11A, 204
reflectometry **7**, 96
refractile bodies *see* inclusion bodies
refractory soil organic matter *see* humic substances
registration, of biomedicinal products, regulations
    12, 221
– of biotechnology products **5A**, 498ff
regualtory concerns *see also* contaminated site
    management
regular expression grammars **5B**, 137
regulation, of biotechnology products **5A**, 519
– of human gene therapy **5A**, 517
regulatory concerns, contaminated soil areas **11B**,
    5
– – action thresholds **11B**, 21f, 38f
– – conflict management **11B**, 28f, 34, 40
– – enforcement **11B**, 26f, 34f, 40
– – financing **11B**, 31f
– – liability **11B**, 13f, 31f, 36f
– – non-polluting sites **11B**, 30
– – organization **11B**, 26f, 34f, 40
– – polluting sites **11B**, 31ff
– – priority setting **11B**, 23f, 39
– – remediation targets **11B**, 24f, 33f, 39
– – verification of clean-up **11B**, 24f
– worldwide **11B**, 5ff
regulatory proteins **5A**, 38ff
– cell surface receptors **5A**, 38
– signal transduction pathways **5A**, 40
– transcription factors **5A**, 40
RegulonDB **5B**, 410
Reichstein–Grussner synthesis **10**, 328f
relaxation time, of processes in microorganisms
    1, 138
relaxin **5A**, 133
release, of GMOs **12**, 183ff, 555
– – Council Directive **12**, 555ff
– – permits **12**, 251, 269f
– – regulations **12**, 149
– – risk assessment **12**, 184ff
– – sites in the U.S. **12**, 251
– – transgenic plants **12**, 198ff, 251f
remediation methods, classification of **11B**, 306
remote stereocenters, of alcohols **8A**, 97
– of carboxylic acids **8A**, 109
renal failure **5A**, 164
rennet **9**, 374
– microbial – **9**, 680
ReoPro (Abciximab, c7E3 Fab) **5A**, 130, 230, 315
– approval process **5A**, 366, 370
– blinding **5A**, 369
– CAPTURE trial **5A**, 376ff
– clinical development **5A**, 365ff
– clinical trials **5A**, 367
– coronary interaction **5A**, 367

– coronary interventions **5A**, 371
– cost-effectiveness **5A**, 370f, 373
– effect on blood clotting time **5A**, 374
– EPIC trial **5A**, 367ff
– EPILOG trial **5A**, 374ff
– high-risk patients **5A**, 368ff, 375
– in bleeding events **5A**, 369, 374, 376
– indication **5A**, 366
– intention-to-treat analysis **5A**, 368f, 377
– ischemia **5A**, 367ff, 377
– low-risk patients **5A**, 374f, 375
– myocardinal infarction **5A**, 367ff, 377f
– phase II trials **5A**, 366
– platelet receptor blockage **5A**, 367
– PTCA **5A**, 376
– reduction in hospitalizations **5A**, 370ff
– sales **5A**, 534
– stent implantation **5A**, 375
– treatment of, acute coronary syndromes **5A**, 371
– – coronary diseases **5A**, 366ff
– – intracoronary thrombus **5A**, 371, 377
– use in, atherectomy **5A**, 375f
– – balloon angioplasty **5A**, 375
– – coronary stenting **5A**, 375
– – saphenous vein grafts **5A**, 371f
– – unstable angina **5A**, 376
repeated sequences **5B**, 392
repeat identification **5B**, 312
repeats finder **5B**, 392
repetitive DNA sequence **2**, 143
– highly repetitive **2**, 143
– long interspersed repeat sequence (LINE) **2**, 143
– moderately repetitive **2**, 143
– short interspersed repeat sequence (SINE) **2**, 143
repetitive extragenic palindromic sequence (REP) **2**, 239
repetitive sequence **5B**, 132
reporter gene **2**, 558
– applicability **2**, 585
– *cat* gene **2**, 558
– for *E. coli* **2**, 384
– β-glucuronidase (*gus*) gene **2**, 558, 585
– in plants **2**, 558f
– in transgenic animals **2**, 779
– luciferase gene **2**, 558
– transcription rate **2**, 585
reporter gene assays **10**, 49f
– in drug screening **10**, 49f
reporter genes *see* bioreporter systems
representational display **5B**, 390
Republic of China *see* Taiwan
Research Institute for Biological Diversity **12**, 443
reservoirs, treatment of water from **11C**, 381ff
– – degree of trophication **11C**, 382

– – process schemes **11C**, 382
residence time distribution, biological test systems **7**, 125
– gas **7**, 121f
– liquid **7**, 123
– measurement **7**, 121ff
– – flow-follower techniques **7**, 124f
– – tracer techniques **7**, 121ff
– of gas bubbles **4**, 111f
– time-of-flow (TOF) measurement **7**, 139ff
residence times, biofilters **11A**, 370
residential waste **3**, 782
– microbiological activity **3**, 783
residue formation in soils, by covalent bonds **11B**, 112
– by physical entrapment **11B**, 112
– by sorption **11B**, 111
resins, biodegradable – **10**, 423ff
resistance, against bacteria, introduction into plants **2**, 647f
– against fungi, introduction into plants **2**, 647f
– introduction into plants **2**, 638ff
– – potato **2**, 642ff
– – tobacco **2**, 642ff
– – tomato **2**, 642ff
– to plant pathogens **5B**, 146, 170
resistance genes, in aminoglycoside producers **7**, 440
resistance mechanisms, in bacteria **7**, 440ff
resolution, of racemic amino acids **6**, 453ff
– – enzymatic methods **6**, 544ff
Resomer® **10**, 433ff
resonance energy transfer **5B**, 189
resorcinol, anaerobic degradation of **11B**, 180f
Resource Conservation and Recovery Act (RCRA) **3**, 772
respiration **1**, 484
– anaerobic **1**, 382
– energy transduction **1**, 64
– heat evolution **1**, 142
respiratory activity, long-term regulation **7**, 395
respiratory chain **1**, 57, **5A**, 20, **9**, 188f
– branched, in *Escherichia coli* **1**, 59
– electron flow **1**, 58
– redox potential **1**, 58
respiratory quotient (*RQ*) **3**, 324
– determination **3**, 226
– ethanol production, by yeasts **3**, 329
respirometer **7**, 36
response surface design, in medium optimization **3**, 135f
– – contour plots **3**, 136
restenosis **5A**, 131f
– medical applications of recombinant proteins **5A**, 131
– monoclonal antibody therapy **5A**, 316
resting cells, in yeast biotransformations, C=C double bond reduction **8A**, 369

restriction analysis **2**, 401
restriction endonuclease **2**, 376ff
– *BamH1* restriction site **2**, 380
– cross index for combinations **2**, 379
– inhibition **2**, 59
– reaction conditions **2**, 377
– specific recognition sequences **2**, 377
– type I **2**, 376f
– type II **2**, 376ff
– type III **2**, 376f
restriction enzyme **1**, 559
restriction enzymes **5B**, 394
– target sites **5B**, 317
restriction fragment length polymorphisms
    (RFLP) **5B**, 43, 78, 167
– in microbial community analysis **11A**, 87
restriction mapping **2**, 161ff, **5B**, 317
restriction site, mapping **2**, 401f
– occurrence frequency **2**, 377
resveratrol, chemical structure **10**, 342
Reteplase **5A**, 130, 168
reticuline **8A**, 338f
– biosynthesis **8A**, 339f
retinitis pigmentosa **5B**, 110
retinoblastoma **5B**, 70
– gene **5B**, 82
retro-aldol reaction, use of catalytic antibodies
    8B, 468
retro-Henry reaction, use of catalytic antibodies
    8B, 468
retrovirus **2**, 725ff, 763ff
– amphotropic **2**, 766
– chromosomal integration **2**, 767
– ecotropic **2**, 766
– handling **2**, 729
– host range **2**, 726f
– integrated provirus **2**, 767
– life cycle **2**, 725, 766
– long terminal repeat (LTR) **2**, 766
– mouse embryo infection **2**, 768
– murine leukemia virus (MLV) **2**, 766
– replication competent – **5A**, 403f
– – generation of **5A**, 404
– restrictions **2**, 729
– RNA structure **2**, 767
retrovirus vectors **5A**, 397ff
– cell entry **5A**, 400
– clinical studies **5A**, 405f
– design of **5A**, 400
– gene expression **5A**, 399
– genome **5A**, 400f
– host range **5A**, 401
– host responses **5A**, 405
– host tropism **5A**, 400
– in HIV gene therapy **5A**, 405, 480f
– integration **5A**, 400
– in therapy of inoperable gliomas **5A**, 405
– in tumor therapy **5A**, 405f
– Kozak sequence **5A**, 402
– life cycle **5A**, 399
– manufacturing **5A**, 402ff
– – complementation systems **5A**, 402f
– packaging cells **5A**, 403
– packaging limit **5A**, 401
– particle structure **5A**, 401
– persistence **5A**, 400
– preclinical studies, *in vitro* **5A**, 404
– – *in vivo* **5A**, 404f
– producer cells **5A**, 402
– properties **5A**, 401
– purification **5A**, 404
– RCR contamination **5A**, 403f
– resistance to human complement – **5A**, 401
– specificity **5A**, 401
– stability in therapy **5A**, 401
– structure **5A**, 399
– subfamilies **5A**, 399
– transgene expression cassettes, design **5A**, 402
– – promoters **5A**, 402
– with two genes **5A**, 402
Revasc **5A**, 130, 169
reveromycins, interruption of eukaryotic cell
    growth **7**, 682
reversed micellar extractions **3**, 606ff
– affinity cosurfactants **3**, 609
– effects of, protein size **3**, 608f
– – salt concentration **3**, 608
– – salt type **3**, 608
– – surfactant concentration **3**, 608
– protein extraction, affinity-based – **3**, 609
– solvents **3**, 607
– surfactants **3**, 606f
reversed micelles **3**, 582, 606ff
– for amino acid extraction **3**, 583
– – interfacial partition coefficient Pcs **3**, 583
reverse genetics **5B**, 151
reverse gyrase, of hyperthermophiles **10**, 72
reverse micelles, lipases in **8A**, 48f
– product recovery from **8A**, 49
– stabilization by anionic surfactants **8A**, 48f
reverse osmosis **11A**, 227f
– desalination by **11C**, 405
– membrane structure **11C**, 406f
reverse osmosis (RO) **3**, 440, 487
– in production water treatment **3**, 756
reverse transcriptase inhibitors, from basidiomy-
    cetes **7**, 509ff
– – structures **7**, 510f
revolving tubular reactors *see* rotating drum reac-
    tors
Reye syndrome **5B**, 69
Reynolds number **7**, 11f, 369, 619, **11A**, 499
– impeller type effects **3**, 406
– membrane aeration **3**, 248
– turbulence in bioreactors **7**, 115

RFLP *see* restriction fragment length polymor-
    phisms
rhamnolipids **6**, 662f, **7**, 180f
– *Pseudomonas* sp. **6**, 663
– structure **7**, 181
– surfactant formation by *Candida* spp. **7**, 180
rhamnulose-1-phosphate aldolase, overexpression
    of **8B**, 14
rhamsan **6**, 620
rhamsan gum, gel formation **6**, 643
rheumatoid arthritis (RA) **5A**, 140, 296, 318f
– animal models **5A**, 344f
– antibody therapy **5A**, 344
– CDP571 **5A**, 347ff
– medical applications of recombinant proteins
    5A, 143
– monoclonal antibody therapy **5A**, 315, 318
– role of tumor necrosis factor **5A**, 344, 346
– treatment with CDP571 **5A**, 346ff
– – repeated dose studies **5A**, 347ff
– – results **5A**, 349f
– vaccines **5A**, 138
rhizobacteria **11B**, 394
– in weed control **10**, 490
– production of siderophores **10**, 488
rhizobactin DM4, carboxylate-type siderphore of
    Rhizobium meliloti **7**, 223
– structure **7**, 223
rhizobactin 1021, dihydroxamate siderophore of
    Rhizobium meliloti **7**, 221
Rhizobium heteroglycans, structure **6**, 650
Rhizobium leguminosarum **2**, 444ff
– genetic map **2**, 448
– physical DNA map **2**, 448
Rhizobium meliloti, rhizobactin DM4 **7**, 223
– rhizobactin 1021 **7**, 221ff
Rhizobium sp., biopolymer formation **6**, 617
rhizodeposition, in wetland systems **11A**, 244f
rhizoferrin **7**, 229f
– fungal siderophores **7**, 229
– structure **7**, 229
rhizofiltration, definition of **11B**, 386
– of metals **11B**, 392ff
– – improvement of **11B**, 394
Rhizomucor lipases, resolution of secondary alco-
    hols **8A**, 77
Rhizomucor miehei, chymosin formation **9**, 680
Rhizomucor miehei lipase (RML) **8A**, 42f
– enantioselective reactions catalyzed by second-
    ary alcohols **8A**, 85
– general features **8A**, 44
Rhizopus **1**, 522
– amyloglucosidase **9**, 745
Rhizopus arrhizus, enzymatic reduction by **8B**,
    371f
– hydroxylation of progesterone **8A**, 477
Rhizopus chinensis, in lao-chao fermentation **9**,
    525

– in sufu fermentation **9**, 524
Rhizopus microsporus, in oncom fermentation **9**,
    535
– in tempeh fermentation **9**, 531ff
Rhizopus nigricans, esterase of **8A**, 219
Rhizopus oryzae, in lao-chao fermentation **9**, 525
– in tempeh fermentation **9**, 530
– lactic acid formation **6**, 302
Rhizopus oryzae lipase (ROL) **8A**, 41ff
Rhizopus sp., fumaric acid production **6**, 370
– glucoamylase **6**, 71
Rhizopus stolonifer, infection of wine grapes **9**,
    500
rhizosphere **11A**, 242ff
– aerobic microbial processes **11B**, 378
– biodegradation of pollutants by bacteria **11B**,
    377ff
– interactions of roots/rhizomes with the soil ma-
    trix **11A**, 247
– in wetland systems **11A**, 242ff
– microbial community of **11A**, 248
– oxygenation of **11A**, 245f
– plant-pseudomonad interaction **1**, 422
rhodamine family, fluorescent dyes **5B**, 188, 191
rhodanese, folding **3**, 539, 542
– hydrophobicity **3**, 549
– refolding, temperature sensitivity **3**, 547
rhodium polypeptide complex, stereoselective hy-
    drogenation by **8B**, 507f
rhodococci, nitrilase activity **8A**, 295
– nitrile hydratase activity **8A**, 284ff
Rhodococcus, isoprene degradation by **11B**, 204f
Rhodococcus coprophilus, Baeyer–Villiger bio-
    transformations **8A**, 560
Rhodococcus erythropolis, trehalose lipid forma-
    tion **6**, 663
Rhodococcus erythropolis carbonyl reductase,
    applications **8A**, 456
– characteristics of **8A**, 456
Rhodococcus fasciens, Baeyer–Villiger biotransfor-
    mations **8A**, 560
Rhodococcus ruber, PHA synthesis from lignite
    depolymerization products **10**, 171f
Rhodococcus sp., bio-de-emulsification **6**, 702
Rhodophyta, characteristics of **10**, 111
Rhodopseudomonas palustris **1**, 23
rhodoquinone **1**, 404
Rhodospiridium toruloides, fatty acyl composi-
    tion **7**, 152
Rhodospirillum rubrum, pathways of PHA biosyn-
    thesis **6**, 415
Rhodotorula, in ogi fermentation **9**, 545
Rhodotorula glutinis, lipid accumulation patterns
    7, 140f
– phenylalanine-ammonia lyase **8B**, 139ff
– production of L-phenylalanine **8B**, 141
rhodotorulic acid, fungal siderophores **7**, 228
– structure **7**, 228

ribavirin, formation, by *Enterobacter aerogenes* **6**, 596f
– – by *Erwinia carotovora* **6**, 596f
ribitol, formation by yeasts **6**, 222
riboflavin, production **3**, 473
– structure **8A**, 536
riboflavine 5′-phosphate (FMN) **1**, 71
ribonuclease (RNase), storage, after freeze-drying **3**, 704
ribonucleic acid *see* RNA
Ribosomal Database Project **5B**, 284
ribosomal RNA *see* rRNA
ribosomal RNA (rRNA) **1**, 98
– superfamily **1**, 403
ribosome **1**, 99, **2**, 245
– aggregation **2**, 680ff
– eukaryotic **2**, 245, 680f
– prokaryotic **2**, 245
– signal recognition particle **2**, 691
– subunits **1**, 99
ribostamin family, structure of – **7**, 467
ribostamycins **7**, 434
ribozyme **2**, 355, 592
– anti-sense effect **2**, 593
– hammerhead formation **2**, 592
ribozymes, and virus resistance in plants **10**, 493
– in HIV gene therapy **5A**, 474ff
ribulose monophosphate pathway **9**, 179f
– of methylotrophic bacteria **9**, 180
ribulose monophosphate (RuMP) cycle **1**, 84, 276
– dissimilatory **1**, 275
rice, as brewing adjunct **9**, 429
– genome **5B**, 147f
ricin **5A**, 34
ricinine, bioconversion to carboxylic acid compounds **8A**, 297
rifamycin, chemical structure **10**, 342
right not to know **5B**, 420
Rio "Earth Summit" **12**, 627f
risk, definition **12**, 101f, 160
– perceptions **12**, 658
– – governmental **12**, 471
– – public **12**, 471
risk assessment **2**, 839, **12**, 349f, 453f, 464
– concepts **12**, 39ff
– factors for recombinants **2**, 845
– – donor characteristics **2**, 845
– – environmental considerations **2**, 846
– – health considerations **2**, 845
– – modified organism characteristics **2**, 845
– – parental characteristics **2**, 845
– – recipient characteristics **2**, 845
– field release of GMOs **12**, 184ff, 528
– for GMO products **12**, 563ff
– GMAG **12**, 516f
– governmental **12**, 486f
– horizontal approach **12**, 58f
– in contaminated soil areas **11B**, 19f, 32f, 38

– of genetic manipulation **12**, 574
– of rDNA products **12**, 157ff
– research, in Europe **12**, 540, 608
risk classification system **2**, 840ff
– for microorganisms **2**, 840ff
risk groups, bacteria **12**, 73f
– fungi **12**, 77
– viruses **12**, 75f
risk scenarios **12**, 463ff
ristocetin A **7**, 372f
– structure **7**, 373
RITE (Research Institute for Environmental Technology for the Earth) **12**, 386
river bank filtration, effect on BOM **11C**, 463
river water, treatment of **11C**, 379
– turbidity of **11C**, 380
RML *see Rhizomucor miehei* lipase
RNA **2**, 193ff, 234, 320
– anti-sense **2**, 589f, 643
– – action mechanism **2**, 590, 643
– capping **2**, 676
– – effect on translation efficiency **2**, 684
– cleavage **2**, 260
– editing **2**, 243f
– fingerprint **2**, 260
– *in vitro* selection **2**, 350
– oligonucleotide fractionation **2**, 260
– satellite- **2**, 643
– – modification of virus **2**, 643
RNA decoys, for HIV gene therapy **5A**, 476
RNA degradation **2**, 675ff
– of transferrin/ferritin receptor mRNA **2**, 679
– signals **2**, 678
RNA–DNA hybridization **5B**, 228
RNA editing **5A**, 49
RNA hybridization **2**, 404
RNA inhibitors, in HIV gene therapy, antisense RNA **5A**, 474f
– – decoy RNA **5A**, 476
– – ribozymes **5A**, 474f
RNA labeling **2**, 260, 397
– indirect labeling **2**, 262
RNA measurement, flow cytometry **7**, 204f
– sample-flow analysis **7**, 205
RNA operon, stable RNA gene organization **2**, 238
– stable RNA processing **2**, 238
RNA polymerase **1**, 105, 393, **2**, 234f
– core enzyme **2**, 234f
– σ-factor **2**, 234
– – prokaryotic **2**, 235f
– – specificity **2**, 235
– holo-enzyme **2**, 234
– in eukaryotes **2**, 234
– in prokaryotes **2**, 234
– in yeast **1**, 492
– of *E. coli*, σ^2 subunit **7**, 64
– σ-factor **1**, 105, 393

– subunits **2**, 234f
RNA polymerase I **2**, 234
RNA polymerase II **2**, 234, 666ff, 670f
– cofactors **2**, 670f
– DNA control elements **2**, 667
RNA polymerase III **2**, 234
RNA polymerization **1**, 97ff
RNA processing **2**, 237f, 675f
– in prokaryotes **2**, 237f
– polyadenylation **2**, 677
RNA replicase **1**, 554
RNAse **2**, 238, 260
– types in *E. coli* **2**, 239
RNA sequencing **2**, 259, **5B**, 185
– history **2**, 259
RNA splicing **2**, 676, **12**, 164ff
– dependence on transcription promoter **2**, 676
– splice acceptor **2**, 677
– splice donor **2**, 677
RNA synthesis **2**, 328
– 2′-protecting groups **2**, 328
RNA transport **2**, 675f
robotics **5B**, 432
– in DNA sequencing **5B**, 202
ROL *see Rhizopus oryzae* lipase
roller bottle, for mammalian cell cultures **3**, 111f
– oxygen concentration **3**, 240
– surface aeration **3**, 240
Romano cheese **9**, 378
root culture, hairy root culture **1**, 601
– of *Hyoscyamus albus* **1**, 601
– secondary metabolite production **1**, 601f
– transformation effect **1**, 601f
root systems, exudates **11B**, 373
Roquefort cheese **9**, 377
Rosenbrock method **7**, 458
rosmarinic acid, from plant cell cultures **7**, 598ff
– – yields **7**, 599
– production by *Coleus blumei* cultures **1**, 590f
– – sucrose effect **1**, 594
– structure **7**, 599
Rossman-like folds **5B**, 351
rotamases **7**, 547f
Rotameter **7**, 59
rotary shear filtration, enzyme transmission rate **3**, 498
rotary vacuum filter, baker's yeast dewatering **9**, 339f
rotated drum reactor **1**, 587f
rotating biological contactors **7**, 429f, **11A**, 346f
– design guidelines **11A**, 346f
– hydrodynamics of the liquid film **7**, 430
– process development **11A**, 346
– wastewater treatment, modelling **7**, 429
– – with nitrification **11A**, 346
– – with nitrification and denitrification **11A**, 346f
– – without nitrification **11A**, 346

rotating disc filters **11A**, 346f
– denitrification rates **11A**, 347
rotating disc reactors **11A**, 9
rotating drum *see also* bioreactors
– for solid state processing **11B**, 333
– – scheme **11B**, 333
rotating drum composting **11C**, 144
rotating drum reactor, use in composting **11C**, 123
rotating drum reactors **11B**, 343f
rotifers, occuring in biofilms **11A**, 357
rotor biofilters **11C**, 346f
– scheme **11C**, 347
rotting processes *see* composting
Rous sarkoma virus (RSV) **2**, 787
Royer report **12**, 587ff
rRNA **2**, 238
– comparative sequence analysis **11A**, 86
– dot blot hybridization **11A**, 91
– evolutionary history **11A**, 86
rRNA-based analysis **11A**, 87ff
– amplified ribosomal DNA restriction analysis **11A**, 87
– genetic fingerprints **11A**, 87
– restriction fragment length polymorphisms **11A**, 87
rRNA-based methods, denaturing gradient gel electrophoresis **11A**, 88ff
– detection of *Zoogloea ramigera* **11A**, 101
– dot blot hybridization **11A**, 97f
– for analyzing of microbial communities in wastewater treatment **11A**, 86ff
– hybridization **11A**, 91ff
– *in situ* hybridization **11A**, 98ff, 103f
– population analysis **11A**, 101
– scheme **11A**, 97
– terminal restriction fragment length polymorphisms (T-RFLP) **11A**, 90f
rRNA databases **11A**, 91
rubber, biodeterioration of **10**, 308
Rubia akane, commercial purpurin production **1**, 589, 604
rubiflavin (vitamin B$_2$), microbial production of **10**, 320ff
Ruppel number **7**, 12
Rushton turbine impeller *see* impeller
Russia, biotechnology regulations **12**, 605
Rutgers process **11C**, 138f
Rutgers strategy, of composting **11C**, 47
rye, analysis of **6**, 63
– as raw material for ethanol production **6**, 63
– autoamylolytical quotient **6**, 63
– autoamylolytic enzyme system **6**, 75, 96
– processing in ethanol production **6**, 92
– production of **9**, 243
– stillage, nutrient content **6**, 106
rye bread *see* sour dough
rye flour doughs, acidification **9**, 264f, 285ff, 299f

S

SABRE™ *see* J. R. Simplot Anaerobic Bioremediation
saccharide metabolization, in clostridia **1**, 298
saccharide transport **1**, 75, 385f
– in bacilli **1**, 385f
saccharide uptake system, in clostridia **1**, 298
saccharification, of liquefied starch **9**, 669, 745, 750ff
Saccharomyces **2**, 121f
– – *see also* yeast
– genetic engineering **2**, 508ff
Saccharomyces bayanus, wine yeast strains **9**, 338
Saccharomyces carlsbergensis see brewers' yeast
Saccharomyces cerevisiae **1**, 115, 197ff, 470ff, 478ff, 525, **2**, 77, 508ff, **12**, 86ff
– – *see also* baker's yeast, brewer's yeast, wine yeast, yeast
Saccharomyces cerevisiae see als yeast
– – see bakers' yeast
Saccharomyces cerevisiae, accumulation of sterols **7**, 172
– adaptive optimization **4**, 552ff
– alcohol dehydrogenase activity **1**, 238
– anaerobic and aerobic catabolism **6**, 124
– as biosuspension model **3**, 194f
– – mass transfer coefficients **3**, 198
– as host cell **3**, 285, 530
– ATP formation **6**, 599
– biological features **12**, 87
– biological properties **4**, 387ff
– biotechnological application **2**, 508
– brewer's yeast **9**, 435
– budding **1**, 115f
– carbohydrate metabolism **1**, 478ff
– – regulation **1**, 486ff
– cell homogenization **3**, 509, 516
– chromosome telomeres **5B**, 21
– continuous culture **4**, 340
– cosmid libraries **5B**, 21
– covalent binding effect **1**, 238
– diaux growth **1**, 154
– diauxic batch growth, metabolic regulator model **4**, 296
– entire genome **5B**, 20
– ERG9-mutation **10**, 581
– expression vectors **9**, 80f
– fermentation of sugars **6**, 123
– formation of ergosterol **7**, 172
– galactose metabolism, regulation **1**, 493
– gene duplications **5B**, 23
– genes **5B**, 22
– genetic map **5B**, 20
– genome project **5B**, 20ff
– glycerol production, with free cells **6**, 210ff
– – with immobilized cells **6**, 213f
– glycolytic enzymes, genes **1**, 480
– glycosylation of heterologous proteins **9**, 85
– glyoxylate cycle, genes **1**, 484
– growth conditions **4**, 150
– growth kinetics **1**, 133
– growth modeling **3**, 87
– growth simulation, metabolic regulator model **4**, 295
– gylcerol production, economics **6**, 223
– hexokinase activity **1**, 236
– high-resolution physical maps **5B**, 21
– immobilization effect **1**, 230ff
– – metabolic changes **1**, 236
– in grape must **9**, 466ff
– in ogi fermentation **9**, 545
– in puto fermentation **9**, 528
– in sour dough **9**, 297
– in sour dough starters **9**, 255ff
– integral genetic redundancy **5B**, 23
– invertase release, lytic enzyme-dependent – **3**, 518
– in waries production **9**, 540
– karyotypes **5B**, 20
– killer plasmids **5B**, 22
– life cycle **1**, 472
– macromolecular composition **1**, 235f
– metabolism **1**, 478ff
– metallothionein **10**, 239
– micrograph **5B**, 20
– mitochondrial genome **5B**, 22
– Mn^{2+} transport system **10**, 237
– occurrence of poly(3HB) **6**, 412f
– open reading frames (ORFs) **5B**, 22
– oxygen consumption **1**, 228
– pentose phosphate pathway **1**, 483
– periodicity of the GC content **5B**, 23
– phosphofructokinase activity **1**, 237f
– postgenome era **5B**, 23f
– production of **1**, **4**, 539
– – bacterial proteins **1**, 498
– – eukaryotic proteins **1**, 498
– – fungal proteins **1**, 498
– – human proteins **1**, 496f
– – mammalian proteins **1**, 497
– – viral proteins **1**, 498
– productivity **6**, 135f
– promoters **9**, 82ff
– – constitutive – **9**, 83
– – hybrid – **9**, 84
– – regulated – **9**, 83f
– protein modification **2**, 508
– pyruvate decarboxylase **8B**, 50ff, 54
– quality control **5B**, 21
– recombinants, in fermentation processes **3**, 285
– repeated sequences **5B**, 23
– respiration **1**, 484
– secretion of heterologous proteins **9**, 84
– secretory pathway **2**, 508
– seeding culture **6**, 152

– selectable markes **9**, 81f
– species- or genus-specific functions **5B**, 23
– starch degradation **1**, 200
– strain improvement **1**, 197ff
– strains for wine/beer production **1**, 496
– subtelomeric regions **5B**, 23
– TCA cycle **1**, 484
– transformation **12**, 91f
– transformed –, screening of steroid receptors **7**, 111
– transposable elements (Tys) **5B**, 23
– tryptophan pathway **9**, 122f
– – activities of enzymes **9**, 123
– wine yeast strains **9**, 338
– xylose metabolization **1**, 198f
– 2m plasmid **1**, 476
Saccharomyces chevalieri, in olive fermentation **9**, 607
Saccharomyces Genome Database **5B**, 285
Saccharomyces rouxii, formation of polyols **6**, 220
– glycerol production **6**, 216f
Saccharomyces uvarum, ethanol formation, kinetic parameters **6**, 132
– – specific ethanol production rate **6**, 132
– fermentation of sugars **6**, 123
Saccharomyces uvarum (*carlsbergensis*), brewer's yeast **9**, 435
Saccharomycodes ludwigii, in wine spoilage **9**, 496
Saccharopolyspora, production of antibiotics **1**, 445
Saccharopolyspora erythraea, polyketide synthase **7**, 33
sacrificial anodes **10**, 298
SADA (Autonomous Service for the Environment Development of Amazon Federal Territories) **12**, 439, 443
S-adenosylhomocysteine, microbial production of **10**, 331
S-adenosyl methionine **9**, 32
– production of, by yeast fementation **10**, 331
safety, in biotechnology **2**, 836ff
– – basic concepts **2**, 839
– – industrial process containment **2**, 850
– – precautions **2**, 849ff
– objectives **2**, 839
– review **2**, 839
safety aspects, preclinical testing of antibodies **5A**, 303ff
safety categories, for GMOs **12**, 49ff
safety education **2**, 840
safety factor, of raw sugar **6**, 10
safety measures **2**, 839
– biological containment **2**, 839
– technical containment **2**, 839
safety precaution system **2**, 848ff
– containment category **2**, 848
– precaution measure **2**, 850

saframycin A, antitumor agent, structure **7**, 662
SAGB (Senior Advisory Group for Biotechnology) **12**, 586, 634ff
– Yellow Book **12**, 652f
salivaricin A, prepeptide structure **7**, 345
– primary structure **7**, 332
– structure **7**, 335
Salmonella, binary fission **1**, 114
– in composting material **11A**, 71
Salmonella newport, specific growth rate **1**, 130
salmonellas, fecal –, in biogas plants **11A**, 228ff
Salmonella typhimurium **2**, 209
– α,β-dihydroxyacid dehydratase **8B**, 112
– strictly aerobic mutant **2**, 210
– strictly anaerobic mutant **2**, 210
salt, use in baking **9**, 252
sample channels **5B**, 201f
sampled-data control system **4**, 252ff
– configuration of **4**, 253
– controllability region **4**, 253
– state-space vector **4**, 253
sample distortion **11B**, 487
sample-flow analysis **4**, 199f
– autoanalyzer **4**, 199, 207f
– cell-viability determination **4**, 202
– DNA measurement **4**, 205
– FIA system **4**, 199, 207f
– pH_i measurement **4**, 203f
– protein monitoring **4**, 207f
– quasi-on-line methods **4**, 199
– RNA measurement **4**, 205
– thermistor **4**, 208
sample injection techniques, classification of **11B**, 502
sample introduction techniques **5B**, 256
sample preservation, parameters **11B**, 487
sample pretreatment **4**, 56ff
sample processing, particle size reduction **11B**, 497f
sampling, for monitored natural attentuation **11B**, 404
– of solid matter **11B**, 477ff
– optimization of **4**, 252ff
– – sensor allocation **4**, 262
– theorem **4**, 252
– time **4**, 252ff
sampling module **4**, 152f
– Bio-filtrator **4**, 152
– Biopem **4**, 152f
– discus-shaped **4**, 155
– Millipore **4**, 152f
– rod-shaped **4**, 153, 155f
sampling period, determination of **4**, 253
sampling points **11B**, 484ff
– localization of epicenters **11B**, 486
sampling protocol **11B**, 488f
sampling strategy **11B**, 482ff
– judgement-based – **11B**, 483f

– statistical – **11B**, 483f
sand, as carrier materials of biofilters **11A**, 466
sanger dideoxy sequencing **5B**, 185f
sanguinarine, biosynthetic pathway **8A**, 338ff
– from plant cell cultures, yields **7**, 606
– production in cell culture **1**, 593
sanitary landfill gas, composition of **11A**, 199
sanitary landfilling, control **11A**, 180f
– – of leachate discharge into the environment
  11A, 180
– – of waste input **11C**, 174
– – of waste pretreatment **11C**, 174ff
sanitary landfills **11C**, 6ff
– – *see also* landfills
– aging of **11C**, 11
– – mass loss **11C**, 12
– – relative metabolite concentration **11C**, 12
– biochemical processes in **11C**, 169ff
– biogas from **11C**, 13f
– – composition of **11C**, 13f
– biogas production in **11C**, 14
– classes **11C**, 7ff
– control, landfill reactor **11C**, 178
– – water input **11C**, 178
– cost per ton **11C**, 196
– deposition of garbage **11C**, 11ff
– design features **11C**, 198
– emission of methane **11C**, 15
– environmental implications **11C**, 191ff
– gas management **11C**, 199
– leachate composition **11C**, 169ff, 200
– leachate constituents **11C**, 172f
– – effect of landfill age **11C**, 173
– – effect of pH **11C**, 173
– – effect of waste composition **11C**, 173
– leachate management **11C**, 199f
– leachate recycle **11C**, 200
– leachate treatment **11C**, 14
– location requirements **11C**, 197
– long-term emissions **11C**, 185
– long-term problems with leachates **11C**, 167ff
– long-term stability **11C**, 191ff
– microbial reactions **11C**, 13
– of solid wastes **11C**, 167ff
– operating criteria **11C**, 197
– overall mass flow **11C**, 13
– regulations for **11C**, 6f
– stabilization, stages of **11C**, 201
– U.S. regulations **11C**, 197
sanitary sewer, biological waste disposal **3**, 783
– – public health implications **3**, 784
sanitation, of sewage sludge **11C**, 219ff
sanitation concepts **11C**, 250ff
– based on biogas plants **11C**, 254f
– differentiation of **11C**, 250
– integrated – **11C**, 251ff
– – pilot project **11C**, 251ff

sanitation systems, comparison of innovative and
  conventional, emissions **11C**, 254
– – energy consumption **11C**, 254
– – material intensity **11C**, 254
– innovative – **11C**, 251
saphenous vein grafts **5A**, 371f
– treatment with ReoPro **5A**, 371
saponification number **10**, 165
saponin aglycone, from plant cultures **7**, 611f
saponins, from plant tissue cultures **7**, 611f
saprobe index **11A**, 8
sarcomere **5A**, 31
Sargrastim **5A**, 157
Sartina albida, thymidine production **6**, 594
Saruplase **5A**, 168
SASP *see* small acid-soluble DNA binding protein
  satellite RNA
– – *see* small acid soluble protein
satellite RNA **2**, 572
satellite virus **2**, 572
satelloid **2**, 572
sauerkraut **9**, 639ff
– production statistics **9**, 632
– sensory properties **9**, 644
– spoilage **9**, 644f
– – discoloration **9**, 645
– – off-flavor **9**, 644
– – softening **9**, 645
sauerkraut fermentation **1**, 333, **3**, 312
– flora **3**, 312
sausage fermentation **1**, 332, **3**, 311
– additives **3**, 311
– broth formulation **3**, 311
scale factor **5B**, 389
scale-up **4**, 301f
– of bioprocesses *see* bioreactor
– of control strategies for fermentation processes
  4, 341
– of stirred tanks, common criteria **4**, 301
– – hypothetical concepts **4**, 303
Scandinavia, biotechnology regulatory policy **12**,
  602ff
scanning electron microscopy **1**, 9
– conventional – **1**, 12
– principle **1**, 12
– sample preparation **1**, 9
– transmission – **1**, 12
scatchard plot **9**, 48f
scenario, assessment **12**, 17, 463ff
Scenedesmus, in human diet **10**, 126
schizokinen **7**, 220f
schizokinen A, structure **7**, 221
schizonellins, structure **7**, 504
schizophrenia **5B**, 71, 80
Schizophyllum commune, scleroglucan production **6**, 645f
Schizosaccharomyces pombe **1**, 473, **2**, 77, 522
– fusion product **9**, 326

– malate metabolism **9**, 490
Schmidt number **4**, 369, 619
Schwanniomyces occidentalis **1**, 474
– α-amylase gene (*AMY1*) **1**, 201
– glucoamylase gene (*GAM1*) **1**, 201
science policy, governmental, history **12**, 28ff
– industrial **12**, 26ff
– information **12**, 33f
– interactive **12**, 30ff
– OECD activities **12**, 29f
– origins **12**, 26ff
– U.S. Congress **12**, 7ff
Scientific and Technical Research Committee *see*
   CREST
scintillation proximity assay, in drug screening
   10, 45
scleroglucan, fungal polymer, precipitation with
   polar solvents **6**, 626
– production by fungal species, *Schizophyllum
   commune* **6**, 645
– – *Sclerotium rolfsii* **6**, 645
– structure **6**, 635, 646
– suggested uses **6**, 646
Sclerotium rolfsii, scleroglucan production **6**, 645f
SCOPE (Scientific Committee on Problems of the
   Environment) **12**, 46
scopolamine, biosynthesis of **8A**, 332ff
– from hairy root cultures **7**, 620
– from plant cell cultures **7**, 613
– production in transgenic *Atropa belladonna*
   plants **7**, 625f
– structure **7**, 624
scorodonin **7**, 502f
– antifungal activity **7**, 503
– structure **7**, 502
scorpions, as sources of nonribosomal peptides **7**,
   314
– as sources of ribosomal peptides **7**, 301
screen filtration **3**, 440
screening **8A**, 31, **10**, 44
– for enzymes, in extremophiles **10**, 73f
– of bioactive metabolites, from basidiomycetes
   7, 496
– of compost **11C**, 115f
– of dalbaheptides **7**, 381
– of lipases **8A**, 149f
– of natural products **7**, 3ff
– – compilation of – **7**, 4ff
– of plant secondary metabolites **7**, 595
– of receptor-active compounds **7**, 107ff
screening libraries **10**, 55ff
– diversity design **10**, 55f
– tools for library tuning **10**, 55
screening methods **10**, 41ff
screening of abzymes **8B**, 445
screw mills, disintegration of organic wastes **11C**,
   114f
– function of **11C**, 115

scrubbing, of soil **11B**, 310
SDS/gel electrophoresis **5B**, 272
sea hares, as sources of nonribosomal peptides **7**,
   314
seals, for bioreactors **3**, 421
search methods, profile-based – **5B**, 135
– sequence-based – **5B**, 135
seaweeds **10**, 445
– production of, agar **10**, 466
– – alginates **10**, 466
– use as food **10**, 462
secondary alcohols, acyclic –, resolution by PPL
   8A, 76
– as lipase substrates **8A**, 68ff
– – binding site **8A**, 70
– cyclic –, enantioselective reactions catalyzed by
   Pseudomonas lipases **8A**, 82ff
– – resolution by CRL **8A**, 74
– – resolution by PPL **8A**, 77
– enantioselective reactions catalyzed by RML
   8A, 85
– enantioselectivity of, CAL-B **8A**, 71f
– – CRL **8A**, 72ff
– kinetic resolution of, by penicillin acylase **8A**,
   230ff
– – by pig liver esterase **8A**, 202
– phenyl acetyl esters of **8A**, 231
– pyridyl acetyl esters of **8A**, 232
– resolution of acyclic –, by CAL-B **8A**, 72
– – by CRL **8A**, 73
secondary alcohols (racemic-), fastest reacting enantiomers with cholesterol esterase **8A**, 216
secondary metabolic aminotransferases **7**, 422
secondary metabolism, evolution of – **7**, 16
– functions **1**, 444
– general aspects **7**, 1ff
– negative feedback control **7**, 30
– of plant cell cultures **1**, 589ff, 595ff
– overproduction of enzymes, in transgenic plant
   cell cultures **7**, 627
– pathway redesign **1**, 214
– peptide bond formation **7**, 36
– regulation of – **7**, 17ff
– – in eukaryotes **7**, 22f
– regulatory molecules, induction of cytodifferentiation **7**, 16
secondary metabolite formation, effect, of nutrient
   depletion **7**, 24ff
– – of stress conditions **7**, 24ff
secondary metabolites, adaptation to metabolic
   imbalances **7**, 13
– agricultural uses **7**, 9ff
– – plant growth regulators **7**, 9
– as a signal of interspecific communication **7**,
   13, 15
– as drugs **7**, 2ff
– – antimicrobial agents **7**, 4ff
– – antitumor agents **7**, 6f

– as endogenous signals **7**, 13
– as metabolic reserves **7**, 13, 15
– biological role **7**, 493
– biosynthesis, biosynthetic modifications **7**, 37f
– – use of cell-free enzymes **7**, 37
– biosynthetic genes clusters **7**, 17ff
– enhancement by metabolic engineering **7**, 623ff
– – nicotine **7**, 624f
– – scopolamine **7**, 625f
– – serotonin **7**, 624
– expression of regulatory mechanisms **7**, 23ff
– formation from a single precuror **7**, 31f
– function **1**, 165
– genetic instability **7**, 26f
– immunosuppressors **7**, 7
– incorporation into cellular structures **7**, 15
– main biosynthetic pathways **7**, 31ff
– overproduction of – **7**, 29f
– pharmacological activities **7**, 7ff
– precursor pools of – **7**, 29f
– producing organisms **7**, 4ff
– production in plant cell cultures **7**, 593ff
– properties **7**, 4ff
– regulation in streptomycetes **7**, 443ff
– roles in producing organisms **7**, 11ff
– structural classification **7**, 38ff
– structural type **7**, 4ff
– sugar components in – **7**, 397ff
secondary pathways, single site manipulations **7**, 626
secondary pollutants, treatment of **11A**, 12f
secondary structure *see* protein structure
– prediction of **5B**, 337f
second messenger assays, in drug screening **10**, 49f
second registrations **12**, 301
secretion **1**, 76
– in corynebacteria **1**, 454
– metabolite transport system **1**, 76
sedimentation, effects of, density **3**, 475
– – viscosity **3**, 475
– inclined sedimenters, for cell retention **3**, 115
– in drinking water treatment **11C**, 386ff
– – settling tanks **11C**, 387f
– in lamella settlers **3**, 476
– rate **3**, 475
– tank, throughput **3**, 477
– theory **3**, 475
sediments, contaminated –, treatment of **11B**, 330ff
seepage trench, scheme **11C**, 431
segregating population *see* marker segregating population
segregating populations **5B**, 166
seldomycin family, structure of – **7**, 469
seldomycins, biosynthetic pathway **7**, 433
selectable marker, in fungi **1**, 536
selective marker **2**, 49

selenides, oxidation to selenoxides by cyclohexanone monooxygenase **8A**, 574
selenium, as electron acceptor for anaerobic respiration **11A**, 43
– phytoextraction of **11B**, 389, 391
– removal by phytoextraction **11B**, 389
selenocysteine **2**, 243f
– regulation of incorporation **2**, 244
selenoxide, *syn*-eliminations, parameters for **8B**, 421
– – substrate desolvation **8B**, 419ff
SELEX_DB **5B**, 410
self-fertilizing plants, for gene linkage studies **2**, 153
self-heated hay **11C**, 66
self-heated material, microorganisms in **11C**, 61ff
Semiconductor, gas analysis **4**, 69f
semi-continuous culture **3**, 28
semimetals, in municipal wastewater **11A**, 179f
sensor allocation, tower loop reactor **4**, 261f
sensors *see* bioreactor instrumentation, fermentation instrumentation
separation membranes *see* filtration membranes
separation techniques *see* cell separation, protein purification, product purification
separators, for downstream processing **3**, 474ff
sepsis **5A**, 134f, 330f
– animal models **5A**, 335f
– anticytokine therapy **5A**, 333
– clinical trial design **5A**, 335ff
– corticosteroid therapy **5A**, 334
– definitions **5A**, 336f
– medical applications of recombinant proteins **5A**, 134
– monoclonal antibody therapy **5A**, 316f, 329ff
– – anti-endotoxin monoclonal antibodies **5A**, 331f
– – anti-tumor necrosis factor **5A**, 332f
– – clinical experience **5A**, 330
– – clinical trials **5A**, 333
– – rationale for **5A**, 330f
– preclinical studies **5A**, 335f
– therapy **5A**, 296
– use of polyclonal antibodies **5A**, 330
septicemia, in mice, effect of dalbaheptides **7**, 380
septic shock **5A**, 344
– treatment with CDP571 **5A**, 346f
sequence alignment **5B**, 202
– algorithms **5B**, 332
sequence alphabet abstractions **5B**, 402
sequence ambiguities **5B**, 393
sequence analysis *see* proteins sequence
sequence-based signals **5B**, 336
sequence cluster databases **5B**, 289
sequence comparison methods **5B**, 307ff
sequence conservation **5B**, 150
sequence data, quality control **5B**, 126
sequence databases **5A**, 53ff, **12**, 325ff

– and bioinformatics **12**, 331
– description of **12**, 326f
– – EMBL (EBI) Data Library **12**, 326
– – GenBank **12**, 326
– – PIR International Protein Sequence Database **12**, 326f
sequence homology searches **5B**, 366f
– tools **5B**, 367
sequence masking **5B**, 311
sequence motifs, pattern-based – **5B**, 334
– profile-based – **5B**, 334
sequence retrieval system **5B**, 291ff
– advanced features **5B**, 296ff
– application integration **5B**, 293f
– book marklets **5B**, 298
– data integration **5B**, 292
– free text description **5B**, 298
– linking **5B**, 292f
sequences, complexity of **5B**, 310
– detection of patterns **5B**, 313
sequence similarity **5B**, 133, 332
– analysis **5B**, 129
sequence-tag **5B**, 261
sequence tagged sites (STSs) **5B**, 90
sequencing batch reactor (SBR), aeration **11A**, 325f
– application of experience from continuous flow plants **11A**, 331f
– applications **11A**, 329ff
– comparison with continuous flow reactors **11A**, 328ff
– concentration profiles **11A**, 324
– current development status **11A**, 323ff
– cycle design **11A**, 325f
– decanter mechanisms **11A**, 327
– definition of **11A**, 322
– design **11A**, 330ff
– – based on mass balances for SBR **11A**, 332f
– historical development **11A**, 323
– inherent process dynamics **11A**, 332
– in municipal wastewater treatment **11A**, 321ff
– mass balances, equation **11A**, 329
– on-line control **11A**, 327f
– on-line measurement **11A**, 328
– operation of **11A**, 325
– phases **11A**, 323ff
– scheme of a plant **11A**, 324
– unaerated mixing **11A**, 325f
sequencing factories **5B**, 436
sequencing technology **5B**, 183ff
sequential kinetic resolution, lipase-catalyzed – **8A**, 64
sequential state/parameter estimation (SSPE) **4**, 238ff
– application, bioreactor contamination **4**, 244ff
– – yeast fermentation **4**, 241ff
– Gauss-Newton iterative searching method **4**, 239

– least square error method **4**, 238
– recursive least square error algorithm **4**, 239f
– – Gradient matrix **4**, 239ff
– – Hessian matrix **4**, 239ff
– schematic diagram **4**, 241
sequest **5B**, 261
Serial Analysis of Gene Expression (SAGE) **5B**, 149
D-serine dehydratase **8B**, 113
L-serine dehydratase **8B**, 113
L-serine hydroxymethyltransferase **6**, 531ff
– reactions catalyzed by – **6**, 531
serine pathway **1**, 277, **9**, 178f
serine production, by methylotrophic microorganisms **1**, 279
L-serine production **6**, 514
– by *Hyphomicrobium* strains **6**, 493
serine protease, heat-stable – **10**, 85ff
serine proteases **5A**, 206f
– characteristics **8A**, 248
serotonin **5A**, 34
– biosynthesis in plant cell cultures **7**, 624
– structure **7**, 624
Serpens, phylogenetic status **1**, 404
serpentine, from plant cell cultures **7**, 608f
– – yields **7**, 608
Serratia marcescens, histidine production **6**, 513
– proline production **6**, 513
– vanillin formation by **8B**, 336
serum, use in cell culture media **10**, 29f
serum-free media **5A**, 256f
sesquiterpenes *see also* terpenes
– biotransformation of **10**, 407ff
– dihydroxylation of **8A**, 513f
– hydroxylation of **8A**, 492f
setpoint optimization **4**, 523ff
– constant **4**, 523
– trajectory **4**, 524
settling tanks, in drinking water treatment **11C**, 387f
sewage *see also* municipal sewage
– bacterial pathogens in **11C**, 216ff
– biological waste content **3**, 783
– contact stabilization system **4**, 433ff
– fungi in **11C**, 217
– parasites **11C**, 217
– pathogenic yeast **11C**, 217
– secondary treatment methods **3**, 783
– viruses in **11C**, 217
sewage compounds, of natural origin **11A**, 169
– – synthetic – **11A**, 169
sewage particle sizes **11A**, 169
sewage pipelines **10**, 299
– corrosion tests, simulation of **10**, 302ff
sewage sludge **11A**, 161ff, 181ff
– aerobic glucose respiration **11A**, 21ff
– aerobic thermophilic stabilization **11C**, 221
– anaerobic glucose degradation **11A**, 23f

– and biowaste, joint fermentation of **11C**, 224
– bacterial pathogens in **11C**, 216ff
– bioleaching of **10**, 216
– chemical constitution of **11A**, 182ff
– cofermentation with biowaste **11C**, 23
– composition of **11C**, 130
– composting of, in reactors **11C**, 223
– – in windrows **11C**, 222f
– – optimal temperature **11C**, 42
– council directive on **11C**, 219f
– definition of **11A**, 181f
– disinfection of **11C**, 220ff
– disposal of **11A**, 187
– formation of **11A**, 182
– fungi in **11C**, 217
– German ordinances on **11C**, 220
– heavy metals, accumulation factors **11A**, 184
– – prescribed limits **11A**, 183f
– irradiation of **11C**, 223
– long-term storage of **11C**, 224
– organic compounds, concentration ranges **11A**, 183
– parasites **11C**, 217
– pasteurization of **11C**, 221
– pathogenic yeasts **11C**, 217
– processing in reed beds **11C**, 224
– raw sludge **11A**, 182
– safe sanitation of **11C**, 225
– sanitation of **11C**, 219ff
– stabilization **11A**, 21ff
– – aerobic glucose respiration **11A**, 21ff
– – anaerobic glucose degradation **11A**, 23f
– – biogas formation **11A**, 23
– thermal drying of **11C**, 223
– treating methods **11A**, 184ff
– treatment with lime **11C**, 222
– use of **11A**, 187
– viruses in **11C**, 217
sewage slurry, effects of anaerobic treatment **11C**, 231
– problems of **11C**, 231
sewage treatment **11A**, 399
sewage treatment plants, connecting grade **11A**, 143
sewer, effect on municipal sewage **11A**, 167
sewer biofilms, concentration of microorganisms **11A**, 72f
sewer networks, for wastewater collection **11A**, 7
sex chromosome, evolution **2**, 178
– in plants **2**, 178
– structure **2**, 178
sex factor **2**, 52
sexual hormone production, inhibitors of –, treatment of hormone-dependent tumors **7**, 685
sexual recombination **2**, 75ff
SFC *see* Supercritical fluid chromatography
shaker flask, mass transfer coefficients **3**, 240
– surface aeration **3**, 240

Shanghai Research Center of Biotechnology **12**, 398
SHE *see* Standard hydrogen electrode
shear stress, effect on plant cells **1**, 586
– effects on, microcarrier cultures **3**, 228
– – suspension cultures **3**, 230
– in animal cell cultures **3**, 228
– – bubble-induced damage **3**, 232f
– turbulent eddy dissipation **3**, 228
SheepBASE **5B**, 285
β-sheet **5A**, 59ff
Sherwood number **4**, 369, 619, **11A**, 498
– membrane aeration **3**, 248
shigellosis, as waterborne disease, outbreats in the 19th and 20th century **11C**, 415f
shii-take *see* Lentinus edodes
shikimic acid, from plant cell cultures **7**, 603
shikonin production, by cell cultures **1**, 589, 592
– copper effect **1**, 594
shikonins, from plant cell cultures **7**, 601f
– – yields **7**, 601
Shine–Dalgarno motifs **5B**, 388
Shine-Dalgarno sequence (SD) **2**, 246, 412
Shine–Delgarno sequences **5B**, 365
shock, clinical definition **5A**, 337
shoot culture, of *Apocynum cannabinum* **1**, 601
shredder machine, mobile – **11C**, 116
sialic acid, structure of **8B**, 16
– synthesis by sialyl aldolase **8B**, 257ff
sialic acid aldolase **8B**, 15ff
sialic acids, structure **6**, 651
sialyltransferases **8B**, 259f
sib pairs **5B**, 80
sickle cell anemia **5B**, 64, 110
– diagnosis with PCR **2**, 340
sideramines, ferric ion-chelating compounds **7**, 493
siderochelin A, catecholate siderophore of *Nocardia* sp. **7**, 208f
– structure **7**, 208
siderophore **1**, 180
– production by pseudomonads **1**, 423
– receptors in pseudomonads **1**, 423
– synthesis regulation **1**, 180
siderophore-mediated iron transport **7**, 233ff
siderophores *see also* bacterial siderophores, fungal siderophores
– as growth inhibitors **10**, 488
– binding of metals and radionuclides **10**, 235
– biological properties **7**, 226
– carboxylate-type **7**, 223ff
– catecholate-type **7**, 201ff
– citrate hydroxamates **7**, 220ff
– ferric uptake regulation proteins **7**, 200
– history **7**, 201
– hydroxamate-type **7**, 209ff
– iron(III) binding ligands **7**, 201

– metal-to-ligand charge transfer bands, coloration **7**, 200
– mycobactins **7**, 217ff
– peptide-type **7**, 212ff
– role in iron nutrition **7**, 235
sidrophores **7**, 14, 199ff
Siemens KWU process **11C**, 26ff
– energy balance **11C**, 30
– input and output data **11C**, 29
– scheme **11C**, 28ff
sieving gel matrix **5B**, 200
signal cascades, in *Bacillus subtilis*, surfactin synthesis **7**, 68
signaling molecules, from bacteria and fungi **7**, 14f
– – structures **7**, 14
– ppGpp **7**, 61
signal peptidase **2**, 249
signal recognition particle (SRP) **3**, 531
signal-to-noise ratio **5B**, 213
signal transduction, domains **5B**, 137
– of graviresponse **10**, 513ff
silage **1**, 333, **9**, 775ff
– additives **9**, 776ff
– – enzymes **9**, 778f
– – microbial inoculants **9**, 776ff
– ensiling process **9**, 771f
– fermentation **1**, 341
silage production **3**, 310
silanol groups **5B**, 271
silicone diffusion membranes *see* diffusion membranes
silicone oil centrifugation **1**, 191
silicones, structure **6**, 676
– use in cosmetics **6**, 674f
simian virus 40 (SV40) **2**, 786
similarity analysis scores **5B**, 387
similarity searches, in protein databases **5A**, 56
Simple Sequence Repeats (SSRs) **5B**, 152
simulation programs for activated sludge plants **11A**, 425
simultaneous bidirectional sequencing **5B**, 198
Singapore, biotechnology **12**, 418ff
– – research centers **12**, 419
– – universities **12**, 419
Singapore Biotechnology Directory **12**, 419
single base mutations, detection using DNA ligases **10**, 90f
single-cell analysis, of proteins **5B**, 275
single cell oils **7**, 135ff
– advantage over single cell protein **7**, 185
single cell protein *see* microbial biomass
– break-even sugar price **6**, 52
– conversion cos **6**, 50
– market prices **6**, 51
single cell protein (SCP), production **3**, 9, 83, 473
single cell protein (SCP) production, by methylotrophic microorganisms **1**, 278

single-level filters, closed – **11C**, 327f
– open – **11C**, 327
– scheme – **11C**, 327
single-locus analysis **5B**, 168
single molecule detection, in drug screening **10**, 47
single nucleotide polymorphisms (SNPs) **5B**, 44, 90f, 112, 153, 177f, 209, 228
single nucleotide primer extension assays **5B**, 116
single photon detection systems **5B**, 433
single sequence repeats (SSP) **5B**, 168
single-spore technique, for strain improvement **6**, 316
single strand conformational polymorphism (SSCP) **5B**, 168
single-strand conformation analysis **5B**, 114
sintered glass, as carrier materials of biofilters **11C**, 466
Siphoviridae **1**, 550
SIRS *see* systemic inflammatory response syndrome
SISIR (Singapore Institute of Standards and Industrial Research) **12**, 420
site-selective mutagenesis, for optimization of enzyme activity **8A**, 26
Sixth Amendment **12**, 547ff
size exclusion chromatography, in enzyme purification **9**, 52f, 56, 104ff
size exclusion chromatography (SEC) *see* chromatography
slab gels **5B**, 200, 207
– ultrathin – **5B**, 203
slaughtering **11A**, 361f
– wastes **11A**, 219
– – amounts of **11A**, 222
– waste treatment, use of submerged fixed bed reactors **11A**, 361f
SL medium, selective for *Lactobacillus* **1**, 347
slow sand filters **11C**, 446
– bacteria in **11C**, 433
– filter layers, biology of **11C**, 431
– scheme **11C**, 429
slow sand filtration **11C**, 379
– change of environmental conditions **11C**, 434f
– effect on BOM **11C**, 463
– removal capacity **11C**, 439
– removal of, microoganisms **11C**, 438ff
– – organic matter **11C**, 438
– – pathogens **11C**, 438ff
– – trace elements **11C**, 438
– – xenobiotics **11C**, 438
sludge *see also* activated sludge
– – *see also* sewage sludge
– aerobic/anaerobic treatment of **11C**, 213ff
– – hygienic considerations **11C**, 213ff
– hydrolysis **11A**, 301f
– – basic principle **11A**, 302
– hygienization, mesophilic – **11A**, 409

– – thermophilic – **11A**, 405f
– liquor **11A**, 60f
– – nitrogen elimination **11A**, 60f
– pathogens in **11C**, 216ff
– spreading of, sanitary aspects **11C**, 225
sludge bed reactors, in anaerobic wastewater treatment **11A**, 517ff
sludge production, chemical phosphorus removal **11A**, 313ff
– – comparison of theoretical calculations and laboratory tests **11A**, 314
– EBPR process **11A**, 303ff
sludges, de-emulsifying activity **6**, 711
Sludge settling **4**, 435ff
sludge stabilization, mesophilic anaerobic – **11A**, 409
– thermophilic autothermic – **11A**, 405f
– – process parameters **11A**, 406
sludge treatment, gravity thickeners **11A**, 306f
– mechanical thickening **11A**, 306
– – thermophilic aerobic – **11A**, 403ff
– processes **11A**, 373ff
sludge volume index (SVI) **11A**, 267
Slug flow, definition **4**, 353
slurry, aerobic treatment **11A**, 232ff
– – inactivation of Enterobacteria **11A**, 232
– – thermophilic processes **11A**, 232f
– anaerobic treatment **11A**, 226ff
– – digester systems **11A**, 226
– animal viruses, inactivation times **11C**, 229
– indicator bacteria in, decimation times **11C**, 229
– pathogenic bacteria in, decimation times **11C**, 229
slurry bioreactors **11B**, 332ff, 472
– – *see also* bioreactors
– batch operation **11B**, 335ff
– – batch curve **11B**, 335ff
– configurations of **11B**, 332
– continuously stirred rank reactor (CSTR) **11B**, 335
– continuous operation **11B**, 338ff
– DMT-BIODYN process **11B**, 337f
– – configuration of **11B**, 337
– FORTEC process **11B**, 338
– full scale batch processes **11B**, 337f
– Huber process **11B**, 338
– OMH process **11B**, 338f
– sequential batch operation **11B**, 338ff
– – cascade of sequencing batch reactors **11B**, 338f
– slurry decontamination **11B**, 338ff
– slurry processing **11B**, 334f
– – set-up **11B**, 334f
slurry decontamination process **11B**, 338ff
– continuous plug flow system **11B**, 338ff
– degradation of, mineral oil **11B**, 340ff
– – polycyclic aromatic hydrocarbons **11B**, 340ff

– dual injected turbulent separation (DITS) reactor **11B**, 339ff
– interconnected suspension bioreactor (ISB) cascade **11B**, 339ff
– scheme **11B**, 336
slurry processing **11B**, 334f
small acid-soluble DNA binding protein (SASP) **2**, 206
small acid soluble protein (SASP), in *Bacillus* endospores **1**, 388f
small-inter-ORF features **5B**, 390
SMART databank **5B**, 138, 289
S_{min} model, application to, biofilters for dinking water treatment **11C**, 473
– – completely mixed reactors **11C**, 472f
– – plug flow reactors **11C**, 473
Smithsonian Institution **12**, 440, 443
Smith–Waterman algorithm **5B**, 130
Smith–Waterman method **5B**, 303
snakes, as sources of ribosomal peptides **7**, 302
SNPs *see* single nucleotide polymorphisms
socioeconomic discrimination **5B**, 427
sodium, effect on microbial growth **1**, 131
sodium bis(2-ethylhexyl)-sulfosuccinate (AOT), as reversed micellar medium **3**, 606
sodium gluconate, production figures **6**, 357
– uses **6**, 357
sodium gluconate process, with *Aspergillus niger* **6**, 353f
– – production medium **6**, 353
– – product recovery **6**, 357
– – sporulation medium **6**, 357
softwood, pulp composition **10**, 526
soil, authochthonous microflora **11B**, 95
– availability of oxygen **11B**, 97
– bacterial communities in **11B**, 374
– bioassays **11B**, 473f
– bioavailability of pollutants **11B**, 469
– – coal particles **11B**, 71
– – increase of the desorption process **11B**, 79ff
– – increase of the mass transfer rate **11B**, 79ff
– – influencing factors **11B**, 67ff
– – long-term contaminated sites **11B**, 69ff
– biodegradation of pollutants **11B**, 71ff
– – kinetics **11B**, 71ff
– – model of the overall process **11B**, 75ff
– – rate **11B**, 74
– biologically enhanced washing of **11B**, 293f
– bioremediation with heap technique **11B**, 319ff
– – monitoring of degradation **11B**, 323
– – use of additives **11B**, 322
– biotransformations in, kinetic data **11B**, 73ff
– carbon pools **11B**, 114
– catalytic capacity **11B**, 98f
– degradation of organic compounds **11B**, 96ff
– – biogenic molecules **11B**, 97f
– – depolymerizing exoenzymes **11B**, 98
– effective porosity, ranges **11B**, 48

– enzymes in **11B**, 98f
– *ex situ* treatment, physical properties **11B**, 57f
– formation of bound residues **11B**, 109ff
– heavy metals in, extraction/elution procedures **11B**, 498
– humic substances, formation of **11B**, 106ff
– *in situ* bioremediation **11B**, 56f, 349ff, 399ff
– – activated zones **11B**, 361ff
– – bioaugmentation **11B**, 408ff
– – bioscreens **11B**, 361ff
– – bioslurping **11B**, 350f
– – biosparging **11B**, 259f
– – biostimulation **11B**, 406ff
– – bioventing **11B**, 356f
– – degradability of contaminants **11B**, 468ff
– – effect of pH **11B**, 430
– – effect of temperature **11B**, 429
– – effect of water activity **11B**, 430
– – features **11B**, 473
– – funnel and gate **11B**, 363f
– – hydraulic circuits **11B**, 357f
– – hydraulic permeabilities **11B**, 56
– – limits **11B**, 465ff
– – monitoring of **11B**, 366ff
– – natural attenuation **11B**, 363f, 401ff
– – perspectives **11B**, 474f
– – preliminary investigations **11B**, 468ff
– – treatment plants in Germany **11B**, 466f
– – use of bioaugmentation **11B**, 356
– – with genetically engineered microorganisms **11B**, 441ff
– interaction with xenobiotics **11B**, 91
– mass transfer in, kinetic data **11B**, 73ff
– metabolic activity **11B**, 98
– metal contaminated – **10**, 243
– – bioremediation of **10**, 243, 245
– microbial activity in **11B**, 96ff
– microorganisms in **11B**, 95f
– – concentration of **11B**, 95
– moisture, effect on microbial activity **11B**, 430
– non-extractable residues, definition of **11B**, 109
– – formation of **11B**, 109
– particle size distribution **11B**, 331
– permeability **11B**, 46ff
– – coefficient **11B**, 471
– – ranges **11B**, 48
– persistence of organic compounds **11B**, 96
– pollutant diffusion in **11B**, 74
– porosity **11B**, 45ff
– – ranges **11B**, 48
– predominant bacteria in polluted – **11B**, 147
– reuse of treated soil **11B**, 315f, 473
– – bioassays **11B**, 473f
– saturated –, *in situ* respiration test **11B**, 367
– – monitoring of remediation **11B**, 367
– – treatment of **11B**, 357
– saturated flow **11B**, 47
– stability of organic compounds **11B**, 105

– threshold concentrations of xenobiotics **11B**, 96
– types of interstices **11B**, 46f
– unsaturated –, *in situ* respiration test **11B**, 367
– – monitoring of remediation **11B**, 367
– – treatment of **11B**, 356f
– unsaturated flow **11B**, 46f
– water activity **11B**, 430
– water content **11B**, 97
– water potential **11B**, 97
– zymogenic microflora **11B**, 95
soil aggregates **11B**, 93f
soil bacteria **11B**, 95
soil bed biofilters, for VOC control **11C**, 363
soil biofilters **11C**, 360
soil bioremediation, by microbial consortia **6**, 689ff
– cost **6**, 692
– effect of biosurfactants **6**, 694ff
– *in-situ* treatment **6**, 692f
– naphthalene contamination **6**, 699
– on-site treatment **6**, 692f
– PAH contamination, use of sophorolipids **6**, 700
– petroleum contamination, use of nutrients **6**, 693
soil carbon pools, turnover rates **11B**, 114
soil components **11B**, 91ff
– aggregation of **11B**, 92
– – aggregate size **11B**, 94
– black carbon **11B**, 92f
– clay minerals **11B**, 93
– clay-organic complexes **11B**, 94
– humic substances **11B**, 93
– physical properties **11B**, 93
soil contaminants, adsorption coefficients **11B**, 69
– biodegradability of **11B**, 468ff
– biodegradation of, use of genetic engineering **11B**, 444
– cometabolic degradation of **11B**, 432f
– ecotoxic effects **11B**, 133
– geochemical models **11B**, 55f
– sorption capacities **11B**, 55
soil contamination, aging of **11B**, 70
– definitions **11B**, 36f
– ecotoxicological assessment **11B**, 127ff
– hazards **11B**, 490
– long-term contaminated sites **11B**, 69ff
– with highly viscous tar oil or tar pitch **11B**, 70
soil decontamination, bacterial activity enhancement **11B**, 425ff
– inoculation of microorganisms **11B**, 416
– nutrient supplements, nitrogen **11B**, 431
– use of immobilized microorganisms **11B**, 415ff
– – results **11B**, 418ff
soil fungi **11B**, 95
soil microorganisms **11B**, 95f
– microenvironments **1**, 244

soil moisture, effect on microbial activity **11B**, 430

soil pores, accessibility for microorganisms **11B**, 68

soil remediation *see also* contaminated site management
- biological processes **11B**, 313ff
- – application of **11B**, 315
- – treatment plants in Germany **11B**, 467
- chemical/physical processes **11B**, 310ff
- – application of **11B**, 312f
- – *ex situ* **11B**, 310ff
- – *in situ* **11B**, 312
- compost addition **11B**, 343
- effect of geology **11B**, 43ff
- effects of hydrology **11B**, 43ff, 56ff
- *in situ* technologies **11B**, 351
- methods **11B**, 306ff
- prices **11B**, 327
- regulations **11B**, 327
- – international approaches **11B**, 5ff
- scrubbing **11B**, 310f
- thermal processes **11B**, 305ff
- – application of **11B**, 309f
- – *ex situ* **11B**, 307f
- – *in situ* **11B**, 308f
- – technologies **11B**, 307f
- – treatment plants in Germany **11B**, 467
- treatment techniques **11B**, 35f
- use of heap technique **11B**, 319ff
- – BTEX aromats **11B**, 326
- – explosives **11B**, 326
- – PAH **11B**, 326
- – petroleum hydrocarbons **11B**, 325
- – phenols **11B**, 326

soil remediation by plants *see* phytoremediation

soil samples, bioassays **11B**, 129
- ecotoxic potential of **11B**, 128
- preparation of aqueous extracts **11B**, 130
- requirements for **11B**, 483

soil sampling **11B**, 480ff

soil structure **11B**, 93f

soil suspension, use of bio-de-emulsifiers **6**, 698

solenoid **2**, 674

solerone, structure of **8B**, 27

solid matter, analysis of samples **11B**, 496f
- analytical investigation program **11B**, 495ff
- – activity parameters **11B**, 496
- – analysis parameters **11B**, 495f
- – assessment parameters **11B**, 496
- – digestion procedures **11B**, 498
- – purification and enrichment methods **11B**, 498
- – stepwise procedures **11B**, 496
- collection of samples **11B**, 487ff
- – boring procedure **11B**, 487f
- – excavation **11B**, 488
- – ram core boring **11B**, 488

- evolution tests **11B**, 490ff
- investigation of **11B**, 477ff
- – steps **11B**, 482
- sampling of **11B**, 477ff, 482ff

solid-phase extraction capillary zone electrophoresis **5B**, 256, 260

solid-phase phosphoramidite chemistry **5B**, 230

solid retention time (SRT), in activated sludge **11A**, 260ff

solid state bioreactors **11B**, 343f, 472
- batch operation **11B**, 343
- composting **11B**, 343
- process configuration **11B**, 343
- rotating drum bioreactor **11B**, 343f
- semi-continuous operation **11B**, 343f

solid-state fermentation, with Basidiomycetes **1**, 521

solid state fermentations **11B**, 343
- fixed bed reactor **11B**, 333f
- reactor configurations **11B**, 332ff
- rotating drum bioreactor **11B**, 333f

solid waste *see also* contaminated soil, contaminated solids
- – *see also* waste
- recycling of, product categories **11B**, 330f
- – sediments **11B**, 330ff
- – sludges **11B**, 330ff
- – soils **11B**, 330ff
- – targets **11B**, 330f
- sanitary landfilling of **11C**, 167ff
- types of, disposal of **11C**, 10
- – reutilization of **11C**, 10

solid waste stream, composition of, comparison data **11C**, 194

solid waste treatment **10**, 215
- biotechnology of **11C**, 5ff
- pyrotechnology of **11C**, 5

solubilization, of coal **10**, 157ff

solute transport systems, in *Lactococcus* **1**, 351
- in *Lactococcus see also* membrane transport

solution diffusion membranes *see* diffusion membranes

solution structures, of lantibiotics **7**, 338ff

solvation effects, on enzyme activity **8A**, 252

solvation energy, of proteins **5A**, 71

solvent engineering **8A**, 226

solvents *see* acetone/butanol
- Alamin 336 **3**, 569
- carbon-bonded-oxygen-bearing extractants (CBO) **3**, 566f
- for liquid-liquid extraction **3**, 560ff
- – for alcohols **3**, 560
- – for antibiotics **3**, 563
- – for carboxylic acids **3**, 561f
- – for ethanol **3**, 560
- – for penicillin G **3**, 563
- high-boiling point **3**, 561
- long chain aliphatic amine extractants **3**, 567

– low-boiling point – **3**, 561
– phosphorous-bonded oxygen donor extractants (PBO) **3**, 566
– strong solvating extractants **3**, 566
– TOPO **3**, 566f
somaclonal variation **1**, 603
somatic cell genetics **2**, 156ff
– mammalia **2**, 156ff
somatic cell technology **5B**, 42
somatic embryogenesis **1**, 603
sophorolipids, structure **7**, 180
– surfactant formation by *Pseudomonas* spp. **7**, 180
sophorose lipids, from *Torulopsis bombicola* **6**, 669f
– production cost **6**, 669
– yeast-derived, use in cosmetics **6**, 675
sorbates, as bread preservatives **9**, 315
sorbic acid, as wine preservative **9**, 488
sorbistins **7**, 439ff
– biogenesis of – **7**, 439
– structure **7**, 472
sorbitol **6**, 42f
– composition **6**, 43
– formation by yeasts **6**, 222
– in vitamin C production **6**, 42
D-sorbitol dehydrogenases **8B**, 306
sorbitol syrup **6**, 43
sorghum, as brewing adjunct **9**, 430
– as raw material for ethanol production **6**, 65
– composition **6**, 142
– in industrial ethanol production **6**, 141
– – cost analysis **6**, 173
– processing in ethanol production **6**, 94
– stillage, nutrient content **6**, 108
sorghum malt, as brewing adjunct **9**, 432
sorghun grain, analysis of **6**, 66
sorivudine, antiviral action **6**, 566f
– structure **6**, 567
sorption, Freundlich isotherm **11B**, 53f
– Henry isotherm **11B**, 53f
– Langmuir isotherm **11B**, 54
– terminology **10**, 228
sorption isotherms, for water **3**, 709
sorption processes, and soil contamination **11B**, 52ff
sour cream, artificial – **9**, 414
– definition **9**, 394
– dip **9**, 414f
– manufacture of **9**, 414ff
– quality problems **9**, 417
sour dough **1**, 333, **9**, 253ff, 285ff
– basic sour overnight **9**, 288f
– Berlin sort sour dough process **9**, 290f
– carboxylic acid degradation **9**, 298
– chemical acidification **9**, 299f
– Detmold single stage process **9**, 290f
– freeze-dried – **9**, 293

– full sour **9**, 289f
– growth of bacteria **9**, 296
– growth of yeasts **9**, 296
– lactobacilli in – **9**, 293ff
– Monheim salt-sour process **9**, 292f
– process stages **9**, 287ff
– protein hydrolysis **9**, 299
– starter cultures **9**, 254ff, 287
– – lactobacilli **9**, 255ff
Southeast Asia **12**, 436ff
Southern blot **5B**, 228
Southern blot analysis, of transgenic animals **2**, 757ff
SoyBase **5B**, 411
soybean fermentation **3**, 307
– products **3**, 307
soybean hull peroxidase, modification of lignite **10**, 168
soybean lipoxygenase **8B**, 365
– commercial availability **8A**, 482
– hydroperoxide formation **8A**, 502f
– synthesis of chiral alcohols by **8A**, 503
soybean milk, fermentation of **9**, 549f
soybean oil, enzymatic extraction **9**, 726
– fatty acid composition **7**, 136
– genetic engineering of – **7**, 138
soy products, use in baked goods **9**, 260
soy protein hydrolysates **9**, 695f, 701
soy sauce **9**, 507ff
– aflatoxin in – **9**, 552
– composition **9**, 513
– flavor compounds **9**, 517
– production of **9**, 513ff
– – koji process **9**, 515f
– – mash stage **9**, 516f
– – pasteurization **9**, 517f
– – preparation of soybeans **9**, 513ff
– – preparation of wheat **9**, 515
– types **9**, 513
space, biotechnology in **10**, 507ff
– exobiology in **10**, 516f
– radiation biology in **10**, 516f
– variations of cellular functions in **10**, 509ff
spaceflights **10**, 508
space stress **10**, 510ff
space travel **12**, 9
SPAD (Signaling Pathways Database) **5B**, 410
sparged aeration **3**, 33, 238, 255ff
– bubble, birth size **3**, 261
– – coalescence **3**, 262
– – dispersion **3**, 265
– – entrainment **3**, 243
– – formation **3**, 261
– – size **3**, 261
– direct sparging **3**, 263f, 270ff
– gas jet regime **3**, 261
– indirect sparging **3**, 239, 265f
– mass transfer, relation **3**, 255

– mechanical stress **3**, 273
– of stirred tank bioreactors **3**, 263ff
– subsurface impellers **3**, 243
– velocity **3**, 261
– vibromixers, power input **3**, 266
– – sorption characteristics **3**, 267
SPA (strategic project assessment) **12**, 217
species cross-reativity **5A**, 295ff
species-specificity, of genes **12**, 78ff
specific growth rate **9**, 203
– baker's yeast fermentation **4**, 237
– estimation **4**, 546f
– identification with SSPE **4**, 245
– indirect measurement **4**, 543
– in recombinant fermentation, effect on plasmid
  content **4**, 498f
– – effect on plasmid stability **4**, 500f
– *Klebsiella pneumoniae* **4**, 282
– – influence of temperature **4**, 282
– microbial growth **9**, 189
– recombinant fermentation, effect on gene expression rate **4**, 499f
– relation to product formation rate **4**, 534
spectinomycins, streptomycin-related aminoglycosides **7**, 427
– structure **7**, 464
spectra, absorption **5B**, 187
– emission **5B**, 187
– shifts **5B**, 187
spectral karyotyping (SKY) **5B**, 92
spectral overlap **5B**, 189
spectroscopic methods, IR spectroscopy **4**, 171f
spenolimycin, structure **7**, 464
spermidine, in pseudomonads **1**, 405ff
sperm-mediated gene transfer **2**, 771f
– target species **2**, 771f
– techniques **2**, 772
– to chicken **2**, 811
spheroblast **2**, 103
spheroids *see* cell spheroids
spheroplast fusion, brewer's yeast **9**, 439
sphingolipids, of yeasts, structure **7**, 183
Sphingomonas paucimobilis, biopolymer formation **6**, 617
– gellan formation **6**, 629, 640
spicamycin, antitumor activity **7**, 647
spiculisporic acid, structure **7**, 181
spiders, as sources of nonribosomal peptides **7**,
  314
– as sources of ribosomal peptides **7**, 301
spina bifida **5B**, 84
spiramycin formation, regulatory gene of in *Streptomyces ambofaciens* **7**, 90
Spirulina, in health food markets **10**, 129
– in human diet **10**, 126f
– production of, in Myanmar **10**, 120
Spirulina maxima, as source of polyunsaturated
  fatty acids **7**, 167

Spirulina platensis, as source of polyunsaturated
  fatty acids **7**, 167
– β-carotene content **7**, 172
spliced cDNA sequences **5B**, 302
Spliceorf **5B**, 373
splice site pattern **5B**, 365
sponge dough process **9**, 273ff
– flow sheet **9**, 274
– formulations **9**, 275
sponges **10**, 445f
– antiinflammatory compounds of **10**, 456
– antiparasitic compounds of **10**, 458
– antiviral compounds of **10**, 460
– aquaculture of **10**, 449
– as sources of nonribosomal peptides **7**, 312f
– cell culture of **10**, 449
– compounds of **10**, 446, 458f
– cultivation of **10**, 449
– cytostatic/antitumor compounds of **10**, 452
spongiform encephalopathies, as risk factors in
  biological products **5A**, 513
spore propagation, in citric acid production **6**,
  318
spores *see also* endospores
– fungal **1**, 517
Sporolactobacillus inulinus, lactic acid formation
  6, 303
Sporosarcina **1**, 368
sporulation, genes **1**, 390
– of bacilli **1**, 390ff
– of clostridia **1**, 292
– of streptomycetes **1**, 437
spray-drying **3**, 708ff
– drying time **3**, 709
– effects on pharmaceuticals **3**, 711
– equipment **3**, 709
– heat transfer **3**, 709
– mass transfer **3**, 709
– nozzle atomizer **3**, 710
– plant components **3**, 710
– rotating disk atomizer **3**, 710
– temperature distribution **3**, 711
squalene synthase (SQS), ERG9 mutation in *Saccharomyces cerevisiae* **10**, 581
SRS **5B**, 291ff
SRS CORBA API **5B**, 298
SRS Objects **5B**, 297
SRS Perl API **5B**, 297
SRS PRISMA **5B**, 296
SSCP *see* single strand conformational polymorphism
SSPE *see* Sequential state/parameter estimation
SSTC (State Science and Technology Commission
  of China) **12**, 397
stabilization, of enzymes **9**, 110ff
– – by disulfide bonds **9**, 17
– – by hydrogen bonding **9**, 16
– – by hydrophobic effects **9**, 16ff

stabilizers, in cultured dairy products **9**, 403
STACK **5B**, 284
staged sludge anaerobic reactor (SSAR) **11A**, 524
staining **1**, 10
– Gram – **1**, 10
– negative – **1**, 11
– ruthenium red – **1**, 10
standard data representations **5B**, 403
standard hydrogen electrode (SHE) **4**, 18
standard Operating Procedures (SOPs) **12**, 227
standards, for biotechnology, in Europe **12**, 641f
Stanton number **4**, 373
Staphylococcus, conjugative transposon **2**, 489
– – Tn916 of *S. faecalis* **2**, 489
– genetic engineering **2**, 488
Staphylococcus aureus **2**, 458, 488
– drug-resistance plasmid **2**, 458ff
– – pTA1060 **2**, 460
– – pT181 **2**, 461
– – pUB110 **2**, 461
– genetic exchange **2**, 56f
– plasmid replication **2**, 458ff
– – rolling circle (RCR) **2**, 461
– – single-stranded DNA (*ss* plasmid) **2**, 458
– specific growth rate **1**, 130
Staphylococcus aureus RN4220 (pC200)/(pC300), metal bioreporters **11B**, 455f
Staphylococcus carnosus **2**, 488
Staphylococcus epidermidis, in tempeh fermentation **9**, 530
Staphylococcus hyicus, staphyloferrin A **7**, 223
staphyloferrin, carboxylate-type siderophore of *Staphylococcus hyicus* **7**, 223f
staphyloferrin A **7**, 223f
– structure **7**, 224
staphyloferrin B, structure **7**, 224
staphylokinase, use in therapy **5A**, 130
starch **6**, 32ff
– amylopectin fraction **9**, 742
– amylose fraction **9**, 742
– anaerobic degradation of **11A**, 31f
– composition of **6**, 32, 34f
– determination with biosensors **4**, 78
– gelatinization temperature **9**, 742
– industrial usage **6**, 615
– in industrial ethanol production **6**, 140
– microbial metabolism **9**, 185
– retrogradation **9**, 311
– substrate for biomass production, Adour-Speichim process **9**, 212
– – ENSA-INRA process **9**, 212
– – SETA process **9**, 212
– – Symba process **9**, 211f
– technical amylolysis **6**, 66ff
– uses of liquefied – **6**, 35
starch-based raw materials *see* raw materials
starch-based sweeteners *see* sweeteners

starch content, of grain **6**, 109
– of potatoes **6**, 109
starch degradation, by autoamylolysis **6**, 73ff
– by improved cultures **1**, 200
– in clostridia **1**, 297
– – enzymes **1**, 310
– in *Saccharomyces cerevisiae* **1**, 200
starch-degrading enzymes, in clostridia **1**, 310
starch derivatives, suppliers **6**, 44
starch-derived syrups, process constraints **6**, 43f
– selection **6**, 43
starch factory wastewater, anaerobic treatment **11A**, 203
– concentrations **11A**, 203
– corn starch production **11A**, 202
– potato starch production **11A**, 202
– specific amounts **11A**, 202
– wheat starch production **11A**, 202
starch hydrolysis, in membrane bioreactors **3**, 454
starch liquefaction **5A**, 193f, **6**, 35ff
– – *see also* starch degradation
– enzymatic – **6**, 67ff
– increase in dextrose equivalents **5A**, 195
– use of α-amylases **5A**, 193
starch modification, enzymes **9**, 667ff, 742ff
– liquefaction **9**, 668ff, 744f, 748ff
– – acid process **9**, 750
– – dual enzyme processes **9**, 749
– – EHE process **9**, 749
– – low temperature process **9**, 749
– – process scheme **9**, 668, 748
– – thermal process **9**, 750
– saccharification **9**, 669ff, 745ff, 750
starch processing enzymes **10**, 74ff
– from hyperthermophiles **10**, 79f
– from thermophiles **10**, 79f
starch production **6**, 32ff
– general scheme **6**, 33
– modification **2**, 651f
– – amylose-free starch **2**, 652
– raw materials **6**, 32
starch saccharification **6**, 68ff, 142f
– α-amylase **6**, 68, 143
– β-amylase **6**, 68, 143
– glucoamylase **6**, 71, 143
– use of enzyme combinations **6**, 71ff
– – corn mash **6**, 72ff
– – triticale mash **6**, 78
starter culture, basic medium **1**, 349
– *Bifidobacterium* **1**, 346
– for food production **1**, 349
– for lactic fermentation **1**, 328
– preparation **1**, 349
– preservation **1**, 349
– propagation **1**, 349
– storage **1**, 349
starter cultures, for cultured dairy products **9**, 390f, 394ff

– – production of **9**, 399ff, 402
– for vegetable fermentations **9**, 651ff
– – enumeration of **9**, 653f
– – growth kinetics **9**, 654f
– – selection of **9**, 651f
– in cheese making **9**, 371ff
– in wine making **9**, 469ff
– – properties **9**, 470
– – use **9**, 470f
– sour dough – **9**, 254ff, 287
STA (Science and Technology Agency) **12**, 253ff, 378, 594f
state estimation, of bioprocesses **3**, 379ff
– – linear system **3**, 380
– – non-linear system **3**, 381
state observer, for bioprocess analysis **3**, 379ff
– – BR model **3**, 379
– – CP model **3**, 379
state variables, estimation by EKF **4**, 231ff
static pile composting **11C**, 138f
stationary bed reactors, specific limitations **11A**, 512
stationary growth phase, of *Bacillus subtilis* **7**, 67
statistical sampling **11B**, 484
statoliths, gravity perception **10**, 513
staurosporin, inhibition of protein kinase C **7**, 677
steady states, in anaerobic wastewater processes **4**, 459ff
steam sterilization **3**, 164ff, 779
– implications on public health **3**, 784
stearic acid, increase in yeasts **7**, 150ff
– – by direct feeding **7**, 150f
– – by inhibition of stearoyl desaturase **7**, 151f
– – by metabolic manipulation **7**, 154
– – by mutation **7**, 152ff
stearoyl desaturase, inhibition of – **7**, 151f
stearoylphytosphingosine, structure **7**, 184
stearoylvelutinal, antimicrobial activity **7**, 495
stem cell factor **5A**, 158
Stenotrophomonas 3664, surfactant resistant bioreporter **11B**, 456
stent implantation, therapy with ReoPro **5A**, 375
stentorin, structures **10**, 141
– therapeutic potential **10**, 140f
STEPI (Science and Technology Policy Institute) **12**, 388
Steptomyces lavendulae, formycin production **7**, 77
stereocenters, of alcohols, non-carbon – **8A**, 98
– – quaternary – **8A**, 94f
– – remote – **8A**, 97
– of carboxylic acids **8A**, 103ff
– – quaternary – **8A**, 109
– – remote – **8A**, 109f
– – sulfur – **8A**, 109f
stereoisomeric alcohols, separation of, antibody mediated – **8B**, 432

stereoselectivity pocket of lipases **8A**, 48
stereospecificity, of acylases **8B**, 356
sterigmatocystin, biosynthetic clusters **7**, 21
sterile filtration, absolute – **3**, 161
– depth – **3**, 160
– efficiency, filter length effects **3**, 179
– – particle size effects **3**, 178f
– – velocity effects **3**, 178
– filter design **3**, 179f
– of air **3**, 177ff
– size exclusion **3**, 161
– types **3**, 160
sterile insect release **10**, 496
sterilization **1**, 129, **3**, 157ff, **10**, 288f
– of callus cultures **1**, 583
– thermal – **3**, 162ff
– – continuous – **3**, 170
– – cool down **3**, 165f
– – cycle design **3**, 166
– – Damköhler number **3**, 173f
– – death kinetics **3**, 162f
– – direct steam injection **3**, 169
– – heat up **3**, 165f
– – high temperature short time (HTST) treatment **3**, 170
– – holding time **3**, 165f
– – nutrient degradation **3**, 166
– – temperature effect **3**, 164ff
– – temperature-time profile **3**, 165
– – thiamine destruction **3**, 169f
sterilization agents, requirements for **10**, 286
sterilizer *see* continuous sterilizer
steroidal compounds, from plant cell cultures **7**, 611f
steroid conversion, by clostridia **1**, 311f
steroid hormone, hydroxylation by fungi **1**, 529
steroid hydroxylases **8A**, 505
steroid manufacture **8A**, 11f
steroid nucleus, biotransformation by whole-cell fermentations **8A**, 13
– microbial oxidation of **8A**, 12
steroid receptors, screening with transformed *Saccharomyces* **7**, 111
steroids, Baeyer–Villiger reaction **8A**, 541
– biosynthesis of **8B**, 28ff
– C—H bond of, microbial oxidation of **8A**, 12
– chlorinated – **8B**, 186
– combined chemical and microbiological synthesis, of estrone **8A**, 14
– hydroxylation of **8A**, 19, 479, 488ff, 505, **8B**, 381
– – by plant cell cultures **8A**, 489
– natural –, used in the manufacture of pharmaceuticals **8A**, 12
– prochiral reduction of **8A**, 19
– side chain oxidation of **8A**, 19
– synthesis of, potential use of hydroxysteroid dehydrogenases **8A**, 448

steroid synthesis, microorganism reductions **8B**, 378
sterols **7**, 170ff
– accumulation in yeasts **7**, 171f
– commercial extraction of eukaryotes **7**, 170ff
– structures **7**, 171
Stevenson-Wydler Technology Innovation Act **5A**, 522
Stickland reaction **11A**, 33
stigmasterol, biotransformation of **8A**, 13
stillage, anaerobic treatment **11A**, 226
– as by-product in ethanol fermentation **6**, 169ff
– as feed **6**, 104ff
– as fertilizer **6**, 106ff
– composition **6**, 170
– – corn **6**, 108
– – wheat **6**, 108
– content of, ethanol **6**, 115
– – fermentable sugars **6**, 115
– – starch **6**, 115
– nutrient content, barley **6**, 107
– – corn **6**, 107
– – manioc **6**, 108
– – potatoes **6**, 105
– – rye **6**, 106
– – sorghum **6**, 108
– – wheat **6**, 106
– recovery processes **6**, 171
– uses **6**, 170
stillage recycling process **6**, 87f
– scheme **6**, 87
stirred tank, activated sludge system **4**, 431ff
– – mass balance equations **4**, 431
– aeration **4**, 311f
– – effect on circulation time **4**, 312
– – effect on mixing times **4**, 311ff
– circulation paths **4**, 310f
– circulation time estimate **4**, 308
– continuous, overall reaction rate **4**, 339
– energy distribution **4**, 304
– macromixer region **4**, 313ff
– – oxygen concentration **4**, 335
– mass distribution **4**, 304ff, 321f
– – five-compartment model **4**, 321
– micromixer region **4**, 313ff
– – oxygen concentration **4**, 334ff
– modelling **4**, 301ff
– recycle models **4**, 304ff
– two-impeller systems, circulation-time dustribution **4**, 315ff
stirred tank bioreactor **3**, 12, 27f, 96ff, 111ff, 120, 404
– aeration **3**, 12, 239, 263ff
– – bubble dispersion **3**, 265
– – impeller geometry **3**, 263
– – indirect sparging **3**, 265
– – mass transfer coefficients **3**, 263ff, 265
– – power input **3**, 263f
– – sorption characteristics **3**, 263f
– – spin filter system **3**, 265f
– – vibromixers **3**, 266
– batch operation, mass balances **3**, 97
– cell retention, devices **3**, 113f
– cleaning **3**, 12
– continuous – (CSTR) **3**, 431
– – membrane reactor **3**, 432
– continuous perfusion operation **3**, 113f
– dissolved oxygen, concentration gradient **3**, 13
– fed-batch operation **3**, 100
– gas supply **3**, 83
– heterogeneity **3**, 13
– ideal – **3**, 96
– macromixer regime **3**, 13f
– mammalian cell culture, hydrodynamic damage **3**, 112
– mass balances **3**, 96
– mass transfer coefficient, volumetric – **3**, 196
– micromixer regime **3**, 13f
– mixing **3**, 12, 206
– – *see also* bioreactor mixing
– operation modes **3**, 113
– operation scale **3**, 12
– oxygen supply **3**, 113
– parameter values **3**, 91
– scale-up **3**, 189
– stirring **3**, 83
– temperature control **3**, 83
– with microcarriers **3**, 108
stirred tank models **4**, 301ff
– applications **4**, 331ff
– coupling of mass and energy distributions **4**, 304
– impulse response, as function of impeller position **4**, 306ff
– influence of scale, on energy dissipation **4**, 334
– – on oxygen uptake **4**, 334
– measurement of circulation time and distribution, flow-follower techniques **4**, 307ff
– – stimulus-response methods **4**, 304ff
– micromixing **4**, 338f
– multi-compartment models, coupling of oxygen transfer, mixing and kinetics **4**, 320ff
– – five-compartment model **4**, 321
– – multiple multi-phase compartment model **4**, 324f
– – multi-turbine model **4**, 322
– – recycle-backmix circulation model **4**, 323
– – stage model **4**, 323
– – two-agitator model **4**, 323
– – two compartment model **4**, 321
– multiple-impeller systems **4**, 336f
– – oxygen transfer capacity **4**, 336f
– oxygen profiles **4**, 322ff
– scale-up **4**, 301ff
– time constants, for conversion processes **4**, 302
– – for transport processes **4**, 302

– turbulence models **4**, 327ff
– turbulent flow **4**, 328
stirred tank reactor **1**, 587f
– anaerobic wastewater process models **4**, 445ff
– oxygen transfer rate **1**, 125
stirrer *see* impeller
STKE (Signal Transduction Knowledge Environment) **5B**, 410
STOA (Science & Technology Options Assessment) **12**, 612
Stokes' law **3**, 475, **4**, 612
Stoke's theory **11A**, 555
stone-washing, use of industrial cellulases **5A**, 204
stop codons **5B**, 388
storage polymer production **1**, 174
storage proteins **5A**, 33
straight dough process **9**, 273ff
– formulations **9**, 275
strain improvement **2**, 30ff
– bacilli **1**, 396
– corynebacteria **1**, 456
– degradation of, aromatics **1**, 202f
– – starch **1**, 200
– – xenobiotics **1**, 210ff
– ethanol production **1**, 204f
– filamentous fungi **1**, 532, 536f
– lactose metabolization **1**, 195
– L-amino acid production **1**, 206ff
– microbial product modification **1**, 204ff
– mutational screening **2**, 32f
– of solvent-producing bacteria, isolation of mutants **6**, 247
– – recombinant DNA techniques **6**, 247
– of solvent-producing clostridia **6**, 246f
– plant cell **2**, 37
– plant cell culture **1**, 592
– product yield **2**, 31
– streptomycetes **1**, 449ff
– xylose metabolization **1**, 196f
strain selection, for microbial biomass production **9**, 172f
– in citric acid production **6**, 315ff
– – passage method **6**, 316ff
– – single-spore method **6**, 316
– of baker's yeast **9**, 323
– of wine yeast **9**, 323
streaming potential **4**, 95f
– dual-flow cell **4**, 95
Strecker reaction, cyclic – dipeptide catalyzed – **8B**, 496f
– during bread baking **9**, 307
– use of titanium dipeptide complex catalysts **8B**, 508ff
streptidine, biosynthetic pathway **7**, 417, 421
streptobacteria **1**, 329
streptococci **1**, 170f
– fecal –, in biogas plants **11C**, 228ff

– homolactic fermentation **1**, 172
streptococcin A-FF22, prepeptide structure **7**, 345
– primary structure **7**, 334f
Streptococcus **1**, 329, 344f
– binary fission **1**, 114
– habitats **1**, 344
– hemolytic – **1**, 344f
– – isolation of **1**, 349
– in olive fermentation **9**, 606
– oral –, isolation media **1**, 349
– – species **1**, 345
– pathogenicity **1**, 344f
Streptococcus dysgalactiae, in tempeh fermentation **9**, 530
Streptococcus equi, hyaluronic acid formation **6**, 644
Streptococcus faecalis see Enterococcus faecalis
Streptococcus faecalis see Enterococcus faecalis
– in idli fermentation **9**, 539
– in miso fermentation **9**, 520
– in puto fermentation **9**, 528
– tyrosine decarboxylase **8B**, 83
– tyrosine-phenol lyase **8B**, 95ff
Streptococcus faecium, as direct feed additives **9**, 781
– in tempeh fermentation **9**, 530
Streptococcus kefir, in poi production **9**, 548
Streptococcus mutants **1**, 344
Streptococcus thermophilus **1**, 332, 335, 344f
– habitats **1**, 344
– in cheese making **9**, 369f
– in cultured dairy products, characteristics **9**, 397
– in Swiss cheese production **9**, 376f
– isolation media **1**, 348
Streptococcus zooepidemicus, hyaluronic acid formation **6**, 644
streptokinase, as therapeutic product **3**, 63
– use in therapy **5A**, 130
Streptomyces **2**, 107f, 465
– actinophages **1**, 568
– antibiotic production by **10**, 365
– antibiotic resistance gene **2**, 467
– biology **2**, 465
– chemotaxonomic marker **1**, 435
– chromosome **1**, 438f
– – architecture **1**, 439
– – silent region **1**, 439
– colony form **2**, 466
– cytodifferentiation, A-factor **7**, 27
– – signal cascade **7**, 27
– G + C content **1**, 435, 440
– genetic engineering **2**, 467ff
– – marker genes **2**, 470
– hybrids **2**, 108
– hydroxamate siderophores, ferrioxamine type **7**, 211f
– industrial applications **2**, 465

– life cycle **2**, 466
– morphology **1**, 435ff
– mycelium **2**, 466
– production of antibiotics **1**, 445
– recombinant stability **2**, 108
– recombination **2**, 106f
– RNA polymerase **2**, 474
– secondary metabolites **2**, 466
– silent antibiotics gene **2**, 475
– – detection method **2**, 475
– – polyketide synthase gene **2**, 476
– strain improvement **2**, 475
– temperature-sensitive plasmids **2**, 469
Streptomyces ambofaciens **7**, 92
– spiramycin production in – **7**, 90
Streptomyces antibioticus **7**, 72ff
– oleandomycin production, growth-phase dependency **7**, 72
– phenoxazinone synthase **7**, 73
Streptomyces aureofaciens, bromoperoxidase of **8B**, 195
– microcolonies **1**, 232
Streptomyces cattleya, formation of fluoroacetate **8B**, 199f
Streptomyces clavuligerus, β-lactam biosynthesis **7**, 251
– clustering of β-lactam biosynthetic genes **7**, 259ff
– penicillin biosynthetic enzymes **7**, 254ff
Streptomyces coelicolor **1**, 438, **2**, 466, 473f, **7**, 71ff, 81, 83
– antibiotic regulatory genes **7**, 85ff
– biosynthetic genes **7**, 447
– fertility factors **2**, 55
– methylenomycin production **7**, 74f, 91f
– – on minimal medium **7**, 75
– ppGpp production **7**, 75ff
– regulatory cassettes for antibiotic export **7**, 92
Streptomyces fradiae, tylosin resistance **7**, 93
Streptomyces glaucescens, biosynthetic genes **7**, 447
– regulatory cassettes for antibiotic export **7**, 92
Streptomyces griseus, A-factor cascade **7**, 87f
– A-factor-induced regulation **7**, 446
– candicidin production **7**, 71f
– – phosphate repression **7**, 74
– decision-making growth phase **7**, 447
– decision phase **7**, 445
– exponential growth phase **7**, 445
– stationary growth phase **7**, 445
– streptomycin biosynthetic genes **7**, 91
– streptomycin production, regulation of – **7**, 445
– streptomycin synthesis regulation **1**, 441
Streptomyces hygroscopicus, bialaphos production **7**, 91
– rapamycin production **7**, 546f
Streptomyces lividans 66 **1**, 451
Streptomyces murinus, glucose isomerase **9**, 672

Streptomyces peuceticus, daunorubicin production **7**, 91
Streptomyces pilosus, ferrioxamines **7**, 209f
Streptomyces sp., formation of anticancer drugs **7**, 660
Streptomyces spp., as host cell **3**, 284
Streptomyces tsukubaensis, FK506 production **7**, 546
Streptomyces venezuelae, bromoperoxidases **8B**, 182
Streptomyces virginiae **7**, 78f
Streptomycetes **1**, 433ff, **2**, 465ff
– actinophages **1**, 440
– antibiotic formation, factors determining onset of – **7**, 86
– antibiotic production **7**, 70ff
– – genes **1**, 439
– – regulation of – **7**, 70
– antibiotic resistance, genes **1**, 447
– arthrospores **1**, 437
– attractant synthesis **1**, 444
– biotechnological application **1**, 434, 447ff
– biotransformation of antibiotics **1**, 452
– γ-butyrolactones, A-factor **7**, 78
– cell-cell interaction **1**, 437
– conidia **1**, 437
– conjugative plasmid **1**, 440
– culture **1**, 436f
– – genetic stability **1**, 439
– differentiation **1**, 437
– diffusible autoregulator **1**, 441
– – A-factor **1**, 441
– drug design **1**, 451
– ecology **1**, 436f
– extracellular enzyme production **1**, 451f
– gene regulation **1**, 440
– genetic engineering **1**, 438, 442, 450
– genetic instability **2**, 473
– genetics **1**, 438ff
– genetic screening **1**, 450
– growth conditions **1**, 438
– heterologous product secretion **2**, 476ff
– – soluble CD-4 (sCD-4) **2**, 478
– – tendamistat **2**, 478
– hormone synthesis **1**, 444
– human protein production **1**, 451
– induction of antibiotic resistance **7**, 92ff
– industrially applied species **1**, 447
– isolation **1**, 436ff
– β-lactam pathway, improvement **1**, 450
– life cycle **1**, 436
– metabolite overproduction **1**, 450
– modular gene product **1**, 446
– mutasynthesis **1**, 450
– mycelium **1**, 437
– nitrogen cycle **1**, 450
– pathway engineering **1**, 450
– phosphate regulation **1**, 443

– precursor formation improvement **1**, 451
– primary metabolism **1**, 442
– products **1**, 447
– recombination-deficiency **2**, 473
– regulation of antibiotic production, interfering metabolites **7**, 73
– regulation of secondary metabolite production **7**, 443ff
– regulatory codon **1**, 442
– regulatory genes **7**, 27
– repellant synthesis **1**, 444
– RNA polymerase, σ-factor **1**, 440
– secondary metabolism **1**, 443ff
– – regulation of **1**, 447
– secondary metabolites **1**, 443ff
– strain improvement **1**, 449
– substrates **1**, 436f
– taxonomy **1**, 434
– thermophilic **11C**, 60
– transcription promoter **1**, 440
– – RpoD-like sequence **1**, 440
– vectors **1**, 442
streptomycin, biosynthesis, gene cluster **1**, 441
– – regulation **1**, 441
– biosynthetic genes **7**, 417
– – in *Streptomyces griseus* **7**, 90
– pathway for its release **7**, 426
streptomycin biosynthesis, regulation of – **7**, 88
streptomycin production, enzymes, in biosynthetic pathway **7**, 418f
– gene clusters **7**, 420
– gene products **7**, 418
– in *Streptomyces griseus*, regulation of – **7**, 445
streptomycins **7**, 415ff
– basic pathway formula **7**, 463
– biosynthetic pathway **7**, 416
– condensation of subunits **7**, 425
– export of – **7**, 425
– processing of subunits **7**, 425
streptomycin sulfate, clarification of bacterial homogenate **9**, 94
streptonigrin, cytotoxic antitumor agent **7**, 656
Streptoverticillium **1**, 436
stress, influence on cell growth in biotechnological processes **10**, 574
stress resistance, of crop plants **5B**, 146, 170
striatals **7**, 493f
– biosynthesis **7**, 494
– structure **7**, 504
striated muscle **5A**, 31
striatins, cytotoxic activity **7**, 507
– structure **7**, 504
strictosidine, biosynthesis of **8A**, 352
strictosidine synthase **8A**, 352
string matching **5B**, 332
strobilurins **7**, 496ff
– blockage of respiration in eukaryotes **7**, 497
– fungistatic activity **7**, 497

– mode of action – **7**, 497ff
– possible functions **7**, 501
– producing organisms **7**, 499
– structures **7**, 498
stroke **5A**, 130ff, **5B**, 68
– medical applications of recombinant proteins **5A**, 130f
strontium, removal by rhizofiltration **11B**, 392f
structural families, of proteins **5A**, 71ff
structural genomics **5B**, 107
structural proteins **5A**, 32ff
structural signatures **5B**, 350f
structured lipids, production of **8B**, 289f
structured triacalglycerides (STs) **8A**, 135
structured triglycerides, production of **8B**, 289
strychnine structure **8A**, 353
styrene, degradation of, in biological waste air purification **11C**, 286
– degradation rate in bioscrubbers **11C**, 309
– scrubbing liquid quantities for the absorption of **11C**, 312
styrene emissions, control of **11C**, 366
styrene oxidation, by *Pseudomonas putida* **8B**, 375
subcellular components, separation of **5B**, 247
sublimation rate, freeze-drying **3**, 701
submerged fixed-bed reactors *see also* biofilm reactors
– aeration **11A**, 354f
– applications **11A**, 351ff, 358ff
– biofilms in, composition of **11A**, 356f
– – development of **11A**, 356f
– – structure of **11A**, 356ff
– – thickness of **11A**, 356
– BOD elimination **11A**, 354
– cascade arrangement **11A**, 355
– definition **11A**, 350f
– in wastewater treatment **11A**, 349ff
– maximum height **11A**, 355
– nitrification in **11A**, 359
– phosphorus removal in **11A**, 359
– principle **11A**, 350f
– reactor volume **11A**, 355f
– scheme **11A**, 351
– substratum **11A**, 351ff
– – extruded grid type **11A**, 352
– – materials **11A**, 352f
– – mushroom-type biofilm **11A**, 353
– – surface structure **11A**, 352
– use in, detergent and cosmetics production plant **11A**, 361f
– – ice-cream plants **11A**, 359f
– – slaughterhouses **11A**, 361f
– – vegetable plants **11A**, 359f
submerged process, in citric acid production **6**, 322f
submicellar separation **5B**, 274
subsoil, biogeochemical conditions in **11B**, 403

– pollutant degrading microbes in **11B**, 403
substance P receptor antagonists **7**, 120
substrate concentration, control **3**, 365
– kinetics **3**, 332
– on-line calculation, anaerobic wastewater process models **4**, 470ff
– profiles in biofilm **1**, 227
substrate consumption, in anaerobic wastewater processes **4**, 457ff
substrate inhibition **1**, 132f
– kinetics **3**, 331
– of alcohol dehydrogenases **8A**, 429f
substrate level phosphorylation **1**, 56
substrate modification, improvement of enzyme selectivity **8A**, 380f
substrates, for cell cultures **1**, 195ff
– for clostridia **1**, 287ff, 308
– for corynebacteria **1**, 454
– for lactic acid bacteria **1**, 334ff
– for methanogens **1**, 255
– for methylotrophs **1**, 266f
– for pediococci **1**, 340
– for streptomycetes **1**, 436f
– spectrum expansion **1**, 195, 201
– uptake, elemental balance **3**, 373
– – gaseous – **3**, 366
– – non-volatile – **3**, 365
– – rates **3**, 364ff
– – volatile – **3**, 366
substrate tuning **8A**, 231
substrate uptake, effect in microbial growth **1**, 139
– in plant cell cultures **1**, 586
– saturation constant **9**, 191f
– – calculation methods **9**, 191f
– specific rate **1**, 132
– transport processes **9**, 190f
substrate uptake kinetics, in cell models **4**, 274f
substratum **11A**, 352ff
– in submerged fixed bed reactors **11A**, 351ff
– – extruded-grid **11A**, 352
– – materials **11A**, 352
– – mushroom-type biofilm **11A**, 353
– – surface structure **11A**, 352
subsurface, chemical-physical conditions **11B**, 50ff
– geochemical processes in **11B**, 49ff
– lithology **11B**, 50
– matrix diffusion **11B**, 50
– soil properties **11B**, 50
– structure of transport matrix **11B**, 49f
– transport of contaminants **11B**, 49ff
subsurface infiltration **11C**, 428, 430f
– use of, horizontal infiltration galleries **11C**, 430
– – infiltration wells **11C**, 430
– – seepage trenches **11C**, 431
subsurface water *see also* groundwater
– composition of **11B**, 51

– divisions of **11B**, 45
– effect of geochemical processes **11B**, 56f
subtiligase, cyclization of linear peptides **8A**, 256
subtilin, prepeptide structure **7**, 345
– primary structure **7**, 331ff
– solution structure, in aqueous solution **7**, 339
– – in lipophilic solvents **7**, 339
– structural gene **7**, 343f
subtilisin **5A**, 206, 208, **8A**, 221ff
– acylation of, castanospermine **8A**, 225
– disaccharides **8A**, 124
– – sucrose **8A**, 225
– esterase activity of **8A**, 221ff
– peptide bond formation, kinetically controlled
  – **8A**, 251
– preferred cleavage sites **8A**, 249
– regioselective hydrolysis of a pyrimidine nucleoside **8A**, 223
– selective acylation of carbohydrates **8A**, 123f
– solvent engineering **8A**, 226
– stability in anhydrous organic solvents **8A**, 224
– variants **8A**, 226
subtilisin Carlsberg **8A**, 251
– acylation of primary amines **8A**, 257
– kinetic resolution of β-sulfonamidopropionic acid esters **8A**, 222
subtractive methods **5B**, 149
succinate, formation by *Actinobacillus succinogenes* **10**, 581f
succinate metabolization, in clostridia **1**, 303
succinic acid, chemical synthesis **6**, 372
– fermentative production of **10**, 425
– – polymers **10**, 425f
– formation in wine making **9**, 477
succinic acid-based products, commodity chemicals **10**, 582
– microbial pathways **10**, 582
– specialty chemicals **10**, 582
succinoglycan, bacterial – **6**, 646f
– formation by *Agrobacterium radiobacter* **6**, 647
– structure **6**, 647
succinoglycan biosynthesis, gene sequence for formation of – **6**, 619
sucrose, acylation by subtilisin **8A**, 225
– determination with biosensors **4**, 78
– effect on plant cell culture **1**, 594
– quality parameters **6**, 9ff
– – EC specifications **6**, 9f
– – ICUMSA methods **6**, 9
sucrose analysis **4**, 90
sucrose esters, as nonionic surfactants **6**, 673
sufu **9**, 523ff
– production of **9**, 524f
sugar, analysis of raw – **6**, 10
– break-even price **6**, 50
– crystallization **6**, 7ff
– – products **6**, 8
– liquid –, specifications **6**, 11

– quality parameters, crystal sugar **6**, 9ff
– – liquid sugar **6**, 11
sugar acid metabolization, in clostridia **1**, 298
sugar-based raw materials *see* raw materials
sugar beet, in industrial ethanol production, typical composition **6**, 137
sugar beet processing **11A**, 200
sugarcane, in industrial ethanol production, energetics **6**, 185f
– – productivity **6**, 151
– – typical composition **6**, 136
sugar components, attachment to cyclitols **7**, 405ff
– basic pathways **7**, 402ff
– biogenesis **7**, 402ff
– in secondary metabolites **7**, 397ff
sugar concentration, effect in wine making **9**, 485
sugar content, beet sugar process materials **6**, 12, 16
sugar derivatives, monomeric – **7**, 436f
sugar esters, food use as nonionic surfactants **8A**, 120
– use in cosmetics **6**, 677
sugar factory wastewater, components **11A**, 200f
– flow fractions **11A**, 200f
– treatment of **11A**, 201ff
– – anaerobic contact sludge processes **11A**, 201f
– – long-term batch processing **11A**, 201
– – soil treatment **11A**, 201
sugar feedstocks, raw material prices **6**, 52ff
sugar processing **6**, 6ff
– – *see also* beet sugar processing
– microflora **1**, 335
– products from, crystallization **6**, 7ff
– – evaporation **6**, 7
– – extraction **6**, 7
– – purification **6**, 7
– schematic diagram **6**, 7
sugars, as protein folding aids **3**, 542
– lipase-catalyzed acylations of, modified – **8A**, 121ff
– – unmodified – **8A**, 119
– protein stabilization effects **3**, 705
– synthesis of, by one-pot multiple enzyme catalysis **8B**, 18
– – by two sep multiple enzyme catalysis **8B**, 19
sugar transport *see* saccharide transport
suicide reagents **9**, 42f
sulfate reducers **11A**, 35f
– competition with methanogenes in methane reactors **11A**, 35f
– oligonucleotide probes **11A**, 93
sulfate-reducing bacteria **10**, 277
– anaerobic cometabolism **11B**, 433
– metal corrosion by **10**, 294ff
– – mechanism **10**, 295
– metal precipitation by **10**, 240ff
– use in biofilm bioreactors **10**, 244

sulfhydryl oxidase, use in baking **9**, 675
sulfide *see also* sulfur
– metal corrosion **10**, 277f
– monitoring with biosensors **11A**, 128f
– oxidation of **10**, 277f
– – scheme **10**, 278
sulfide oxidation, wastewater treatment **11A**, 398f
sulfides, biotransformation by isolated enzymes **8A**, 576
– electrochemical enantioselective oxidation of, use of poly-L-valine **8B**, 506
– oxidation to sulfoxides by cyclohexanone monooxygenase **8A**, 572
sulfite, monitoring with biosensors **11A**, 128
sulfite oxidation, dissolved oxygen quantification **3**, 196
sulfite waste liquor, substrate for biomass production **9**, 211
Sulfolobus acidocaldarius **10**, 65
Sulfolobus shibatae, phage SSV1 **1**, 548
Sulfolobus solfataricus **10**, 65
– dialysis fermentation of **10**, 71
– MAGPIE project **5B**, 380
– 2D gel electropherogram profile **5B**, 250
β-sulfonamidopropionic acid esters, kinetic resolution by subtilisin **8A**, 222
sulfonates, degradation of **11B**, 186
sulfoxidations *see also* asymmetric sulfoxidations
– active-site model **8A**, 570ff
– asymmetric –, flavin dependent monooxygenase enzymes **8A**, 569ff
– proposed mechanism for **8A**, 570
sulfur *see also* sulfide
– in coal, removal of **10**, 172ff
– in wastewater **11A**, 197
– physiological function **1**, 175
– supply, in plant cell culture **1**, 581
sulfur assimilation **10**, 277
sulfur compounds, degradation by genetically engineered microorganisms **11B**, 452
– in beer **9**, 443
– in municipal wastewater **11A**, 178f
– metal corrosion by **10**, 277f
– oxidation of **10**, 277f
– – by bacteria **10**, 278
– removal from burn exhaust gases **11C**, 350f
sulfur cycle **1**, 93f, **10**, 276ff, 300f
– microorganisms involved in metal corrosion **10**, 295ff
– scheme **10**, 301
sulfur dissimilation **10**, 277
sulfur etherocycles, improvement of reduction efficiency in yeast **8A**, 381
sulfuric acid, biogenic corrosion **10**, 281f, 299
sulfur limitation, effect, on *Escherichia coli* **1**, 173
– – on metabolism **1**, 172ff
– – on pyruvate production **1**, 176
sulfur metabolism **1**, 93

– assimilatory sulfate reduction **1**, 94
– desulfurylation **1**, 94
– dissimilatory sulfate reduction **1**, 94
– sulfur oxidation **1**, 95
– sulfur respiration **1**, 95
sulfurous acid, effect in wine making **9**, 485
– formation in wine making **9**, 480
sulfur oxidizing bacteria **11A**, 398f
sulfur stereocenters of carboxylic acids **8A**, 109
sulindac, redox reaction **8A**, 576
summation relationships **9**, 128f
sunflower, cell wall polysaccharides **9**, 707
sunflower oil, enzymatic extration **9**, 726
– genetic engineering of – **7**, 138
sunflower protoplasts, electrofusion of **10**, 515f
– exposition to microgravity **10**, 514f
sunscreens, of aquatic organisms **10**, 468
supercontigs **5B**, 385
supercritical CO_2 reactor, diagram of **8A**, 62
supercritical fluid chromatography (SFC) **4**, 172f
– capillary **4**, 172
supercritical fluid extraction (SCF) **3**, 579
– fluids **3**, 580
– – critical density **3**, 580
– – critical pressure **3**, 580
– – critical temperature **3**, 580
– of ethanol **3**, 580
supercritical fluids, lipase activity in **8A**, 60ff
superoxide dismutase, purification **3**, 602
superoxide dismutase activity, in clostridia **1**, 293
supports **11A**, 499f
– in activated sludge aeration tanks **11A**, 257ff
– in fixed film stationary bed reactors **11A**, 502f
– in fluidized bed reactors **11A**, 499f, 502ff
– – characteristics of **11A**, 499f
– – fluidization of **11A**, 507
– – loss of **11A**, 507
– in stationary fixed film reactors **11A**, 499
suppressor **2**, 11
– intergenic **2**, 11
– intragenic **2**, 11
surface aeration **3**, 237ff
– bubble entrainment **3**, 243f
– design **3**, 244
– effects of additives **3**, 241, 243
– impeller design **3**, 245
– impeller height effects **3**, 241
– in agitated vessels **3**, 240
– in laboratory vessels **3**, 240
– mass transfer coefficients **3**, 239, 241, 244, 269
– scale-up **3**, 245
– sorption characteristics **3**, 242
– with subsurface impellers **3**, 240
– with surface impellers **3**, 241
surface infiltration **11C**, 428ff
– flooding of grassland **11C**, 428f
– – scheme **11C**, 429
– recharge through overgrown soil **11C**, 428f

– use of infiltration basins **11C**, 429
– use of trough-trench systems **11C**, 429f
surface moistening model **11C**, 341
surface-plasmon resonance, of protein complexes
   5A, 92
surface process, in citric acid production **6**, 320ff
surfactant-coated lipases **8A**, 53
surfactant groups **6**, 671
surfactants **6**, 659ff
– – *see also* biosurfactants
– affinity cosurfactants **3**, 609
– as adjuvants for vaccines **6**, 680f
– – trehalose dimycolate **6**, 681
– blood substitutes **6**, 678ff
– cosmetics emulsions **6**, 671ff
– effect on bioavailability of pollutants **11B**, 78ff
– for reversed micellar extractions **3**, 606f
– in foods, properties **6**, 684f
– liquid emulsions for drug delivery **6**, 680
– mechanisms in microbial hydrocarbon uptake
   11B, 67
– non-ionic –, degradation of **11B**, 172f
– oral emulsion systems **6**, 681f
– parenteral emulsions **6**, 677f
– prepared by lipase-mediated reactions **8A**, 120
– pulmonary **6**, 683
– stabilization of reserve micelles **8A**, 48f
– topical emulsion systems **6**, 682
– Triton **3**, 612
surfactin, from *Bacillus subtilis* **6**, 664
surfactin synthesis **7**, 67f
– by *Bacillus subtilis* **7**, 67ff
– – signal cascade **7**, 68
susceptibility loci **5B**, 68
susceptibility to disease **5B**, 93
suspended growth reactor, completely mixed –
   11A, 480f
– – mean cell residence time (MCRT) **11A**, 480
– – scheme **11A**, 480
– – types of suspended solids **11A**, 481
suspended solids, concentration in wastewater
   11A, 114
suspension adaption **5A**, 256f
suspension cells **3**, 107
suspension culture *see* plant cell culture
– aeration **3**, 237ff
– effects of, agitation **3**, 230f
– – shear stress **3**, 230
suspension cultures *see* plant cell suspension cultures
sustainable development **12**, 142
sustainable technology **12**, 138, 144
Svedberg relation **5A**, 91
SVI *see* sludge volume index
SWALL **5B**, 287
Sweden, biotechnology regulatory policy **12**, 604f
sweeteners **9**, 674
– carbohydrate-based – **9**, 737ff

– – consumption **9**, 741
– – definition **9**, 739
– – history **9**, 739ff
– – production **9**, 741ff
– – terminology **9**, 739
– in cultured dairy products **9**, 403
– proteinaceous – **10**, 577ff
sweet potato, in industrial ethanol production **6**, 141
Swiss cheese **9**, 376f
SWISS-PROT **5B**, 286
Switzerland, biotechnology regulations **12**, 606
SYFPEITHI **5B**, 411
symba process, yeast biomass production **9**, 211f
Symbiobacterium sp., tyrosine-phenol lyase **8B**, 95ff
symport, over membranes **1**, 74
synchrony, in plant cell cultures **1**, 584
syncytium **2**, 96
syn-eliminations, of selenoxide **8B**, 419ff
synteny **2**, 158, **5B**, 148, 176
– homology **2**, 172ff
synthase *see* ligase
synthases, systematic names **8B**, 6
synthetic enzymes **8B**, 491ff
– artificial peptides in stereoselective synthesis **8B**, 491ff
– cyclic dipeptides **8B**, 492ff
– poly-leucine **8B**, 499ff
– use of linear dipeptides **8B**, 508ff
– use of tetrapeptides **8B**, 511ff
– use of tripeptides **8B**, 510ff
– use stereoselective synthesis **8B**, 492ff
synthetic gene **2**, 342ff, 392
synthetic gene expression **2**, 346ff
– fusion-protein expression **2**, 346
– transport expression **2**, 347
synthetic organics, in compost **11C**, 145
syntrophic bacteria **11A**, 457f
syntrophic interaction, of fermentative and acetogenic bacteria **11A**, 20
Syntrophococcus sucromutans, glucose metabolism **11A**, 462
Syringa vulgaris, production of verbascoside **7**, 599f
syrups **9**, 758ff
– applications **9**, 760
– composition **9**, 759
– production of **9**, 669ff, 758f
– – dextrose – **9**, 753
– – fructose **9**, 756
– – maltose – **9**, 758
– – oligosaccharide – **9**, 758
– properties **9**, 759
systematic base calling errors **5B**, 202
systemic acquired resistance, artificial induction of **10**, 494
– elicitors **10**, 494f

– in plants **10**, 494f
– signaling compounds **10**, 494f
systemic inflammatory response syndrome (SIRS) **5A**, 316f
– monoclonal antibody therapy **5A**, 316
systemic vasculitis **5A**, 318f
– monoclonal antibody therapy **5A**, 318
system identification *see also* Bioreactor identification
– parameter vectors **4**, 229f, 238f
– schematic description **4**, 227
– state vectors **4**, 229f, 238
– theory **4**, 226ff
system noise, in DNA sequencing **5B**, 205f
systems Biology **5B**, 438

T

TAB (Technology Assessment Board) **12**, 21, 24
tagatose-1,6-diphosphate aldolase, overexpression of **8B**, 14
tagatose-6-P pathway **1**, 353
tag sequence studies **5B**, 24
Taiwan, bio-products **12**, 405
– biotechnology **12**, 403ff
– – companies **12**, 405
– – industry **12**, 404ff
– – universities **12**, 404
– biotechnology regulatory policy **12**, 408
Taiwan Bioindustry Development Association *see* BIDEA
Tanabe process, of diltiazem synthesis **8A**, 147
tandem mass spectrometry **5B**, 255
tandem processor systems **4**, 570
tangential filtration **3**, 440f
tangential flow filtration (TFF) *see* crossflow filtration (CFF)
Tanimoto coefficient **10**, 55f
tapé, production of **9**, 528f
tapioca *see* cassava (*Manihot esculenta*)
tapioca starch, composition **6**, 32, 34f
Taq DNA polymerase **2**, 269, 334, 394, **5B**, 196
– error rate **2**, 270
– properties **2**, 336
Taq polymerase **10**, 87ff
TAR decoys, antiviral action **5A**, 476
tarhana, production of **9**, 547
tar oil, soils contaminated by – **11B**, 116
tartaric acid, from wine making residues **6**, 374
tartaric acid production **6**, 373f
– by *Gluconobacter* sp. **6**, 374
tartrate metabolization, by pseudomonads **1**, 422
tartronate semi-aldehyde synthase **8B**, 85f
– acyloin condensation of glyoxalate **8B**, 86
tautomycin, inhibition of protein kinase C **7**, 677
taxol **7**, 665ff

– biosynthetic pathway **7**, 617
– from plant tissue cultures **7**, 614ff
– – yields **7**, 616
– from *Taxus brevifolia* **8A**, 330
– preparation of, using poly-leucine catalyzed
  epoxidation **8B**, 501f
– structure **7**, 615, 668
taxol synthesis, use of lipases **8B**, 385f
taxonomy, of bacilli **1**, 368ff
– of bacteriophages **1**, 548ff
– of corynebacteria **1**, 434, 453
– of lactic acid bacteria **1**, 328ff
– of methanogens **1**, 255ff
– of methylotrophs **1**, 268ff
– of pseudomonads **1**, 402, 404ff
– of streptomycetes **1**, 434f
taxonomy databases **5B**, 283
Taxus brevifolia **7**, 615, **8A**, 330
Taylor cone **5B**, 254
TBADH see Thermoanaerobium brockii alcohol
  dehydrogenase
TCA cycle *see* tricarboxylic acid (TCA) cycle
TCE *see* trichloroethylene
T cell activation, cytokine receptor expression **7**,
  545
– gene **7**, 545
– inhibition of –, by rapamycin **7**, 547
– – role of immunophilins **7**, 548f
– release of cytokines **7**, 545f
– role of calcineurin **7**, 549
– signal recognition **7**, 542ff
– – binding of antigen **7**, 542f
– – costimulatory signals **7**, 542ff
– signal transduction **7**, 544f
TCTD *see* tetrachlorothiophene dioxide
TDC *see* tyrosine decarboxylase
T-DNA **5B**, 156
Teceleukin **5A**, 151
technological developments, controllability **12**,
  17f
– predictability **12**, 17
technological innovation, biological revolution
  12, 8
– electronic revolution **12**, 8
– nuclear revolution **12**, 8
– use of plastics **12**, 8
technology, benefits **12**, 7ff
– – economic welfare **12**, 8
– – maintaining peace **12**, 9
– – prestige **12**, 9
– – social welfare **12**, 8
– critical evaluation **12**, 9ff
– – costs **12**, 10f
– – social relevance **12**, 10
– information **12**, 33f
– interactive assessment **12**, 22ff
– potential dangers **12**, 11
– social consequences **12**, 32

technology assessment (TA) **12**, 5ff
– failure **12**, 15ff
– history **12**, 11ff
– methodology **12**, 12ff
– office **12**, 13ff
– predictions **12**, 17
technology planning, institutionalization **12**, 35f
– – decentralized **12**, 36f
– – multiform **12**, 36f
technology policy, governmental, history **12**, 28ff
– industrial **12**, 26ff
– information **12**, 33f
– interactive **12**, 30ff
– OECD activities **12**, 29f
– origins **12**, 26ff
– U.S. Congress **12**, 7ff
technology transfer offices **5B**, 98
Tectiviridae **1**, 550
teicoplanin **7**, 375f
– effect in mouse septicemia **7**, 380
– heptapeptide, amino acid composition **7**, 384
– structure **7**, 376
teicoplanin aglycone, stereo model **7**, 377
Tellus model **11A**, 572
TEMED *see* N,N,N',N'-tetramethylethylene diamine (TEMED)
tempeh **9**, 529ff
– nutritional characteristics **9**, 533
– production of **9**, 530ff
– – preparation of soybeans **9**, 530
temperature, control **4**, 515
– effect on, enzyme activity **9**, 146f
– – microbial growth **9**, 198
– – wine fermentation **9**, 481ff
– effect on microbial activity **11B**, 429f
– influence on microbial growth **10**, 273
– instruments for determination of **4**, 10
– measurement **4**, 10
– – accuracy of **4**, 21
– optimization **4**, 526ff
– – anaerobic wastewater process models **4**, 479f
– – computer-controlled **4**, 479
– – in batch bioreactors **4**, 528ff
– optimum **4**, 6f
– use in process control **4**, 6f
temperature control, of specific gene expression
  rate **4**, 501f
– – in a two-stage continuous culture system **4**,
  501ff
temperature measurement, in production bioreactors **3**, 322
template-primer-polymerase complex **5B**, 197f
tenderization, of meat **9**, 698
tentative consensus sequences **5B**, 178
teratocarcinoma **2**, 760
– development **2**, 760
– embryoid body **2**, 760
teratoma **2**, 760

terminal restriction fragment length polymorphisms (T-RFLP), in microbial community analysis **11A**, 90f
terminal transferase **2**, 378, 381
terminator **5B**, 388
terpenes **10**, 376
– – *see also* diterpenes, monoterpenes, sesquiterpenes
– – *see also* sesquiterpenes
– bioconversion of, microbial transformation reactions **10**, 375
– biosynthesis **7**, 35
– chlorinated – **8B**, 185
– cyclic –, hydroxylation of **8A**, 492
– enzymatic biosynthesis of **8B**, 365
– hydroxylation of **8A**, 479, 491ff
terpenoid biosynthesis **8B**, 28ff
terpenoid flavor and fragrance compounds, biotechnological production of **10**, 373ff
– – future aspects **10**, 412ff
terpenoid flavor compounds, anticarinogenic activity **10**, 415
– antimicrobial activity **10**, 415
terpenoid flavor formation, bioprocess modifications **10**, 414
– genetic engineering approaches **10**, 413
terpenoid fragrance compounds, anticarinogenic activity **10**, 415
– antimicrobial activity **10**, 415
terpenoid fragrance formation, bioprocess modifications **10**, 414
– genetic engineering approaches **10**, 413
terpenoids *see also* monoterpenoids
– biotransformations of, with flavor and fragrance impact **10**, 376ff
– enantioselective biotransformation of **10**, 402ff
– microbial transformation reactions **10**, 375
– natural substances class **10**, 413
– structural class **7**, 39
terpenyl esters *see also* monoterpenyl esters
– *in vivo* production of **10**, 399
– laboratory-scale production of **10**, 401
terphenyl, hydroxylation of **8A**, 500
terpineol, product of limonene bioconversion **10**, 391ff
terraferm technology, biological soil treatment by – **11B**, 321
Terranox system **11B**, 344
terrestrial ecotoxicity test system, assessment of contaminated soils **11B**, 135ff
– – setup and assessment of test results **11B**, 136f
– – use of bacteria **11B**, 136
– – use of earthworms **11B**, 136
– – use of plants **11B**, 136
tertiary alcohols, kinetic resolution by pig liver esterase **8A**, 202

– lipase-catalyzed enantioselective reactions **8A**, 94
tertiary structure *see* protein structure
testosterone, hydroxylation of **8A**, 489
testosterone 5α-reductase inhibitors **7**, 686
TETRA *see* tetrachloromethane
tetracenomycin, biosynthetic pathway **10**, 349
tetrachlorobenzene, aerobic degradation **11B**, 258
tetrachloroethene, biodegradation, anaerobic – **11B**, 255
tetrachloroethene (PER, PCE), biodegradation **11B**, 254f
tetrachloromethane degradation **1**, 313
tetrachloromethane (TETRA), biodegradation pathways **11B**, 252f
tetrachlorothiophene dioxide (TCTD), Diels–Alder cycloaddition **8B**, 418
tetracycline, effect on transcription regulation **2**, 634
tetracyclines, structures **7**, 34
tetrad analysis, of yeast spores **1**, 473
tetrafibricin, fibrinogen binding antagonists **7**, 115
Tetragenococcus **1**, 340
tetrahydrobiopterin **9**, 32
tetrahydrofolate **9**, 32
tetrahydrofolic acid **1**, 72
tetrahydrofuran, bacterial degradation of **11B**, 205f
Tetrahymena **10**, 145f
– expression of foreign genes **10**, 147
– growth kinetics **10**, 145
– use in bioassays **10**, 140f
Tetrahymena thermophila, enzyme production **1**, 239
– formation of acid hydrolases **10**, 142ff
– formation of lysosomal enzymes **10**, 143
– growth in bioreactor **10**, 143
– high cell density cultivation of **10**, 142f
– transformation of **10**, 147
N,N,N',N'-tetramethylethylene diamine (TEMED), in polyacrylamide gel electrophoresis **3**, 720
tetrapeptide catalysts, D-proline based –, acylation of alcohols **8B**, 512f
– L-proline-based, acylation of alcohols **8B**, 512
tetrapeptides, use in stereoselective synthesis **8B**, 511ff
textile desizing, protective layer **5A**, 194
– stages **5A**, 194f
– use of α-amylases **5A**, 194f
textile industry, use of cellulases **5A**, 204f
textiles, biodegradation of **10**, 308
T-flask, oxygen concentration **3**, 240
– surface aeration **3**, 240
Thailand, biotechnology **12**, 420ff
– forest conservation **12**, 438f, 443

Thalictrum minus, auxin effect in suspensions **1**, 591
– berberine production **1**, 598f
Thalictrum sp., accumulation of protoberberines **7**, 604f
thaumatin, recombinant – **10**, 579
Thaurea aromatica, degradation of benzoate **11B**, 177
thebaine, transformations by *Trametes sanguinea* **8A**, 344
thelephoric acid, iron-binding compounds **7**, 232
– structure **7**, 232
The Netherlands, biotechnology regulatory policy **12**, 597ff
– contaminated site management, financing schemes **11B**, 15
– – legal development **11B**, 10
– – principles **11B**, 12
– – registration **11B**, 17
– – risk assessment **11B**, 19
– – technical development **11B**, 10
– guidelines, for foods containing rDNA-encoded materials **12**, 177ff
– identification of contaminated sites **11B**, 17
– public opinion survey, on biotechnology **12**, 193, 602
– regulatory concerns, action thresholds **11B**, 21
– – conflict management **11B**, 28
– – enforcement **11B**, 27
– – in soil contamination **11B**, 7ff
– – liability rules **11B**, 14
– – oranization **11B**, 27
– – priority setting **11B**, 23
– – remediation targets **11B**, 24
– – verification of clean-up **11B**, 24
theobromine, in cocoa cotyledons **9**, 563
therapeutic antibodies *see also* antibodies, monoclonal antibodies
– glycosylation **5A**, 255
– manufacture of, antibody purification **5A**, 263ff
– – bioreactor types **5A**, 261f
– – cell banking **5A**, 257
– – cell culture media **5A**, 256
– – cell line **5A**, 248f, 257
– – cell line stability **5A**, 258f
– – cell removal **5A**, 263
– – choice of host cell **5A**, 255
– – cloning techniques **5A**, 256
– – copy numbers **5A**, 252f
– – culture conditions **5A**, 255, 258, 262f
– – debris removal **5A**, 263
– – DHFR expression systems **5A**, 250
– – economic aspects **5A**, 266
– – gene amplification **5A**, 251f
– – gene expression systems **5A**, 250f
– – glutamine synthetase expression systems **5A**, 250
– – good manufacturing practice **5A**, 268f
– – high expression **5A**, 251ff
– – integration into genome **5A**, 252f
– – key parameters **5A**, 260
– – Kozak sequence **5A**, 253
– – limits **5A**, 260
– – methods **5A**, 252f
– – methotrexate **5A**, 253f
– – pilot scale fermentations **5A**, 260
– – process consistency **5A**, 256
– – process definition **5A**, 259f
– – process monitoring **5A**, 261
– – product consistency **5A**, 254f
– – production systems **5A**, 261f
– – product recovery **5A**, 262f
– – reproducibility of bioprocessing **5A**, 255
– – selectable marker **5A**, 253
– – selectable marker genes **5A**, 250
– – selection **5A**, 256f
– – selection of suitable cell lines **5A**, 251ff
– – serum-free media **5A**, 256f
– – stability of expression **5A**, 254
– purification of **5A**, 263ff
– – DNA contamination **5A**, 265
– – protein contaminants **5A**, 264f
– – purification methods **5A**, 267f
– – requirements **5A**, 264f
– – virus contamination **5A**, 265
– robustness **5A**, 265f
therapeutic applications, of antibody fragments **5A**, 313
– of monoclonal antibodies **5A**, 311ff
– – autoimmune disorders **5A**, 318f
– – clinical applications **5A**, 315f
– – immunosuppression **5A**, 317f
– – infection **5A**, 316
– – rheumatoid arthritis **5A**, 318f
– – sepsis syndrome **5A**, 316
– – targets for intervention **5A**, 314
– – transplantation **5A**, 317f
therapeutic cloning **5B**, 420
therapeutic products **3**, 9, 57ff
– – *see also* peptide drugs, polypeptide drugs, protein drugs
– drug operation, scale **3**, 8
– insulin **3**, 58, 70
– process development **3**, 18
– proteins **3**, 10
– purification **3**, 57ff
– recombinant – **3**, 67
– second-generation peptides **3**, 69
– sources **3**, 41, 57ff
– tissue plasminogen activator (tPA) **3**, 68
therapeutic proteins *see also* proteins
– antibodies **5A**, 311ff
– – clinical applications **5A**, 311ff
– – engineering and expression **5A**, 219ff
– – for sepsis **5A**, 329ff
– – in chemotherapy **5A**, 371ff

– – in chronic therapy **5A**, 359ff
– – manufacture of **5A**, 245ff
– – preclinical testing **5A**, 289ff
– – ReoPro **5A**, 343ff
– blood products **5A**, 21
– classes **5A**, 21
– economics **5A**, 531
– functions **5A**, 21ff
– growth factors under development **5A**, 156
– in humans and animals **5A**, 125ff
– in tumor therapy **5A**, 135ff
– regulations **5A**, 495ff
– regulatory factors **5A**, 21
– under development **5A**, 171
– use as vaccines **5A**, 21, 137ff
– use in immunosuppression **5A**, 141
thermal bio-waste treatment **3**, 779
thermal mass flow meter, capillary **4**, 59
– immersible **4**, 59
thermal soil treatment **11B**, 305ff
– *ex situ* **11B**, 307f
– *in situ* **11B**, 308f
– plants in Germany **11B**, 467
thermistor **4**, 85, 208
– application of **4**, 89f
– glucose oxidase activity **4**, 88
Thermoacetobacter ethanolicus, ethanol formation **6**, 132
Thermoactinomyces **1**, 368
thermoalkaliphiles **10**, 69
Thermoanaerobacter, amylases **9**, 745
Thermoanaerobium brockii alcohol dehydrogenase (*TBADH*), applications of **8A**, 441f
– Baeyer–Villiger biotransformations **8A**, 568
– secondary – **8A**, 403
– thermostability of **8A**, 441
– tolerance to organic solvents **8A**, 441
thermobacteria **1**, 329
thermodynamics **1**, 52ff
– first law **1**, 52
– free energy **1**, 53
– internal energy **1**, 52
– irreversible **1**, 52ff
– of microbial growth **1**, 141
– second law **1**, 52
– standard free energy of formation **1**, 53
thermodynamic water activity, control methods **8A**, 59f
thermogram, cultivation of *Kluyveromyces fragilis* **4**, 190
thermolysin **3**, 452, **5A**, 207
– peptide synthesis **8B**, 355f
– preferred cleavage sites **8A**, 249
– production of aspartame **8B**, 286f
thermometer, Pt-100 resistance thermometer **4**, 10
thermophiles **10**, 273f
– cellulase, xylanases and chitinase from **10**, 83

– characteristics of **10**, 66
– production of fuel alcohol **6**, 131f
– proteolytic enzymes from **10**, 86
– starch hydrolyzing enzymes from **10**, 79f
thermophilic actinomycetes *see* actinomycetes
thermophilic bacilli *see* bacilli
thermophilic bacteria, in wastewater treatment **11A**, 403ff
thermophilic fungi *see* fungi
thermophilic microorganisms, bioleaching by **10**, 215
– specific growth rate **1**, 128
thermoplasts **10**, 308f
thermoselect process **11C**, 26ff
– energy balance **11C**, 30
– input and output data **11C**, 29
– scheme **11C**, 28ff
thermo Sequenase **5B**, 197
thermosome **10**, 72
thermostability, of lipases, directed evolution of **8A**, 149f
thermostable enzymes **10**, 72f
– amylases **10**, 74ff
– cellulase **10**, 81f
– chitinases **10**, 84f
– DNA processing enzymes **10**, 87ff
– proteases **10**, 85ff
– pullulanases **10**, 74ff
– xylanases **10**, 82ff
Thermotoga maritima **10**, 65, 81
Thermus aquaticus, Taq DNA polymerase **5B**, 196f
Thermus thermophilus, fumarase **8B**, 109
thiamin, cell-based biosensor assay **4**, 91
thiamine, thermal destruction, kinetics **3**, 169f
thiamine diphosphate (TPP) **1**, 72
thiamine pyrophosphate **9**, 34
– structure of **8B**, 50
thiamin (vitamin B_1), chemical production of **10**, 321, 331
Thiele modulus **4**, 332f, **11A**, 498f
– generalized – **3**, 94
Thiobacillus ferrooxidans **10**, 65
– characterization of **10**, 201, 204
– fly ash leaching **10**, 196, 211
– in bioleaching **10**, 194ff
Thiobacillus sp. **11A**, 398f
Thiobacillus species, in bioleaching, pH effects **10**, 207
Thiobacillus thiooxidans, fly ash leaching **10**, 196, 211
– in bioleaching **10**, 194ff
– ornithine-containing lipid formation **6**, 664
Thiocapsa pfennigii, PHA synthase, gene structure **6**, 418
thioesterase domain, of polyketide synthases **10**, 362
thioguanine, antitumor activity **7**, 647

thiolactone inhibitors, of cyclohexanone monooxygenase **8A**, 573
thiol protease **10**, 86
thiols, lipase-catalyzed enantioselective reactions
of **8A**, 102f
thionine, as redox mediator in bioelectrochemical
fuel cells **10**, 13ff
thiophenes, chlorinated – **8B**, 187
thioproteases, characteristics **8A**, 248
Thiosphaera pantotropha, coupled nitrification/de-
nitrification **11A**, 40
thiosulfate reductase **10**, 242
thiouridine, antitumor activity **7**, 647
thraumatin, proteinaceous sweetener **10**, 577ff
thraustochytrids **10**, 445
– cultivation of **10**, 448
– PUFA production by **10**, 463
threading **5B**, 338f, 347, 352f
– 2D – **5B**, 339
– 3D – **5B**, 339
three-dimensional (3-D) structures, of proteins
5B, 346
three-locus analysis **5B**, 169
threonine aldolase **8B**, 19f, 89f
– from mammalian tissues **8B**, 19
– synthesis of amino acids **8B**, 20
L-threonine dehydratase **8B**, 113
L-threonine, aspartate derived amino acids, in corynebacteria **8B**, 323
– bacterial production of, pathway engineering
8B, 321ff
– biosynthetic pathway **6**, 509
– feed-grade – **6**, 482ff
– producing mutants **6**, 483f
– producing strains, recombinant – **6**, 486f
L-threonine acetaldehyde lyase **6**, 532
L-threonine production **6**, 483ff, 508ff
– strain development **6**, 482
threonine production **1**, 209f
threshold factors, of biotechnology **12**, 344ff
thrombopoietin **5A**, 157f
– in therapy **5A**, 157
thrombosis, treatment with ReoPro **5A**, 366, 371
– urokinase therapy **5A**, 373
thromboxane **5A**, 34
thymide, production by *Sartina albida* **6**, 594
thymine **2**, 194, 320
thyroid stimulating hormone **5A**, 165
thyroxine, thyroid metabolite **8B**, 191
tiamulin, antibiotic action **7**, 496
– structure **7**, 495
tight junction **1**, 40
tightly integrated units **5B**, 438
tiling path **5B**, 47
time-of-flow distribution **4**, 139f
tissue cultures, of marine organisms **10**, 448
tissue culture technology *see* plant cell cultures,
plant tissue cultures

tissue organization **1**, 7
tissue plasminogen activator (tPA), recombinant
–, production **3**, 68, 473
– use in therapy **5A**, 129f
tissue samples, in proteomics **5B**, 245
tissue-type plasminogen activator, *in vitro* disulfide bond formation **5A**, 115f
– *in vitro* folding **5A**, 115ff
TISTR (Thailand Institute of Scientific and Technological Research) **12**, 421
titanium dipeptide complex **8B**, 508
T lymphocytes **5A**, 35
– CD4+ **5A**, 472f
TNF *see* tumor necrosis factor
TNT *see* trinitrotoluene
– – *see* 2,4,6-trinitrotoluene
tobacco mosaic virus (TMV) **2**, 565
– as gene expression vector **2**, 566
– infection cycle **2**, 566
tocopherol *see* vitamin E
α-tocopherol, biocatalytic synthesis, by pyruvate
decarboxylase **8B**, 27
– enzymatic synthesis of **8B**, 388ff
tofu **9**, 524
toluene **11C**, 309
– anaerobic degradation of **11B**, 183f
– bacterial degradation of **11B**, 153
– bioremediation of **11B**, 427
– direct ring cleavage **11B**, 214
– extraction of **11B**, 499
– metabolism of **11C**, 282f
– oxidation of, by genetically engineered microorganisms **11B**, 449
– scrubbing liquid quantities for the absorption
of **11C**, 312
– side chain attack of **11B**, 214
toluene degradation, by pseudomonads **1**, 416
– gene regulation **1**, 411
– in biological trickling filters **11C**, 342f
– rate in bioscrubbers **11C**, 309
toluene dioxygenase, in industrial biotransformations **3**, 459
tolypocin, iron-binding compounds **7**, 232
– structure **7**, 232
tomato, cell wall polysaccharides **9**, 707
tonoplast **1**, 36
topaquinone **9**, 36
topinambur *see* Jerusalem artichoke
topioca starch, production **6**, 34
topoisomerase I **2**, 207f
– effect on (an)aerobic growth **2**, 210
– gene expression regulation **2**, 209
topoisomerase II **2**, 207
– – *see also* gyrase
topoisomerases, inhibition by anticancer drugs **7**,
653ff
topoquinone (TPQ) **1**, 274
torulene **7**, 175

– structure **7**, 174
Torulopsis **1**, 269
– in miso fermentation **9**, 520
Torulopsis bombicola, production of sophorose lipids **6**, 669
Torulopsis candida, in idli fermentation **9**, 539
– in olive fermentation **9**, 612
– nitrilase activity **8A**, 299
Torulopsis holmii, in idli fermentation **9**, 539
– in sour dough **9**, 297
– in sour dough starters **9**, 255ff
Torulopsis magnoliae, glycolipid formation **6**, 664
TOSCA (Toxic Substances Control Act) **12**, 548, 571, 619
total organic carbon (TOC), in wastewater **11A**, 259ff
– of municipal sewage **11A**, 173ff
total turnover number, definition **8A**, 399
tower filters **11C**, 328
tower loop reactor **4**, 259ff
– c_{OF} sensor position **4**, 261f
– modelling of aerobic growth **4**, 259ff
– – cell concentration **4**, 260
– – oxygen concentration **4**, 260
– schematic diagram **4**, 259
tower reactor, back flow cell model **4**, 370, 373
– operation policies **4**, 378f
– process models **4**, 372ff
tower reactor models **4**, 351ff
– air lift reactor flow models **4**, 360ff
– air lift reactors **4**, 374
– axial dispersion model **4**, 364f
– bubble column flow models **4**, 354ff
– bubble column reactors **4**, 372ff
– classification based on flow characteristics **4**, 372
– flow configuration **4**, 352
– flow regime transitions **4**, 353f
– – homogeneous bubbly flow-slug flow **4**, 353f
– – slug flow-churn-turbulent flow **4**, 354
– – slug flow-dispersed bubble flow **4**, 353f
– mass transfer modelling **4**, 368ff
toxic compounds, in fertilizer **11C**, 18
– in soil, biodegradation of **6**, 689ff
toxicity, quantification of **11B**, 132
toxicity control, by biosensors **11A**, 135f
– – automated – **11A**, 136
toxicity equivalency **11A**, 573
toxicological assessment, of soil samples, methodologies **11B**, 129ff
toxic substances, in municipal wastewater **11A**, 180f
– removal in the activated sludge process **11A**, 267
toxin, gene transfer, by phages **1**, 545
– of *Bacillus thuringiensis* **1**, 377
Toxoplasma gondii, elimination from drinking water **11C**, 420ff

– transmission via drinking water **11C**, 414ff
– – listing of outbreaks **11C**, 418f
tPA *see* tissue plasminogen activator
TPL *see* L-tyrosine phenol-lyase
trace elements, for microbial growth **9**, 200
– removal during artificial groundwater recharge **11C**, 438
trace generation **5B**, 212ff
trace metals, effect on citrate accumulation **6**, 313
tracer **4**, 121f
– dyes **4**, 123f
– – bromocresol green **4**, 123
– fluorescent substances, coumarin **4**, 123
– for circulation measurement **4**, 304ff, 313ff, 319f
– – technetium isotope **4**, 319
– heat **4**, 139f
– noble gases **4**, 121
– pseudostochastical addition **4**, 121f
– sampling **4**, 121
trade secret **12**, 301
Trametes sanguinea, transformations of, neopinone **8A**, 344
– – thebaine **8A**, 344
Trametes versicolor, lignin-degrading white-rot fungi **11B**, 101
transaldolase, gene (*TAL*) **1**, 480
– use in multienzyme systems **8B**, 20
transaminase, reaction scheme **8B**, 324
transcription **2**, 234ff
– *cis*-acting elements **2**, 667ff
– factor of inversion stimulation (FIS) **2**, 253
– feedback control **2**, 254
– in animals **2**, 666
– – silencer **2**, 667
– inhibitory antibiotics **2**, 248
– regulation by DNA conformation **2**, 214, 217f
– start site determination **2**, 404f
transcription boundaries **5B**, 366
transcription control, by DNA supercoiling **2**, 208f
– – promoter orientation **2**, 208
transcription enhancer **2**, 667
– function **2**, 669
– inducible **2**, 669f
– – inducing agents **2**, 670
– structure **2**, 669
transcription factor **2**, 670ff
– TAT protein **2**, 673
– *trans*-acting factor **2**, 670f, 721
– – action mechanism **2**, 671
transcription factor binding sites **5B**, 317
transcription initiation **2**, 235ff, 665ff
– Goldberg-Hogness box **2**, 666ff
– initiator (INR) **2**, 667
– preinitiation complex **2**, 671
– signals **2**, 666ff
– site **2**, 236f

– TAT-box **2**, 666ff
– upstream promoter sequence (UPS) **2**, 667ff
– – negative **2**, 667, 670
– – positive **2**, 667
transcription level, of heterologous genes **2**, 780
transcription promoter **1**, 97, 106, **2**, 234ff, 669f, 779, 806
– determination **2**, 404
– *E. coli lac* **2**, 412
– *E. coli tac* **2**, 412f
– *E. coli trp* **2**, 412
– inducible **2**, 669f, 720ff
– – inducing agents **2**, 670
– in fish **2**, 813
– in mammals **2**, 719
– in plants **2**, 562f, 633f
– – artificial **2**, 562, 633
– – expression specificity **2**, 633
– – for overexpression **2**, 562
– – for vectors **2**, 570
– – inducible **2**, 563
– – intron insertion **2**, 563
– – of cauliflower mosaic virus 35S RNA **2**, 562
– – promoter-trapping **2**, 561
– – regulated **2**, 562f
– – species specificity **2**, 633
– – strength improvement **2**, 563
– – test vector **2**, 561
– in streptomycetes **1**, 440
– investigation strategy **2**, 587
– in yeast **1**, 489, 499
– *lac* **2**, 237
– – protein interaction sites **2**, 237
– lambda p_1, p_r **2**, 411ff
– metabolic-response regulated **2**, 411
– metallothionein **2**, 779
– of cauliflower mosaic virus 35S RNA **2**, 633f
– of heat shock gene **2**, 593
– oxygen-responsive **2**, 412
– phage T7- **2**, 413, 721
– phosphate-responsive **2**, 412
– phylogenetic specificity **2**, 593
– Pribnow-box **2**, 236
– regulation system **2**, 411
– strong promoters **2**, 719
– structure **2**, 236
– synthetic **2**, 722
– temperature regulated **2**, 411
– tissue-specific **2**, 720
– "topological coupling" **2**, 209
transcription promoters, metallothionein – **3**, 32
transcription regulation **2**, 250ff, 590, 665ff
– – *see also* gene expression, gene activation
– A elements **2**, 780
– bacteriophage genes **1**, 554f
– effect of scaffold-associated regions (SARs) **2**, 674
– external modification **2**, 779

– heterologous genes **2**, 718ff
– in plants **2**, 633ff
– – modification by bacterial repressors **2**, 634
– locus control region (LCR) **2**, 675
– modification by bacterial repressors **2**, 563
– of heterologous genes, in tobacco **2**, 591
– – in tomato **2**, 591
– "switching" **2**, 563f, 590, 634
– transcription quality **2**, 593
– transcription quantity **2**, 593
– yeast genes **1**, 491f
transcription repressor **2**, 670
transcription termination **2**, 237
– rho protein **2**, 237
– signals **2**, 668
transcription terminator **2**, 237
transcription unit, structure in eukaryotes **2**, 750
transcript processing, of heterologous genes **2**, 594
transducer, detectors **4**, 169f
– receptor-transducer systems **4**, 169f
transducers for biosensors, electrochemical detectors **4**, 83f
– – amperometric detectors **4**, 84
– – fuel cells **4**, 84f
– – potentiometric detectors **4**, 84
– integration **4**, 81ff
– optical detectors **4**, 85f
– thermometric detectors, thermistor **4**, 85
transduction **1**, 545, **2**, 48, 50, 59, **12**, 80, 162f
– abortive **2**, 50
– complete **2**, 50
– generalized – **1**, 559, **2**, 59
– in bacilli **2**, 460f
– in Streptomycetes **2**, 472
– specialized – **1**, 559, **2**, 61
transduction of an extracellular signal, protein kinases, effect on neoplastic cell growth **7**, 675
transesterification, of racemic hydroxyacid esters **8A**, 416ff
TRANSFAC **5B**, 284, 411
transfection **2**, 705ff, 725
– – *see also* transformation
– electrotransfection **2**, 99
– of animal cells, bicistronic expression vectors **2**, 722f
transfection systems **5A**, 429ff
– anionic liposomes **5A**, 434
– calcium phosphate precipitation **5A**, 430f
– cationic liposomes **5A**, 432ff
– – efficiencies **5A**, 432
– – transfection levels **5A**, 432
– direct gene transfer **5A**, 429ff
– electroporation **5A**, 431
– liposome-HVJ complexes **5A**, 434
– microinjection of DNA **5A**, 430
– neutral liposomes **5A**, 434
– non-targeted cationic polymers **5A**, 431

– particle bombardment, DNA immunization **5A**, 432
– receptor-mediated endocytosis **5A**, 434f
– – DNA-binding **5A**, 435
– – DNA-compaction **5A**, 435
– – efficiencies **5A**, 437
– – endosome release **5A**, 436
– – targeting ligands **5A**, 436
– – targeting moiety **5A**, 435
transferase **1**, 66
transferases **5A**, 25f, **9**, 8
– carbon–carbon bond formation **8B**, 6
transferrin **5A**, 19, 30
transferrin receptor *see* ferritin/transferrin receptor
transfer RNA *see* tRNA
transfer RNA (tRNA) **1**, 98
– aminoacyl-tRNA **1**, 100
transformation **2**, 48, 51, 62, 381ff, **12**, 80, 162
– bacterial genera **2**, 62
– chromosomal integration, detection techniques **2**, 757ff
– – hybridization pattern **2**, 757f
– frequency, in competent cells **2**, 381
– in bacilli **2**, 460f
– in Corynebacteria **2**, 483f
– in Streptomycetes **2**, 472
– – electroporation **2**, 472
– – PEG-mediated **2**, 472
– in yeasts **2**, 508, 523
– mechanism in *Pneumococcus* **2**, 51
– methods **1**, 477
– of animal cells **2**, 665ff, 705ff
– – application **2**, 666
– – bicistronic expression vectors **2**, 710, 723
– – calcium phosphate method **2**, 705f
– – DEAE-dextran method **2**, 705f
– – efficiency **2**, 705ff
– – electroporation **2**, 705f
– – germ line cells **2**, 748ff
– – lipofection **2**, 705f
– – micro-injection **2**, 705f
– – protoplast fusion **2**, 705f
– – strategies **2**, 705ff
– of bacilli **1**, 395
– of embryonic stem cells **2**, 762
– of fungi **2**, 530ff
– – limitations **2**, 530
– – mutant recipient **2**, 531
– – techniques **2**, 530
– – wild-type recipient **2**, 531
– of leaf discs **2**, 573
– of plant organ cultures **1**, 601
– of plants **2**, 557ff
– of yeasts **1**, 476f
– plant cell transformation, *Agrobacterium*-mediated **2**, 630
– – cereals **2**, 632

– stability **2**, 714, 725
– transient gene expression **2**, 724
transformation of plants **2**, 557ff
– – *see also* plant cell transformation
transforming growth factor β **5A**, 132, 160f
– animal experiments **5A**, 161
– binding proteins **5A**, 161
– in therapy **5A**, 140, 161
transfusion **2**, 114
transgene **2**, 584
– – *see also* heterologous gene
– chromosomal integration **2**, 584
– – DNA rearrangement events **2**, 585
– – integration site **2**, 584
transgene expression *see* heterologous gene expression
transgene inheritance **2**, 584
transgenes **5B**, 153
– expression of **5A**, 389f
– persistence of **5A**, 390
– size **5A**, 389
transgenic animals **2**, 747ff, **3**, 35, **5A**, 539, **12**, 359
– breeding program **2**, 777
– effect of gene integration locus **2**, 776f, 781
– future possibilites **2**, 814f
– generation **2**, 749ff
– – cell type suppression **2**, 784
– – domestic animals **2**, 770
– – mice **2**, 781
– – purposes **2**, 783f
– – steps **2**, 749
– – transgenic ovary transplantation **2**, 774
– – yields in mice **2**, 769
– hemizygous **2**, 776
– insertion mutation **2**, 781
– line establishment **2**, 775ff
– – time schedule **2**, 778
– public opinion **12**, 453
– Southern blot analysis **2**, 757f
transgenic chicken **2**, 809f
– irradiated sperm insemination **2**, 811
– transferred genes **2**, 810
transgenic engineering, of plants **5B**, 155
transgenic fish **2**, 811f
– species **2**, 813
– transferred genes **2**, 813
transgenic livestock **2**, 793ff
– application **2**, 793ff
– carcass composition **2**, 793
– growth regulation **2**, 793
transgenic modifications, of crop plants **5B**, 153ff
transgenic offspring **2**, 775ff
transgenic plant cell cultures **7**, 623ff
– overexpression of enzymes **7**, 627
transgenic plant-derived vaccinogens **5B**, 155
transgenic plants **2**, 549ff, 629ff
– APHIS regulations **12**, 203ff
– development **12**, 197ff, 359ff

– – domestic concerns **12**, 207
– – international needs **12**, 207ff
– herbicide resistance of **10**, 493f
– in developing countries **12**, 362
– marketing **12**, 253
– production of compounds against phytopathogens **10**, 491
– production of PAHs **10**, 430f
– production of plantibodies **10**, 493
– regulations **12**, 197ff
– release **12**, 198ff, 251f
– transformation target species **2**, 631
– virus resistance of **10**, 492f
transglucosidase **9**, 746
transglutaminase **9**, 705
transglycosylation, chitooligosaccharide synthesis **8B**, 251
– diastereoselectivity of **8B**, 248
– glycopeptide synthesis **8B**, 252
– simple alcohols as acceptor substrates **8B**, 247f
trans-isopiperitenol, product of limonene bioconversion **10**, 396f
transition state analog, chemical synthesis of **8B**, 407ff
– chorismate mutase reaction **8B**, 416
– conjugation to carrier proteins **8B**, 409
– for acyl cleavage reaction via tetrahedral transition states **8B**, 411
– generation of catalytic antibodies **8B**, 410ff
transketolase **8B**, 20ff
– from spinach leaves **8B**, 21
– gene (*TKT*) **1**, 480
– synthesis of a bark beetle pheromone **8B**, 22
– synthesis of deoxysugars **8B**, 23
– synthesis of fagomine **8B**, 22
– use in organic synthesis **8B**, 20f
translation **1**, 97, **2**, 245ff, 412ff
– feedback control **2**, 254
– fidelity **2**, 248
– hybrid released **2**, 399
– inhibitory antibiotics **2**, 248
– initiation site **2**, 412
– *in vitro* **2**, 406
– overlapping genes **2**, 690
– polycistronic mRNA **2**, 688f
– – reinitiation **2**, 688f
– regulation mechanisms **12**, 165f
– – frameshifting **12**, 165f
– – hopping **12**, 166
– ribosomal frameshifting **2**, 688ff
– stringent response **2**, 248
– termination suppression **2**, 688f
translational entropy, in enzyme catalysis **8B**, 417ff
translation efficiency **2**, 682ff
– effect of hairpin structures **2**, 685
– influence of RNA sequence **2**, 682, 684
– Kozak sequence **2**, 684

translation elongation **2**, 247, 679
– factor EF-Tu **2**, 247
translation inhibition, anti-sense concept **2**, 325
translation initiation **2**, 245ff, 679ff
– cap-binding complex (CBP) **2**, 683
– cap-independent **2**, 687
– cardiovirus **2**, 687
– effect of untranslated regions (UTRs) **2**, 685
– eIF factor **2**, 680
– enterovirus **2**, 687
– initation factors **2**, 245f, 680ff
– – eIF modification **2**, 683
– – modification by viruses **2**, 683f
– in picornavirus **2**, 687
– melting model **2**, 682
– start codon selection **2**, 682
translation machinery, eukaryotic **2**, 679ff
– prokaryotic **2**, 680
translation regulation **1**, 106, **2**, 679ff
– – *see also* enzyme synthesis regulation
– attenuator model **1**, 106
– by mRNA structure **2**, 684ff
– by synthetic control units **2**, 347
– concerned genes **2**, 686
– ferritin/transferrin **2**, 681
– GCD/GCN system **1**, 495
– in yeast **1**, 494
– of phage genes **1**, 556
translation termination **2**, 247, 680
– release factor **2**, 247
transmembrane distillation **3**, 584
transmembrane helix prediction **5B**, 336
transmembrane proteins **5A**, 29, **5B**, 348
transmission disequilibrium test **5B**, 90
transmission electron microscopy **1**, 10, 12
– conventional **1**, 12
– cytochemical sample preparation **1**, 10
– immunocytochemical sample preparation **1**, 11
– low temperature **1**, 12
– – sample preparation **1**, 10
– shock freezing **1**, 10
– ultrathin sectioning **1**, 10
– – chemical fixation **1**, 10
TRANSPATH **5B**, 411
transplantation **5A**, 314, 317f
– mAb products **5A**, 222
– monoclonal antibody therapy **5A**, 317f
– role of tumor necrosis factor **5A**, 344
transplant rejection therapy, use of cyclosporin A **7**, 539f
transport proteins **5A**, 26ff
– as transmembrane proteins **5A**, 29
– ATPases **5A**, 29
– erythrocyte glucose transporter **5A**, 28
– facilitated transport **5A**, 27
– functions **5A**, 26ff
– receptor-mediated endocytosis **5A**, 30
transport systems, in clostridia **1**, 298

transposable element **2**, 29, 56, 440
– methylation **2**, 596
– RP4::Mu **2**, 434
– RP4::Mu hybrid **2**, 57
– Tn10, use in gene regulation **2**, 634
transposon **2**, 29, 56f, 440ff
– DNA sequencing using Tn1000 **2**, 297
– DNA sequencing using Tn5 derivatives **2**, 297
– DNA sequencing using Tn9 derivatives **2**, 296f
– for Gram negative bacteria **2**, 443f
– – characteristics **2**, 444
– – with reporter genes **2**, 443f, 449
– for *Streptomyces* **2**, 470
– – Tn4560 **2**, 471
– – Tn5060 **2**, 471
– in filamentous fungi **2**, 540
– Tn5 derivatives **2**, 443ff
– – with promoterless indicator genes **2**, 443
– Tn916 **2**, 56
transposon mutagenesis **2**, 441
– fragment specific **2**, 441f, 448
– in bacilli **2**, 462
– random **2**, 441f, 446
transposons, in pseudomonads **1**, 410f
– pWWO plasmid **1**, 411
– stability **12**, 170f
trehalosamine family, structure of – **7**, 470
trehalosamines **7**, 437f
– as glycosidase inhibitors **7**, 437
trehalose, in yeast **1**, 485
– protein stabilization effects **3**, 706
trehalose dimycolate **6**, 681
trehalose lipids, *Rhodococcus erythropolis* **6**, 663
trehazolin, structure **7**, 472
– trehalase inhibitor **7**, 439
TrEMBL **5B**, 287
trend analysis, of bioprocesses **3**, 389
– – degree of certainty **3**, 393ff
– – feature extraction **3**, 389ff
– – real-time profile **3**, 392ff
– – sequence extraction **3**, 395
– – ShapeDescriptor **3**, 393
– – template shape library **3**, 394
– – trend detection **3**, 389
triacylglycerides, containing medium chain fatty
   acids (MLMs) **8A**, 136
– containing PUFAs **8A**, 137
– glycerolysis to 1(3)-MAGs **8A**, 141
– hydrolysis or alcoholysis to MAGs **8A**, 141
– lipase-catalyzed reactions **8A**, 135
– structured – (STs) **8A**, 135
triacylglycerols **7**, 146ff
– biosynthetic pathway **7**, 144
– glucose conversion, overall stoichiometry **7**,
   145
– pathway of synthesis **7**, 143ff
tricarboxylic acid (TCA) cycle **1**, 85, 484, **11A**, 20

1,1,1-trichloroethane (TRI), abiotic degradation
   11B, 250f
trichloroethylene (CTE), biodegradation of, by
   Burkholderia cepacia **11B**, 410
– – by genetically engineered microorganisms
   11B, 449ff
– – by plants **11B**, 378
– – by *Pseudomonas mendocina* **11B**, 410
– – degradating microorganisms **11B**, 409f
– – use of bioaugmentation **11B**, 434
– bioreporter systems **11B**, 454
– degradation by cometabolism **11B**, 152f, 432f
– – cometabolic dechlorination by oxygenases
   11B, 254
– model target for bioaugmentation **11B**, 409
– phytotransformation of **11B**, 375f, 382f
trichloroethylene (TCE), degradation pathways
   11C, 285
– – in waste air cleaning **11C**, 285f
trichloromethane, cometabolic degradation of
   11B, 155
2,4,5-trichlorophenoxyacetic acid degradation **1**,
   201
trichloropropanes, biodegradation of **11B**, 255
Trichoderma, as biocontrol agent **2**, 120
– fusion products **2**, 120
– strain improvement **2**, 120
Trichoderma reesei, cellulolytic system of **10**, 526f
– growth kinetics **1**, 133
Trichoderma viride, cellulase of, glycosylation **8B**,
   255
Trichosporon cutaneum, growth kinetics **1**, 133
– in pozol fermentation **9**, 549
Trichosporon pullulans, in idli fermentation **9**,
   539
trichostatins, phosphoinositol kinase inhibition **7**,
   681
Trichothecium roseum, infection of wine grapes
   9, 500
trickle bed reactors **11C**, 346
– – *see also* trickling filters
– Biotex technique **11C**, 347
– variations **11C**, 347
trickling, biological –, fundamentals of **11C**, 339ff
– – modeling of **11C**, 339ff
trickling filter, wastewater treatment, modelling
   4, 426
trickling filters **11A**, 337ff, **11C**, 362, 393ff
– – *see also* biofilm reactors
– biofilms in, structure of **11C**, 279
– biological –, in practice **11C**, 342
– – large-scale – **11C**, 342
– – packing material **11C**, 335f
– – scheme of **11C**, 335
– – technical scale – **11C**, 336
– – toluene degradation **11C**, 342f
– clogging in **11C**, 337ff
– – avoidance of **11C**, 338f

– concentration of germs, in clean gas **11C**, 294
– construction guidelines **11C**, 336f
– definition of **11C**, 334
– degradation of, alkylated and polyalkylated benzenes **11C**, 286
– – dichloromethane **11C**, 285
– – diethylether **11C**, 286
– – *m*-cresol **11C**, 286
– – methyl-*t*-butyl ether **11C**, 286
– design guidelines **11A**, 339ff
– development of **11A**, 8f
– filter media **11A**, 338f
– growth of biofilm **11C**, 338f
– in waste gas purification **11C**, 261
– mass transfer **11C**, 340f
– modeling of **11C**, 341f
– operating parameters **11C**, 265
– operational guidelines **11C**, 336f
– operation of, cocurrent – **11C**, 337
– – countercurrent – **11C**, 337
– – crosscurrent – **11C**, 337
– packing material, wetting of **11C**, 336f
– pollutant concentrations **11C**, 265
– pollutant degradation in, kinetics **11C**, 341f
– – limiting step **11C**, 340ff
– pre-denitrification **11A**, 341ff
– principle of **11C**, 335f
– process development **11A**, 337
– process efficiency **11A**, 337f
– removal of surplus biomass **11C**, 338f
– scheme **11C**, 262, 359, 393
– second stage-, surface related nitrification rates **11A**, 340
– semi-technical –, distribution of biomass **11C**, 340
– simultaneous denitrification **11A**, 341
– specific surface **11A**, 338f
– treatment of waste gas pollutants in **11C**, 333ff
– use in drinking water treatment **11C**, 393
– wastewater treatment, with nitrification **11A**, 340f
– – with nitrification and denitrification **11A**, 341ff
– – without nitrification **11A**, 339
tridecanone, degradation of **8A**, 542
trifluoramphenicol, hydrolysis of, use of catalytic antibodies **8B**, 439f
triglycerides *see* triacylglycerols
– enantioselectivity of lipases **8A**, 91ff
– production of structured – **8B**, 289
trihydroxybenzenes, anaerobic degradation of **11B**, 181f
trihydroxybenzoates, anaerobic degradation of **11B**, 181
trinitrotoluene (TNT), as the sole nitrogen and carbon source **11B**, 283f
– biodegradability of **11B**, 468f
– biodegradation of **11B**, 315

– – by aerobic bacteria **11B**, 283f
– – by fermenting bacteria **11B**, 286f
– – by fungi **11B**, 289f
– – by *Phanerochaete chrysosporium* **11B**, 289f
– – by plants **11B**, 378f
– – by sulfate reducing bacteria **11B**, 286f
– bioremediation with the heap technology **11B**, 326
– conversion to triaminotoluene by anaerobic bacteria **11B**, 186
– decontamination, by mobilization **11B**, 293f
– degradation of **11B**, 474
– – degradation graph **11B**, 327
– microbial transformation of **11B**, 290ff
– remediation by immobilization **11B**, 292
– transformation in hybrid poplar tissues **11B**, 377
– worldwide production **11B**, 283
2,4,6-trinitrotoluene (TNT) **11A**, 383
triol, structure of **8B**, 22
triolein, structure **8A**, 194
triosephosphate isomerase, gene (*TPI*) **1**, 481
tripeptide benzamide, catalysis of enantioselective acylation **8B**, 510ff
tripeptides, use in stereoselective synthesis **8B**, 510ff
triple quadrupole mass spectrometer **5B**, 255
triptolide, structure **7**, 615
trisaccharides, synthesis by glycosylation **8B**, 253f
– synthesis by immobilized glycosyltransferases **8B**, 258
Tris-Cycle process **11A**, 276
trisomy **2**, 8, 158f
– in humans **2**, 8
trisomy X **5B**, 75
trisomy 13 **5B**, 76
trisomy 18 **5B**, 76
trisomy 21 **5B**, 75
triticale, as raw material for ethanol production **6**, 63
– autoamylolytical quotient **6**, 63
– autoamylolytic enzyme system **6**, 75, 95
– composition of **6**, 64
– processing in ethanol production **6**, 92
triticale malt **6**, 70
triticale mash, saccharification of **6**, 78
triton, as surfactant, for reversed micellar extractions **3**, 612
tRNA **2**, 238
– cistron **2**, 240
– gene frequency **2**, 240
– mutated suppressor tRNA **2**, 243f
– rare **2**, 248
– wobble hypothesis **2**, 241f
tropane alkaloids **8A**, 331ff
– biosynthesis of, hyoscyamine **8A**, 332ff
– – scopolamine **8A**, 332ff
– – tropic acid **8A**, 333

– from plant cell cultures **7**, 613f
– from plant hairy root cultures **7**, 619
– metabolic pathways, atropine **8A**, 335f
– – cocaine **8A**, 337
– microbial metabolism of **8A**, 335
– parasympathetic inhibition **8A**, 331f
– structures **8A**, 332
trophication, of lakes **11C**, 382f
– of reservoirs **11C**, 382f
tropic acid, biosynthesis of **8A**, 333
tropical forests, biodiversity **12**, 435ff
– – hypothesis of disturbance **12**, 435f
– conservation **12**, 437ff
– – by biotechnology **12**, 440ff
– products **12**, 437f
– utilization **12**, 437ff
– – by biotechnology **12**, 438ff
– – sustainability **12**, 442
tropical regions, classification **12**, 436
– genetic resources **12**, 442
– organizations **12**, 438ff
– research facilities **12**, 438ff
– state of biotechnology **12**, 441
tropinol dehydrogenase **8A**, 335ff
tropinone reductases **8A**, 334
trough-trench systems **11C**, 429f
– scheme **11C**, 430
TRRD – Transcription Regulatory Regions Database **5B**, 411
truffle **1**, 532
Trypanosoma brucei, ornithine decarboxylase **8B**,
   78
trypsin, preferred cleavage sites **8A**, 249
trypticase-yeast-extract-cystine (TYC) agar **1**, 349
tryptophan, biosynthesis, main pathway **3**, 287f
– effect on plant cell culture **1**, 596
– production in *Saccharomyces cerevisiae* **9**, 122f
– solubility **3**, 292
– uptake, by *Escherichia coli* **3**, 291
tryptophanase **6**, 523ff, **8B**, 91ff
– structure **6**, 523f, **8B**, 92
L-tryptophan, analogs **8B**, 94
– as feed additive **6**, 484ff
– biosynthesis, in *Escherichia coli* **6**, 489
– biosynthetic pathway **6**, 511
– cleavage of, by tryptophanase **8B**, 91
– conversion to indole-3-acetic acid **8B**, 89
– decarboxylation of, by aromatic L-amino acid
   decarboxylase **8B**, 83
– feed uses **6**, 468
– production by antimetabolic resistant mutants
   6, 488
– production strains, recombinant – **6**, 490
– substitution **8B**, 94
– tryptophanase catalyzed cleavage of, mechanism **8B**, 92
L-tryptophan production **6**, 512

– direct fermentation by overproducing mutants
   6, 485
– fermentation technology **6**, 484ff
– strain development **6**, 485ff
tryptophan pathway, cooperative feedback inhibition **9**, 130
tryptophan production, by recombinant *Escherichia coli* **3**, 287f
– – hyperproduction **3**, 292f
– by recombinant microorganisms, cultivation
   conditions **3**, 293
– – organism characteristics **3**, 293
– crystalline fermentation **3**, 292ff
– culture technique effects **3**, 289
– fed-batch design **3**, 292
– indole accumulation **3**, 292
– plasmid stability **3**, 288
– time course **3**, 292
– yield **3**, 288f
tryptophan synthase **6**, 525ff
– activity, in recombinant *Escherichia coli* **3**, 289
– inhibition **3**, 292
– reactions catalyzed by – **6**, 525
– S-substituent donors **6**, 526
tryptophan tryptophyl quinone (TTQ) **1**, 274
tryptophylquinone **9**, 36
TSCA **12**, 462ff
Tuber **1**, 532
Tuber melanosporum **1**, 525
tubular bioreactor **3**, 96, 101
– plug flow – **3**, 101
tubular centrifuge, grade efficiency curve **3**, 478
tubulin **5A**, 4
– contractile properties **5A**, 32
tumor *see also* cancer
– biodistribution studies **5A**, 361f
– gene therapy with adenovirus vectors **5A**, 413,
   421
– gene therapy with retroviruses **5A**, 405f
– monoclonal antibody therapy **5A**, 315, 320ff
– radiotherapy **5A**, 320
– treatment of **5A**, 361
tumor cells **3**, 107
– – *see also* cancer cells
– efflux of cytotoxic drugs **7**, 672
– multidrug resistance **7**, 540, 551
– redifferentiation of – **7**, 683
tumorigenesis **2**, 786ff
– *myc* gene **2**, 788f
tumor necrosis factor (TNF) **5A**, 283, 291f, 295f,
   332f
– antibodies **5A**, 293ff
– – in therapy **5A**, 296
– antibodies against –, CDP571 **5A**, 345ff
– function **5A**, 344
– half-life in blood **5A**, 334
– in Crohn's disease **5A**, 346
– in inflammatory bowel disease **5A**, 346

– in monoclonal antibody therapy, anti-tumor necrosis factor **5A**, 332f
– in tumor therapy **5A**, 136
– in ulcerative colitis **5A**, 346f
– response to inflammatory stimuli **5A**, 334
– testing of antibodies **5A**, 291
– TNFα, antibodies to – **5A**, 135
– – receptor **5A**, 135
– – use in therapy **5A**, 130, 134, 144
tumor specific antigens, in tumor therapy **5A**, 136
tumor suppressor protein, in tumor therapy **5A**, 136
tumor therapy **5A**, 135ff, 294ff
– antibody fragments **5A**, 294f
– hematopoietic growth factors **5A**, 137
– medical applications of recombinant proteins **5A**, 135ff
– use of monoclonal antibodies **5A**, 135f, 294f
tunicates **10**, 445f
– antiviral compounds of **10**, 460
– cytostatic/antitumor compounds **10**, 453
– substances of **10**, 446
tunnel composting **11C**, 142
turbidity, as function of cell concentration **4**, 182
– measurement methods **4**, 182ff
– of drinking water **11C**, 375, 422
– of river water **11C**, 380
turbulence, in stirred reactors, mass transfer coefficients **3**, 191
turbulence models **4**, 327ff
– single-phase flow **4**, 330
– two-phase flow, gas/liquid **4**, 328
turbulent flows, in bioreactors **4**, 115
β-turn **5A**, 62
Turner syndrome **5B**, 75
turning machines, aerated pile composting **11C**, 140f
turnover, definition of **8A**, 398f
turnover number, definition **8A**, 31
turpentine oil, constituents **10**, 404
twin studies **5B**, 80
two-dimensional gel electrophoresis **3**, 725
– – *see* 2D gel electrophoresis
two-factorial design, in medium optimization **3**, 134
two-hybrid assays, in drug screening **10**, 51
two-locus analysis **5B**, 169
two-phase extraction **3**, 593ff
two-phase flow **4**, 328ff
– gas/liquid, governing equations **4**, 328f
two-stage continuous culture system **4**, 497ff
– for recombinant fermentation processes **4**, 497ff
– schematic diagram **4**, 497
tylosin resistance, in streptomycetes, induction of – **7**, 93
tyndallization **10**, 287
type II diabetes **5B**, 117

typhoid fever, as waterborne disease, outbreaks in the 19th and 20th century **11C**, 415f
tyromycin, structure **7**, 519
tyrosinase, modification of coal **10**, 169
β-tyrosinase **6**, 522f
– reaction mechanism **6**, 522
tyrosine decarboxylase (TDC) **8B**, 82f
– of *Streptococcus faecalis* **8B**, 83
L-thyrosine, biosnythetic pathway **6**, 511
L-thyrosine production **6**, 510ff, 522
L-tyrosine, decarboxylation of, by tyrosine decarboxylase **8B**, 82
– synthesis of, by tyrosine phenol-lyase **8B**, 94
tyrosine phenol-lyase reaction, mechanism **6**, 524
tyrosine phenol-lyase (TPL), mechanism of action, with β-substituted L-alanines **8B**, 97
– – with L-tyrosine **8B**, 96
– of *Citrobacter freundii* **8B**, 95ff
– of *Erwinia herbicola* **8B**, 95ff
– of *Streptococcus faecalis* **8B**, 95ff
– of *Symbiobacterium* sp. **8B**, 95ff
– products from TPL catalyzed reactions **8B**, 94
tyrosine protein kinase inhibitors **7**, 680
L-tyrosine-related amino acids, enzymatic synthesis **6**, 523
tyrosyl-arginine production **3**, 453

U

UASB *see* upflow anaerobic sludge bed reactor, upflow anaerobic sludge blanket reactor
ubiquinone **1**, 71, 404
ubiquinone Q (coenzyme Q), microbial production of **10**, 336
ubiquitin, activities in protoplasts under stress **10**, 511f
UCT process, in enhanced biological phosphorus removal **11A**, 298
UHT process **10**, 288
– – *see* uperization
UK House of Lords Select Committee on Science and Technology **12**, 56
UK 63.052, antitumor agent, structure **7**, 651
ulcerative colitis, monoclonal antibody therapy **5A**, 319
– role of tumor necrosis factor **5A**, 346
– treatment with CDP571 **5A**, 346f
Ulcophyceae, characteristics of **10**, 112
ultrafiltration **3**, 440, 487ff
– boundary layer **3**, 489f
– capillary module, scheme **11C**, 408
– effects of solids concentration **3**, 495
– flux **3**, 489
– membrane structure **11C**, 406f
– transmembrane pressure **3**, 489
– use in water treatment **11C**, 405ff

ultramicrotomy **1**, 10
– cryo ultramicrotomy **1**, 10
– freeze etching **1**, 10
– freeze facturing **1**, 10
ultrasound **5B**, 83
ultrasound Doppler technique **4**, 112f, 117, 134ff, 141
– measurement of bubble velocity distribution **4**, 134ff
ultrasound pulse reflection methods, determination of interfacial areas **4**, 137
ultrasound transmission methods, determination of interfacial areas **4**, 136
umami taste **6**, 466
– definition **6**, 563f
UMP *see* uridine monophosphate
UNCTAD (United Nations Conference on Trade and Development) **12**, 294
undercooling, effects on freeze-drying **3**, 699
UNEP (United Nations Environment Program) **12**, 439, 627f
UNESCO, Universal Declaration on the Human Genome and Human Rights **5B**, 419
UN Food and Agriculture Organization **12**, 144
UNIDO (UN Industrial Development Organization) **12**, 625f
uni-Gene **5B**, 284
unigenes **5B**, 47
union Chemical Laboratories (UCL) **12**, 404
uniport, over membranes **1**, 74ff
uniquegene Cluster **5B**, 178
United Kingdom, biotechnology regulatory policy **12**, 573ff
– – history **12**, 516
– contaminated site management, financing schemes **11B**, 15
– – legal development **11B**, 10
– – principles **11B**, 12
– – registration **11B**, 17
– – risk assessment **11B**, 19
– – technical development **11B**, 10
– guidelines, on the assessment of novel foods and processes **12**, 175f
– identification of contaminated sites **11B**, 17
– – conflict management **11B**, 28
– – enforcement **11B**, 27
– – in soil contamination **11B**, 7ff
– – liability rules **11B**, 14
– – organization **11B**, 27
– – priority setting **11B**, 23
– – regulatory concerns, action thresholds **11B**, 21
– – remediation targets **11B**, 24
– – verification of clean-up **11B**, 24
– public opinion survey, on biotechnology **12**, 126, 193, 662
United Nations (UN), agencies **12**, 618ff
United States, biotechnology history **12**, 461ff

– biotechnology regulatory policy **12**, 51ff, 198ff, 241ff, 349, 461ff, 566ff
– – agriculture **12**, 57f, 461ff
– – biopharmaceutical sector **12**, 57, 461ff
– – Coordinated Framework **12**, 568ff
– – history **12**, 513ff
– patent laws **12**, 290f, 302ff
– – Continuation-in-Part (CIP) **12**, 315
– – swearing-back **12**, 316
– public opinion survey, on biotechnology **12**, 133f, 146ff, 473, 658
United States Environmental Protection Agency (USEPA), biowaste regulations **3**, 772
United States of America, contaminated site management, financing schemes **11B**, 15
– – legal development **11B**, 10
– – principles **11B**, 12
– – registration **11B**, 18
– – risk assessment **11B**, 20
– – technical development **11B**, 10
– identification of contaminated sites **11B**, 18
– – action threshold **11B**, 22
– – conflict management **11B**, 29
– – enforcement **11B**, 27
– – liability rules **11B**, 14
– – organization **11B**, 27
– – priority setting **11B**, 23
– – regulatory concerns **11B**, 7
– – remediation targets **11B**, 24
– – verfication of clean-up **11B**, 25
Universal Movement for Scientific Responsibility *see* MURS
uperization (UHT process) **10**, 288
upflow anaerobic sludge bed (UASB) reactor **11A**, 519ff
– granulation **11A**, 522
– in the potato processing industry **11A**, 206
– technology **11A**, 519ff
– treatment of agricultural wastewater and slurry **11A**, 226f
– treatment of starch factory wastewater **11A**, 203
upflow anaerobic sludge blanket reactor (UASB) **11A**, 471
– effect of gas recirculation **11A**, 551ff
– laboratory-scale – **11A**, 551f
upflow sludge bed *see also* expanded granular sludge bed reactor (EGSB), upflow anaerobic sludge bed reactor (UASB)
UPOV Convention **12**, 284ff, 292
– New Act **12**, 286f
UPOV (International Union for the Protection of New Varieties of Plants) **12**, 284, 638f
upstream activating site (*UAS*), in yeast **1**, 492f
upstream repressing site (*URS*), in yeast **1**, 492f
uranium, as electron acceptor for anaerobic respiration **11A**, 43
– bacterial reduction of **10**, 240

– bioleaching of **10**, 213
– biosorption by fungi **10**, 229f
– removal by rhizofiltration **11B**, 392ff
urea, estimation methods **1**, 126
urea fuel cell **10**, 11
uridine, production, by *Bacillus subtilis* **6**, 592f
– – by *Brevibacterium ammoniagenes* **6**, 592
– – by *Microbacterium lacticum* **6**, 592f
uridine monophosphate (UMP), fermentative production **6**, 579
uridine triphosphate (UTP), and phosphodiester formation **8B**, 233
uridylate biosynthesis, pathways **8B**, 445
urokinase, in therapy of thrombus **5A**, 373
U.S. Congress **12**, 514ff
– technology assessment (TA) **12**, 12ff
USDA (U.S. Department of Agriculture) **12**, 198f, 207, 241, 359, 461ff, 570ff)
U.S. Departments of Interior and Agriculture **12**, 42
U.S. Environmental Protection Agency *see* EPA
use patents **12**, 309
user interfacing, in automation systems **4**, 578ff
– X-windows-based **4**, 579f
U.S. Food and Drug Administration *see* FDA
U.S. Human Genome Program **12**, 331
U.S. National Academy of Science *see* NAS
U.S. National Institutes of Health *see* NIH
U.S. National Research Council *see* NRC
U.S. Office of Science and Technology Policy *see* OSTP
U.S. Patent and Trademark Office *see* PTO
U.S. Plant Genome Research Program **12**, 331
U.S. science policy **12**, 7ff, 461
U.S. Supreme Court, Chakrabarty case **12**, 284, 290
U.S. technology policy **12**, 7ff, 461
UV absorbance measurements **5B**, 272
UV radiaton, extraterrestrial spectrum **10**, 516

V

vaccination, plasmid vector design **5A**, 452
– use of plasmid DNA **5A**, 444
vaccine production **3**, 11, 24
– conjugate vaccines **3**, 11
– foot-and-mouth disease vaccine **3**, 25
– hepatitis B vaccine **3**, 11
– polio vaccine **3**, 24
vaccines **5A**, 21, 137ff
– AIDS **5A**, 138
– bacterial **5A**, 139ff
– cancer **5A**, 135, 138
– hepatitis virus **5A**, 139
– herpes simplex **5A**, 138f
– malaria **5A**, 139

– melanoma **5A**, 138
– plasmid DNA **5A**, 465
– recombinant proteins **5A**, 137ff
– viral – **5A**, 137f
vaccinia virus **2**, 729
vacuole **1**, 36, 578
vacuum biogas systems, integration into a new settlement **11C**, 252
– in urban areas, pilot project **11C**, 251ff
vaginal flora **1**, 331
valencene, biotransformation of **10**, 409
validamycin family, biogenesis of – **7**, 438
– structure of – **7**, 471
validamycins **7**, 438f
L-valine production **6**, 510
– by direct fermentation **6**, 492
valves, for bioreactors **3**, 423
vanadium haloperoxidases **8B**, 193
vancomycin **7**, 373f
– biotransformation of – **7**, 390
– effect in mouse septicemia **7**, 380
– structure **7**, 373
van der Waals energy **5A**, 88f
vanilla flavor, from plant cell cultures **7**, 617f
– natural – **8B**, 334
Vanilla planifolia, plant cell cultures **7**, 618
vanillin **7**, 617
– fermentative production of **8B**, 334ff
– – by hydrolysis of curcumin **8B**, 335
– – by plant cell culture **8B**, 335
– – ferulic acid as intermediate **8B**, 336ff
– – from ferulic acid **8B**, 338ff
– – from glucose **8B**, 335ff
– – from isoeugenol **8B**, 335f
– – with dioxygenase **8B**, 335f
– – with lipoxygenase **8B**, 335f
– production by *Amycolatopsis* **8B**, 339f
vanillin formation, by *Serratia marcescens* **8B**, 336
van't Hoff equation **5A**, 90
variable number of tandem repeats (VNTRs) **5B**, 43, 78, 90
varicella-zoster virus, monoclonal antibody therapy **5A**, 317
variegatic acid, iron-binding compounds **7**, 232
– structure **7**, 232
variegatin, iron-binding pigments **7**, 231
– structure **7**, 231
varnishes, biodeterioration of **10**, 311
vascular endothelial cell growth factor **5A**, 133f
VCI (Chemical Industry Association of Germany) **12**, 586
VCOGEM (Ad Hoc Committee for Recombinant DNA Activities) **12**, 600ff
vector **2**, 268, 382ff, 724
– – *see also* cloning vector, expression vector, viral vector
– disarmed *Agrobacterium tumefaciens* – **2**, 630
– for animals **2**, 717

– – characteristics **2**, 726
– for Corynebacteria **2**, 480f
– – bacteriophages **2**, 483
– – broad host range plasmid RP4 **2**, 484
– – endogenous plasmids **2**, 480f
– – integration – **2**, 482
– – pBL1 **2**, 480f
– – pHM1519 **2**, 480f
– – pNG2 **2**, 480
– – suicide – **2**, 482
– for filamentous fungi **2**, 531ff
– – chromosomal integration **2**, 531
– – endogenous plasmid **2**, 533
– – replicating – **2**, 533
– – reporter gene **2**, 537
– – stability **2**, 534
– for methylotrophic yeasts **2**, 523
– for multiplex DNA sequencing **2**, 286
– for plants **2**, 550ff
– – classes **2**, 561
– – expression – **2**, 561
– – transformation – **2**, 561
– for *Saccharomyces* **2**, 509ff
– – break introduction **2**, 510
– – chromosomal integration **2**, 510f, 522f
– – stability **2**, 510
– – transformation frequency **2**, 510
– for *Streptomyces* **2**, 467, 468ff
– – cassette insertion **2**, 470
– – cosmid **2**, 468
– – integration – **2**, 472
– – pGM **2**, 469
– – pGM11 restriction map **2**, 471
– – phage **2**, 467
– – pIJ702 restriction map **2**, 467
– – plasmid pIJ702 **2**, 467
– – restriction **2**, 472
– in plants **2**, 630
– intron-containing **2**, 722
– pAS1 **2**, 143
– pGP1-2 **2**, 143
– pJF119 HE **2**, 413
– pT7-7 **2**, 143
– shuttle – **2**, 723
– transfer – **2**, 729f
vector design, in HIV gene therapy **5A**, 479ff
– – adeno-associated virus based – **5A**, 480f
– – amphotropic murine leukemia virus based –
   5A, 479f
– – novel retroviral and based – **5A**, 480f
– – optimization of – **5A**, 484
vector plasmid *see* plasmid vector
vectors, design for gene therapy **5A**, 387ff
– – non-viral – **5A**, 427ff
– dihydrofolate reductase (DHFR) expression –
   3, 31f
– food grade – **12**, 171
– for streptomycetes **1**, 442

– GILSP recommendations **3**, 776
– in *Lactococcus* **1**, 362
– integration **12**, 171
– in yeast **1**, 499
– phage lambda **1**, 562
– phage Mu **1**, 562
– plasmid – **12**, 91f
– stability **12**, 171
– viral – **5A**, 395ff, **12**, 98ff
vegetable fermentations **9**, 629ff
– – *see also* indigenous fermented foods, olives
– culture development **9**, 651ff
– lactic acid bacteria in – **9**, 635
vegetable oil industry, wastewater treatment of
   11A, 203ff
vegetable plant **11A**, 359ff
– use of submerged fixed bed reactors in **11A**,
   359ff
vegetable processing, typical wastes **11A**, 224
vegetable protein *see* protein
vegetable raw material, extraction **9**, 706ff
vegetative incompatibility **2**, 79f
– *Podospora anserina* **2**, 79
Veillonella, anaerobic wastewater process **4**, 447
verbascoside, from plant cell cultures, yields **7**,
   600
– structue **7**, 599
Verlinden process, preparation of distillers'
   mashes **6**, 84
vermicomposting **11C**, 139
vertical flow composting **11C**, 144f
vertical flow reactors **11C**, 123
vessel *see* fermentation vessel
viability, of bioreactor biomass **3**, 386
viability assay, metabolic activity assay **4**, 202
– plate counting **4**, 202
– sample-flow analysis **4**, 202
– vital staining **4**, 202
viable plate count **1**, 120
Vibrio anguillarum, anguibactin **7**, 206f
vibriobactin **7**, 205f
– catecholate siderophore of *Vibrio cholerae* **7**,
   205
– structure **7**, 205
Vibrio cholerae, vibriobactin **7**, 205f
vibrioferrin, carboxylate-type siderophore of *Vibrio parahaemolyticus* **7**, 224
– structure **7**, 224
Vibrio fischeri, light emission **7**, 66
– OHHL **7**, 65f
– – structure **7**, 65
Vibrio fluvialis, fluvibactin **7**, 206
Vibrio parahaemolyticus, vibroferrin **7**, 224
Vibrio vulnificus, vulnibactin **7**, 206
Vicia faba protoplasts, exposition to microgravity
   10, 511f
Viehoff report **12**, 540ff
vinasse **6**, 333

vinblastin, from plant cell cultures **7**, 614
– inhibition of the microtubular system **7**, 668
– structure **7**, 615, 668
vinblastine, structure **8A**, 353
– treatment of leukemia **8A**, 354
vinca alkaloids **7**, 665
vincamine, structure **8A**, 353
– use as vasodilator **8A**, 352
vincristin, from plant cell cultures **7**, 614
– inhibition of the microtubular system **7**, 668
– structure **7**, 668
vincristine, structure **8A**, 353
– treatment of leukemia **8A**, 354
vindesin, inhibition of the microtubular system **7**, 668
– structure **7**, 668
vindoline, biosynthetic pathways, in *Catharanthus* **8A**, 354f
vinegar **9**, 579ff
– – *see also* acetic acid
– additives **9**, 589
– adulteration with added acetic acid **9**, 585
– cider vinegar **6**, 383
– constituents of, cider vinegar **9**, 584
– – malt vinegar **9**, 584
– – rice vinegar **9**, 584
– – spirit vinegar **9**, 583
– – whey vinegar **9**, 584
– – wine vinegar **9**, 583
– filtration, continuous ultrafiltration **9**, 582
– – traditonal – **9**, 582
– food law provisions **9**, 588ff
– – European Code of Practice **9**, 588ff
– legal definitions **9**, 581
– malt vinegar **6**, 383
– pasteurization **9**, 582
– properties **9**, 581
– rice vinegar **6**, 383
– spirit vinegar **6**, 383
– sulfiting **9**, 582
– uses **9**, 583
– wine vinegar **6**, 383
vinegar analysis **9**, 584ff
– determination of, acetic acid **9**, 585f
– – ethanol **9**, 586f
– – extract **9**, 587f
– – lactic acid **9**, 588
– – sugars **9**, 587
– – sulfurous acid **9**, 588
– – total acids **9**, 587
vinyl guaiacol, formation from ferulic acid **8B**, 346f
violaxanthin, structure **7**, 173
viral-based vector systems, in HIV gene therapy **5A**, 479ff
viral gene expression **2**, 786ff
– phenotypical effects **2**, 786ff
– tissue specificity **2**, 787

viral oncogene **2**, 786ff
– virus types **2**, 786ff
viral pathogens, in household wastes **11C**, 232
viral plant inoculation **2**, 574
viral protein production, by *Saccharomyces cerevisiae* **1**, 498
viral vector **2**, 564ff, 725ff, 767ff
– – *see also* gene expression vector, plant viral vector, vector
– animal transformation **2**, 768
– – gene expression **2**, 770
– – mice **2**, 769f
– – risks **2**, 770
– baculovirus **2**, 726, 729f
– bovine papillomavirus **2**, 724f
– – transfection stabilization **2**, 725
– for random insertion mutagenesis **2**, 781
– packaging (helper) cell line **2**, 728, 767
– ping-pong amplification **2**, 728
– production strategy **2**, 767f
– retroviral chicken transformation **2**, 809f
– retrovirus **2**, 766ff
– – active from regeneration **2**, 770
– – capacity **2**, 770
– – in domestic animals **2**, 770
– – *see also* retrovirus
– – target cells **2**, 768
– – yield of transgenic mice **2**, 769
– retrovirus systems **2**, 725ff
– – genomic structure **2**, 727
– subretroviral particle **2**, 770
– titer **2**, 728
– vaccinia virus **2**, 726, 729f
viral vectors, adeno-associated – **5A**, 397f, **12**, 100
– adenovirus **5A**, 397ff, 406ff
– derived from adenoviruses **12**, 99f
– derived from BPV-1 **12**, 100f
– derived from HSV-1 **12**, 100
– derived from poxvirus **12**, 98
– derived from retroviruses **12**, 98f
– for gene therapy **5A**, 395ff
– retrovirus **5A**, 397ff
virion **1**, 516
viroid **2**, 572
– as plant vector **2**, 572
– infection cycle **2**, 572
virtual cell **5B**, 438
virtual screening, of protein drug targets **5B**, 353f
virulence gene, of Salmonellae **2**, 212
– regulation **2**, 212
virus **1**, 546
– – *see also* bacteriophage
– classification scheme **1**, 548ff
– genome structure **1**, 549
– mRNA formation **1**, 551
– nucleic acid replication **1**, 551
viruses **3**, 58, **12**, 74ff
– – *see also* baculovirus

– classification, risk groups **12**, 75f
– contamination of cell cultures **12**, 97
– horizontal transmission **12**, 75
– in biological waste **3**, 783
– in contaminated ground water **3**, 783
– in plant protection **10**, 496f
– in sewage **11C**, 217
– in sewage sludge **11C**, 217
– live – **3**, 58
– plant pathogenic –, in organic wastes **11C**, 234
– recombinant – **9**, 88
– removal of, during artificial groundwater re-charge **11C**, 440
– vertical transmission **12**, 75
virus inactivation **2**, 96
virus receptor binding, inhibition, by chloropep-tins **7**, 124ff
– – by isochromopilones **7**, 124f
– – of HIV entry **7**, 124ff
virus resistance, in plants **2**, 641ff, **10**, 492f
– – against tobacco mosaic virus (TMV) **2**, 642ff
– – mechanisms **2**, 632
– introduction into plants **2**, 641ff
– – against RNA viruses **2**, 590
– – anti-sense RNA **2**, 590
– – anti-sense RNA-mediated protection **2**, 643
– – coat protein-mediated protection **2**, 641f
– – into tobacco **2**, 591
– – resistance degree **2**, 643
– – satellite RNA-mediated protection **2**, 643
– *Mx* gene **2**, 806
virus vectors, adeno-associated – **5A**, 413ff
viscosimeter **4**, 13f
– capillary **4**, 13, 192
– coaxial cylinder **4**, 14
– Couette **4**, 14
– falling sphere **4**, 14
– Searle **4**, 14
– torsional pendulum **4**, 14
viscosity, apparent –, of highly viscous fluids **3**, 202
– – shear rate effects **3**, 204
– biomass estimation **4**, 192
– effects on mass transfer coefficients **3**, 202
– measurement **4**, 192
viscosity methods, for biomass determination **4**, 192
vitamin, determination, by lactic acid bacteria **1**, 364
– precursor production **1**, 211
– synthesis, in yeast **1**, 490f
vitamin A, structure of **10**, 332
vitamin A, structure **7**, 174
vitamin B$_1$ *see* thiamin
vitamin B$_{12}$, fermentative production of **10**, 328
vitamin B$_2$ *see* riboflavin
vitamin B$_6$ *see* pyridoxine
vitamin C *see* ascorbic acid

– – *see* L-ascorbate
– synthesis of, use of *Gluconobacter oxydans* **8B**, 297
vitamin C (L-ascorbic acid), chemical synthesis of **8A**, 10
– manufacture of **8A**, 8
vitamin content, of bread **9**, 247
vitamin D, production using ultraviolet radiation **10**, 333
vitamin D compounds, structures of **10**, 334
vitamin D$_2$, formation in yeasts **7**, 172
vitamin D$_3$ hydroxylation **8A**, 485
vitamin E (tocopherol), chemicoenzymatic synthe-sis of **10**, 334
vitamin K compounds **10**, 335f
– structures of **10**, 336
vitamin K$_1$, chemicoenzymatic synthesis of **10**, 334
vitamins **4**, 91f
– as nutrients, for mammalian cell cultures **3**, 146
– biosensor assay **4**, 91f
– definition of **10**, 320
– fat-soluble – **10**, 324, 332ff
– – biosynthesis of **10**, 332ff
– in yeasts **9**, 225f
– microbial production of **10**, 319ff
– of marine origin **10**, 466
– physiological function **3**, 146
– production of **10**, 321
– – microbial and enzymatic processes **10**, 322f
– thermal destruction, activation energy **3**, 164
– water-soluble – **10**, 320ff
VLDL **5A**, 15
VNCI model **11A**, 572
volatile acids, gas analysis in bioreactors **4**, 60
volatile halogenated hydrocarbons, extraction of **11B**, 499
volatile organic compounds (VOC), control of em-issions **11C**, 363
– treatment of emissions **11C**, 365
volume, measurement, accuracy of **4**, 21
volumetric rates, biofilters **11A**, 370
Volvariella volvacea **1**, 534f
von Recklinghause disease **5B**, 69
vulnibactin **7**, 205f
– catecholate siderophore of *Vibrio vulnificus* **7**, 206
– structure **7**, 205
vulpinic acid, iron-binding compounds **7**, 232
– structure **7**, 232

W

waries, production of **9**, 540
waste *see also* biowaste, garbage, municipal solid waste, solid waste

– – *see* biological waste
– – *see* solid waste
– anaerobic fermentation of **11C**, 157ff
– – procedures **11C**, 157ff
– as a supplementary fuel for production processes **11C**, 30
– avoidance **11A**, 157
– bioconversion of **11C**, 15
– biogenic –, pathogens **11C**, 228
– classification of **11C**, 250
– co-incineration in power plants **11C**, 30
– collection systems **11C**, 9f
– composition of, in Europe **11C**, 204
– – influence on landfill leachate **11C**, 173
– composting of **11C**, 15ff
– disposal **11A**, 157
– generation of **11C**, 8f
– handling of **11C**, 8ff
– heavy metal content in **11C**, 104f
– in sanitary landfills, aerobic degradation phases **11C**, 169
– – anaerobic degradation phases **11C**, 169ff
– land disposal **11C**, 195ff
– liquid agricultural –, value of **11C**, 225
– pathogens in **11C**, 233ff
– plant pathogens in **11C**, 234f
– processing, costs **11C**, 30
– – management **11C**, 30f
– recycling of **11C**, 8ff
– rotting of **11C**, 15ff
– sieving of, screen underflow **11C**, 206f
– stabilization of, anaerobic – **11C**, 19
– starting material **11C**, 43f
– – biochemical composition of **11C**, 43
– – lignin content **11C**, 43
– suitability for, aerobic digestion **11C**, 157
– – anaerobic digestion **11C**, 157
– – composting, biochemical composition of **11C**, 43
– thermal conversion process, by Noel **11C**, 27
– – by Siemens **11C**, 28
– thermal treatment, comparison of processes **11C**, 29
– – energy balances **11C**, 30
– thermal treatment of **11C**, 25ff
– with high sulfate levels, anaerobic digestion **11A**, 461
waste activated sludge (WAS) **11A**, 303f
waste air, pollutant concentrations in, fluctuations of **11C**, 314
waste coke degassing **11C**, 26ff
waste disposal, with mechanical-biological treatment, concepts of **11C**, 205f
waste gas, biochemically oxidizable hydrocarbons in **11C**, 283
– components, organic –, water solubility of **11C**, 323
– composition, analysis of **11C**, 326

waste gas filtration plants, emission of germs from **11C**, 294f
– – aerobiological methods **11C**, 294
waste gas purification, activated carbon **11C**, 288
– applications for **11C**, 362
– bioaugmentation **11C**, 269
– biological – **11C**, 259ff
– – membrane procedures **11C**, 348ff
– – membrane reactors **11C**, 349
– – two-stage plant **11C**, 319
– – use of and alternative techniques **11C**, 345ff
– – use of membrane processes **11C**, 345ff
– commercial applications **11C**, 357ff
– handling of mixtures of compounds **11C**, 288f
– humification **11C**, 283f
– inoculation of reactors **11C**, 268f
– in trickling filters **11C**, 333ff
– membrane reactors, treatment of sparingly soluble compounds **11C**, 288
– microbial degradation of pollutants **11C**, 267ff
– microbial degradative reactions **11C**, 280f, 291
– microbial metabolism, influence of toxicity on **11C**, 289
– – of alkylated and polyalkylated benzenes **11C**, 286
– – of ammonia **11C**, 287
– – of α-pinene **11C**, 287
– – of dichloromethane **11C**, 285f
– – of diethylether **11C**, 286
– – of fluorobenzene **11C**, 284f
– – of gasoline vapors **11C**, 286
– – of hydrogen sulfide **11C**, 287
– – of *m*-cresol **11C**, 286
– – of methyl-*t*-butyl ether **11C**, 286
– – of styrene **11C**, 286
– – of trichloroethylene **11C**, 285f
– – of xenobiotics **11C**, 282ff
– – problemes of **11C**, 282ff
– microbial uptake of dissolved compounds **11C**, 290f
– microbiological aspects of biological – **11C**, 275ff
– microorganisms in, as biochemical agents **11C**, 280f
– needs **11C**, 358f
– performance parameters **11C**, 261ff
– process engineering of **11C**, 259ff
– process principles **11C**, 265ff
– reactor control **11C**, 271
– reactor performance **11C**, 269ff
– technology **11C**, 260f
– use of, biofilters **11C**, 260f, 322ff
– – bioscrubbers **11C**, 261, 305ff
– – membrane reactors **11C**, 261
– – membranes **11C**, 348f
– – trickling filters **11C**, 261
waste gas reactors, application of a designed oligospecies culture system **11C**, 292

– as open systems **11C**, 291f
– bacteria, used for inoculation **11C**, 278
– clogging in **11C**, 292
– fungi, used for inoculation **11C**, 278
– inoculation of **11C**, 291f
– microbial communities in **11C**, 276ff
– monitoring of mixed cultures **11C**, 292f
– – by molecular genetic probing **11C**, 293
waste gas stream, characteristics of **11C**, 264f
– composition of **11C**, 264
– pollutants in **11C**, 264
waste gas treatment, in trickling filters, in practice **11C**, 342
– use in trickling filters **11C**, 333ff
waste gas treatment facilities, active mixed cultures in **11C**, 281
– microbiology of, practical aspects **11C**, 289ff
waste generation, future settlements with minimized – **11C**, 249ff
waste incineration **11C**, 195
– off-gases, purification of **11C**, 28
– plants **11C**, 26ff
– – off-gas composition **11C**, 25f
– products of **11C**, 29
– scheme **11C**, 27
– slags, biogas production from **11C**, 29
– – microbial colonization of **11C**, 28f
waste legislation **11A**, 154ff
– in Germany **11A**, 155f
– – administrative enforcement of **11A**, 157
– – distribution of legislative responsibilities **11A**, 156
– – structure of **11A**, 156
waste management, aerobic – and anaerobic –, comparison of **11C**, 155
– aerobic – composting **11C**, 156
– anaerobic –, characteristics of **11C**, 160
– – commercial processes **11C**, 163f
– – continuous operation **11C**, 161f
– – discontinuous operation **11C**, 161f
– – dry and wet fermentation **11C**, 160f
– – mass flow **11C**, 16
– – mesophilic operation **11C**, 162
– – one-stage process **11C**, 160
– – processes **11C**, 156ff
– – thermophilic operation **11C**, 162
– – two-stage process **11C**, 160
– anaerobic digestion **11C**, 156
– – biogas production **11C**, 207
– biological – **11C**, 207f
– – basic aspects **11C**, 152ff
– – microbial risks **11C**, 238f
– by incineration **11C**, 26ff
– for final disposal **11C**, 10f
– in agriculture **11A**, 225ff
– integrated – **11C**, 193f
– – definition of **11C**, 193f
– legally defined priorities **11C**, 10f

– mechanical-biological, BIOPERCOLAT® procedure **11C**, 208ff
– – combined with incineration **11C**, 206
– – concept **11C**, 204ff
– – mechanical pretreatment **11C**, 206f
– – pilot plant studies **11C**, 209f
– plants, health of personnel **11C**, 243f
– practices, incineration **11C**, 195
– – landfilling **11C**, 195
– – recycling **11C**, 195
– rotting systems **11C**, 207
waste materials, industrial –, bioleaching of **10**, 215f
waste plant materials *see* plant residues
waste recycling **11A**, 157
wasterwater, definition of **11A**, 6
wastes, nitrile containing –, bioremediation of **8A**, 312f
waste sulfite liquor, components **6**, 139
– in industrial ethanol production **6**, 138f, 168f
waste treatment *see also* solid waste treatment
– aerobic –, mass flow **11C**, 16
– anaerobic digestion **11A**, 455ff
– laws, statutory orders and directives **11A**, 141ff
waste utilization, in agriculture **11A**, 225ff
wastewater *see also* agricultural wastewater, industrial wastewater, municipal wastewater
– aerobic/anaerobic treatment of **11C**, 213ff
– – hygienic considerations **11C**, 213ff
– aerobic biological processes **11A**, 115ff
– ammonia concentrations, sources **11A**, 395
– analysis **4**, 92, **11A**, 435
– and infectious diseases **11C**, 218ff
– carbon removal **11A**, 259ff
– classification of **11C**, 250
– composition of **11A**, 259, 434
– definition of **11A**, 19
– disinfection of **11C**, 219f
– disposal of, decentralized – **11C**, 227
– – in rural areas **11C**, 226ff
– fecal –, discharge of **11C**, 227
– from bakers' yeast production **11A**, 538
– highly loaded –, sources **11A**, 395
– infectious diseases **11A**, 7
– land application of **11A**, 9
– limiting values for drinking water **11A**, 435f
– lipid containing – **11A**, 520
– maximum permissible concentrations **11A**, 435f
– mechanical treatment **11A**, 115
– monitoring of organic substances, betaine **11A**, 534f
– pathogens in **11C**, 216ff
– quality, influence on phosphate removal **11A**, 299ff
– settling tanks **11A**, 7f
– sewer networks **11A**, 7ff
– sources **11A**, 433f
– specific amount **11A**, 434

– titration **4**, 466
– treatment in fluidized bed reactors **11A**, 504ff
– treatment in stationary bed reactors **11A**, 504f
– use in fish ponds **11A**, 9f
– with high sulfate levels, anaerobic digestion **11A**, 461
wastewater characterization, for modeling **11A**, 422
– physical parameters, flow **11A**, 113
– – suspended and settleable solids **11A**, 114
– – temperature **11A**, 114
wastewater components **11A**, 196ff
– cleaning agents **11A**, 198
– corrosion inducing substances **11A**, 198
– disinfectants **11A**, 198
– hazardous substances **11A**, 197f
– impact of, pH value **11A**, 196
– – temperature **11A**, 196
– lubricants **11A**, 198
– nutrient salts **11A**, 197
– organic substances **11A**, 197
wastewater constituents, assumptions for modelling **4**, 409
wastewater generation, future settlements with minimized – **11C**, 249ff
waste water management, in olive fermentations **9**, 621f
wastewater process **4**, 475
– anaerobic, biomass concentration **4**, 450
– – gas flow measurement **4**, 453ff
– – measuring techniques **4**, 450ff
– – methane formation **4**, 448ff
– – parameter identification for model-based controllers **4**, 475
– – pH course **4**, 465
– – regulation of gas production **4**, 477
– – scheme of experimental setup **4**, 455
– – substrate regulation **4**, 477f
– – thermodynamics **4**, 448f
wastewater process models, aerobic, assumptions **4**, 409f
– – balanced growth conditions **4**, 411, 432
– – basic unstructured model **4**, 409ff, 432
– – biofilm kinetics **4**, 418ff
– – complexity versus utility **4**, 416f
– – empirical **4**, 417f
– – IAWPRC model **4**, 410ff, 433
– – mass balance equations **4**, 415f
– – mechanistic **4**, 417f
– – rate equations **4**, 410f
– – simple structured model **4**, 414, 432f, 436
– – yield equations **4**, 411ff
– anaerobic **4**, 456ff
– – calculation of biomass concentration **4**, 470
– – calculation of substrate concentration **4**, 470ff
– – CO_2 transport **4**, 466
– – gas production **4**, 461

– – growth kinetics **4**, 457ff
– – non-steady states **4**, 462f
– – observation of state variables **4**, 472f
– – pH calculation **4**, 463ff
– – steady states **4**, 459ff
– – substrate consumption **4**, 457ff
wastewater purification, anaerobic, advantages **4**, 445ff
– – limitations of model approaches **4**, 445
– – metabolic pathways **4**, 447ff
– – microorganisms **4**, 447
– – process control **4**, 474ff
wastewater samples, estimation of pollutants in **11B**, 497
waste water treatment **3**, 83
– – *see also* agricultural wastewater, industrial wastewater, municipal wastewater
– activated sludge process **11A**, 10ff, 253ff
– activated sludge system, modelling **4**, 430
– adsorption technology **11A**, 566
– aerobic **4**, 409ff
– aerobic and anaerobic systems **11A**, 24f
– aerobic oxidation of inorganic substances **11A**, 394ff
– – nitrification **11A**, 394ff
– – sulfide oxidation **11A**, 398f
– aerobic processes, activated sludge process **11A**, 400ff
– – fixed bed reactor **11A**, 400f
– – fluidized bed reactor **11A**, 401
– aerobic thermophilic – **11A**, 403ff
– – kinetic and stoichiometric coefficients **11A**, 405
– anaerobic –, immobilized bacterial aggregates **11A**, 520
– – reactor types **11A**, 469ff
– anaerobic digestion **11A**, 13, 455ff
– anaerobic sludge bed reactors **11A**, 517ff
– – application **11A**, 522f
– – reactor technology **11A**, 519ff
– – start-up **11A**, 522f
– bacterial metabolism in – **11A**, 17ff
– biodegradation of, chlorinated *n*-alkanes **11A**, 377ff
– – chlorobenzenes **11A**, 380ff
– – chlorophenols **11A**, 382
– – DCA **11A**, 378f
– – DCM **11A**, 378f
– – dyes **11A**, 392ff
– – first-order theory **11A**, 10
– – naphthalene sulfonic acids **11A**, 390ff
– – nitroaromatics **11A**, 383ff
– charging of contributions and fees **11A**, 154
– chemical phosphorus removal **11A**, 308ff
– comparison of aerobic and anaerobic degradation **4**, 445ff
– – of acetic acid **4**, 446
– – of carbon **4**, 446f

– constituent microbial processes, rate equations **4**, 412f
– constructed wetlands **11A**, 242f, 247
– decentralized – **11A**, 358
– decomposion of carbon compounds **11A**, 19ff
– denitrification **11A**, 118f
– effects of detergents **11A**, 12
– elimination of persistent substances **11A**, 567ff
– – kinetics **11A**, 569f
– – membrane reactor **11A**, 568ff
– – substrate-specific residence times **11A**, 567ff
– end-of-pipe technologies **11A**, 563
– enhanced biological phosphorus removal **11A**, 288ff
– – by activated sludge **11A**, 290ff
– excess sludge **11A**, 570
– extraction module for resource recovery **11A**, 576
– fluidized-bed bioreactors, modelling **4**, 427
– historical development **11A**, 5ff
– in the food processing industry, abattoirs **11A**, 206ff
– – beverage industry **11A**, 209f
– – breweries **11A**, 210f
– – dairy industry **11A**, 208f
– – distilleries **11A**, 211ff
– – fruit juice **11A**, 209
– – oil and fat production **11A**, 203ff
– – potato processing **11A**, 205f
– – starch factories **11A**, 202f
– – sugar factories **11A**, 200ff
– laws, statutory orders and directives **11A**, 141ff
– membrane technology **11A**, 565f
– – influence on chemical potential **11A**, 565
– – nanofiltration **11A**, 565f
– metal ion removal **11A**, 42ff
– microbial communities in, use of nucleic acid-based techniques **11A**, 85ff
– monitoring of, basic tools **11A**, 111ff
– – data assessment **11A**, 111ff
– nitrification **11A**, 117f
– nitrogen removal **11A**, 13, 37f
– – ammonia oxidation **11A**, 40, 56
– – ammonification **11A**, 37f
– – combined nitrification/denitrification **11A**, 39f, 60f
– – denitrification **11A**, 39f, 57f, 59f
– – nitrification **11A**, 38f, 58f
– nucleic acid-based techniques, denaturing gradient gel electrophoresis **11A**, 88ff
– – phylogenetic classification **11A**, 88
– oligonucleotide probes specific for microorganisms involved **11A**, 92ff
– organic carbon removal **11A**, 115ff
– oxidation pond **4**, 430
– phosphorus removal **11A**, 13, 41f, 119
– production-integrated – **11A**, 561, 564ff
– – EPS model **11A**, 572

– – evaluation by ecopoints **11A**, 572
– – future aspects **11A**, 561ff
– – materials management **11A**, 570ff
– – method of cumulative energy expenditure (KEA) **11A**, 573
– – methodological approaches to the implementation of an environmental compatibility assessement **11A**, 573f
– – methodology of material intensity per product and service unit (MIPS) **11A**, 573
– – model of critical air and water volumes **11A**, 571f
– – model of critical loads **11A**, 572
– – quantitative methods to assess ecological impacts **11A**, 571ff
– – recovery of useful materials **11A**, 564ff
– – techniques **11A**, 567ff
– – Tellus model **11A**, 572
– – toxicity equivalencies **11A**, 573
– – utility analysis **11A**, 571
– – VNCI model **11A**, 572
– – water pinch analysis **11A**, 573
– – weak point analysis **11A**, 570ff, 574f
– reactive extraction method **11A**, 566
– ribosomal ribonucleic acid-based analysis **11A**, 86ff
– rotating biological contactors **4**, 429f
– – modelling **4**, 429
– rotating disc reactors **11A**, 9
– saprobe index **11A**, 8
– secondary pollutants **11A**, 12f
– self-purification **11A**, 8
– sludge anaerobic reactor, staged – **11A**, 524
– sludge stabilization **11A**, 405f
– special aerobic processes **11A**, 373ff
– stages **11A**, 199
– submerged fixed bed reactors **11A**, 349ff
– trickling filter, modelling **4**, 426
– trickling filters **11A**, 8ff
– use of biofilters **11A**, 365ff
– use of rotating biological contactors, denitrification **11A**, 346f
– – nitrification **11A**, 346
– use of trickling filters **11A**, 337ff
– – denitrification **11A**, 341ff
– – denitrification rates **11A**, 342
– – nitrification **11A**, 340f
– water recycling **11A**, 564
– wetland systems **11A**, 241ff
– – aquaculture systems **11A**, 248f
– – hydrobotanical systems **11A**, 248f
– – role of microorganisms **11A**, 248
– – role of plants **11A**, 242ff
– – soil systems **11A**, 248f
wastewater treatment plants, modeling of **11A**, 418
– monitoring during operation, aerobic biological treatment **11A**, 115ff

– – mechanical treatment **11A**, 115
– – parameters for operational decisions **11A**, 121
– monitoring of, analytical paramaters **11A**, 109ff
– – quality control **11A**, 122
– odor control in **11C**, 364
wastewater treatment processes, aerobic – **11A**, 399ff
– carbon/nitrogen ratio **11A**, 14
– process combinations **11A**, 406ff
water **3**, 697ff
– deionization (DI) **3**, 757
– deionized –, use in biotechnological manufacturing **3**, 758
– for biotechnological manufacturing **3**, 755f
– – pretreatment **3**, 756
– – sources **3**, 756
– – system validation **3**, 759
– – treatment **3**, 756
– for brewing **9**, 422f
– for fermentation processes, purification **3**, 145
– influence on microbial growth **10**, 272f
– in macromolecular solutions, adhesive – **3**, 708
– – adsorptive – **3**, 708
– – capillary – **3**, 708
– – constitutional – **3**, 708
– – hydrate – **3**, 708
– – swelling – **3**, 708
– public supply **11C**, 373
– reverse osmosis (RO) **3**, 756
– use in baking **9**, 252
water activity **11C**, 45f
– effect in microbial growth **1**, 130
waterborne diseases **11C**, 414ff
– listing of outbreaks **11C**, 415f, 418f
water for injection (WFI), distillation apparatus **3**, 759
– distribution systems **3**, 759
– production **3**, 758
– storage tanks **3**, 759
water-in-oil microemulsions **3**, 606
water legislation **11A**, 143ff
– administrative water law enforcement **11A**, 151
– in Germany **11A**, 145f
– – distribution of legislative competence **11A**, 146
– – federal legislation **11A**, 146f
– – land legislation **11A**, 147f
– – standardization **11A**, 148
– – structure of **11A**, 148ff
– requirements for, direct wastewater discharge into water bodies **11A**, 151ff
– – indirect discharges **11A**, 153
– – wastewater treatment **11A**, 153
– structure of wastewater charges **11A**, 153
water management **11A**, 148f
water molecules, binding energy **3**, 709
– binding types **3**, 708

– crystal formation **3**, 697
– heterogenous nucleation **3**, 698
– homogeneous nucleation **3**, 697
– nucleation rate **3**, 697
– properties **3**, 697ff
– undercooling **3**, 697
water pinch analysis **11A**, 573
water potential **11C**, 45f
– for microbial growth **10**, 273
water protection **11A**, 149f
water purification *see* wastewater
water quality **3**, 755ff
– monitoring **3**, 757
water runoff, regulation of **11A**, 150
water samples, estimation of pollutants in **11B**, 497
water technology, production-integrated – **11A**, 564ff
water treatment processes, adsorption **11C**, 395ff
– – granular activated carbon **11C**, 397f
– – powdered activated carbon **11C**, 396f
– chemical softening **11C**, 402ff
– deferrization **11C**, 404f
– demanganization **11C**, 404f
– flocculation **11C**, 383ff
– flotation **11C**, 386ff
– gas exchange **11C**, 399f
– membrane filtration **11C**, 405ff
– oxidation **11C**, 400ff
– precipitation **11C**, 383ff
– rapid filtration **11C**, 389ff
– – applications of **11C**, 392ff
– sedimentation **11C**, 386ff
wax esters **7**, 176f
– pathways to a diunsaturated wax ester **7**, 176
Waxmann Hatch legislation **5A**, 503
weak organic acids, membrane proton transport **1**, 181
weak point analysis, in production-integrated wastewater treatment **11A**, 570ff, 574f
web-based database integration systems **5B**, 436
Weber number, of spargers **3**, 261
weeds, biocontrol of, by bioherbicides **10**, 489f
– – by leaf-spot causing bacteria **10**, 490
– – by rhizobacteria **10**, 490
– – use of bacteria **10**, 487f
welan gum, gel formation **6**, 642
wetland systems **11A**, 241ff
– constructed wetlands **11A**, 242f, 247ff
– – retention time of wastewater **11A**, 247
– for wastewater treatment **11A**, 241ff
– role of plants **11A**, 243ff, 246ff
– – metabolism of xenobiotics **11A**, 246f
– – oxygen release via the roots **11A**, 245f
– – rhizodeposition **11A**, 244f
– – storage potential of nutrients **11A**, 243f
– – transpiration **11A**, 247
– technological aspects **11A**, 248f

– – aquaculture systems **11A**, 248f
– – hydrobotanical systems **11A**, 248f
– – soil systems **11A**, 248f
wet milling, cell disruption **3**, 510ff
– – disc selection **3**, 512
– energy dissipation **3**, 514
– operating conditions **3**, 513
– parameters of, bead mills **3**, 512f
– – cell suspensions **3**, 512f
wheat, analysis of **6**, 62
– as raw material for ethanol production **6**, 62
– autoamylolytical enzyme system **6**, 74f, 95
– autoamylolytical quotient **6**, 63
– processing in ethanol production **6**, 91f
– production of **9**, 243
– stillage, composition **6**, 108
– – nutrient content **6**, 106
wheat dwarf virus **2**, 568
– as vector for *Triticum* **2**, 568f
wheat gluten *see* gluten
wheat malt **6**, 70
– as brewing adjunct **9**, 432
wheat starch, as brewing adjunct **9**, 430
– composition **6**, 32, 34f
– production **6**, 33
– wastewater from – **11A**, 202
whey, anaerobic degradation – **11A**, 536f
– – concentration of the different volatile acids
  11A, 537
– Bel process **9**, 210
– composition **6**, 139, **9**, 209ff
– composition of **11A**, 221
– fermented – **9**, 211
– hydrolysis **9**, 210f
– in industrial ethanol production **6**, 169
– protein-enriched – **9**, 210
– substrate for acetone/butanol fermentation **6**,
  255
whey expulsion, from cheese curd **9**, 361f
whisky production, microflora **1**, 334
white Paper **5A**, 526
white-rot fungi **1**, 534f
– – *see also* filamentous funi enzyme production
– biodegradation of nitro compounds **11B**, 288f
– degradation of lignin **11B**, 161ff
– degradation of organo pollutants **11B**, 163
– oxidation of polycyclic aromatic hydrocarbons
  11B, 230ff
WHO **5A**, 445f
– ethical aspects of genome research **5B**, 422
– plasmid DNA, quality issues **5A**, 460
– plasmid DNA therapeutics, guidelines **5A**, 445
– plasmid vectors, guidelines **5A**, 452
whole-cell biotransformations, dehydrogenases
  8A, 396ff
– hydroxylation reactions **8A**, 483ff
– of steroid nucleus **8A**, 13
– practical aspects **8A**, 26ff, 31f

whole-cell catalysts, for dihydroxylation **8A**, 481f
– for hydroxylation **8A**, 481f
whole-cell simulation **5B**, 403
whole-genome random sequencing **5B**, 17
– of *A. fulgidus* genome **5B**, 16f
whole-genome shotgun strategy **5B**, 46
– in Human Genome Project **5B**, 48f
WHO (World Health Organization) **12**, 619ff
– Children's Vaccine Initiative (CVI) **12**, 358
– Expanded Programme on Immunization (EPI)
  12, 357
wild yeasts, in brewing **9**, 447
– in wine making **9**, 468f
Wiliamson Committee **12**, 638
Williams report **12**, 516, 521
windows *see* composting configurations
windrow, geometric shape of **11C**, 119f
windrow composting **11C**, 118f, 138, 140
– airborne microorganisms in **11C**, 241f
– classification of **11C**, 119f
– of sludge **11C**, 222f
– turning machines **11C**, 121f
– with forced aeration, by dearation **11C**, 121
– – by ventilation **11C**, 121
– – turning machines **11C**, 121f
– with natural aeration **11C**, 120f
– – and horizontally growing heap **11C**, 121
– – and one-step buildup **11C**, 120
– – and vertically growing heap **11C**, 120f
– – support of **11C**, 121
windrow turning machines **11C**, 121f
– self-driving – **11C**, 122
– tractor driven – **11C**, 122
wine fermentation, alcohol yields **9**, 473
– by-products **9**, 474ff
– – acetaldehyde **9**, 475
– – acetic acid **9**, 476
– – carbon dioxide inhibition **9**, 483
– – esters **9**, 479
– – glycerol **9**, 475f
– – higher alcohols **9**, 477f
– – hydrogen sulfide **9**, 480
– – lactic acid **9**, 477
– – pyruvate **9**, 475
– – succinic acid **9**, 477
– – sulfurus acid **9**, 480
– – 2-oxoglutarate **9**, 475
– – 2,3-butanediol **9**, 477f
– effect of, must particles **9**, 484
– – sugar concentration **9**, 485
– – sulfurous acid **9**, 485f
– – temperature **9**, 481ff
– ethanol inhibition **9**, 483
– formation of, carbon dioxide **9**, 472ff
– – ethanol **9**, 472ff
– spontaneous – **9**, 468f
wine making, malolactic fermentation **9**, 489ff
– use of, active dry yeast **9**, 471

– – baker's yeast **9**, 471
– – enzymes **9**, 727f
– – liquid pure culture yeast **9**, 471
– volatile acidity **9**, 486
wine preservatives **9**, 487ff
– carbon dioxide **9**, 488
– dimethyldicarbonate **9**, 489
– ethanol **9**, 488
– sorbic acid **9**, 488
wine production **1**, 334
– process improvement **1**, 496
– red wine **1**, 341
– *Saccharomyces* strains **1**, 495
wines **9**, 463ff
– aromatic terpenes in – **9**, 728
– β-glucans in – **9**, 727
– definition **9**, 464
– designations of German quality wines **9**, 473
– fermentation of botrytized musts **9**, 499
– history **9**, 465
– malate-degrading bacteria in – **9**, 490ff
– noble rot **9**, 497ff, 727
– pesticides in – **9**, 487
– spontaneous fermentation **9**, 468f
– starter cultures **9**, 469ff
– – properties **9**, 470
– – use **9**, 470f
wine spoilage **9**, 492ff
– by acetic acid bacteria **9**, 493
– by acrolein **9**, 495
– by lactic acid bacteria **9**, 493
– by *Lactobacillus brevis* **9**, 496
– by *Saccharomyces ludwigii* **9**, 496
– by *Zygosaccharomyces bailii* **9**, 496
– lactic acid tone **9**, 494
– mannitol spoilage **9**, 494f
– mousiness **9**, 495
wine yeast **9**, 466ff
– – *see also* yeast
– activity tests **9**, 346
– commercial active dry yeast **9**, 337ff, 342
– commercial production **9**, 321ff
– history **9**, 322f
– nitrogen assimilation **9**, 486
– quality criteria **9**, 324
– SO₂ tolerance **9**, 485f
– strains **9**, 338
– – maintenance **9**, 325
– – modification methods **9**, 325f
– – selection of **9**, 323ff
WIPO (World Intellectual Property Organization) **12**, 286
WIT project **5B**, 290, 411
wood, glucomannan content **10**, 527f
– in industrial ethanol production **6**, 161ff, 167
– – energetics **6**, 186f
– – flow diagram **6**, 163
– laminated – **10**, 306f

– – biodeterioration of **10**, 306f
– xylan content **10**, 527f
wood chips, cellulase treatment **10**, 534
– microbial treatment of, with lignin-degrading fungi **10**, 535
wood extractives **10**, 530
– action of enzymes on **10**, 532
wood fibers, size of pores **10**, 534
working cell bank **5A**, 452
– quality control **5A**, 453
working Cell Bank (WCB) **10**, 30
working draft of human genome sequence **5B**, 48
World Health Organization **2**, 841, 846
– Biosafety Manual **2**, 841ff
– – exposure **2**, 846
– – instructions **2**, 846
– – practice codes **2**, 846
– – work organization **2**, 846
worldwide amounts of water **11C**, 372
worms, occuring in biofilms **11A**, 357
wort *see* brewer's wort
WTO (World Trade Organization) **12**, 290

X

xanthan **6**, 620
– aseptic production **6**, 649
– consumption **6**, 627
– diversity of carbohydrate chains **6**, 648
– price **6**, 627
– production by *Xanthomonas campestris* **6**, 647f
– – mutant strains **6**, 648
– structure **6**, 647
– uses **6**, 616
– – in drilling muds **6**, 629
– – industrial – **6**, 628f
– – in food **6**, 649
xanthan gum **3**, 11
– production by *Xanthomonas*, use of cane molasses **6**, 26
xanthan gum production, modification **1**, 213
xanthan-like polymers **6**, 648
xanthan solution, apparent viscosity **3**, 205
– shear rate relations **3**, 205
xanthan synthesis, gene sequence for formation of – **6**, 619
Xanthobacter autotrophicus, Baeyer–Villiger biotransformations **8A**, 561, 566
– degradation of dichloroethane **11B**, 253
Xanthomonas, phylogenetic status **1**, 404
Xanthomonas campestris **6**, 616f
– biopolymer formation **6**, 617
– xanthan gum production **1**, 213, **6**, 26
– xanthan-like polymers **6**, 648
– xanthan production **6**, 647f
– – mutant strains **6**, 648

Xanthophyta, characteristics of **10**, 113
xanthosine, production by *Bacillus subtilis* **6**, 585f
xanthylate (5′) (XMP), fermentative production,
   by *Brevibacterium ammoniagenes* **6**, 576f
– flavor compounds **6**, 563
xenobiotics **1**, 201ff, **11A**, 44ff, 246f
– aerobic dechlorination **11A**, 45
– aerobic degradation of **11A**, 44f
– anaerobic dechlorination of **11A**, 45
– anaerobic degradation of **11A**, 44ff, 524
– bacterial degradation of, dioxygenase reaction
   11B, 149
– – monooxygenase reactions **11B**, 149
– bioavailability of **11B**, 61ff
– biodegradation by fungi **11B**, 156ff, 160ff
– complete metabolization and mineralization
   11B, 221
– degradation **1**, 235
– – by clostridia **1**, 311f
– – by pseudomonads **1**, 414ff
– – pathway construction **1**, 201
– – plasmid compatibility **1**, 410
– – strain improvement **1**, 201ff
– degradation of, using bioaugmentation **11A**,
   46ff
– detoxification, by white-rot fungi **1**, 534f
– formation of bound residues **11B**, 110f
– – binding modes **11B**, 111ff
– humification processes **11B**, 90, 101f
– interactions, with black carbon **11B**, 92f
– – with soil **11B**, 91
– metabolism of **11C**, 282ff
– microbial degradation of **11B**, 101f, 221
– – threshold concentrations **11B**, 96
– oxidative coupling reactions **11B**, 113
– release of their transformation products **11B**,
   114
– removal by plants, in wetland systems **11A**,
   246f
– removal of, during artificial groundwater recharge **11C**, 438
– solubility in water **11B**, 64ff
– sorption of hydrophobic – **11B**, 218
xenotransplantation, medical applications of recombinant proteins **5A**, 146
– monoclonal antibodies **5A**, 146
– of parathyroid tissue **10**, 556f, 565
xerocomic acid, iron-binding compounds **7**, 232
– structure **7**, 232
Xeromyces bisporus, minimum water potential
   10, 273
xerulin, inhibitor of cholesterol biosynthesis **7**,
   518
– structure **7**, 519
XML language **5B**, 436
XMP *see* xanthylate (5′)
X-ray radiation, determination of gas holdup **4**,
   137

X-windows **4**, 579ff
– applications **4**, 579
– definition **4**, 579
xylanases, bleaching of paper pulp **10**, 535ff
– from hyperthermophiles **10**, 83
– from thermophiles **10**, 83
– mode of action **10**, 528
– thermostable – **10**, 82ff, 537
– treatment of craft pulp **10**, 84
xylan degradation, in clostridia **1**, 297
xylan degrading enzymes **10**, 82ff
xylans, action of enzymes on **10**, 532
– composition of **10**, 527
– content in wood **10**, 527f
xylem, cells **1**, 580
xylene, degradation rate in bioscrubbers **11C**, 309
– scrubbing liquid quantities for the absorption
   of **11C**, 312
xylene degradation, by pseudomonads **1**, 416
– gene regulation **1**, 411
xylenes, anaerobic degradation of **11B**, 184
– extraction of **11B**, 499
– microbial degradation of **11B**, 214
xylitol, formation by yeasts **6**, 223
xylitol fermentation, redox reactions **6**, 224
xylose, determination with biosensors **4**, 78
xylose isomerase **1**, 196
– thermostable – **10**, 91
xylose metabolization **1**, 205
– by improved cultures **1**, 196f
– cofactor regeneration **1**, 198f
β-xylosidase **10**, 529
xylulose monophosphate (XuMP) cycle **1**, 84, 276

Y

YAC physical mapping **5B**, 172
YAC vectors **5B**, 44
YADH *see* yeast alcohol dehydrogenases
Yarrowia lipolytica **1**, 470ff
– expression of heterologous proteins **9**, 85
– formation of γ-decalactone **8B**, 341f
yeast **2**, 88, 114, 121ff, 508ff
– – *see also* Baker's yeast
– – *see also* bakers' yeast, brewers' yeast
– – *see also* baker's yeast, brewer's yeast, wine
   yeast
– – *see also* baker' yeast, brewers' yeast
– – *see also* Saccharomyces
– – *see Saccharomyces cerevisiae*
– active dry – **9**, 254, 337ff, 342, 471
– alcohol tolerance **9**, 483
– as economical source of enzymes **8A**, 364f
– autonomously replicating sequence (ARS) **2**,
   512
– biomass production, Samba process **9**, 211f

– biotransformation of furanyl acrolein **8B**, 387
– brewery yeast **2**, 122
– – fusion products **2**, 122
– bulk yeast **9**, 253f
– C—C bond formation **8A**, 378
– C=C double bond reduction **8A**, 369ff, 375ff
– – mechanism **8A**, 372
– disintegration plasmid **2**, 512
– – *see also Saccharomyces*
– elemental composition **9**, 199
– enoate reductase, reduction of triple substituted double bonds **8A**, 375
– – substrates for **8A**, 375
– episomal plasmid **2**, 511
– – 2 μm circle **2**, 511
– essential amino acids in – **9**, 224
– expression of cyt. P-450 **8A**, 483
– expression of glycosyltransferases **8B**, 268
– expression systems **9**, 80ff
– fatty acid synthetase cycle **8A**, 373
– genetically modified – **8A**, 383
– hydrolytic enzymes **8A**, 379
– improvement of reduction efficiency, use of sulfur etherocycles **8A**, 381
– instant active dry – **9**, 254
– integrating plasmid (YIp) **2**, 509
– killer protein **2**, 88
– life cycle **2**, 121
– mitochondrial gene **2**, 122
– NADPH regeneration system **8A**, 367
– nutritional requirements, mineral elements **9**, 327
– – vitamins **9**, 327
– oversecreting mutant **2**, 522
– – PMR1 gene **2**, 522
– petite cell **2**, 114
– reducing enzymes, from the fatty acid synthetase complex **8A**, 373f
– reducing enzymes of **8A**, 370ff
– reduction of α-diketones **8A**, 382
– reduction reactions **8A**, 370ff
– replicating plasmid **2**, 512
– – centromeric plasmid (YCp) **2**, 512
– replicating plasmid (YRp) *see also* autonomously replicating sequence (ARS)
– reproducibility of biotransformation **8A**, 365f
– resting cells, biotransformation by **8A**, 369
– secretory pathway **2**, 521f
– selectable marker **2**, 522ff
– sources of **8A**, 31f
– – influence on catalytic activity **8A**, 366
– sterol cyclase, lanosterol cyclization **8A**, 378
– strain improvement **2**, 121ff
– transformation **2**, 522
– use for cell-based biosensor assays **4**, 91f
– use in animal feeds **9**, 780f
– use in baked goods **9**, 273
– use in prostaglandin synthesis **8B**, 377f

– viability **8A**, 29
– virus-like particle (VLP) **2**, 88
– yeast cakes **9**, 253
yeast alcohol dehydrogenase **8B**, 359ff
– reduction of β-keto esters **8B**, 359ff
– reduction of simple ketones **8B**, 359
– synthesis of a pheromone component **8B**, 360
– synthesis of compactin **8B**, 360
– synthesis of mevinolin **8B**, 360
yeast alcohol dehydrogenases (YADH) **8A**, 371, 439ff
– – *see also* alcohol dehydrogenases
– applications of **8A**, 440f
– characteristics of **8A**, 439f
– enantioselective reduction of enones **8A**, 440
– selective reduction of an α,β-unsaturated ketone **8A**, 371f
– sources of **8A**, 439ff
– use for coenzyme regeneration **8A**, 440
yeast artifical chromosome (YAC) **2**, 512f
yeast artificial chromosome (YAC) **1**, 499f, **5B**, 173
yeast biotransformations **8A**, 363ff
– enzymatic activities **8A**, 370ff
– operational conditions **8A**, 366ff
– preparation of chiral building blocks **8A**, 365
– product recovery **8A**, 383f
– reproducibility **8A**, 365f
– selectivity of **8A**, 379ff
– – improvement by selective enzyme inhibition **8A**, 382f
– – improvement substrate modification **8A**, 380f
– use of cell immobilization techniques **8A**, 369
– use of genetically modified yeast **8A**, 383
– use of organic solvents **8A**, 369f
– use of resting cells, C=C double bond reduction **8A**, 369
– using ethanol as energy source, NADP$^+$ regeneration in the reduction of 3-oxo-butyrate to 3-hydroxy-butyrate **8A**, 368
– – NAD$^+$ regeneration in the reduction of acetol to propylene glycol **8A**, 368
yeast cell cycle **4**, 389f
– cycling age intervals **4**, 396
– cycling phase equations **4**, 396
– dynamics **4**, 389
– model structure **4**, 396
– schematic diagram **4**, 389
– simulation **4**, 397
yeast cells, asymmetric reductions by **8A**, 393
yeast cell vacuole, regulation of cytosolic metal ion concentrations **10**, 238
yeast centromer plasmid (YCp) **1**, 499f
yeast chromosomes **1**, 474f
– repetitive DNA sequence **1**, 475
– replication origins (*ARS*) **1**, 475
yeast culture, ethanol production **4**, 70

– measurement of respiratory quotient **4**, 70
yeast episomal plasmid (YEp) **1**, 499f
yeast extracts **9**, 702f
– autolysis **9**, 703
– plasmolysis **9**, 703
yeast fermentation, adaption of parameter estimates, by SSPE **4**, 243
– biomass concentration, by SSPE **4**, 242, 246
– endogeneous consumption of reducing power **8A**, 367
– energy source, endohydrates **8A**, 366f
– – ethanol **8A**, 367f
– *Escherichia coli* contaminated, use of SSPE **4**, 247
– feedback control actions, by SSPE **4**, 244
– feedforward control actions, by SSPE **4**, 244
– glucose concentration, by SSPE **4**, 242
– specific growth rate estimation, by SSPE **4**, 246
yeast genome **5A**, 3
yeast growth *see* microbial growth
yeast harvesting **4**, 608
yeast integrative plasmid (YIp) **1**, 499
yeast linear plasmid (YLp) **1**, 499f
yeast mash, analysis **6**, 114
– for ethanol production **6**, 96f
yeast oxidoreductases, listing of **8A**, 371
yeast phosphatases, hydrolysis of phytic acid **8A**, 379
yeast plasmids **1**, 475
– 2m plasmid **1**, 476
yeast production, baker's yeast **9**, 321ff
– control of, dissolved oxygen **9**, 334
– – fermentation kinetics **9**, 334
– – foaming **9**, 334
– – pH **9**, 334
– – temperature **9**, 333f
– fermentation stages **9**, 329ff
– fermenter design **9**, 331f
– media **9**, 326ff
– process model **4**, 386ff
– quality assurance **9**, 344ff
– – activity tests **9**, 345f
– – microbiological purity **9**, 345
– – physical and chemical parameters **9**, 346f
– – strain integrity **9**, 347
– wine yeast **9**, 321ff
yeast protoplast **2**, 508
yeast protoplast fusion **2**, 121ff
– chromosome segregation **2**, 123
– hybrid stability **2**, 123
– intergenera hybridization **2**, 122f
– interspecies hybridization **2**, 122f
– intraspecies fusion **2**, 121
– metabolic function recombination **2**, 122
– recombinant **2**, 122
yeast pyruvate decarboxylase **8B**, 24, 55
yeast reductions, NAD$^+$ regeneration cycle **8A**, 368

yeast replicative plasmid (YRp) **1**, 499
yeast RNA, use in 5′-nucleotide production **6**, 568ff
yeasts **1**, 32, 469ff
– – *see also* oleaginous yeasts
– amino acid metabolism **1**, 489
– apiculate, in grape must **9**, 467
– ATP production by **10**, 330f
– basal media supplements **1**, 143
– biotechnologically important species **1**, 470
– budding **1**, 115f
– carbohydrate metabolism **1**, 478ff
– – defective mutants **1**, 480
– – regulation **1**, 486ff
– cell compartmentation **1**, 478
– cell envelope organization **1**, 32
– cell lysis **3**, 518
– cell structure **1**, 32
– characteristics **1**, 470
– citric acid production **6**, 331
– – process kinetics **6**, 332ff
– composition of **9**, 171
– Crabtree effect **1**, 133, 154
– cultivation **1**, 471
– DNA binding proteins **1**, 492
– effects of microgravity on **10**, 509
– elemental composition **3**, 131
– essential amino acids in – **9**, 174
– ethanol formation **6**, 123ff
– ethanol inhibition **6**, 124ff
– ethanol metaoblism **9**, 181f
– ethanol production **3**, 305
– expression of fungal glucose oxidase **6**, 350
– expression vector **1**, 499
– extrachromosomal inheritance **1**, 476
– fatty acid composition **9**, 225
– fluorescence monitoring for biomass concentration **4**, 187
– for human consumption, clinical tolerance studies **9**, 230f
– formation of biosurfactants **6**, 664
– formation of polyols **6**, 220ff
– – arabitol **6**, 220
– – erythriol **6**, 220
– – mannitol **6**, 222
– – ribitol **6**, 222
– – sorbitol **6**, 222
– – xylitol **6**, 223
– galactose-inducible gene **1**, 493
– GCD/GCN system **1**, 495
– gene expression regulation **1**, 491ff
– genetic manipulation **1**, 476
– genetic organization **1**, 474
– gluconeogenesis **1**, 484
– glycerol metabolism **9**, 182
– glycerol production **6**, 210ff
– – with osmotolerant – **6**, 216
– growth in sour doughs **9**, 296

– growth in wine making **9**, 480f
– haploid strain production **1**, 473
– heterologous gene expression **1**, 496ff
– hexose metabolism **9**, 182ff
– – Entner-Doudoroff pathway **9**, 183
– – glycolysis pathway **9**, 182f
– – hexose monophosphate pathway **9**, 183
– in bioleaching **10**, 206
– in BOD monitoring **11A**, 134
– in cocoa fermentation **9**, 567f
– in cucumber fermentation **9**, 636
– in cyanide monitoring **11A**, 133
– industrial applications **1**, 495
– in grape must **9**, 466ff
– in olive fermentation **9**, 611, 615
– isolation **1**, 471
– killer phenotype **1**, 476
– life cycle **1**, 472ff
– lipid synthesis **1**, 490f
– mating types **1**, 472
– – *HO* genes **1**, 473
– metabolic compartmentation **1**, 478
– metal removal by – **11A**, 441
– methanol metabolism **9**, 180f
– – *Hansenula polymorpha* **9**, 181
– methanol metabolization **1**, 271, 471
– methylotrophic **1**, 269
– monoterpenoid transformation by **10**, 388f
– occurrence **1**, 471
– pathogenic –, in sewage **11C**, 217
– – in sewage sludge **11C**, 217
– phosphatases **1**, 490f
– – genes **1**, 491
– Pol II mediated gene transcription **1**, 492
– post-transcriptional gene regulation **1**, 494
– production of S-adenosylmethionine **10**, 331
– protein secretion **1**, 500
– pseudomycelia **1**, 115
– pure culture –, in wine making **9**, 470f
– RAS/cAMP signaling pathway **1**, 487f
– *RAS* genes **1**, 487f
– RNA polymerase **1**, 492
– selective marker **1**, 499
– sphingolipids **7**, 183
– sterol contents **7**, 171f
– storage carbohydrates **1**, 485
– surface-active properties of cell walls **6**, 685
– tetrad analysis **1**, 473
– transcription, promoter **1**, 489, 499
– – regulation **1**, 491f
– transformation methods **1**, 476f
– transposable element **1**, 475
– use in foods **9**, 226f
– use of aromatic compounds as growth substrates **11B**, 159
– vitamin content **9**, 225f
– vitamin synthesis **1**, 490f
– xylose metabolization **1**, 196

– yeast "lecithin" **7**, 183
yeast two-hybrid system **5B**, 89, 151
yellamycin B, antitumor agent **7**, 659
– structure **7**, 659
yellow-green algae **10**, 113
yersiniabactin **7**, 207f
– catecholate siderophore of *Yersinia enterocolitica* **7**, 208
– structure **7**, 207
Yersinia enterocolitica, yersiniabactin **7**, 208
Yersinia pseudotuberculosis, biosynthesis of CDP-ascarylose **8B**, 119ff
yield equations, assumptions for modelling **4**, 411ff
– biomass decay parameter **4**, 414f
– yield coefficient **4**, 414
yield improvement, of crops **5B**, 146
yoghurt *see* milk products
yoghurt fermentation **3**, 298, 307
yogurt *see also* frozen yogurt
– definition **9**, 387f
– fruit preparations for – **9**, 403f
– manufacture of **9**, 407ff
– – fruit-flavored – **9**, 407, 409
– – plain – **9**, 407f
– – processing stages **9**, 406
yohimbine, structure **8A**, 353
YPD (Yeast Proteome Database) and WormPD **5B**, 288, 411

Z

Z-DNA **2**, 194ff, 202
– anti-Z-DNA antibody **2**, 204
– biological function **2**, 203ff
– d(AT) **2**, 203
– d(CA) **2**, 203
– d(TG) **2**, 203
– *in vivo* **2**, 205
– methylase inhibition **2**, 204
– probing **2**, 203ff
– restriction endonuclease inhibition **2**, 204
– sequence **2**, 203ff
zealarenone, as estrogen receptor agonist **7**, 112
zearalenone **7**, 123
Zenapax **5A**, 318
ZFIN **5B**, 285
zinc, bioleaching of **10**, 213
– bioprecipitation **11B**, 365
– biosensors for depletion of **11B**, 456
– hyperaccumulator **11B**, 389ff
– removal by phytoextraction **11B**, 389ff
zinc finger motifs **5A**, 19
zink, removal by biosorption **11A**, 42f
Zoogloea, phylogenetic status **1**, 404

Zoogloea ramigera, detection by fluorescent *in situ* hybridization **11A**, 101
– in wastewater treatment **11A**, 102
z-Transform **4**, 252f
Zygomycetales, metabolite production **1**, 522
Zygosaccharomyces bailii, in wine spoilage **9**, 496
Zygosaccharomyces rouxii, in miso fermentation **9**, 520
Zymomonas mobilis, continuous cultivation kinetics **1**, 158
– elemental fluxes **1**, 125
– ethanol formation **6**, 128ff
– – kinetic parameters **6**, 132
– – kinetic parameters for growth **6**, 130f
– – specific ethanol production rate **6**, 132
– ethanol production **1**, 136
– gluconic acid production **6**, 352, 356
– growth kinetics **1**, 133
– immobilized cell systems, ethanol productivity **6**, 135f
– lactose metabolization **1**, 195
– L-alanine production **1**, 210
– pyruvate decarboxylase **8B**, 25, 49f, 53ff
– xylose metabolization **1**, 198f
– ^{31}P NMR spectrum **1**, 192f
zymosterol, structure **7**, 171